[redacted]

DIGITAL PROCESSING OF GEOPHYSICAL DATA

A REVIEW

Roy O Lindseth P. Geoph.
TEKNICA RESOURCE DEVELOPMENT LTD.
600, 633 SIXTH AVENUE SW, CALGARY, ALBERTA, CANADA.

FIRST PUBLISHED : OCTOBER 1968.
REVISED : OCTOBER 1969, JANUARY 1970, DECEMBER 1971, NOVEMBER 1973, AUGUST 1974, FEBRUARY 1976, OCTOBER 1978.
REVISED AND NEW FORMAT : SEPTEMBER 1982.
DESIGN, GRAPHICS AND PRINTING : R. DRAPER, ADVERTISING DEPT.

These course notes have not been edited nor been through peer review by the Society of Exploration Geophysicists. Any questions regarding this material should be directed to the author.

Roy O. Lindseth
Tecknica Resource Development Ltd.
1100, 736 Sixth Avenue SW
Calgary, Alberta, Canada

Library of Congress Catalog No. 87-060964

ISBN 931830-48-6 (Series)
ISBN 931830-50-8 (Volume)

Printed in the United States of America

Contents

Contents

Introduction

This publication originated as a few notes to accompany a basic seminar in digital processing organized by the Canadian Society of Exploration Geophysicists in 1967, and expanded in 1968 into a course of the Continuing Education Program of the Society of Exploration Geophysicists. Over the years the notes have been revised and expanded to include many of the developments in digital data processing. In the rapidly developing field of geophysical exploration, it is a problem to maintain a working knowledge of the state of the art.

It seems now that basic data enhancement techniques have reached a plateau in their development and research is directed increasingly toward the interpretive aspects. Process routines include more direct intervention of the interpreter. Stratigraphic techniques reveal traps not previously seen. New display techniques, many with colour, add yet another dimension to the presentation of the data. The goal of all this activity is to increase our ability to find oil and gas, offsetting the decreasing amount that remains to be found. Some of the world's very large oilfields have been found in the last ten years, testifying to the skill of the explorationist.

This edition of the notes represents an effort to consolidate old and new information about geophysical data processing and to translate that knowledge into an understanding of the physical nature of the process. Understanding the cause and effect relationship between input and output should provide a better understanding of why the results of processing are not always consistent, and how to choose processes and parameters for any given field data.

Many liberties have been taken with the development in order to avoid rigorous mathematics. A twelve hour lecture series directed to a broad spectrum of professional oil finders does not lend itself to comprehensive treatment in any great depth. Fortunately, there are many excellent sources of detailed information with full development of the material. The references cited will be useful in this regard, particularly for specific items. A recommended program for general background reading would begin with the Robinson Treitel Reader, for basic development. The Encyclopedia Dictionary of Exploration Geophysics, by R.E. Sheriff provides a clear definition, and frequently a basic review, of most of the terminology used. Bracewell's book, The Fourier Transform and its Applications, is a good source of understandable material on Fourier transforms and the book by Y.W. Lee on the Statistical Theory of Communication provides a thorough coverage of the field.

Foreword

Geologists and geophysicists share a common goal in their search for petroleum. Both seek to locate those geologic anomalies in the subsurface in which porous reservoir rocks form a trap for hydrocarbons. Finding the play is usually not enough; they must also convince management to drill the play, and hopefully in the end discover a new pool of hydrocarbons.

Although they have the same interests and goals, geologists and geophysicists tend to ply their profession in different ways. Often the skills which each type of explorationist has developed are quite different, due largely to natural differences in the basic data which each uses as the basis for interpretation.

Most exploration today is based on some form of indirect measurement, otherwise known as remote sensing. Subsurface geological studies are not made by direct examination of the rocks but are often derived from borehole log information correlated with analysis of a much smaller amount of cores and samples.

Over wide areas the seismic method may be the only source of information. The seismic survey is, of course, almost the only effective tool for subsurface exploration that does not first require drilling a deep hole into the earth. Hence, the geophysicist tends to rely more and more on the seismic reflection survey as the principal source of his working data. This technique, by which comparative reflection times are measured to some particular reflecting horizon in the section, is unexcelled for mapping structure. For this to be effective, the horizon must be continuous. Therefore, the geophysicist and his allies, the data processors, strive to obtain clear, continuous reflectors which extend over an entire prospect. This allows continuous structure maps to be drawn up, together with isopachs to show changes in interval thickness. If a horizon is not continuous, it may be ignored for mapping, since it would then not be possible to develop a complete prospect map. All of these conditions tend to develop in the geophysicist a concept of layer cake geology which ignores the fact that in any sedimentary basin, depositional conditions rarely extend without change over any great distance.

The geologist works extensively with subsurface sections based on geologic tops and logs from existing boreholes. He develops maps, based on isopach intervals and other criteria, which usually succeed to a high degree in demonstrating the conditions and topography existing at the time the sediments were laid down. He may use samples and cores to determine lithology, or use a suite of logs to estimate lithology with a fair degree of precision. These operations frequently lead to successful conclusions regarding the development of a basin, and to localize regions in which facies changes, important to the accumulation of hydrocarbons, occur across the subsurface section.

1 FUNDAMENTAL WAVE THEORY

Cyclicity Sedimentation

Geologists and geophysicists share the common goal of exploration. They also both seek to measure and describe the same subsurface section. They may measure different physical properties of that subsurface section, but the section remains the same.

Some of their measurements may be more closely related than others. For instance, the sonic log, which is most often used by geologists, measures the transit time of soundwaves through the subsurface section. The seismic reflection method does essentially the same thing, measuring the cumulative transit time from the surface. Some analytical method of comparing the results of different measuring methods is needed to determine the relationships, so that the information contained in one may be expressed in terms of the other, or in absolute terms of geology. After all, it is geology, not geophysics which determines the source and accumulation of sedimentary beds and produces the conditions which gather and trap the hydrocarbons we seek to discover.

A description of the sedimentary geology in a basin is often based on some model or concept of how that particular section was developed. To the uninformed observer, the sedimentary column may appear to be merely a purely random sequence of different rock types. Indeed, one type of seismic signal deconvolution is based on the assumption that the sedimentary section is a completely random sequence. Rarely is this true. Most sedimentary sections are the result of depositional cycles.

At the other extreme, many geophysicists, although they should know better, tend to model the sedimentary section in terms of a layer cake sequence of thick units which have continuous reflections at the boundaries. The seismic markers are used to develop a sequence of structural maps and isopachs which can then be interpreted in terms of the major geological features. Such interpretations are primarily structural and ignore all but the most gross results of sedimentary processes.

A thorough understanding of the nature of the sedimentary deposition may require a nearly continuous stratigraphic analysis of the entire section, including an examination of rock properties and bedding planes to determine the conditions that most likely existed at each stage of the development of the entire section.

Fundamental Wave Theory

It is generally recognized that cyclic phenomena play a major role in the development of any sedimentary section. Cyclothems and related phenomena of both longer and shorter duration are frequently observed over much, if not all, of the section. Such processes continue. Any point on the earth's surface today merely represents the present condition of a continuous geologic process which has gone on since the beginning of time and will continue indefinitely into the future.

The cyclic phenomena which influence deposition are perhaps most visible along any shoreline, where repetitive wave action continues without interruption as long as that shoreline exists. The waves may be a short, gentle chop along the edge of a small lake or may be the much larger and longer period ocean waves that crash on a shore facing the open sea.

The edge of the waves will move to higher and lower levels of the shore in accordance with the tides, moving in and out twice each day regardless of any other events which might occur. A measurement of the combined effects would show the relatively long period of the tidal cycle overridden, or modulated, by the much shorter oscillation of the wave action.

It is a relatively simple matter to extend this thinking to include a great number of other cyclic phenomena which occur at more or less uniform periods. The lunar cycle affects the tides, and the seasonal change brought about by the earth's travel around the sun may result in variation of the rate of sedimentation.

Sun spots appear to move in cycles and these influence rainfall and sedimentation. The process continues on through cycles of glaciation, major geologic ages and presumably through cycles of continental drift. All of these changes have direct effects on deposition, changing the amount and the type of sedimentation. The resulting sequence is a historical record of the particular combination of all the conditions which existed at any particular instant in time. Even some major condition which appears to have occurred only once through this entire history, such as the existence of the earth itself, may be but one cycle of a sequence occurring over such long periods of time that we are incapable of measuring it.

This fundamental model of the sedimentary column representing the sum of several cyclical processes fits known conditions very well and is widely accepted. By good fortune, a particular field of signal theory, harmonic analysis, is very well suited to measurement, study and analysis of such series. Let us now examine this in detail to determine how it may be useful in applications for geology and geophysics.

Any continuing constant cyclic phenomenon may be represented graphically by a sine or cosine wave. Each cycle can be described by two terms; the amplitude and the wavelength. The amplitude is the peak value to which each cycle rises above or falls below some neutral base line. The wavelength is the distance from peak to peak of one full cycle. If the wavelength is measured in time it is called the period of oscillation. The periodicity is often expressed in terms of frequency. For instance, common domestic electric current oscillates at a frequency of 60 cycles per sec, or 60 Hz. Such cyclic phenomena may continue for an indefinite length of time, but for purposes of analysis they are usually measured over some discrete time interval.

If a continuous cosine wave is used to graph the oscillation it is possible to construct the cyclic wave over any length of time from the mere definition of amplitude and frequency.

One more characteristic of the periodic component must be defined in order to describe it precisely for a discrete interval; namely the point in the cycle at which the sequence started. The angular phase position of the cyclic component at some arbitrary reference, usually the start of the signal sequence, is specified.

FIGURE 1.1

FIGURE (a)

Two frequencies can be added together to produce a signal different in appearance from either of the components.

FIGURE (b)

Changing the amplitude of either of the components will immediately affect the appearance of the output.

SIMPLE SUMMATION OF TWO DIFFERENT FREQUENCIES

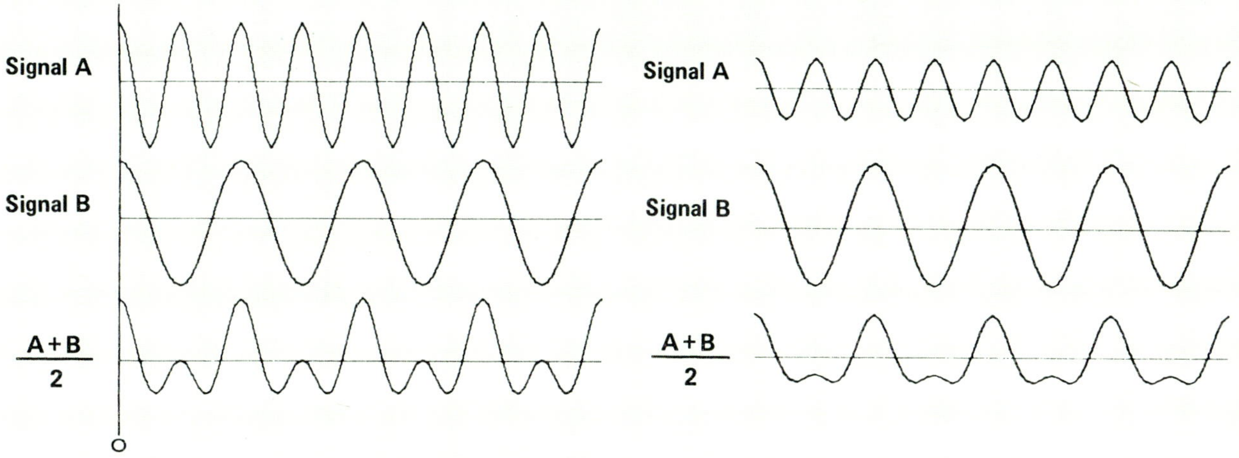

(a) The bottom signal represents the sum of the upper two.

(b) Sum of base frequency and first harmonic at two to one amplitude ratio.

The three coordinates of frequency, amplitude and phase are sufficient to describe each frequency component. If the cyclic sequence is known or believed to continue indefinitely, a sample taken over some length of time long enough to ensure reliable definition can be extended into the future.

This brings up another interesting point of information theory which relates to the information carried in the sample. The sample measured must be long enough to obtain a reliable measurement of amplitude and frequency. The length of a discrete time sample limits the determination of the lowest frequency component that can be measured reliably.

Complex signals are composed of the sum of a number of elementary wave forms, each at some constant amplitude and frequency, starting at some defined phase angle. The fact that a given continuous signal is merely the sum of a number of simple periodic components may not be evident from examination. The summation of even a relatively small number of different frequency components can produce rather complex signals and the nature of the components may be very obscure in the sum.

Two frequencies, for example a base frequency and a harmonic twice the base frequency, can be added together to produce a signal different in appearance from either of the components (**Fig. 1.1a**). Changing the amplitude of either of the components will immediately affect the appearance of the output (**Fig. 1.1b**).

The addition of more frequencies will produce still more complex signals. The sum of frequencies representing two, four, and eight periodic cycles over some interval will itself be periodic, repeating the output wave form twice over the interval (**Fig. 1.2a**). If the frequencies are two, four, and eight cycles per sec, (or more correctly two, four and eight Hertz) the resulting signal will be periodic over a length of time of one-half sec, the shortest time in which all components will complete one or more full cycles. This reveals another fact; the output from the summation of a number of periodic components will itself be periodic over the interval into which all of the component wavelengths will divide evenly.

The summation of such frequencies as, say, six, seven and eight Hz (**Fig. 1.2b**) will be periodic only over a full second, the shortest time in which all of the components complete an exact integral number of cycles. The sum of these three components is not unlike a portion of a seismic trace. Much more complex signals can be obtained by adding a larger number of frequencies and/or varying the amplitude of the components.

In addition to variations in the amplitude and the frequency of the signal components the phase of each component may vary. Any composite signal may be changed substantially in appearance by a relatively minor change of the phase of the components (**Fig. 1.3**). Note however that the changes of phase does not change the periodicity of the signal (**Fig. 1.3 and Fig 1.4**). Characteristics features of the signal (say a peak) may appear to shift in time, but the relative time difference (periodicity) between features will not. Changing the phase of the components of a seismic record may change the reflection times, a source of mis-ties on different seismic surveys in the same area. A phase shift in the unit impulse response is the cause of the typical asymmetrical appearance of the seismic reflection.

A relatively small number of frequency components sum to produce a signal not unlike that of a seismic signal. In fact, any continuous signal can be approximated by simply summing a set of individual frequency components at some specified amplitude and phase relationship. The example in **Fig. 1.5** duplicates a portion of an integrated borehole sonic log. The sum of a set of frequencies produces a good approximation of the log trace. Detail is less than on the original log, because the normal frequency content of a log is much higher than 100 Hz, the highest

FIGURE 1.2

FIGURE (a)
The sum of frequencies representing two, four, and eight periodic cycles over some interval will itself be periodic, repeating the output wave form twice over the interval.

FIGURE (b)
The summation of such frequencies say, six, seven and eight Hz will be periodic only over a full second, the shortest time in which all of the components complete an exact integral number of cycles.

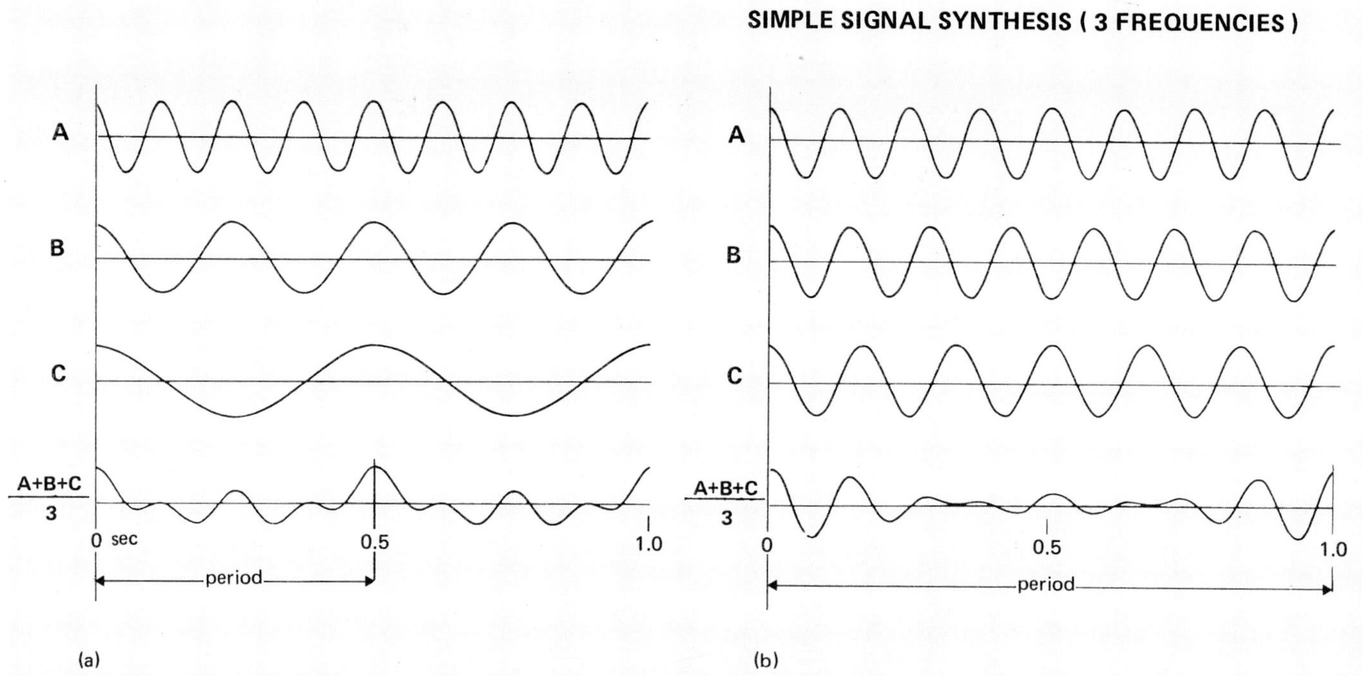

FIGURE 1.3

Any composite signal may be changed substantially in appearance by a relatively minor change of the phase of the components. The change of phase does not change the periodicity of the signal.

EFFECTS OF PHASE SHIFT (2 FREQUENCIES)

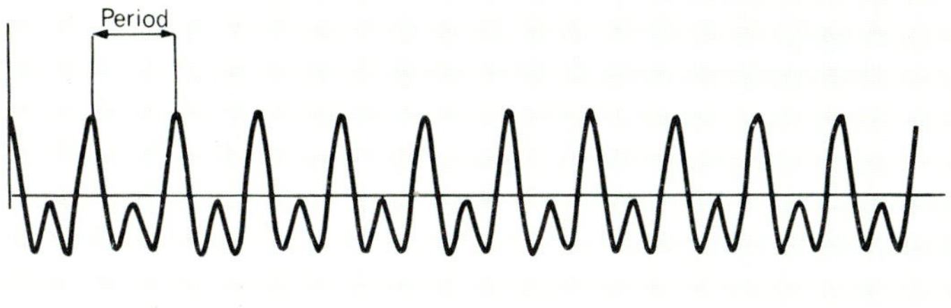

Basic frequency plus harmonic in phase.

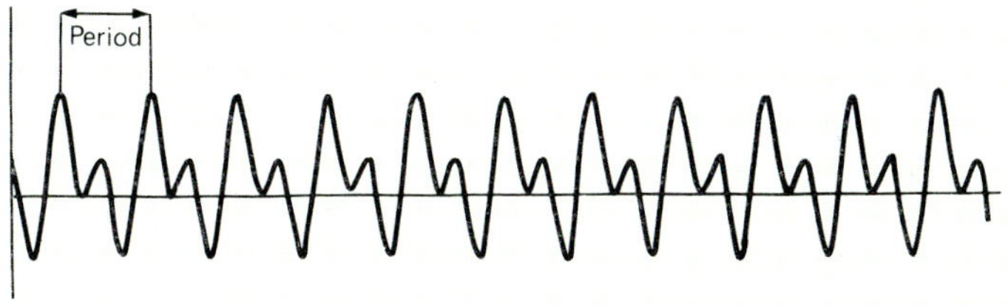

frequency used in the approximation. Also, the range of frequencies is not uniformly spaced. A better approximation to the log would probably have resulted from the addition of the intervening frequencies. Not shown is the DC component which would be necessary in the reconstruction. Seismic information usually oscillates about a zero axis, but on a log trace all values are positve, hence it has a positive non-zero algebraic average, or DC component.

A better example is provided by **Fig. 1.6**. A single unprocessed seismic field trace is recorded at the top. Beneath it is the series of individual frequency components which, when summed, yield the seismic signal trace.

Each frequency component is plotted at the same relative amplitude it has in the field trace (although all amplitudes have been scaled up a constant amount for the illustration). Each frequency component has a relative phase position which can be seen at the zero-time reference line.

FIGURE 1.4

The change of phase does not change the periodicity of the signal.

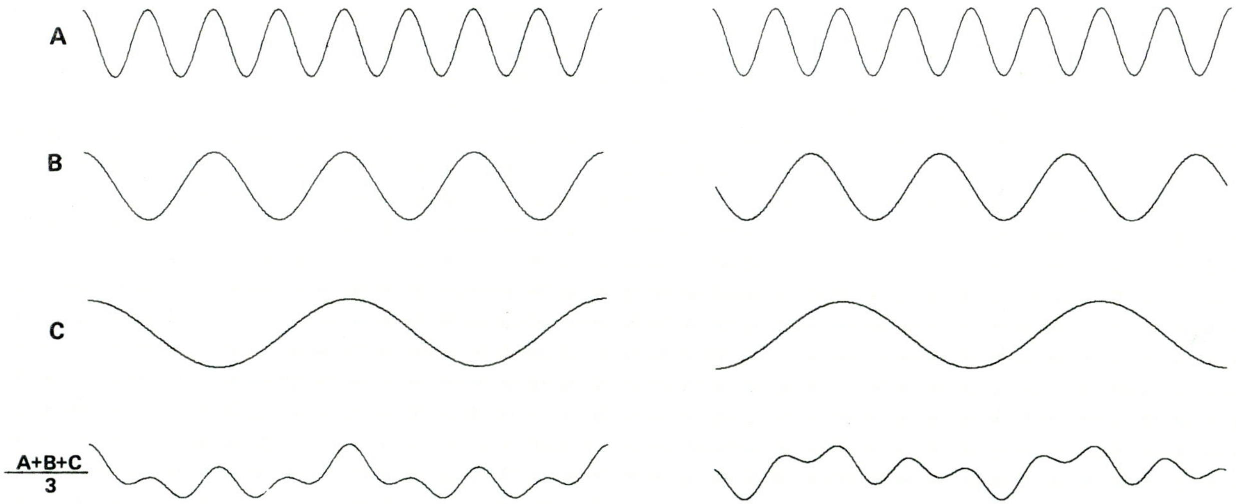

EFFECTS OF PHASE SHIFT (3 FREQUENCIES)

The Fourier Transform

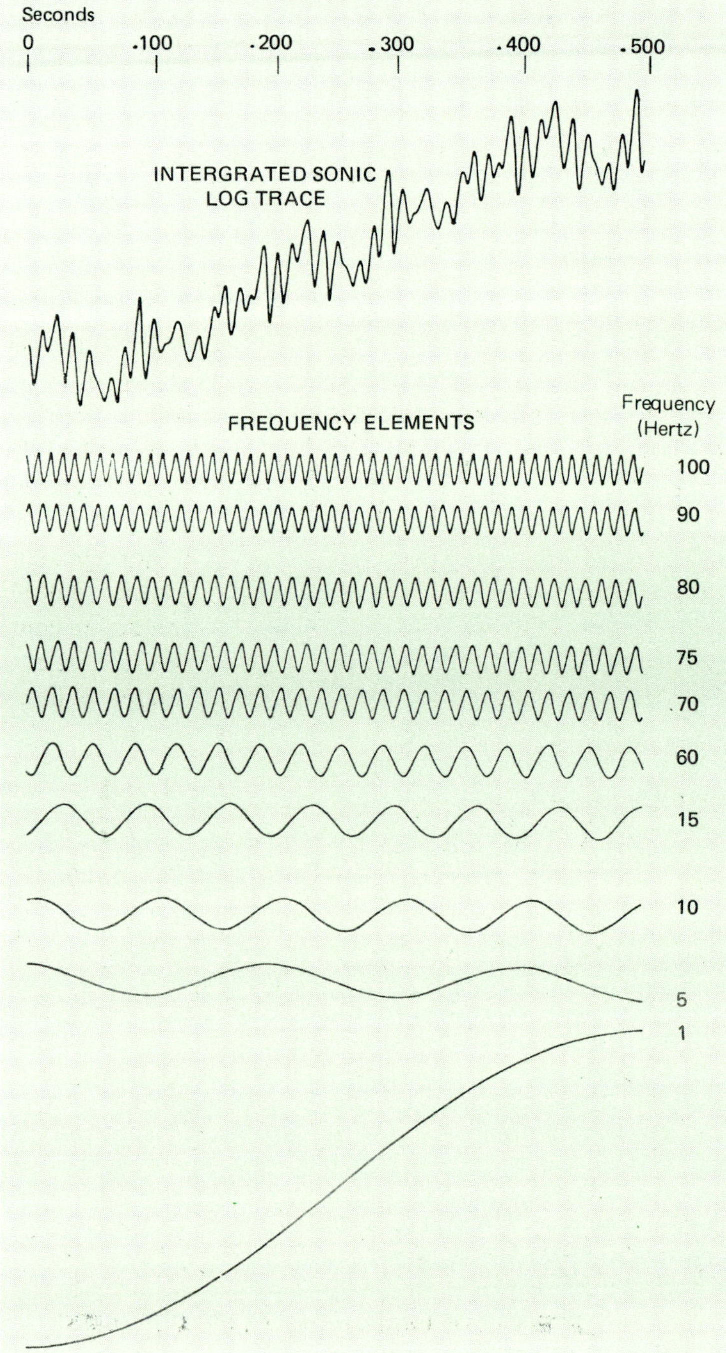

FIGURE 1.5

The sum of a set of frequencies produces a good approximation of the log trace.

When all the individual frequency components are added together at the amplitude and phase relationship indicated the result will be equal to the single trace at the top of the display.

The complete specification of the range of frequencies contained in a given signal, defined in terms of amplitude and phase of each frequency, forms the Fourier transform of that signal. It is a description of the signal in the frequency domain. The computation of the Fourier transform is completely reversible so each side of the transform can be developed from the other. A time-domain description and the frequency-domain description of the same signal are known as a transform pair.

A number of interesting observations can be drawn from the dual presentation of **Fig. 1.6**. With the exception of a few components, frequencies near 30 Hz have the maximum amplitude. Above 30 Hz the amplitude gradually diminishes. Below 30 Hz the amplitude drops at a rather rapid, if irregular, rate. This response is characteristic of seismic information in many areas and illustrates the rather narrow frequency band in which the natural earth response to seismic energy occurs. Note also the absence of frequencies below 5 Hz even though this seismic trace is rich in low frequency (which may be noise). This is probably the result of a field filter or the geophone response. A frequency spectrum of reflection coefficients corresponding to this location would extend from zero to at least 250 Hz.

The limited bandwidth of seismic signals simplifies some of the processing operations, but the lack of high frequencies seriously limits the definition of thin beds.

The band of energy between 5 and 10 Hz in **Fig. 1.6** has rather high amplitude. This illustrates another common problem. The lower

FIGURE 1.6

A single unprocessed seismic field trace is recorded at the top. Beneath it is the series of indivdual frequency components which, when summed, yield the seismic signal trace.

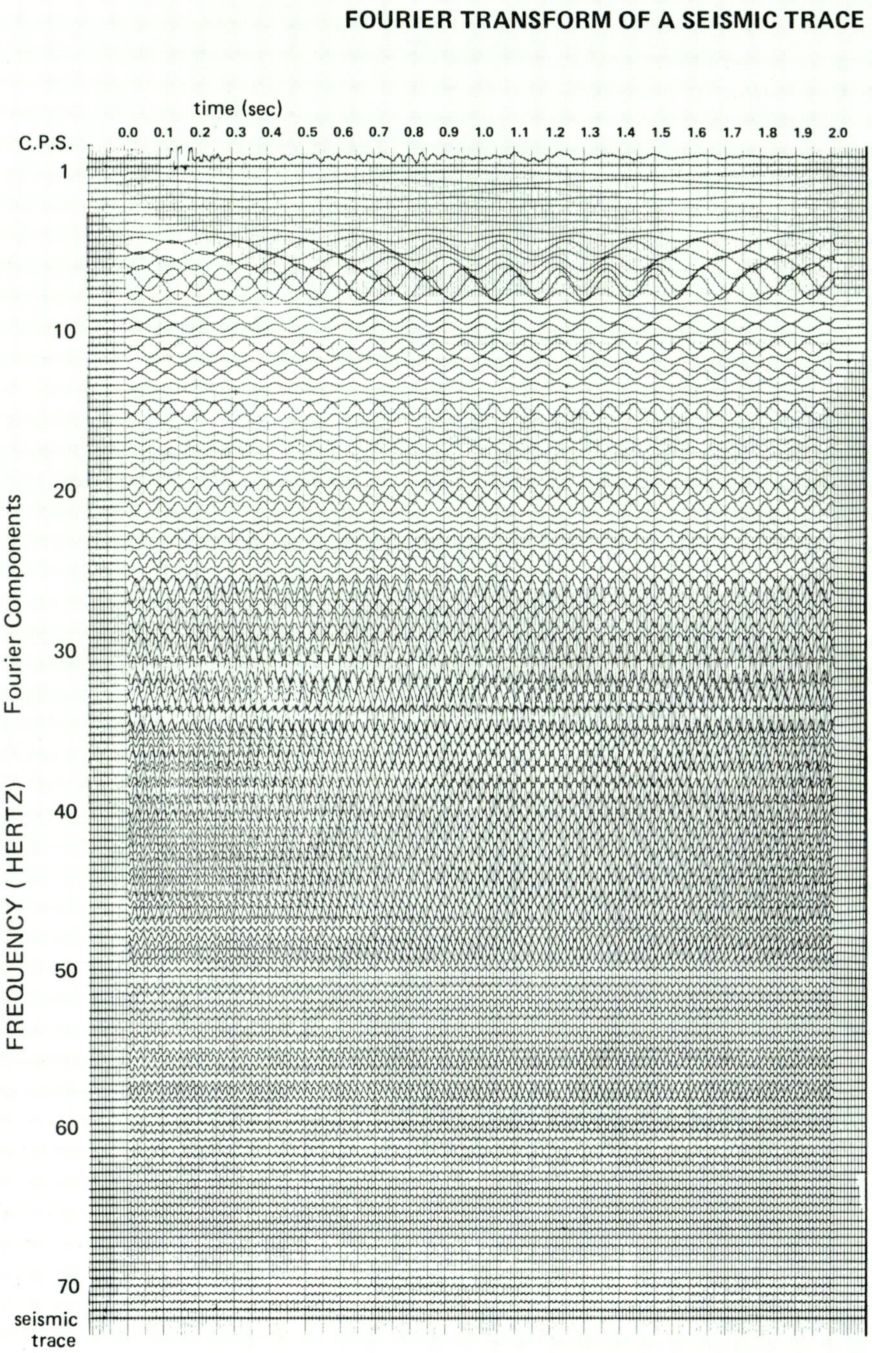

frequencies, which may include a large component of noise, often carry much more energy than do the remaining frequencies on the record. If the dynamic range of the instruments is limited, the signal must be attenuated to a level within the range of the instruments. Should this condition occur, it is likely that the very weak signal components, particularly the high frequencies, will be lost.

Elimination of the low frequencies by analog filtering in the field is one way to ensure that the remainder of the signal is not overwhelmed. On the other hand, the fact that the seismic signal is by nature already severely band-limited makes it imperative to record as broad a band as possible. A better solution is the current trend toward the use of accelerometers rather than geophones. Accelerometer response increases toward high frequencies, while geophone response is usually flat. Any necessary correction of the response can be made by deconvolution, and unwanted frequencies can be removed more selectively by digital processing than by field filters. It is important to increase the bandwidth as well as the dynamic range of the recording equipment to permit recording of all the available frequencies in the field. However, there is no point in recording a bandwidth greater than that of the seismic signal.

FIGURE 1.7

A much more convenient method is a graphical presentation of the amplitude in terms of frequency.

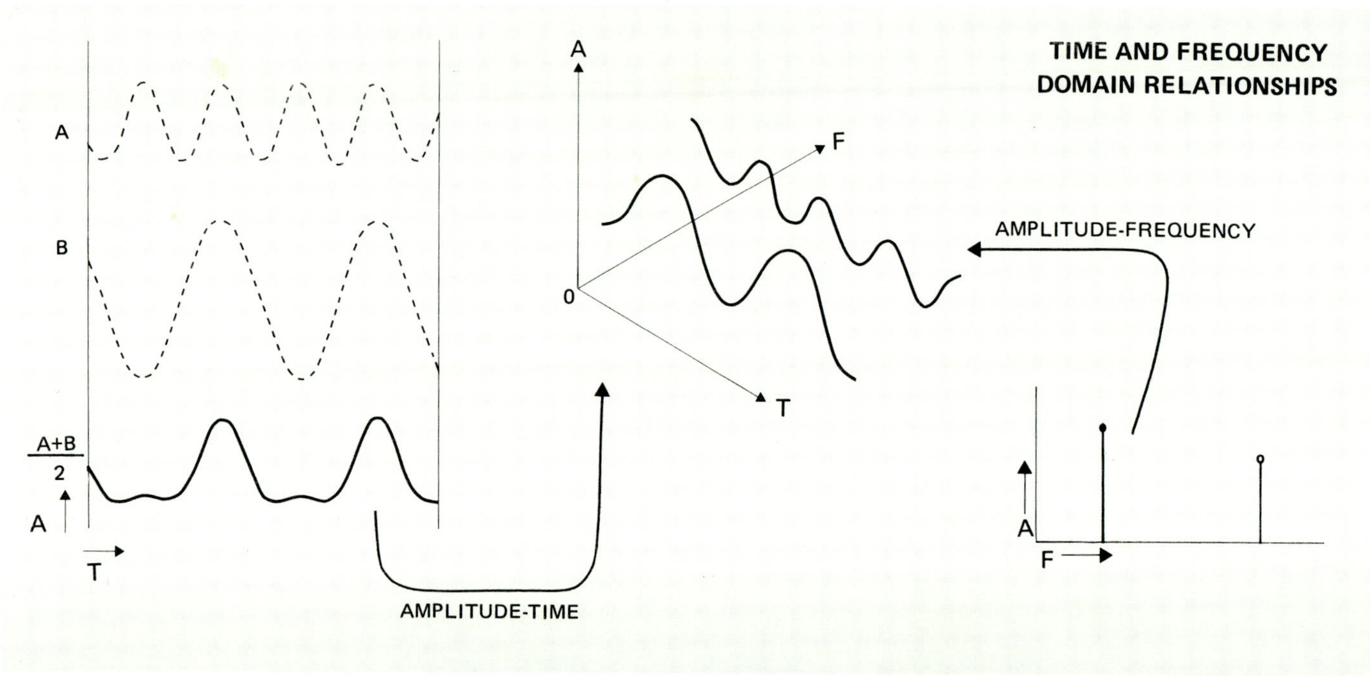

Signal bandwidth is rarely defined sharply, but regardless, energy outside this band will be mostly noise.

It is evident that the amplitude level of the frequency components can vary substantially. Certain frequency components have nearly zero amplitude while adjacent components are much higher. These frequency/amplitude relationships produce a spectrum signature or "character" which is as significant as the character on the time-domain trace.

Understanding the Fourier transform often includes some puzzlement over seemingly paradoxical relationships. For instance, loss of seismic high frequencies with increasing reflection time is an accepted phenomenon. Yet, in the transform, all frequency components have constant amplitude. The answer lies in phase relationships. Although each frequency component has constant amplitude over the length of the signal, low-frequency events, such as those near 1.2 sec, occur on the trace because a number of low-frequency components fall in phase over this portion of the record. Elsewhere, particularly in the shallow events, the same low-frequency components are shifted out of phase and cancel out when summed. On the other hand, the high-frequency components are in phase over the early part of the trace, so they dominate. But they become increasingly out of phase with depth.

This transform provides a fairly coarse approximation to the trace, yet it can define the trace with considerable accuracy. The frequency range examined extends from 1 to only 70 Hz. The frequency increment is approximately one-half Hz. Summing a narrower band of frequencies or at larger frequency steps would still provide a recognizable seismic trace, but one defined with less precision. The Fourier transform is a type of curve fitting routine which provides increasing closeness of fit with increasing numbers of coefficients.

This leads to another curious phenomenon related to the selection of the zero time reference point on the signal to be examined. In this example of a real field trace, the zero reference is the time break. A short time interval, about 130 ms, elapsed between the firing of the shot and the arrival of the first seismic wave at the nearest geo-

FIGURE 1.8

The full transform consists of one spectrum describing the amplitude as a function of frequency and an associated spectrum which describes the phase of frequency components contained in a signal.

AMPLITUDE SPECTRUM OF SEISMIC TRACE

Figure A

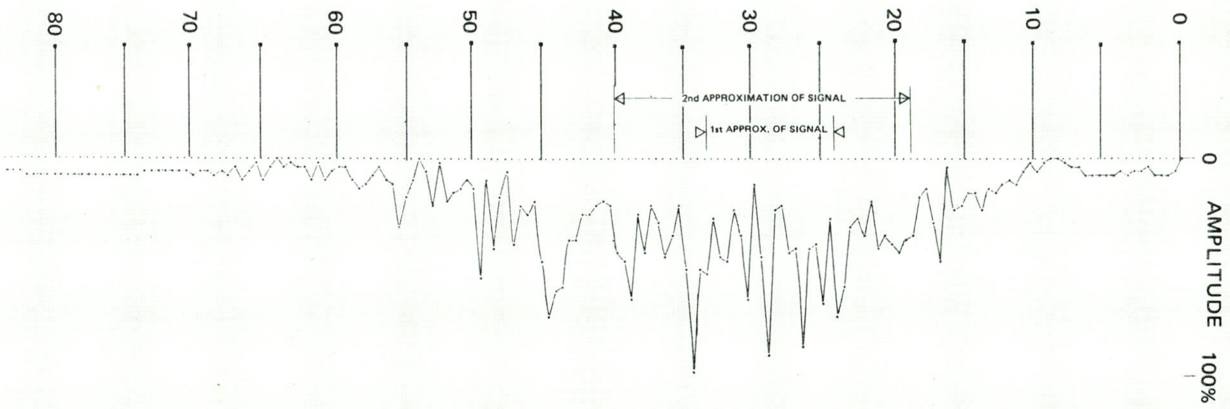

PHASE SPECTRUM OF SEISMIC TRACE

Figure B

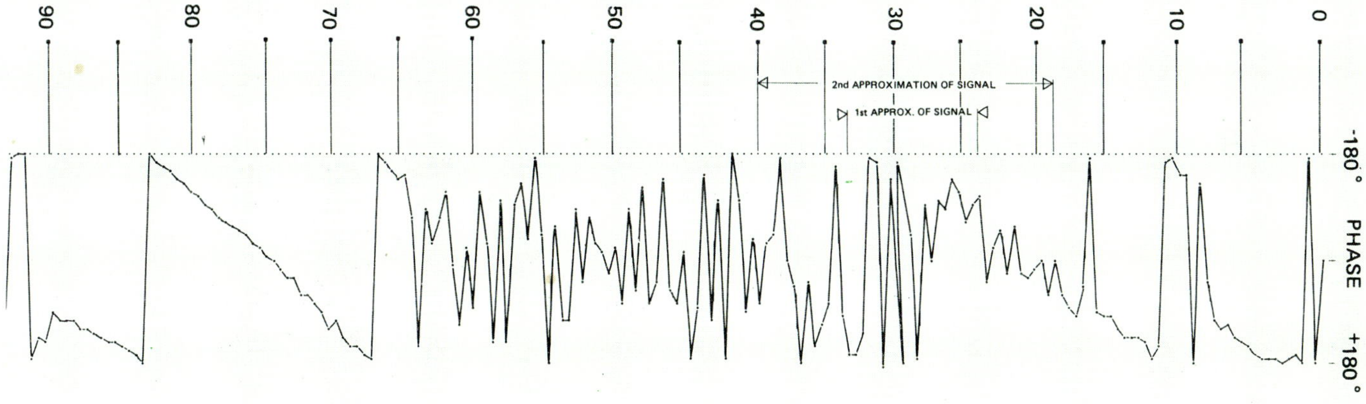

FIGURE 1.9

(a) If the two signals are summed the result will be identical to the source signals in frequency and phase, but double in amplitude. Normalizing (division by two) will make the output and input identical.
(b) If one of the source signals is shifted in phase 180°, with respect to the other, the numerical value of the two signals will be equal but opposite in polarity. When the two signals are summed the resulting amplitude is zero.
(c) If one signal is phase shifted with respect to another of the same frequency and amplitude, the phase of the resulting summed output trace will be the average of the phases of the two input traces. The amplitude of the output will be evidently somewhat less than the sum of the same two traces in phase.

ADDITION OF LIKE-FREQUENCY ELEMENTS AT DIFFERENT PHASE ANGLES

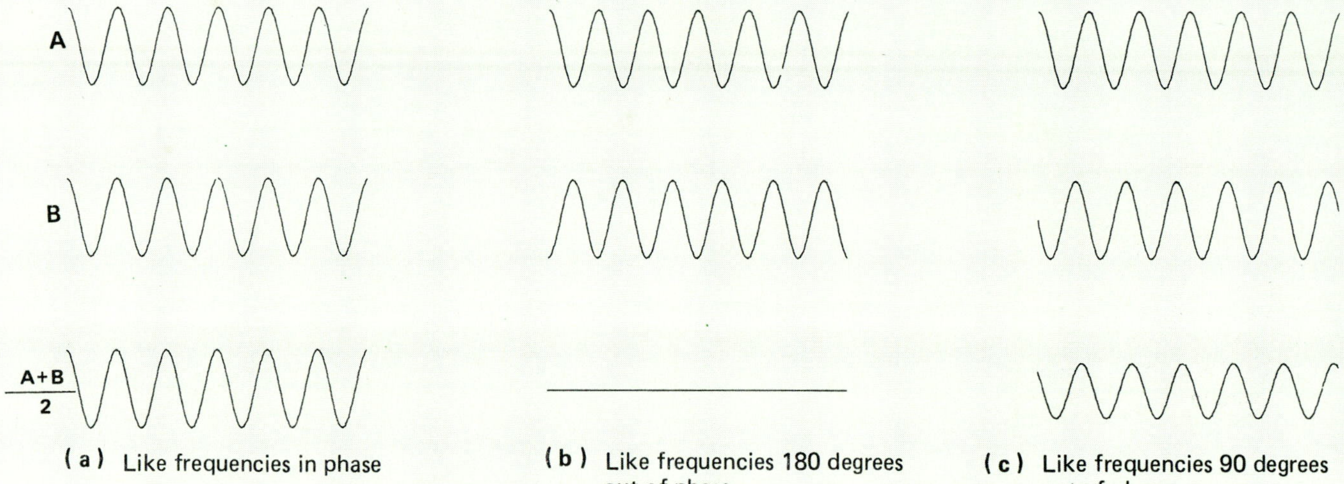

The sum of like frequencies at different phase relationships may be equated to a single wave.

phone. This is confirmed by the zero amplitude of the trace over the interval. Yet each frequency, referenced to the same zero, oscillates with unchanging amplitude over the full time span. Again, the phase relationships are such that all frequencies cancel out precisely to zero in the region prior to the first arrival. If for some reason, some of the frequencies were omitted from the sum, the total would no longer be zero, and there would be some residual oscillation from time zero to the first arrival. This is a common effect of band-pass filtering which removes frequency components from the spectrum. A band-pass filter may remove noise, but it will also likely have some side effects as the result of the fewer frequencies remaining in the sum.

Fourier Spectra

Presentation in the form shown illustrates the nature of the transform rather well but it is otherwise of little practical use. A much more convenient method is a graphical presentation of the amplitude in terms of frequency (Fig. 1.7). The amplitude spectrum represents only one part of the Fourier transform. The full transform consists of one spectrum describing the amplitude as a function of frequency (Fig. 1.8a) (in this example, the amplitude squared, or power, is plotted to accentuate the relationships) and an associated spectrum (Fig 1.8b) describes the phase of frequency components contained in a signal. The two spectra together provide the total frequency-domain description of the signal. The identical signal can be described in the time domain by a function of amplitude versus time.

The ability to separate a signal into its component parts is a powerful tool for analysis and operation. The Fourier transform is completely linear, and specification of any signal in either the time domain or the frequency domain can be equally correct. Any operation in one domain has a counterpart in the other. However, some operations are substantially easier to perform and/or easier to understand in one domain than in the other. It is very important that the relationships existing between transforms be thoroughly understood as they are not always clear, nor are all of the relevant considerations always accounted for.

Note that one and only one value of amplitude and phase defines each frequency component in the transform of the seismic trace. In fact, each frequency component may be expressed in terms of a single vector having a length

FIGURE 1.10

The sum of the two vectors will provide the amplitude and phase of the output frequency components. The frequency will remain unchanged.

VECTOR ADDITION OF LIKE FREQUENCIES

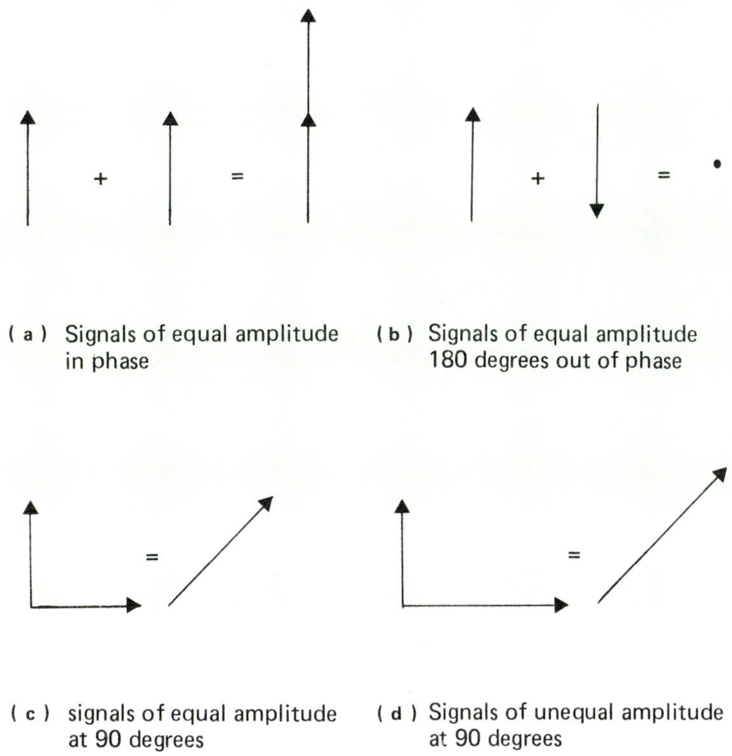

Stacking or summing of signal components may be computed by vector addition

equal to the amplitude directed at the angle of the phase. It is perfectly conceivable that a recorded signal might contain a number of components containing identical frequencies generated by more than one energy source. For example, a trace could contain signals of seismic energy, wind noise, AC pickup and random noise, all having common frequency components.

The range of effects of multiple energy sources on a seismic trace may be illustrated in the following manner. Select two identical sources of energy which have the same frequency and phase. If the two signals are summed the result will be identical to the source signals in frequency and phase, but double in amplitude. Normalizing (division by two) will make the output and input identical (**Fig. 1.9a**)

If one of the source signals is shifted in phase 180°, e.g., the origin is lagged one-half period with respect to the other, the numerical value of the two signals will be equal but opposite in polarity. When the two signals are summed the resulting amplitude is zero (**Fig. 1.9b**).

These two examples represent the extreme limits of phase relationships between two components of the same frequency; either completely in, or out of phase. Any other condition must fall somewhere between. If one signal is phase shifted with respect to another of the same frequency, and amplitude, the phase of the resulting summed output trace will be the average of the phases of the two input traces. The amplitude of the output will be evidently somewhat less than the sum of the same two traces in phase (**Fig. 1.9c**). The resulting amplitude value may be determined by vector operations. To repeat, each frequency component may be considered as a vector of length equal to amplitude, directed to the corresponding phase angle. The sum of the two vectors will provide the amplitude and phase of the output frequency component. The frequency will remain unchanged (**Fig. 1.10**). By extension it will be seen that this applies to any number of components of the same frequency. They may be summed as vectors to produce a single component having a resultant amplitude and phase which will be that determined by Fourier analysis. The amplitude and phase spectra are usually of little or no help in defining the source or nature of the signal examined, nor the relative proportions of signal and noise. Only the final result of all contributions to any frequency is measured.

The entire Fourier transform may be illustrated by a series of vectors in space originating from a frequency axis

FIGURE 1.11

The entire Fourier transform may be illustrated by a series of vectors in space originating from a frequency axis.

VECTOR REPRESENTATION OF FOURIER TRANSFORM

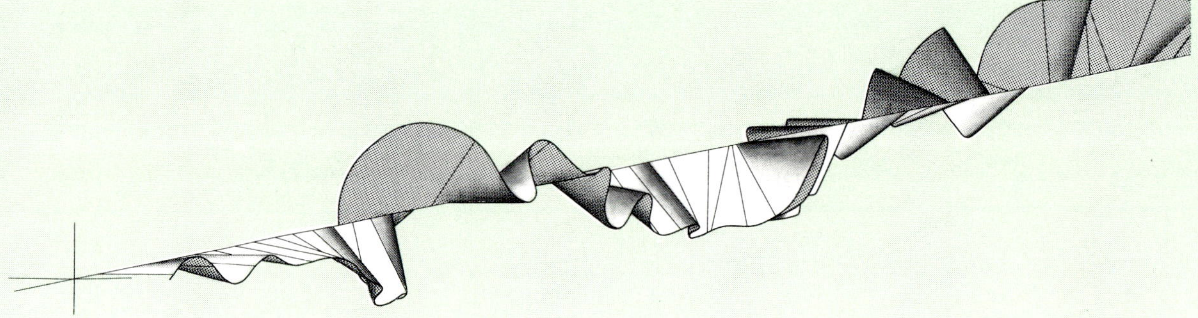

(Fig. 1.11). Phase shifts cause the transform to wind around the axis. The actual coefficients are computed as complex-number pairs representing the sine and cosine components of the frequency vectors. The cosine component is the real term and the sine the imaginary.

The Fourier integral is not new. It has been applied for over 150 years to a wide range of problems. The broad scope of the subject and the wide range of applications tend to cause a theoretical study of Fourier transforms to become very comprehensive and extremely complex. However, as pointed out by Dr. Ron Bracewell in his excellent book "The Fourier Transform and Its Applications", the special mathematical difficulties with the Fourier series are usually associated with their extreme nature in non-physical situations.

The mathematical definition of a pulse, for example, is a signal of infinite amplitude extending over an infinitely short period of time. Yet, for practical purposes it is possible to approximate a pulse with as much precision as the ability to measure it. The limits of mathematical precision required for digital processing are usually pre-ordained by the constraints of the parameters and field equipment. The sampling rate and the length of record immediately fix the range of frequencies and the frequency sampling increment while the amplifiers and recording equipment determine the precision and range of the measurement of amplitude.

Likewise, for most applications simple mathematical expressions may be used in place of cumbersome functions which do not have a physical possibility. This may be considered to have special application to the special case of the seismic signal used for petroleum exploration. The inherent low-frequency band-limited nature of the seismic signal restricts the treatment to a relatively simple case, in much the same manner as a low-order polynomial may be used to describe exactly a simple surface. As a result, most of the more difficult and complex developments do not apply.

The ability to transform the data from one domain to the other is a very useful one since it provides a means to make quantitative analyses and comparisons of several related signals. The amplitude and the phase of each frequency component of each signal can be measured with considerable precision and it is often then very easy to detect the exact nature of differences which may exist between two signals. Also, it is possible to modify signals selectively by removing or changing specified components of the signal in a controlled manner. Another useful application is in determining what changes were produced in a signal as it passes through a given process.

A substantial benefit of familiarity with Fourier transforms and frequency-domain operations is a better understanding of the procedures and relationships in signal processing. Many situations are clarified by reference to the equivalent operation in the transform domain without any need for actually carrying out numerical operations.

Common Relationships

The amplitude spectrum, and to a lesser extent the phase spectrum, are very useful for analytical procedures. Certain conditions in time-domain signals, particularly when combined with others, may be difficult to detect in that domain, but may be clear in the frequency domain. Therefore, a few basic time-domain/frequency-domain relationships should be committed to memory, together with an understanding of their effects and significance.

A sinusoidal wave can be designated by a single point having coordinates of amplitude and frequency on the amplitude spectrum. The representation of the component could be a vertical line bar having a height equal to the scaled amplitude (**Fig. 1.12a**).

A single spike in the time domain contains all frequencies (**Fig. 1.12b**). The amplitude spectrum of the transform is a continuous line parallel to the base, representing a uniform amount of each frequency over the entire frequency range. Incidentally, a spectrum of this nature is termed white, by corollary with the visible spectrum where a uniform amount of each frequency over the visible spectrum is seen as white light.

True random noise measured over a sufficiently long time period will also contain equal amounts of all frequencies. Therefore, the amplitude spectrum of random noise is identical to that of the spike (**Fig. 1.12c**). The important characteristic that separates the two is contained in the phase spectrum. The impulse spike has all components at zero phase. Random noise will have a completely random phase spectrum.

Seismic signals are band-limited. And so is seismic noise. Also seismic noise is white only in certain forms such as wind noise. In general, seismic noise is stronger in the middle of the band and fades towards the edge (**Fig. 1.12d**).

The theoretical seismic pulse should have a white amplitude spectrum, but in practice it also has a Gaussian or bell-shaped spectrum (**Fig. 1.12e**). Noise and signal are additive, so the amplitude spectrum of a field trace represents the sum of signal and noise elements (**Fig 1.12e**). A small amount of practice can develop the knack of mentally converting from the time-domain to the frequency-domain and vice versa.

Data Processing

Seismic data processing is merely a special case of the general field of signal processing. Familiar signal processing devices are such things as the telephone, radio and television. Most of these devices process analog signals through several electronic components.

The majority of electronic circuits can be simulated by digital computer programs, which today frequently replace analog circuitry as a tool for signal analysis and processing. Changes to the digital circuit models are made by merely changing a few program commands or control parameters. Many analytical techniques, in particular the Fourier transform, are performed best by digital computers. Perhaps the major advantage of digital processing is that it can be auto-adaptive, first applying some analytical techniques to the input data and then using the results to modify the process to be applied to improve the output.

Most signals are expressed as continuous analog functions over some interval of time. The digital computer operates on discrete numerical samples. It is therefore necessary to digitize the continuous signal into a series of samples of the amplitude of the signal. For simplicity, the spacing of the samples should be uniform.

FIGURE 1.12

A sinusoidal wave can be designated by a single point having coordinates of amplitude and frequency on the amplitude spectrum. The representation of the component could be a vertical line bar having a height equal to the scaled amplitude.
A single spike in the time domain contains all frequencies. The amplitude spectrum of the transform is a continuous line parallel to the base, representing a uniform amount of each frequency over the entire frequency range.

True random noise measured over a sufficiently long time period will also contain equal amounts of all frequencies. Therefore, the amplitude spectrum of random noise is identical to that of the spike.
Seismic signals are band-limited. And so is seismic noise. Also, seismic noise is white only in certain forms such as wind noise. In general seismic noise is stronger in the middle of the band and fades towards the edge.
The theoretical seismic pulse should have a white amplitude spectrum, but in practice it also has a Gaussian or bell-shaped spectrum.
Noise and signal are additive, so the amplitude spectrum of a field trace represents the sum of signal and noise elements.

FOURIER TRANSFORM PAIRS

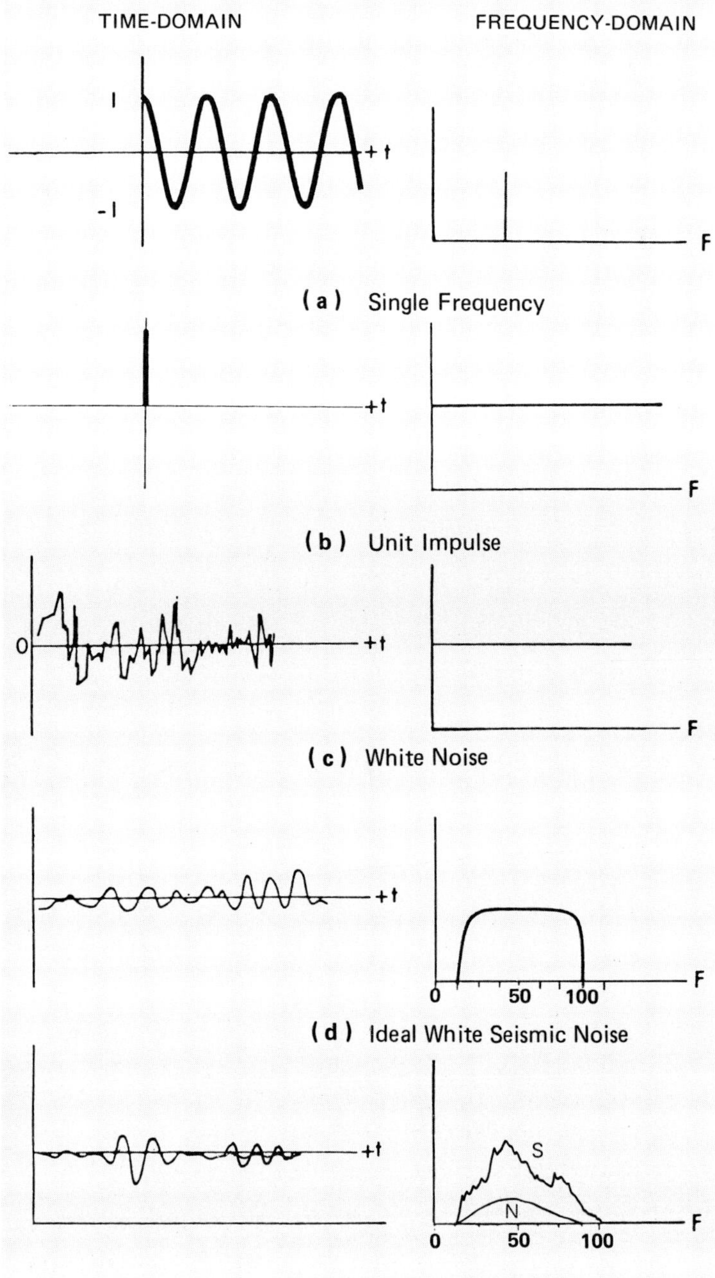

A very important consideration of all sampling is the fact that once any continuous function is sampled all information between the samples is lost. This is not a problem unique to signal processing. A geologist, for example, may use a limited set of information, say well data, to construct a map. Strictly speaking, he has absolutely no knowledge of anything between those points. He may use his experience to map the probable situation, interpolating the continuation of a trend between certain control points to represent a fault rather than steep dip. But if his map is based solely on the sample information from a well file he has absolutely no way of determining what lies between the points of a set of samples.

The addition of factual or interpretive external knowledge to a set of samples is carried out so routinely in so many operations that it may not be recognized as such. For example, the values interpolated between sample points used for signal processing and Fourier analysis, by definition, all fall on continuous sinusoid functions, so the information contained in sine and cosine functions is added to that contained in the basic data set. A geologist or geophysicist mapping subsurface structure almost invariably adds his knowledge of regional trends to the information contained in the data set. Correlation of well logs, particularly in areas of changing interval thickness, is a complex interpolation process which can be performed better and more quickly by a trained person than by machine, since the machine has available only the factual information contained in the data set, plus a relatively limited set of rules for interpolation that have been programmed into the system.

Maps based on sampled data sets can be misleading if the sample set is not a valid statistical representation of the total. Presumably all exploratory tests for hydrocarbons are drilled on anomalies. If so, a map constructed from only

FIGURE 1.13

Signal frequencies outside the limit 'fold back' into the frequency spectrum and add to whatever elements are legitimately there.

FREQUENCY ALIASING

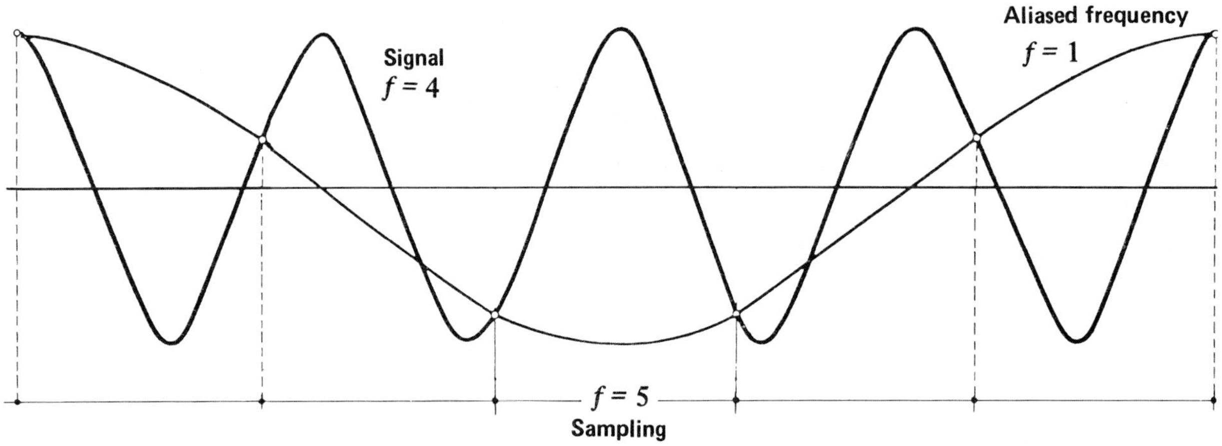

exploratory well data would not truly represent the continuous surface that it was intended to map. For this reason, randomly spaced samples are usually considered to provide the most reliable data set. Regularly spaced samples are acceptable if they do not introduce a systematic bias. The number of samples taken should be sufficient to be representative of the total.

So it is with digital signal analysis. The signal must be sampled often enough to make sure the set of samples truly represents the signal. On the other hand, the cost of computer process operations is strictly a function of the number of calculations made. Therefore it is desirable to perform the minimum number of operations required to obtain a reliable answer. Fortunately, these different needs can be reconciled by certain well defined properties of sampled time series.

A fundamental rule states that at least two samples per cycle must be taken if any given frequency is to be defined properly. Thus the sampling rate must be at least twice the frequency to be sampled. This sets the mathematical limit of the highest frequency that can be measured with a given sample rate. It follows that within mathematical limits a rate of two samples per cycle is sufficient and adequate to define a given frequency component, given reliable sampling equipment. This is an extremely useful fact, for it provides a simple and direct calculation of the sample rate required for any signal for which the upper limit of frequency is known.

A frequent but erroneous belief is that more frequent sampling will always provide more information or greater resolution. This is true only when the detail information, expressed in terms of high frequency, is actually present in the signal. But once the sampling rate is sufficient to define the highest frequency contained in the signal, no additional information can be obtained by more frequent sampling. Increasing the sample rate merely extends the upper limit of the frequency spectrum that can be examined. This upper region of the spectrum will be empty if the high frequencies are not present in the signal. So more frequent sampling is merely a waste of effort, and may be worse (since sampling equipment may produce noise, and the total sampling noise is a direct function of the number of samples). Excessive sampling rates may actually increase the amount of noise in the sampled data set, thus reducing the signal-to-noise ratio of the data.

If the signal to be sampled contains frequencies higher than one-half the sampling frequency, i.e. wavelengths shorter than twice the sample interval, the definition of the higher frequencies will be lost. Worse yet, the lost information will reappear as a false lower frequency distorting the measurement of the true lower signal frequencies (longer wavelengths). This phenomena is known as aliasing, and the highest signal frequency which can be defined correctly by a given sample frequency is known as the aliasing frequency or the Nyquist frequency, F_n . Frequencies higher than the Nyquist frequency corresponding to a given sample rate will appear on the frequency spectrum of the sampled signal as components having a frequency equal to the absolute value of the difference between the signal frequency and the sampling frequency. Signal frequencies outside the limit fold back into the frequency spectrum of the sample series. (**Fig 1.13**). A familiar example of aliasing is the apparent reversal of rotation which occurs to spoked wheels in motion pictures. Slow rotation of a wheel is filmed but as the rate increases, and the position of the spoke advances between camera frames by more than ½ the distance between spokes, the wheel appears to rotate backward. If the wheel continues to speed up, eventually the spokes advance slightly more than one spoke position between camera frames and the rotation once again appears to be correct.

To avoid the aliasing problem in seismic recorders manufacturers place analog high cut frequency filters in the input circuit ahead of the sample and digitize modules in order to remove any frequencies above the Nyquist frequency corresponding to a given sample rate. But electronic filters do not have perfectly abrupt cutoffs, or at least not without introducing other problems, such as phase distortion to the output, so the manufacturers construct their high cut filters with a relatively slow roll-off on the high side, and place the alias high-cut point in their seismic instruments with a 50% safety factor. Hence, most seismic alias filters are designed to cut off at one-half of the theoretical limit. A sample rate of 1 ms has a Nyquist frequency of 500 Hz, but the field alias filters are designed with a high-cut of 250 Hz.

This has important considerations for 4 ms sampling. Conventional alias filters are designed to high-cut the data at 62.5 Hz which is hardly satisfactory for today's broadband shooting. Some manufacturers now supply an alias filter at 75% of the Nyquist limit. A 75% cutoff extends the range to 93.75 Hz (approximately, since filters are not that precise), a much more satisfactory limit.

Sample Rate Δt	Theoretical F_n	Alias Filters F_a	Proposed F_a
1ms	500 Hz	250 Hz	375 Hz
2ms	250 Hz	125 Hz	187.5 Hz
4ms	125 Hz	62.5 Hz	93.75 Hz

At present, due to the 50% cutoff convention, it is common practice in data processing centers to decimate field data to one-half the original sampling rate. Decimation to increase the sample rate can be accomplisher very effectively by merely discarding every other sample of the field trace. This practice is acceptable but is not without risk, and should not be used routinely without an understanding of the effects of aliasing as they apply to the data at hand. If the field alias filter rolls off over one octave, for example, and the spectrum of the field data contains frequencies higher than the Nyquist frequency for the new sample rate, some of the higher frequencies above the limit will merely be attenuated, but not removed from the signal by the alias filter. These will be aliased back into the low frequency end of the spectrum, where the subsequent application of deconvolution may restore them to an appreciable amplitude. Even this may not necessarily distort the output if standard practice is to low-cut filter the output. If the frequencies above the Nyquist limit do not exceed that number by the amount of the low frequency filter, say 15 Hz, they will alias back into the range from 0-15 Hz, and be removed by the low-cut filter.

If decimation is to be done it should include recognition of the above and compensation to avoid aliased frequencies. One remedy is to high-cut filter the input data to the new limit with a sharp digital filter before decimation. This can be done very precisely in the frequency domain by taking the Fourier transform of the input trace, truncating the spectrum to one-half its length, and applying the reverse transform to return to the time domain.

The reverse process, to increase the sample rate, may not be so simple. The output should be a smooth sinusoidal function. Linear interpolation between samples is not acceptable since the intervening point will not follow the true sinusoidal form.

A much better solution is again provided by resampling in the frequency domain. The sample set is first transformed. Then, for each doubling of the sample rate desired, the length of the transform is doubled by adding the required number of samples, at zero amplitude. The inverse transform back into the time domain yields a perfect sinusoidal interpolation.

In the time domain resolution of broad band signals is related to the sample rate, Δt, in time and in the frequency domain to the length of the frequency spectrum.

In the frequency domain, resolution of the frequency spectrum, Δf, is a function of frequency sampling which is not controlled by the sample rate in the time domain, but rather by the length of the time sample, for the following reasons:

The Nyquist frequency defines the highest frequency that can be mapped with a given sample rate. At least two time samples are required to measure each cycle at the Nyquist frequency. If a signal one second long is sampled every millisecond, it will yield 1000 samples, and since at least two are needed to define each cycle of the highest (or any) frequency, the total number of different frequencies that can be measured would be 1000/2 = 500.

The Nyquist frequency sets the upper limit of frequencies that can be sampled at 1 ms to 500 Hz. Therefore, the resolution in F will be the spectrum 500 Hz divided by 500 increments, for a ΔF of 1 Hz. Changing the time sample rate does not affect the spacing in F as may be seen by the following table for a time series one second long.

Δt	Samples = S_n	$S_n/2$	F_n	Δf
1 ms	1000	500	500 Hz	1 Hz
2	500	250	250	1
4	250	125	125	1

But for a signal two seconds long:

Δt	Samples = S_n	$S_n/2$	F_n	Δf
1 ms	2000	1000	500 Hz	.5 Hz
2	1000	500	250	.5
4	500	250	125	.5

Hence, the total length of the time signal determines the rate of sampling in frequency Δf. The longer the signal, the better will be the resolution of the frequency spectrum contained in the signal. A signal one second long will yield frequency information in exact multiples of one cycle per second or one Hz. If it is desired to examine frequencies in step of one-half cycle a sample two seconds long must be available.

Time Sample	ΔF
1.0 sec	1.0 Hz
2.0	0.5
4.0	0.25

There is a fixed reciprocal relationship between time sampling and frequency. The interval between time samples, Δt, determines the total length of the frequency spectrum, F_n; the total length of the time sample, T_n, determines the frequency spacing, Δf. This is typical of the reciprocal relationships which exist between the time domain and the frequency domain.

The sampled time series data set is the primary input to the majority of automated procedures for extracting and enhancing significant information from any set of data. The fundamental constraints that govern harmonic analysis of time series are relatively few and simple. Yet they are extremely valuable as a means for both understanding and utilizing the information in sampled data series, whatever the origin. Complete familiarity and understanding comes only with experience and a partial understanding can be very misleading. They also provide some quantitative measure of the size and form of features that can be resolved by an appropriate process procedure. Certainly much effort is wasted today in attempts to define features which a modest analytical effort would show to be impossible of measurement or nearly so with the data at hand.

Time		Frequency
Δt	⟺	F_n
T_n		Δf

Fast Fourier Transforms

It would seem then that there is no good reason for not using the frequency domain operation as a regular processing procedure. However, during the initial years of development of the seismic processing industry, frequency operations were used primarily as a conceptual or analytical aid to determine the effect of various processes, while normal production operations were done in the time domain. The reason for this was the difficulty of transforming the time domain into the frequency domain. During the initial years of digital seismic processing the Fourier transform operation was an extremely tedious and time-consuming operation, requiring an enormous amount of calculation, doubled if it was desired to apply the reverse transform to step back into the time domain. The very large amount of calculating tended to reduce the precision as round-off errors were significant.

The cumbersome methods for calculating the transform precluded frequency-domain operations during the various stages of digital processing. Those cases for which transforms were essential, such as power-spectrum analysis, derived the spectrum from the autocorrelation function of the trace. This provides a much easier method of computation but at the expense of loss of the phase information. So the method cannot be applied to general transform operations. In many cases the power spectra were computed in gross frequency steps of anywhere from 1 to 5 Hz which provided a very crude approximation of the amplitude spectrum.

Later, the Cooley-Tukey computing algorithm allowed calculation of the Fourier transform with relative ease. The Cooley-Tukey algorithm utilizes an ingenious system of sorting related values to provide an increase in speed of better than two orders of magnitude, depending upon the length of the signal over which the transform is to be taken. Whereas the conventional method involves about n^2 multiplications to analyze a signal n points long, the Cooley-Tukey Fast Fourier transform requires only $n \log_2 n$. For 1024 points the FFT is approximately 100 times faster than the conventional transform. Also, the sums are small so the accuracy increases with little danger of round off errors. The algorithm requires the number of samples to be a power of 2.

At least in theory, every time domain operation has a frequency domain equivalent, and vice versa. In practice, execution of the exactly equivalent operation may be very difficult. For instance, a time domain filter operator usually is much shorter than the signal to which it is applied. This means that the filter operator samples rather coarsely in F. The output of the filter operation will not be quite the same as the output of the usual frequency domain equivalent process, normally done by modifications to the frequency spectrum. For this operation, the time domain equivalent is to filter with an operator of the same length as the signal.

General Fourier transform theory is based on the case of a continuous signal. Use of the discreet Fourier Transform, which is necessary when the signal has been truncated to a finite length, assumes the signal to be periodic over the length of the time sample, and may produce unwanted side effects. For example, the truncation of a continuous signal may produce distortion in the frequency response.

On the other hand, some operations in the frequency domain readily achieve results which are not easily obtained in the time domain. Another consideration may be an added computational cost of transforming the data prior to performing the operation. Yet in some cases, because of the simplicity of an application in the frequency domain, the total cost may be less than the time domain equivalent. New developments in hardware Fast Fourier Transform boxes supplied by several manufacturers increase the use of frequency domain procedures.

The main point of this discussion is to stress that both time domain and frequency domain operations should be available to enable selection of the best procedure for the needs. The Fourier transform is a completely linear process and it can be shown that given equal parameters, the operations apply in equal manner to both domains. The Fast Fourier transform has removed previous limitations of computing the transform. Either the time domain or the frequency domain may be used where best suited. An understanding of the time/frequency relationships and of the operations being conducted will help provide greater knowledge of both the advantages and the limitations in all phases of processing.

Computation of the discrete Fourier transform yields a unique, and repeatable, solution for any given sample, but that solution is totally dependent upon the constraints imposed on the solution by sample length, sampling rate, and the method of computation (which today, for all practical purposes is uniform). But it is a mathematical fit, and a change of any element of the basic parameters, or in the values of the data set, will yield a different solution.

The solution is correct only insofar as our ability to measure and define it, and this limitation necessarily requires certain assumptions to be made. For example, one of the assumptions made is that the signal is periodic over the length of the transform, which patently may be erroneous. The analysis resulting from one second of seismic data may have or may have little relation to the signal content of the next second.

Another very important point to remember is that the basic unit of construction of the sedimentary section is the sedimentary particle, or grain. These combine in groupings of every increasing size and at any boundary, a velocity contrast exists which responds in the manner described by our basic model.

The range of frequencies required to measure precisely the signal content corresponding to that section is very large, ranging from low to very high. Any lesser range of frequencies can only provide an approximation to the sedimentary sequence. The absence of high (or low) frequencies from a seismic section does not mean they do not exist in the sedimentary section; they simply have not been retrieved or measured.

2 BASIC PRINCIPLES

Drivers

Claus

PDCON

WDCON

~~Stack~~

Output Shots Stack.

(Stack)

Signal Theory

Much of the available signal theory, particularly that developed for radar and telephone purposes, is based on the assumption that a pure signal having some well defined shape or characteristic is transmitted from a source and is later received, contaminated by noise, at some distant recording point. The noise is usually assumed to be additive in nature. The problem then becomes one of retrieving the known signal from the noise and, if the signal has been distorted or damaged in transmission, of restoring it to its original shape or character. Techniques which do this extremely well have been developed, and are well known.

Unfortunately, most of the operations in petroleum seismic exploration fail to meet the basic assumption. The source signal, when produced by any impulse energy source such as conventional explosives, is a single, relatively short sharp impulse not readily defined on land work. It lacks any known constant definitive shape or signature and thus can not be extracted in the processing. Even if it could be defined, as the recording of the source signature in marine work is, the subsequent influences of instruments and transmission through the earth modify the pulse to an uncertain form so, inevitably, the processing must proceed under certain assumptions or approximations.

It is not possible, on a routine basis, actually to measure or observe the seismic wavelet at any intermediate stage between the shot and the geophone. We can only input energy into the ground,and observe the end result of the reflection process as it appears on the recorded trace. Special experimental recording is very expensive, therefore it is often necessary to model the process, either by electronic equipment or, often better, by a digital computer model. It is also frequently difficult to determine the actual physical conditions which exist in the field, so modelling is used a great deal in the development of computer programs for seismic processing. The models are relatively inexpensive and can simulate both the passive and active parts of the system, i.e., the sedimentary column and the seismic survey process applied to it. As a result, practically all processing is based on certain assumptions

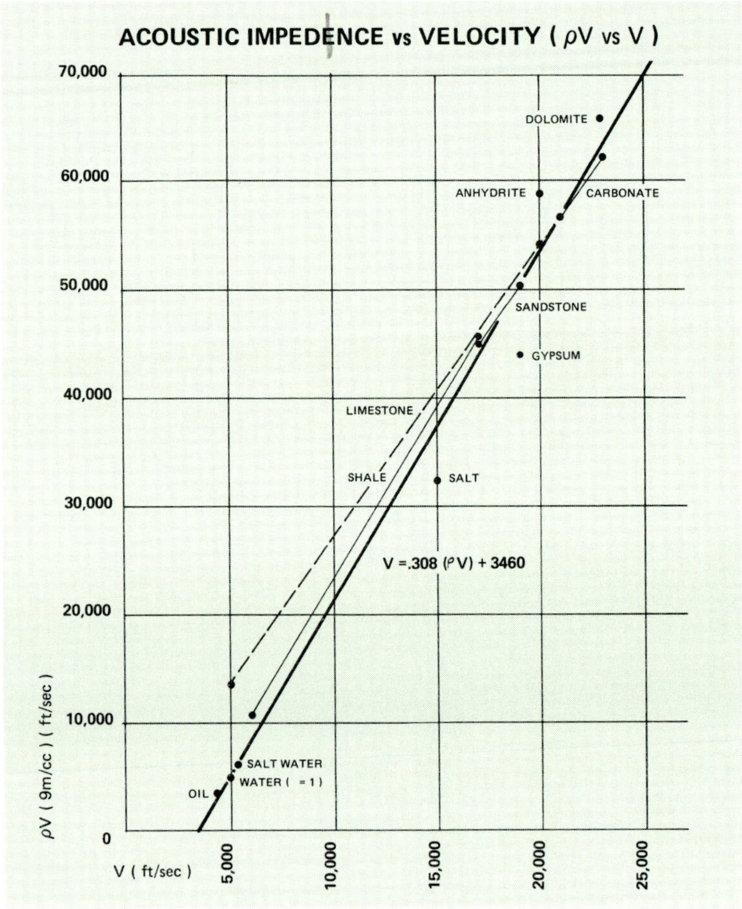

FIGURE 2.1

Acoustic impedence can be shown to approximate a linear function of velocity.

usually derived from a limited amount of measured data. The assumptions are used to model the conditions to be considered in the process sequence. Although the sedimentary section may represent all or part of the result of a continuous sequence of deposition extending over geologic time, it is almost always considered to be made up of a layered sequence consisting of discrete individual units of varying thickness of shale, sand, limestone or some other lithology. In most cases, this is a fairly reasonable assumption if the sampling of units is frequent enough to provide a reliable approximation of the depositional sequence. Although the depositional process may be continuous, it is by no means a constant or steady-state process. Changes in weather, environment and other conditions continually alter the nature and rate of sedimentation. Most of the changes in the sedimentary column are rather small. Sandstones grade into finer-grained material or into shale. Carbonates vary in velocity and density as the result of chemical changes or a variable percentage of some other component. Only occasionally do major changes occur, yet these may be so distinctive that there may be a tendency to ignore all but these boundaries, and this may lead to erroneous conclusions.

When lithology changes, velocity and/or density, which are closely related to lithology or rock type, change also. It is precisely such changes in velocity and density which make the seismic reflection method possible. Sonic energy is reflected from the contact between two units of differing velocity and/or density as a function of acoustic impedance. Acoustic impedance can be shown to approximate a linear function of velocity (**Fig. 2.1**), and often the resulting reflection coefficient will be nearly the same as that produced by the velocity differences alone. Also, the range of density values is generally much smaller in magnitude compared to velocity. For these reasons, and for the fact that only recently have density logs become commonly available, density is often ignored in calculations of reflection coefficients. This will be the case in much of the development to follow. However it is important to remember that density may have an important influence and in some cases, account for reflection amplitudes that cannot be explained by velocity.

Concepts of Models

The simplest model of the reflection process starts with two homogeneous rock layers, one resting on the other. For any physical property which can be measured by the seismic method, the only point at which any change in the response can be expected is at the boundary between the two layers. By definition, this model boundary is a mathematical line, i.e., infinitely thin, and the change in any measurement would occur at a point on the boundary. (Fig. 2.2a).

SONIC LOG MODELS

FIGURE 2.2

FIGURE (a)

A borehole sonic log is a close but not perfect match to the ideal model log. The typical sonic tool has a spacing of 3 ft between the two sound detectors. Therefore, the response of the real logging tool is not instantaneous but is averaged or spread over the 3 ft span as it moves across the boundary.

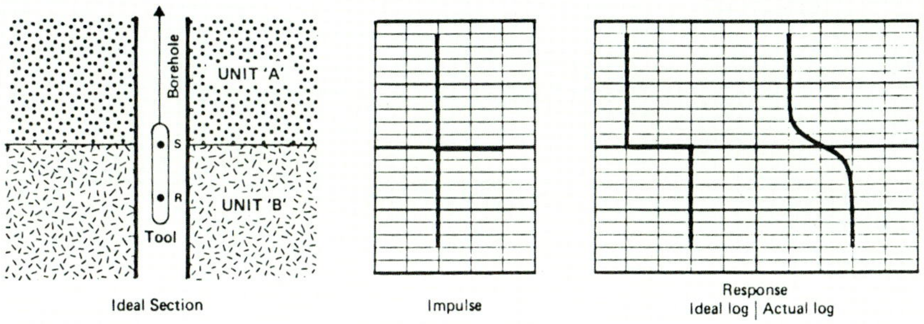

A Single interface develops an 'impulse' which causes a shift in potential, or response, of the log recording.

FIGURE (b)

The logging model may be extended into successively more complicated models first by introducing another interface.

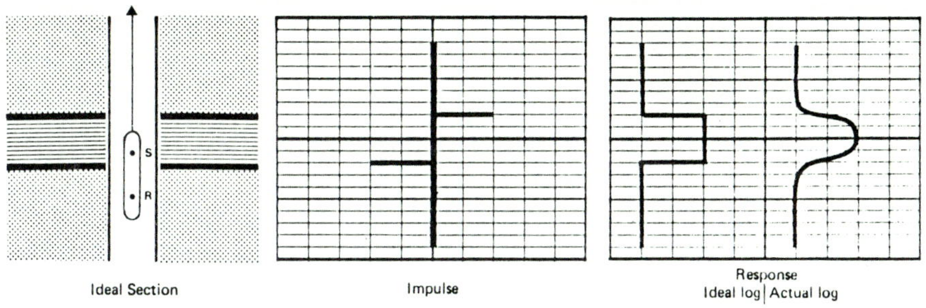

Reversed interfaces develop compensating impulses which return the log curve to its origional position.

FIGURE (c)

Each of these responses could be plotted separately and then subsequently summed to obtain the final response of the tool traversing the sequence of formations.

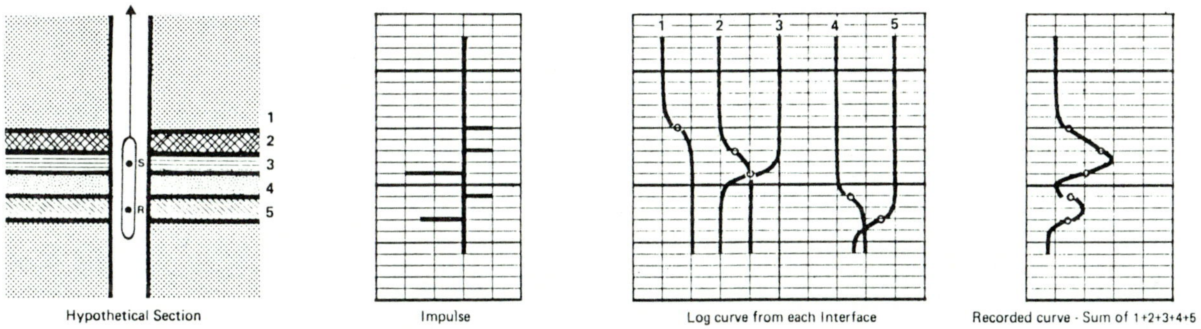

Random interfaces develop many responses which in sum, produce a log pattern which characterizes the geological section.

A borehole logging tool measuring some physical property of the earth would respond with some sort of impulse or potential shift at the boundary, otherwise the response would be flat with no change. The nature of the response might be a simple shift of measured value, say of velocity, or it might be an impulse, as in the case of some potential measured across the boundary. The two types of response could be related. A velocity log measures the velocity of the rocks, but the output could be arranged to measure only the velocity difference across the boundary, i.e., the first derivative. The output would then take the form of a spike representing the actual amount and direction

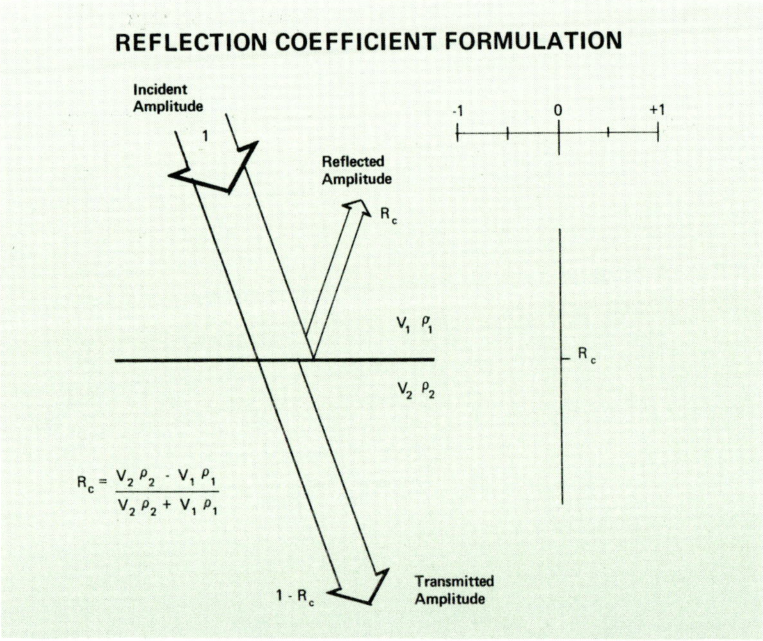

FIGURE 2.3

The fundamental concept of seismic exploration is to send into the earth a short signal which is then reflected back from a boundary between two units.

of change in velocity across the boundary. This is a useful concept for seismic reflection work, since the coefficient of reflection is dependent upon the change in velocity. A numerical model of the velocity-change impulse would have zero at all sample points except at the boundary, where the impulse could be represented by a single spike of a length proportional to the velocity change, i.e., $V_2 - V_1$, where V_2 is the velocity of the deeper unit. A simple integration of this impulse function, plus some constant, would generate the response of a sonic-velocity logging tool.

A borehole sonic log is a close but not perfect match to the ideal model log. The typical sonic tool has a spacing of 3 ft between the two sound detectors. Therefore, the response of the real logging tool is not instantaneous but is averaged or spread over the 3 ft span as it moves across the boundary (**Fig. 2.2a**).

A series of very closely spaced samples might be used in a digital model of the ideal log. Any change in response would be confined to the interval between two adjacent samples. To model the response of a real log would require applying a 3 ft running average over the sample values. If the numerical sequence representing the sedimentary column had been sampled every foot, the running average could be obtained by simply multiplying three adjacent samples by 1/3, 1/3, 1/3, and adding the products to obtain the average value. The running average would be shifted 1 ft and the sum-of-products operation repeated to produce the next output point. If a more accurate representation were warranted, sampling at more closely spaced intervals could provide greater detail.
This type of numerical running-average operation is convolution. It is the manner by which the response of any linear system can be simulated by a numerical model.

The logging model may be extended into successively more complicated models first by introducing another interface (**Fig. 2.2b**), and on to numerous interfaces (**Fig. 2.2c**). Here the amplitude and polarity of each impulse is proportional to the amount of velocity change across the boundary. A logging tool moving across the series of beds will respond to each boundary in an identical manner, with amplitude proportional to the amplitude and polarity of the impulse. Each of these responses could be plotted separately and then subsequently summed to obtain the net final response of the tool traversing the sequence of formations (**Fig. 2.2c**).

Velocity logs and seismic reflections are closely related, so a similar impulse model can be used to model the seismic process.

Reflection Coefficients

The fundamental concept of seismic exploration is to send into the earth a short signal which is then reflected back from a boundary between two units (Fig. 2.3). In this model, the impulse function is not a production of the simple velocity difference but rather of acoustic impedance. This gives rise to the familiar fundamental calculation of the reflection coefficient:

$$R_c = (\rho_2 V_2 - \rho_1 V_1) / (\rho_2 V_2 + \rho_1 V_1) \quad \text{where } \rho \text{ is density.}$$

LOSS OF REFLECTION COEFFICIENT AMPLITUDE
WITH INCREASING VELOCITY (DEPTH)

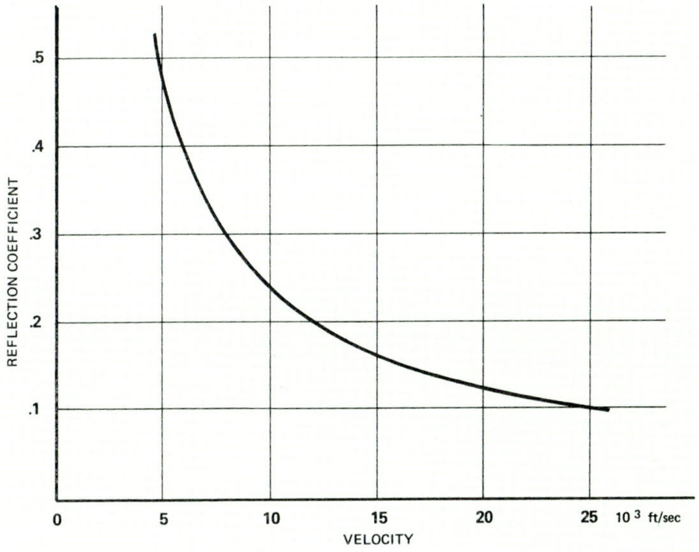

FIGURE 2.4

A velocity shift of 5,000 ft/sec in nearsurface sediments of, say, 6,000 and 11,000 ft/sec will produce a reflection coefficient of about 0.3, but the same shift of 5,000 ft/sec in rocks having velocities of 18,000 and 23,000 ft/sec will produce a reflection coefficient of only 0.12.

Density effects may be ignored for simple cases so the formula can be:

$$R_c = (V_2 - V_1) / (V_2 + V_1).$$

Second-and-third-order factors, in addition to density, could be included but they will not change the result appreciably. For simplicity, only the velocities, which provide a reasonable approximation, will be used in the basic discussion.

In the large, velocity increases with increasing depth, so the velocity change between any two samples might be considered ΔV, the increment. Then the reflection coefficient formula can be written:

$$R_c = \frac{\Delta V}{2V + \Delta V}.$$

Clearly, with increasing depth, the denominator of the equation will become increasingly large, without a corresponding increase in the numerator, so the effect of introducing the velocity sum on the bottom of the equation is gradually to diminish the average amplitude of reflections with depth. A velocity shift of 5,000 ft/sec in near-surface sediments of, say, 6,000 and 11,000 ft/sec will produce a reflection coefficient of about 0.3, but the same shift of 5000 ft/sec in rocks having velocities of 18,000 and 23,000 ft/sec will produce a reflection coefficient of only 0.12 **(Fig. 2.4)**. This is one of several reasons for the decrease in reflection amplitude with depth.

If a signal of unit amplitude arrives at any reflector, a fraction of that signal, equal to the reflection coefficient, will be returned to the surface and $1 - R_c$ will be transmitted through the boundary to the next reflection boundary. **(Fig. 2.3)**. This process continues across all boundaries **(Fig. 2.5)** and it is easy to see that back-scattering produces an exponential decay of the amplitude. Internal reflections further complicate the problem, but these will be ignored in the interests of simplicity at this stage.

The Seismic Model

A model of the seismic process is commonly used in the development of digital data reduction processes (Fig. 2.6). A single spike impulse having an amplitude of unity is the input into the model system. The action of the model system on the input should represent the total effect of all influences on the signal between the pulse origin and the final recorded signal. The output is the unit impulse response of the model system.

The theoretical model of **Fig. 2.6** has certain well defined characteristics. The input has zero amplitude at all points except one, where it is a maximum. There is no output before any input, for in a real system there can be no output prior to some input. The output responds to the unit input impulse in a definite manner, rising immediately to a maximum, and then oscillating with an exponential decay of amplitude.

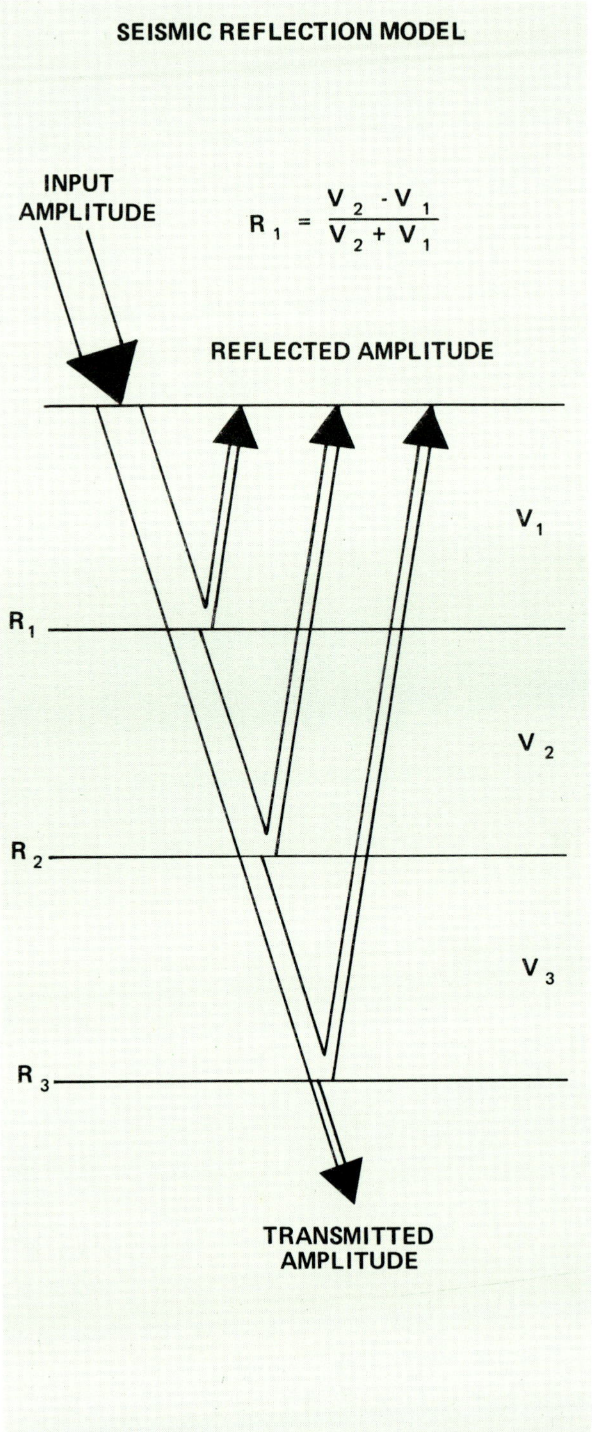

FIGURE 2.5

This process continues across all boundaries.

Systems which respond to a transient impulse in this manner are common in many real physical situations. They have well documented response properties backed by extensive development of theory. But, as is so often the case with seismic data operations, the well developed theory applies only to systems which are better defined or better behaved than the average seismic system. The model of real seismic system may appear more like the seismic model of **Fig. 2.6**. The input is not exactly the instantaneous unit impulse of the theoretical model, nor does the output exactly exhibit pure transient impulse response characteristics.

This fundamental concept of the system model is of utmost importance in the development and understanding of all seismic data processing operations. Nearly every innovation in seismic data processing results from two basic operations:

(1) The development of a system model which accurately represents the real system. Any differences which exist between the two must be corrected by either modifying the model to match the system or modifying the system to match the model. Residual differences will likely remain, and these are tolerable only if they are small enough to have an insignificant effect on the output, or if their nature is well enough understood to permit an effective evaluation of their effects on any given set of data,

(2) Once the model is defined, development of a data enhancement process then becomes (usually) one of reversing the effects of the system; that is, to try to convert the output of a system back to the form of the input, eliminating insofar as possible all of the influences which may have modified the data in its transit through the real system.

It is very important that the model closely represent the real physical system, rather than an idealized system. Seismic data processing operations based on purely ideal systems may not always perform in practice as desired. Many apparently excellent processing programs break down in application simply because the real system does not meet the specifications required by the theoretical system used to develop the process.

The development of the model is complicated by the fact that the response of the real system may be somewhat different from trace to trace and from shot to shot. The same broad-

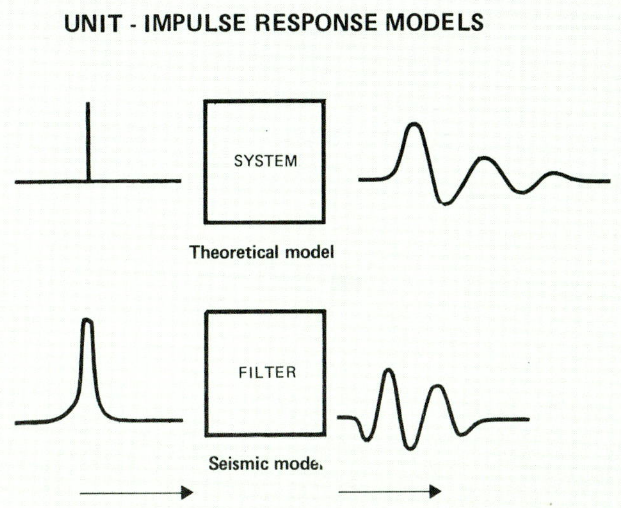

FIGURE 2.6

A model of the seismic process is commonly used in the development of digital data reduction processes.

band pulse may be sent out each time but the conditions of the shothole and the lithology of the section may change in such a manner as to cause some variation in the unit impulse response. Fortunately the majority of the system components, although they cannot all be individually defined, remain constant. Therefore, variations in the system are normally due only to some change in the conditions of the earth (the transmission medium) and the coupling of the earth to the instruments.

The reflecting interface has been defined as a line boundary. If it is to be located precisely in space the transmitted signal pulse must be at least as sharp as the boundary. Otherwise, like the span of the logging tool crossing the boundary, a pulse which spans some period of time will reflect back in a continuous stream from the boundary during the length of time it takes for the transmitted pulse to pass through the boundary (**Fig. 2.7**). This reflection will not be an instantaneous impulse but, like the running average of the logging tool, the response will extend over some period of time. In this model, the spike of the reflection coefficient is replaced by the familiar seismic wavelet, scaled in amplitude and polarity in direct proportion to the amplitude of the reflection coefficient (**Fig. 2.8**).

Probably the sharpest available pulse source is explosive. However, for several reasons the explosive is confined within a borehole in the ground (or under water in marine work) and is surrounded by material which is not completely elastic. As a result, the pulse produced by the explosion, although it might be considered as a spike at its source, expends part of its energy in deforming the somewhat plastic surrounding material with the result that the pulse is deformed into a transient response several milliseconds long. This can be modelled by the simple system of **Fig. 2.6** which shows the spike entering the system and emerging, as through the side of the hole, in the form of an extended time pulse.

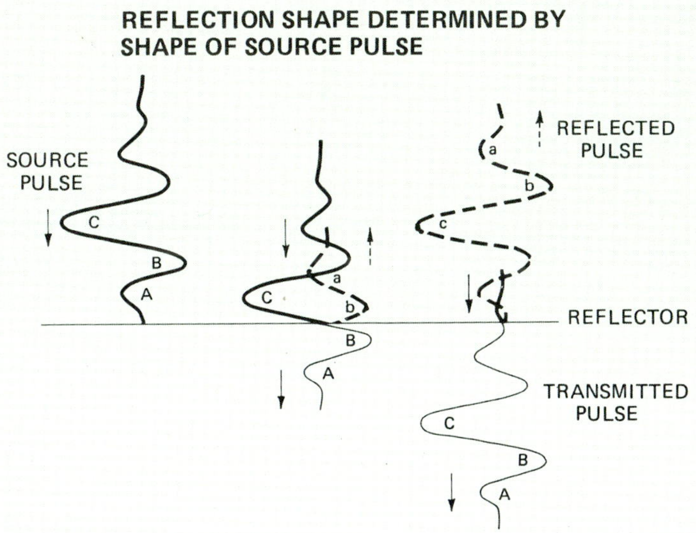

FIGURE 2.7

A pulse which spans some period of time will reflect back in a continuous stream from the boundary during the length of time it takes for the transmitted pulse to pass through the boundary.

In terms of frequency content, an explosive develops a very sharp input pulse which, for all practical purposes, is a completely white broad-band signal containing all frequencies. However it is soon modified substantially by the inelastic coupling of the charge to the earth into a severely

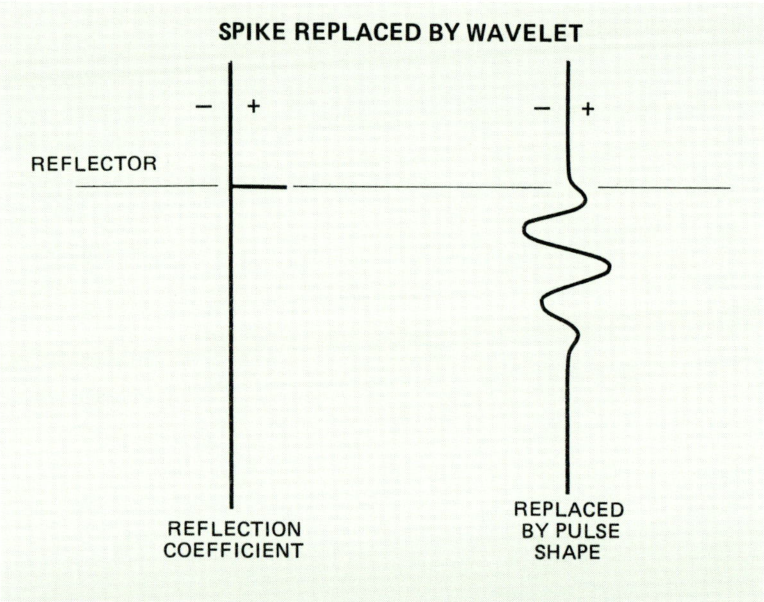

FIGURE 2.8

The spike of the reflection coefficient is replaced by the familiar seismic wavelet, scaled in amplitude and polarity in direct proportion to the amplitude of the reflection coefficient.

band-limited frequency spectrum. The amount and manner of modification invariably is unknown and cannot be readily determined. Hence the shape or signature of the transmitted waveform is unknown. To complicate matters, seismic noise often has properties very similar to those of the signal and the two cannot always be readily distinguished. The input pulse travels through the earth and is reflected from the reflection coefficient at the geologic boundary. At the receiver the response from the reflection coefficient appears to be the waveform rather than the spike of the reflection coefficient. In other words, the reflection coefficient has been replaced by the pulse. The impulse has been replaced by the impulse response at the input, and again at the reflecting boundary. This process is modelled very nicely by convolution of an operator representing the system response with a spike representing the impulse. The returning wavelet may have suffered further changes in transmission. These may include the effects of the earth, the geophones, the amplifiers and the modulation system which places the signal on tape. It may not be possible to determine the individual contributions of each of the components. Fortunately, it is not necessary to separate the response of the components of the earth from that of the instruments in order simply to synthesize or model a seismic signal. It is sufficient to determine the response of the entire system to the unit impulse, but even this may not be easy. The response function is termed the transfer function, and is the model which is used in subsequent work.

The System Model

The simple basic model of the system (Fig. 2.6) includes a single spike or impulse input which is modified by the system to a different form at the output. The effect of the system on the input might be described as a stretching in time and a change of shape, but this is not very precise nor does it permit a quantitative measure of the system effect.

The input has been described as a unit impulse or spike. It has the unique characteristic that when it is sampled, regardless of the sample rate, all values except one will be zero. The exception is the impulse itself, which may have any numerical value, but is usually considered to be the unit input value of 1.

If a range of pure frequencies is generated and a peak on each waveform is aligned on a common point, eg., a cosine wave at zero phase, summation of the sequence will produce a crude approximation of a spike (**Fig. 2.9**). Increasing the number of frequency components will improve the approximation until eventually a sampling of the output results in a time series having all sample values very nearly zero, except one which is a maximum and represents the unit impulse.

The unit impulse at time zero is therefore a broad frequency band signal at zero phase. The pure pulse input to the ideal model represents the zero-phase integration of all frequencies, from zero to infinity, at the same amplitude. Thus, in the frequency-domain, the amplitude spectrum would be a flat line of constant amplitude across all frequencies. The phase spectrum, referred to the time of impulse, would also be flat at zero phase for all frequencies.

FIGURE 2.9

If a range of pure frequencies is generated and a peak on each wave form is aligned on a common point, eg., a cosine wave at zero phase, summation of the sequence will produce a crude approximation of a spike.

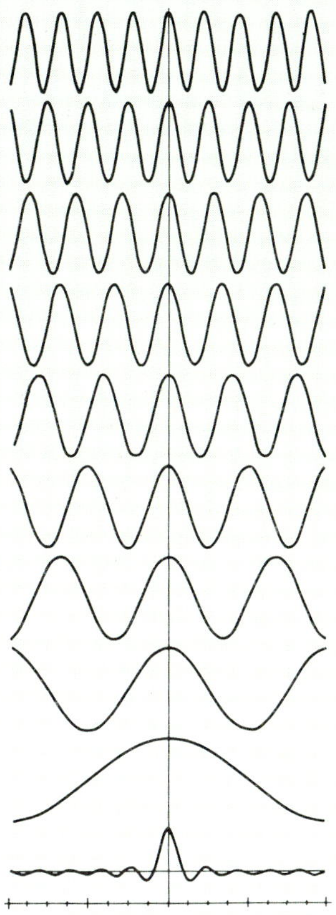

The signal from the model output is not at all like the input but extends over several samples and is not symmetrical **(Fig. 2.6)**. Signals which are not symmetrical are not zero-phase signals. Evidently some phase shifting must have occurred in the transmission of the signal through the system. A phase shift can be related to a time shift so obviously some of the components have been shifted in time in relation to the others. The system is a real time system. That is nothing can come out of the system until something goes into it. Therefore any time shift must be a time delay **(Fig. 2.10)**. Some of the frequency components have been delayed or lagged slightly in time and arrive later than the other components.

The output of the model is a waveform quite different from the input. The information signal which is recorded by a seismic system is composed of a very narrow band of frequencies. For comparison, the range of the human hearing extends from something like 20 to 20,000 Hz, yet the normal seismic signal range rarely exceeds 5 to 100 Hz.

Earth Filter

The seismic signal travels from the source to the geophone through the transmission medium, the earth, which acts as a filter to remove all but a very limited band of frequencies from the broad-band input. Those frequencies that do remain may be attenuated to some degree, particularly those frequencies which lie near the ends of the passband. Only over a very small portion of the band is the response fairly uniform.

The combined effect of the loss of several of the input frequency components and the phase shifting of the remainder produces a wavelet which is quite different from the sharp source pulse. The result is defined as the unit impulse response.

The impulse input to the complete system, earth plus instruments, has a uniformly flat amplitude spectrum and zero phase. Fourier analysis of the recorded signal will define its amplitude and phase spectra. It is therefore possible to define, for each frequency component, the exact alteration it has suffered in passage through the system. It is a relatively simple matter to define the effect of the instruments alone on the input. A single spike is input into the system and the response recorded. Fourier analysis of the response to a single spike input provides the amplitude and phase spectra. The familiar amplifier response curves, published for each set of instruments, show the amplifier system response. The amplitude and phase responses define the changes that have taken place in the input, which had 100% amplitude at all frequencies and zero phase. The time-domain response to a unit impulse is the wavelet (**Fig. 2.11**).

The response of the system results from convolution of the transfer function with the input. The convolution process is in many ways similar to the mathematical operation of multiplication. Since the input is a unit value

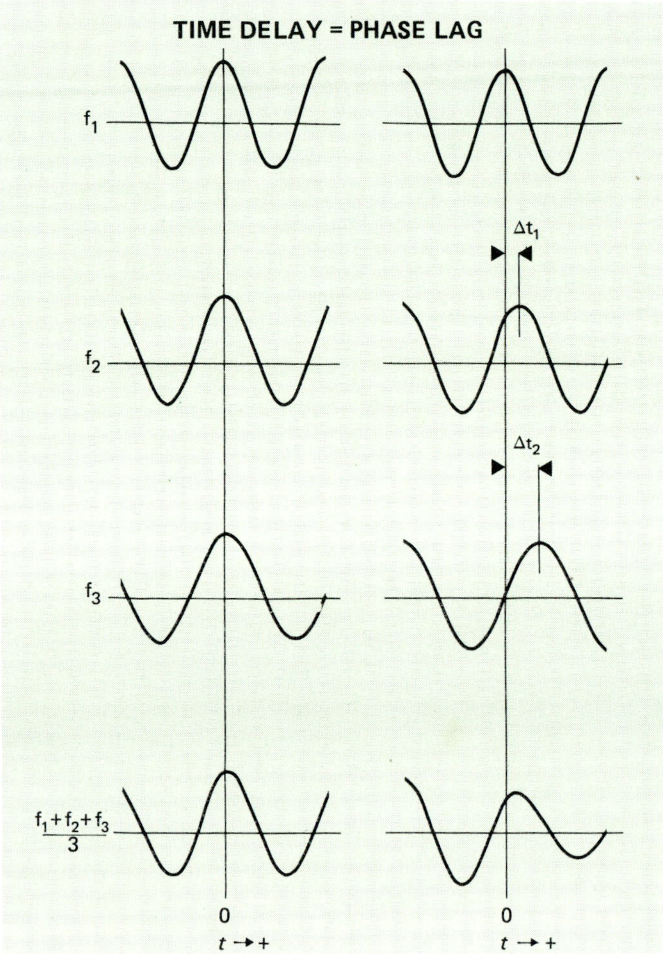

FIGURE 2.10

A phase shift can be related to a time shift.

the output response actually defines the transfer function of the system. In other words, the input having a unit value of 1 is convolved ('multiplied') with the transfer function to obtain the unit-impulse response. In this case, since the 'multiplication' was by 1, the unit-impulse response equals the system transfer function. In any case, if any output is 'divided' by the input, the answer will again be the transfer function.

It may be seen that if any two of the three system components, input, transfer function and response are known the third may be determined. If the response of the system to a single spike can be defined, the resulting model can be used to determine the response to a complex input signal and conversely, the transfer function or unit-impulse response, can be 'divided' out of a complex response to recover a complex input signal. Unfortunately, the unit-impulse response of the total seismic system is extremely difficult to define. The seismic trace contains a collection of unit-impulse responses, one for each reflection, but these are so intimately entwined that it is almost impossible to locate any single response to a unit impulse. A test pulse may be input to a geophone and the response recorded. However, this test will record only the response of the system between the geophone and the recording system output. It will omit a very important part of the total system, that of the earth. A closer approximation might be obtained by firing a light charge, such as a cap buried in a hole to obtain the response through the earth to the geophone. This might provide an improved approximation, but it is immediately contaminated by reflections and noise. Even if a reliable definition of the total system could be obtained at any particular point, a different shot or location a short distance along the seismic line could produce a quite different response.

Basic Numerical Operations

Three basic mathematical operations, convolution, cross-correlation and summing, or stacking, account for the majority of basic digital signal-enhancement processing in the time domain.

Summing, or trace stacking, is the well known simple process of summing corresponding samples from two or more sampled time series corresponding to the same subsurface point. The other processes used in the model systems to simulate the action of the real system are slightly more complicated, but can be readily modelled by computer.

The advantage of a mathematical computer model is that it can be modified easily by changing a few program statements or numerical values. Signals may be represented by uniformly spaced amplitude measurements of the signal sampled over some period of time. Thus the input might appear a 0, 0, 0, 1, 0, 0, 0,............, representing a

RESPONSE CURVES FOR A TYPICAL SEISMIC AMPLIFIER AND FILTER.

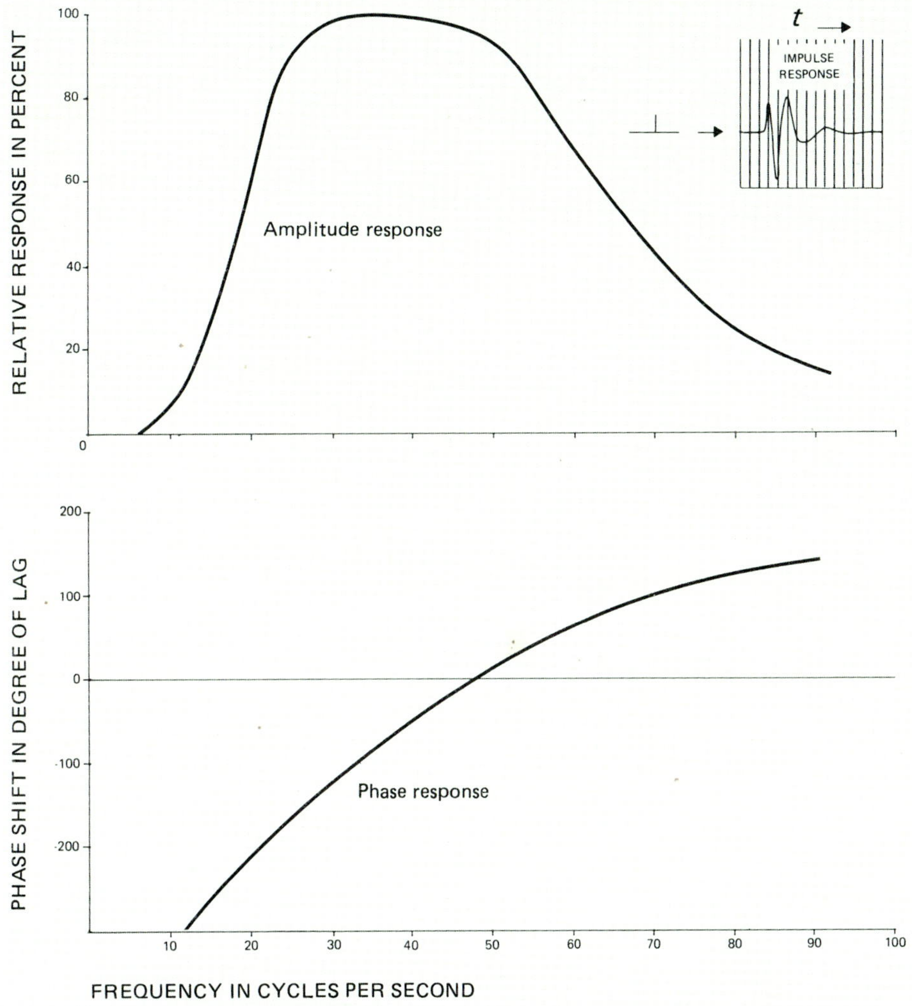

FIGURE 2.11

The familiar amplifier response curves, published for each set of instruments, show the amplifier system response to a pulse containing all frequencies and zero phase. The time-domain response to a unit impulse is the wavelet.

single pulse of unit amplitude occuring at some defined point in time.

The output would be a similarly spaced series of values which, if joined with a continuous line (**Fig. 2.12**) would approximate the shape of the unit-impulse response.

In the simplest case of a single impulse passing through the system, the operation amounts to a mere scaling operation, where each value of the transfer function, (equal to the unit impulse response) is in turn multiplied by the amplitude value of the single input pulse as it passes through the system to create a scaled version of the transfer function. The input is replaced by the scaled transfer function at the output When several closely-spaced pulses are passed through the system, as would be the case for a model of several closely-spaced reflection coefficients, each is replaced by the response, but in this case the response pulses will over-lap in time, and where they overlap, the output will be the sum of the several response pulses (**Fig. 2.13**).

This could be accomplished for any series of input pulses by calculating the response for each impulse and then summing them at the appropriate spacing given by the input. In fact, the same result can be obtained from a continuous multiply-add sequence termed convolution. The process is illustrated in **(Fig. 2.14b).**

The series of samples representing the transfer function is termed the operator. This signal, which is to be shifted in the process, *is first reversed end-for-end* and placed to the left of the series of samples representing the input sequence.

The operator is shifted right one position and each coincident value of the operator and of the signal are multiplied.

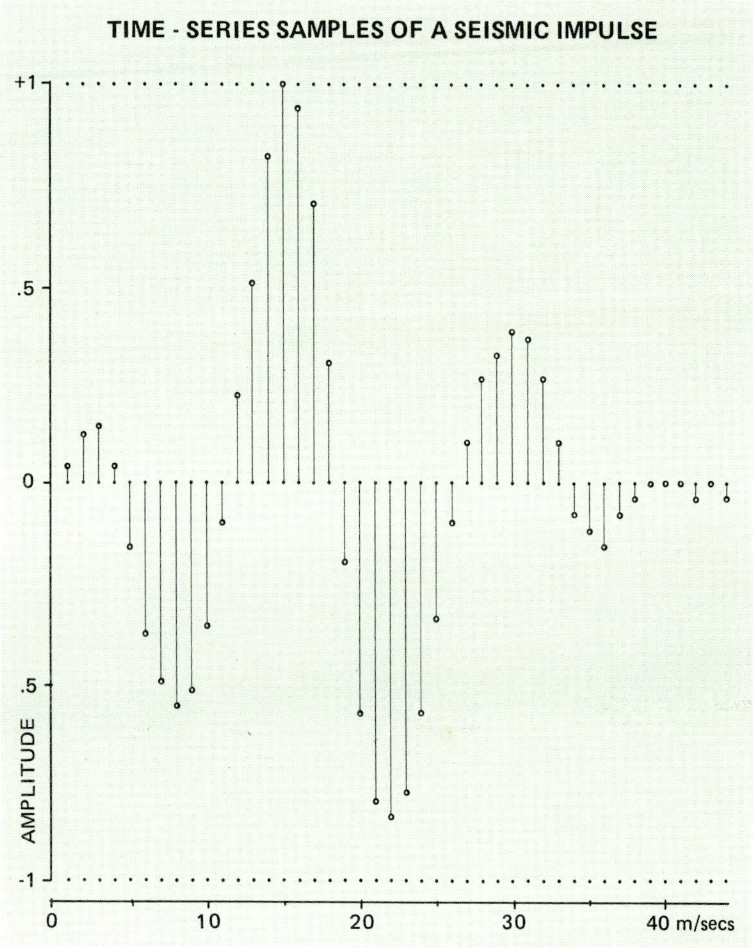

FIGURE 2.12

Signals may be represented by uniformly spaced amplitude measurements of the signal sampled over some period of time.

All the products are then summed to yield one output sample. The operator is shifted right one position and the operation repeated, until all positions have been occupied.

The result is the output series, in this case the sequence reading down the left side of the example.

It will be noticed by turning **Fig.2.14b** sideways, so that the output reads across the bottom of the page, that the output is exactly equal to the sum of a series of output responses scaled to the amplitude of each input impulse. Reading across the product matrix show each row to be a scaled response replacement for each point of the input signal, and the output to be the sum of these.

The only operational difference between convolution and cross-correlation is that the *convolution operator is reversed* end-for-end before application *whereas it is not for cross-correlation* (**Fig. 2.14a**), a seemingly simple difference but one with important properties.

The mechanics of performing the three simple basic operations is routine once the process is understood. However, the properties of the operations are fascinating. They provide power and, in many cases, limitations of signal processing.

Cross-correlation and/or convolution of a pure sine or cosine wave with a signal will extract that same frequency component from the signal (**Fig. 2.15**). Information regarding any other components of the signal will be lost from the output. A separate appropriate cosine frequency operator must be cross-correlated with the signal for each desired frequency component in the output. However, cosine waves of the same amplitude for several frequencies may be added and the sum of these used as a single operator. All of the corresponding components will then be extracted simulateously, and the output will also be the sum of those frequencies.

This property is used to advantage in digital band-pass filtering. A time-domain band-pass filter operator is the sum of all frequencies that fall within the desired band to be passed. Cross-correlation of this operator with any signal will, in a single operation, extract from the signal all frequencies within the band and reject the others. Band-pass filters are completely symmetrical, therefore either cross-correlation or convolution may be used with equal success.

From this may be derived another important property: the product of cross-correlation or convolution contains only those frequency components present in both the operator and the signal.

Convolution has much in common with the mathematical process of multiplication, including the commutative

OVERLAPPING IMPULSE RESPONSES ON A SEISMIC RECORD

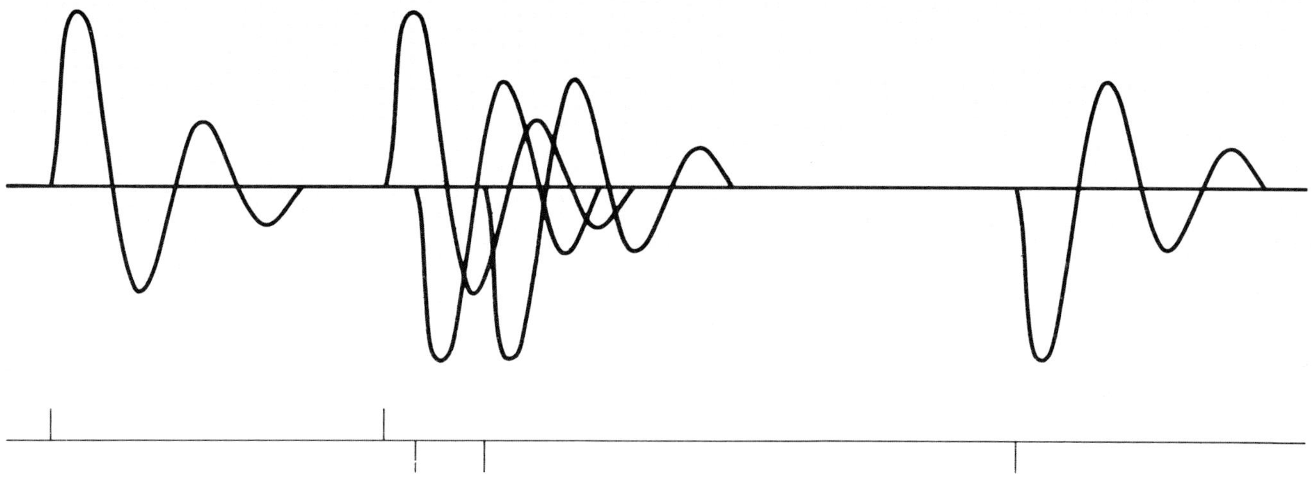

FIGURE 2.13

For a model of several closely-spaced reflection coefficients, each is replaced by the response, but in this case the response pulses will overlap in time, and where they overlap, the output will be the sum of the several response pulses.

property. That is, the order in which any series of convolution operations is done may be changed without affecting the final result. This is not true for cross-correlation.

In the frequency domain, *convolution,* the basic procedure for filter application, is accomplished by cross-multiplying the amplitude values of each frequency of the operator and the signal. The products yield the amplitude spectrum of the convolved output. Simultaneously, the values of the *phase spectrum* of the operator are *added* to the corresponding phase values of the signal to obtain the phase spectrum of the output (**Fig. 2.16a**).

In the time domain, the operations of convolution and cross-correlation are identical with the exception that the operator is reversed for convolution, but not for cross-correlation. Similarly, *cross-correlation* in the frequency domain is identical to convolution, with the exception that the *phase spectrum* of the operator is *subtracted* from, rather than added to, the phase spectrum of the signal (**Fig. 2.16b**). In both cases the product of the amplitude spectrum is computed.

Autocorrelation

Autocorrelation is a special case of cross-correlation in which the signal is correlated with itself. This amounts to squaring each component of the amplitude spectrum and subtracting the phase spectrum from itself, which will always yield a zero-phase output (Fig. 2.16c). For this reason, autocorrelations of all signals which have identical amplitude spectra, regardless of their phase spectra will be identical (Fig. 2.17). The operation of autocorrelation in the frequency domain becomes simply one of squaring the amplitude spectrum and setting the phase spectrum to zero. Squaring the amplitude spectrum also yields the power spectrum.

The autocorrelation is the time-domain equivalent of the power spectrum, and is itself a useful analytical tool. Characteristic phenomena associated with seismic recordings are often revealed when the effects of phase are removed from the time-domain signal. The autocorrelation discards all phase information but consolidates all of the amplitude information in a uniform relationship.

The sum of a broad range of frequencies at zero phase in the time domain will approximate a spike. Since auto-

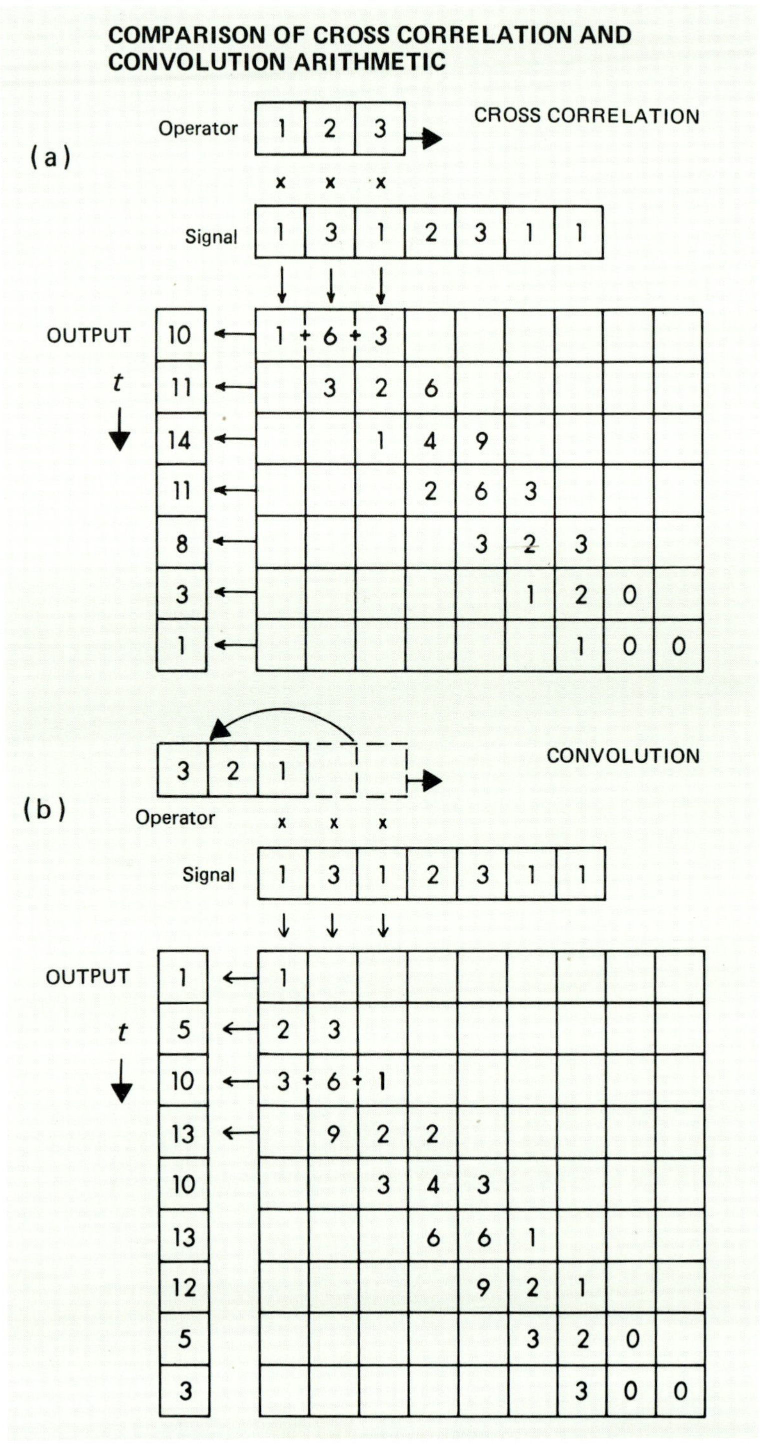

FIGURE 2.14a

The only operational difference between convolution and cross-correlation is that the convolution operator is reversed end-for-end before application whereas it is not for cross-correlation, a seemingly simply difference but one with important properties.

FIGURE 2.14b

The response for each of a series of impulses can be summed at the appropriate spacing given by the input. The same result can be obtained from a continuous multiply-add sequence termed convolution.

correlation forces all frequency components to zero phase, the autocorrelation of a broad-band signal, as might be expected, is a spike (**Fig.2.18b,c**). This is true whether the input signal is itself a spike or random noise.

Neither an autocorrelation nor a power spectrum will distinguish between noise and signal having similar frequency components.

The autocorrelation of the usual seismic signal will be neither the single spike of the broad-band signal nor the continuing cosine wave of the single frequency **(Fig. 2.18e)**, but will be between the two extremes. Seismic signals tend to be band-limited, and narrow-band signals, when autocorrelated, produce an output which oscillates over considerable time, slowly decaying in amplitude. The rate of decay of the autocorrelation provides an indication of the frequency bandwidth. Narrow-band signals decay at a slower rate than broad-band signals. The autocorrelation of any single-frequency wave will produce output of the same frequency **(Fig. 2.18a)**. It will be a spike in the frequency domain.

The addition of broad-band random noise to any signal will cause the center point of the autocorrelation to be elevated by the height of the spike representing the autocorrelation of the noise (**Fig. 2.18b & c**), while the remainder of the output will be identical to that of the autocorrelation of the signal frequency.

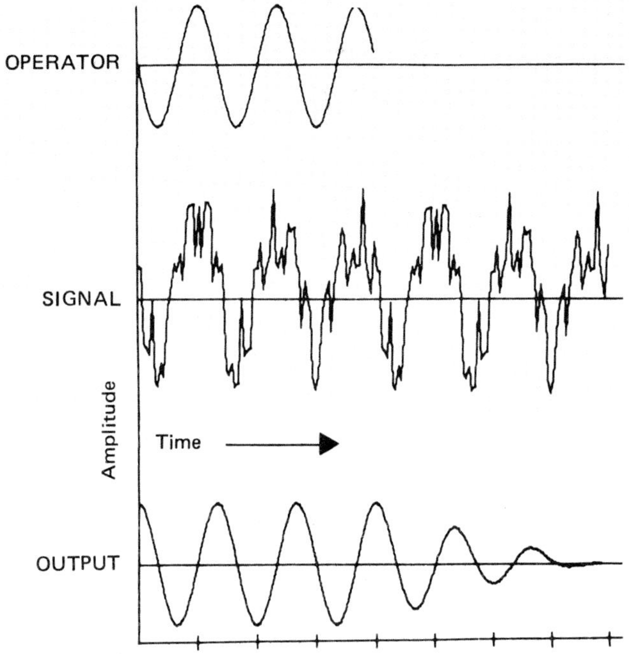

FIGURE 2.15

Cross-correlation and/or convolution of a pure sine or cosine wave with a signal will extract that same frequency component from the signal.

Synthetic Seismograms

The simple basic reflection model, in which the reflection coefficient at a single interface is replaced by the wavelet representing the unit impulse response of the system, has enormous value and potential, for it can be extended easily into much more complicated applications. An immediate and practical example is the production of synthetic seismograms. The basic system response model can be used to synthesize a reliable approximation of a seismic record for any location where a sonic log is available. A synthetic seismogram is nothing more than a model of the total seismic response to a model of the many reflection boundaries in a sedimentary section.

A sonic log is a continuous direct measurement of velocities over the length of the sedimentary section penetrated by a borehole. Each change in velocity can be converted into an approximate reflection coefficient by the simple calculation of the velocity ratios. A more exact model would include density, but for many areas the effects of density are minor. All velocity changes, no matter how small, can be modelled in this manner, converting the absolute measurement of velocities into the equivalent series of velocity ratios referred to as reflection coefficients.

Implementation of this next stage of model development requires certain preparation of the data. Models must observe conventions of units of measurement, and be scaled so the dimensions of the models respond in proportionate manner to the full scale systems they represent. A sonic log is usually presented in demensions of vertical depth and horizontal transit time. Seismic traces are plotted in vertical time. It is a simple matter to integrate a sonic log, summing transit times over small depth-sample intervals until the total amount of transit time equals the time interval of one seismic sample. Most seismic work today, on land at least, is recorded at 2 ms sampling

FIGURE 2.16a

In the frequency domain, convolution, the basic procedure for filter application, is accomplished by cross-multiplying the amplitude values of each frequency of the operator and the signal. The products yield the amplitude spectrum of the convolved output. Simultaneously, the values of the phase spectrum of the operator are added to the corresponding phase values of the signal to obtain the phase spectrum of the output.

FIGURE 2.16b

In the time domain, the operations of convolution and cross-correlation are identical with the exception that the operator is reversed for convolution, but not for cross-correlation. Similarly, cross-correlation in the frequency domain is identical to convolution, with the exception that the phase spectrum of the operator is subtracted from, rather than added to, the phase spectrum of the signal. In both cases the product of the amplitude spectrum is computed.

FIGURE 2.16c

Autocorrelation is a special case of cross-correlation in which the signal is correlated with itself. This amounts to squaring each component of the amplitude spectrum and subtracting the phase spectrum from itself, which will always yield a zero-phase output.

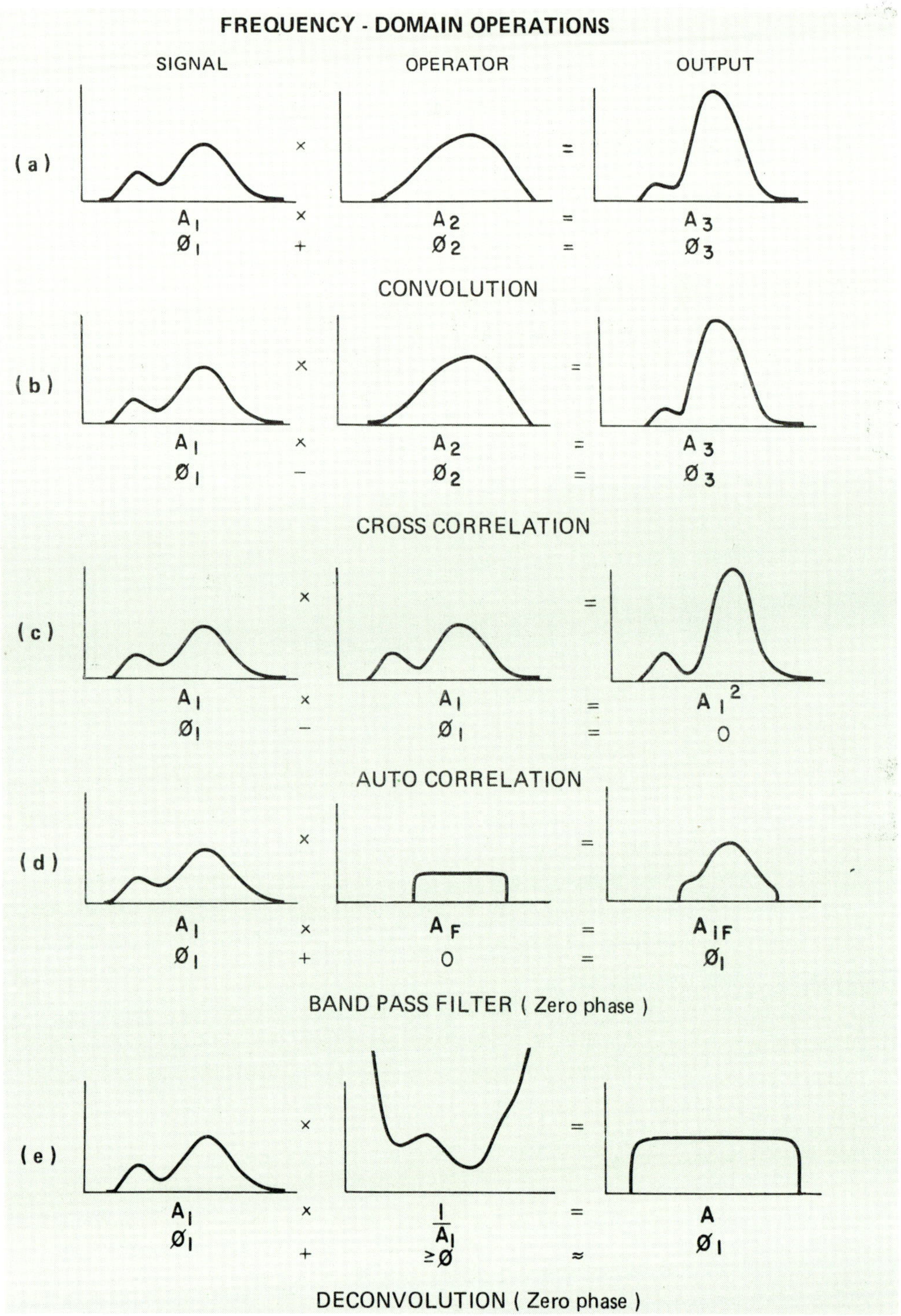

in two-way time; integration of a sonic log into 1 ms one-way time provides the equivalent sampling rate. Integration of a sonic log, digitized in depth to uniformly-spaced time samples also yields the velocity over each time sample increment. The results may be plotted to produce a sonic log sampled in time and velocity instead of the original depth and transit time. The log now represents a model of the earth layered at 1 ms intervals. At the boundary of each layer, the reflection coefficient may be calculated **(Fig. 2.19)**. The reflection coefficients will always be numbers between zero and one, positive or negative, and thus each constitutes a small amplitude spike at each layer boundary. The reflection-coefficient series contains the same information as the integrated sonic log which, within the constraints imposed by the integration process, contains the same information as the original depth sonic log. The operation has merely transformed the presentation of the data from one form to the other. There is a direct sample-to-sample relationship from the time-scaled sonic log to the series of reflection coefficients The basic model provides the means to simulate the modification by the system of the

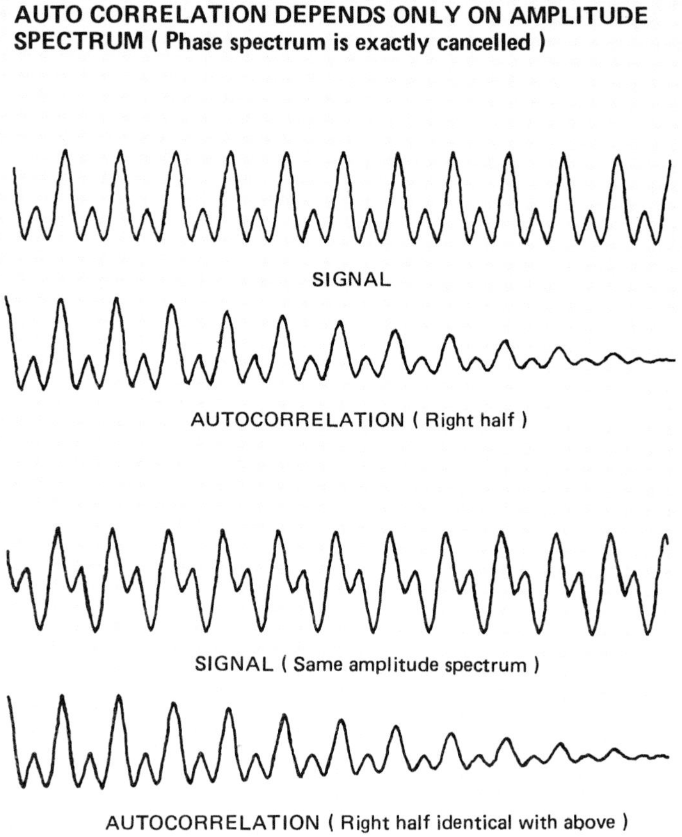

FIGURE 2.17

Autocorrelations of all signals which have identical amplitude spectra, regardless of their phase spectra will be identical.

sharp input energy source pulse into a wavelet.

It also provides the means to simulate the effect taking place in the real field seismic survey when a sharp reflection coefficient spike is replaced by the system wavelet in the seismic reflection process.

Generation of the synthetic seismogram requires each one of the series reflection coefficients to be replaced by the system response wavelet. If the series of spikes is input in continuous sequence to the system model, each of the spikes in turn as it passes through the system will be convolved with the transfer function (the unit-impulse response) converting each spike to a proportionately-scaled response wavelet. Convolution of the transfer function with the reflection coefficients produces an output trace which is a close approximation to a noise-free seismic field record, provided that the transfer function has been correctly defined and that the reflection coefficients computed from the log represent a reliable model of that particular location (**Fig. 2.21**).

The recorded result then represents the convolution of the transfer function with the series of reflection coefficients marking the boundaries of the individual sedimentary members within the earth. The detection of these boundaries and their precise definition, free of noise and distortion, is the conventional end result of seismic data interpretation. Geophysicists and interpreters try to detect these boundaries and to map them as they carry around an area. It is the purpose of this review to define some of the problems that affect the data and what can be done by data processing to correct them.

The use of the model system not only simulates the input and output but the procedure closely parallels the real system process. The model is obtained by convolving a continuous series of spikes (the reflection coefficients) with the unit-impulse response to obtain the output. A direct comparison of the seismic trace to the sonic log is somewhat misleading, since their appearance is quite different. In real life the shot pulse, modified to the input wavelet, passes through the earth and the reflection coefficient at each reflection boundary is replaced by the wavelet in the signal returning to the surface. The reflection coefficients merely present the sonic log information in a form more directly related to the seismic reflection process.

Since the original sonic log has been resampled at the same rate as the seismic trace, all of these three time series can be plotted as continuous signals to the same vertical scale on a conventional seismic plotter (**Fig. 2.22**).

FIGURE 2.18a

The autocorrelation of any single-frequency wave will produce output of the same frequency.

FIGURE 2.18b,c

The autocorrelation of a broad-band signal is a spike.

FIGURE 2.18d

Band-limited or narrow-band signals, when autocorrelated, produce an output which oscillates over considerable time slowly decaying in amplitude.

FIGURE 2.18e

The addition of broad-band random noise to any signal will cause the center point of the autocorrelation to be elevated by the height of the spike representing the autocorrelation of the noise.

COMPARISON OF TIME DOMAIN AND FREQUENCY DOMAIN SIGNAL REPRESENTATION. PHASE SPECTRUM IS NOT SHOWN.

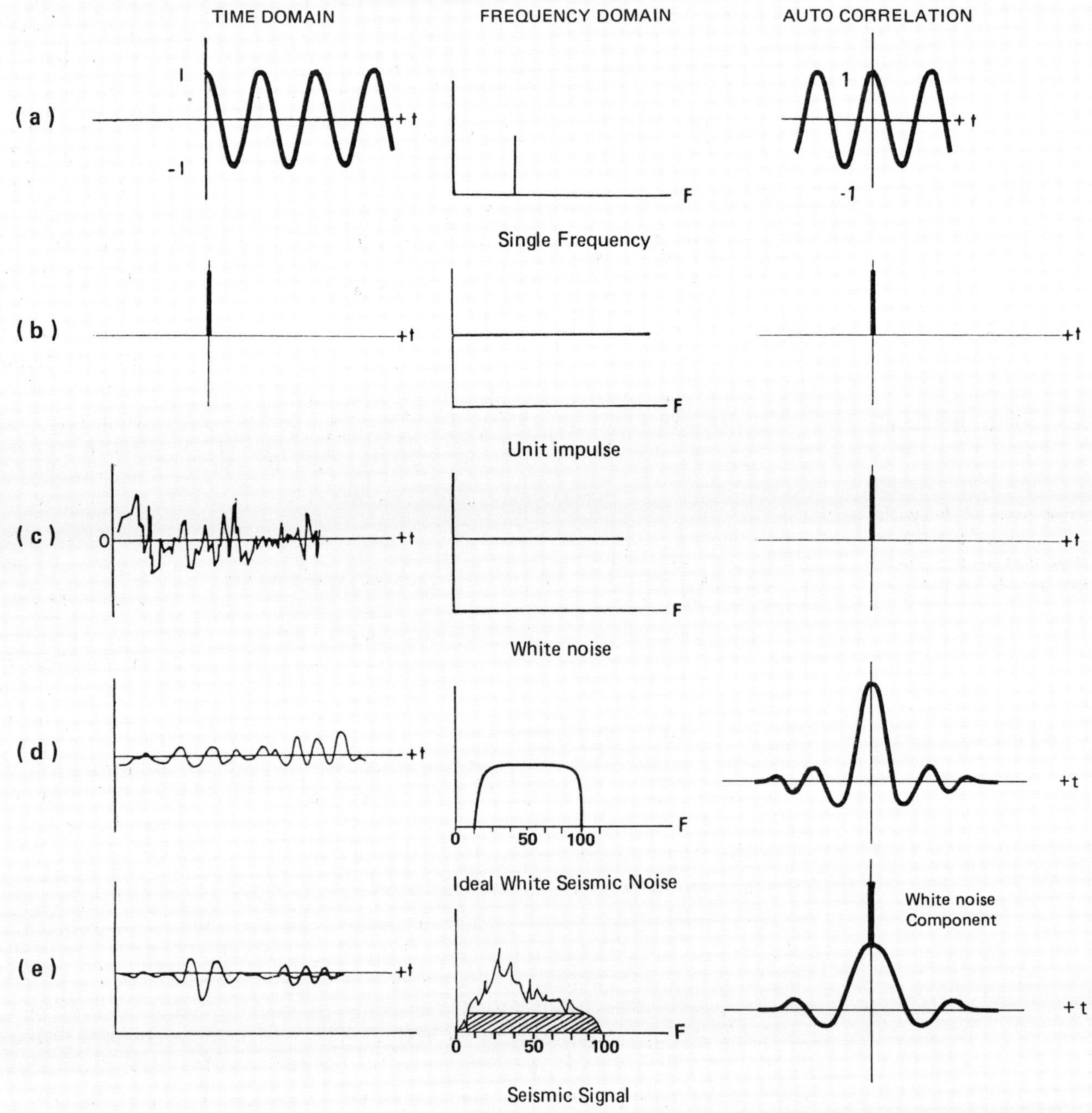

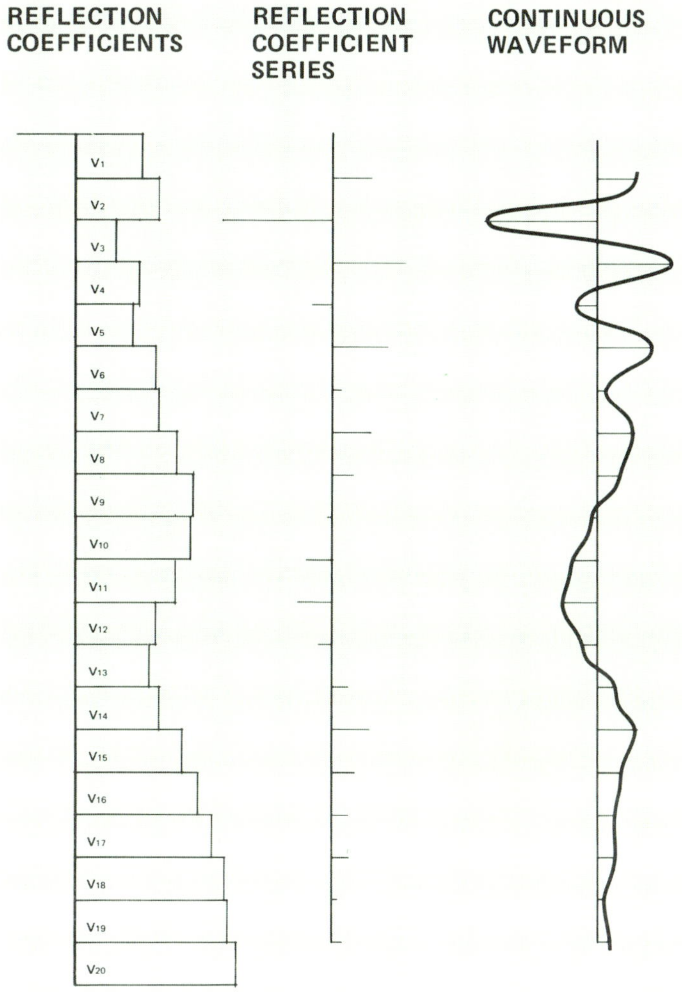

Figure 2.19

Figure 2.20

FIGURE 2.19

A sonic log sampled in time and velocity instead of the original depth and transit time. The log now represents a model of the earth layered at 1 ms intervals. At the boundary of each layer, the reflection co-efficient may be calculated.

FIGURE 2.20

The reflection-coefficient series contains the same information as the integrated sonic log which, within the constraints imposed by the integration process, contains the same information as the original depth sonic log.

$$Rc_1 = \frac{V_2 - V_1}{V_2 + V_1} \quad \text{Basic approximation}$$

$$Rc_1 = \frac{\rho_2 V_2 - \rho_1 V_1}{\rho_2 V_2 + \rho_1 V_1} \quad \text{Better approximation}$$

ρ = Density

This establishes the very important point that like the sonic log and the seismic trace, the reflectivity sequence is a continuous signal. The earth is not correctly modelled by separate, indvidual unrelated reflection coefficients, and certainly not by a few widely spaced reflectors with no information between.

The reflectivity series, or earth signal, represents the ideal reflection sequence. Neglecting density, it models the exact response of the seismic reflection process completely free of transmission losses, noise and distortion and therefore is the ultimate goal of all seismic data processing. This is a clearly defined goal and the success of any part of the seismic data acquisition and processing should be measured in terms of how much it contributes toward achieving this goal.

The earth signal also defines the limits of data processing. It is not possible, from a single conventional seismic trace, to obtain any geological information which is not contained in the earth signal. The earth signal can be manipulated and analyzed in exactly the same manner as any other signal, in particular the seismic trace, so it is possible to make quantitative measurements of the differences which exist between the real and the ideal signals. The spectra of components of two signals can be compared directly to obtain a measurement of the differences that may exist (**Fig.2.23**).

The comparison is very revealing. The earth signal has frequencies distributed over the total bandwidth from 0 to well beyond 100 Hz. The seismic trace on the other hand, has lost most frequencies below 15 Hz and above 90 Hz. The remainder are nearly all attenuated to varying degrees. These changes represent a real loss of information over the missing frequencies, while most of the remainder are severely limited in their capacity to carry information.

FIGURE 2.21

Convolution of the transfer function with the reflection coefficients produces an output trace which is a close approximation to a noise-free seismic field trace, provided that the transfer function has been correctly defined and that the reflection coefficients computed from the log represent a reliable model of that particular location.

SONIGRAM courtesy of DIGITECH LTD.

SONIGRAM
VELOCITY ANALYSIS

LOCATION LSD 6 SEC 12 TWP 27 RGE 1 W 5 M
WELL NAME TIPPERARY ET AL CROSSFIELD
COMPL. DATE 5 JAN 71 DATA SOURCE BHC SONIC LOG
K.B. 3566 G.L. 3553 TD. 9000 TD. corr.
AREA AREA CODE
INTEGRATED FROM 1220 TO 9010 TOTAL 7790
INTEGRATION ADJUSTED TO

GEOLOGIC FORMATION	DEPTH KB	ELEVATION	INTERVAL THICKNESS	INTERVAL TRANSIT TIME SECONDS (TWO-WAY)	INTERVAL TIME R.M.S. VELOCITY	INTERVAL HALF VEL.	REFLECT. TIME SECS (TWO-WAY)
	1220	2346					•000
			510	•090	11•28	5•641	
KNHL TF	1730	1836					•090
			1803	•331	10•97	5•453	
BELLY R	3533	33					•421
			957	•151	11•92	6•353	
LEA PRK	4490	−924					•572
			95	•017	11•69	5•622	
MILK R	4585	−1019					•589
			285	•045	11•97	6•352	
COLO	4870	−1304					•633
			780	•127	12•06	6•163	
CARD ZO	5650	−2084					•760
			86	•014	12•10	6•119	
CARD SS	5736	−2170					•774
			304	•049	12•16	6•210	
2 W S	6040	−2474					•823
			467	•078	12•12	5•970	
B F S	6507	−2941					•901
			67	•013	11•88	5•148	
VIKING	6574	−3008					•914
			270	•038	12•20	7•108	
BLAIR	6844	−3278					•952
			483	•070	12•41	6•893	
L BLAIR	7327	−3761					1•022
			256	•033	12•81	7•849	
SHUNDA	7583	−4017					1•055
			202	•022	13•46	9•087	
PEKISKO	7785	−4219					1•077
			288	•030	14•12	9•681	
BANFF	8073	−4507					1•107
			482	•055	14•45	8•707	
EXSHAW	8555	−4989					1•162
			8	•001	14•43	7•095	
WABAMUN	8563	−4997					1•163
			295	•030	14•88	9•693	
CROSSFD	8858	−5292					1•194
			152	•015	15•32	10•011	
TD INT	9010	−5444					1•209

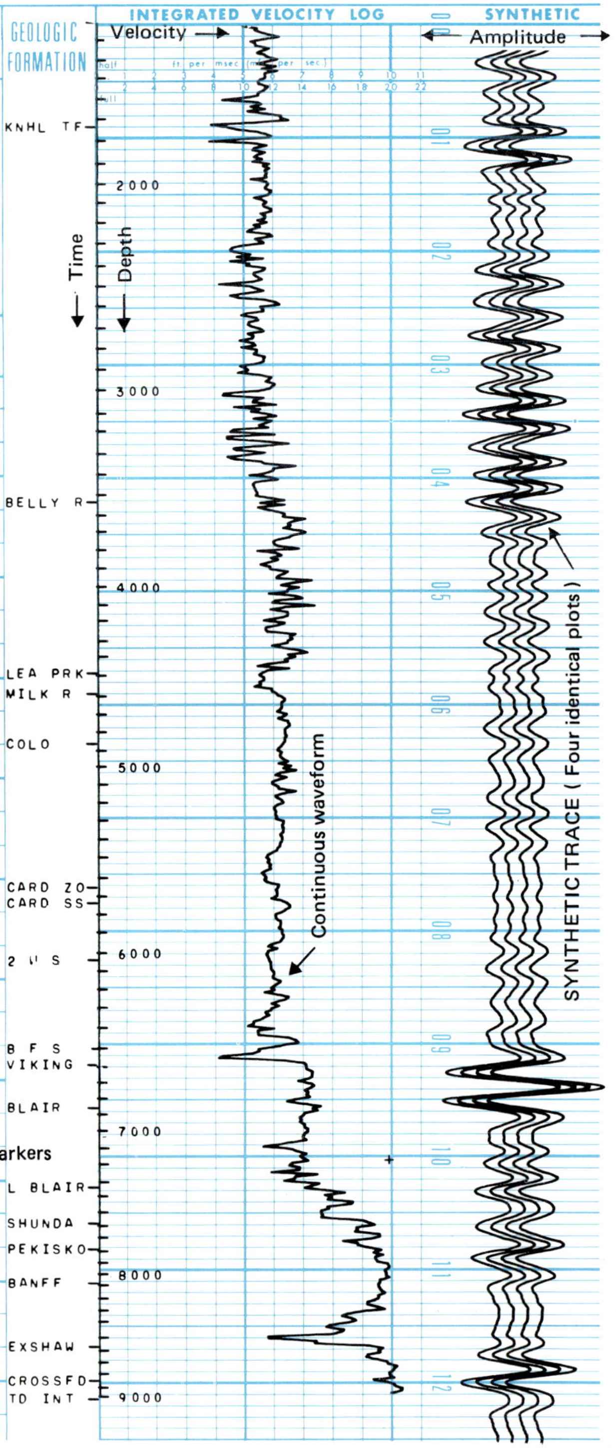

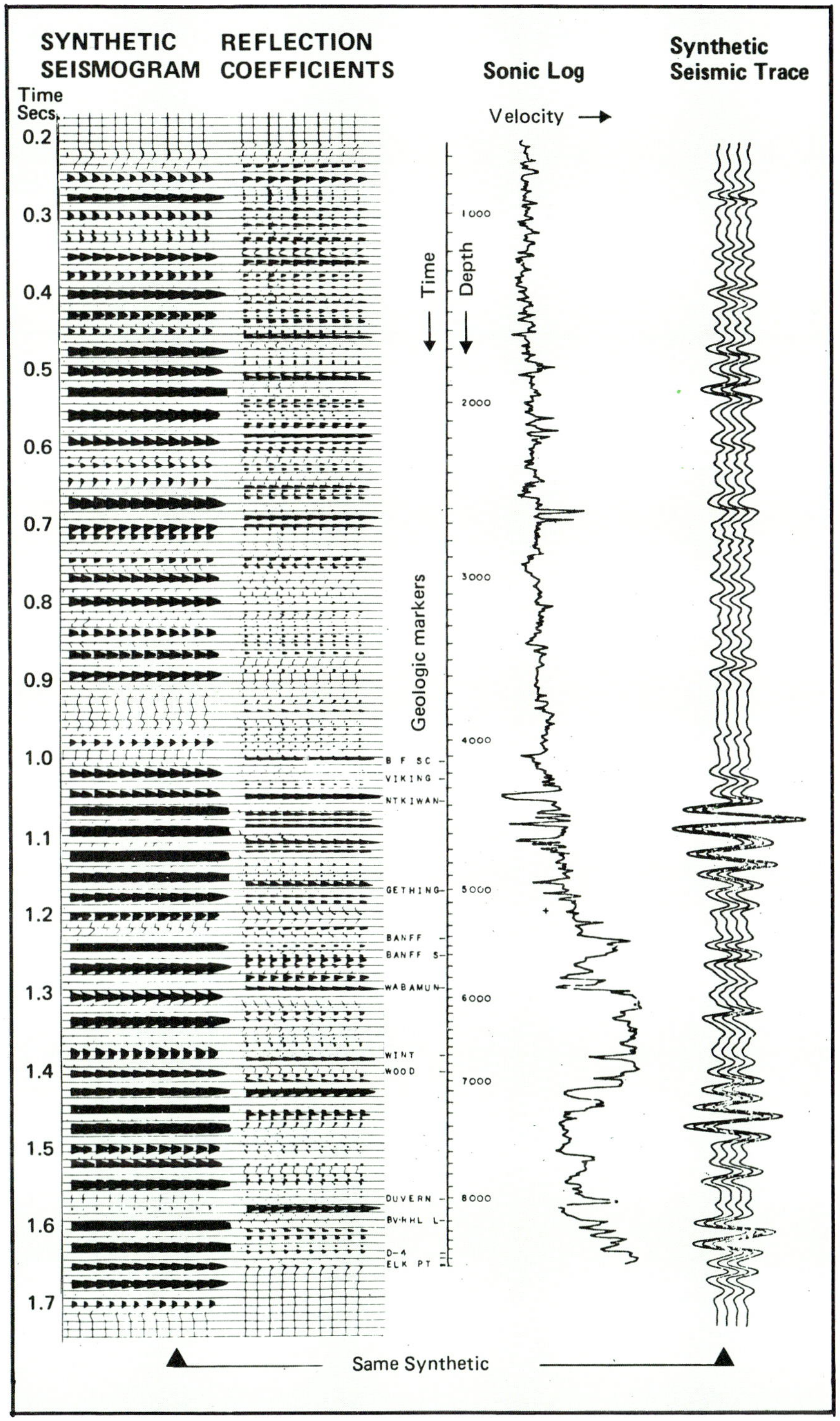

EARTH SIGNAL AND SYNTHETIC SEISMOGRAM

FIGURE 2.22

The reflection coefficients and the sonic log have been sampled at the same rate as the seismic trace, so all of these three time series can be plotted as continuous signals to the same vertical scale on a conventional seismic plotter.

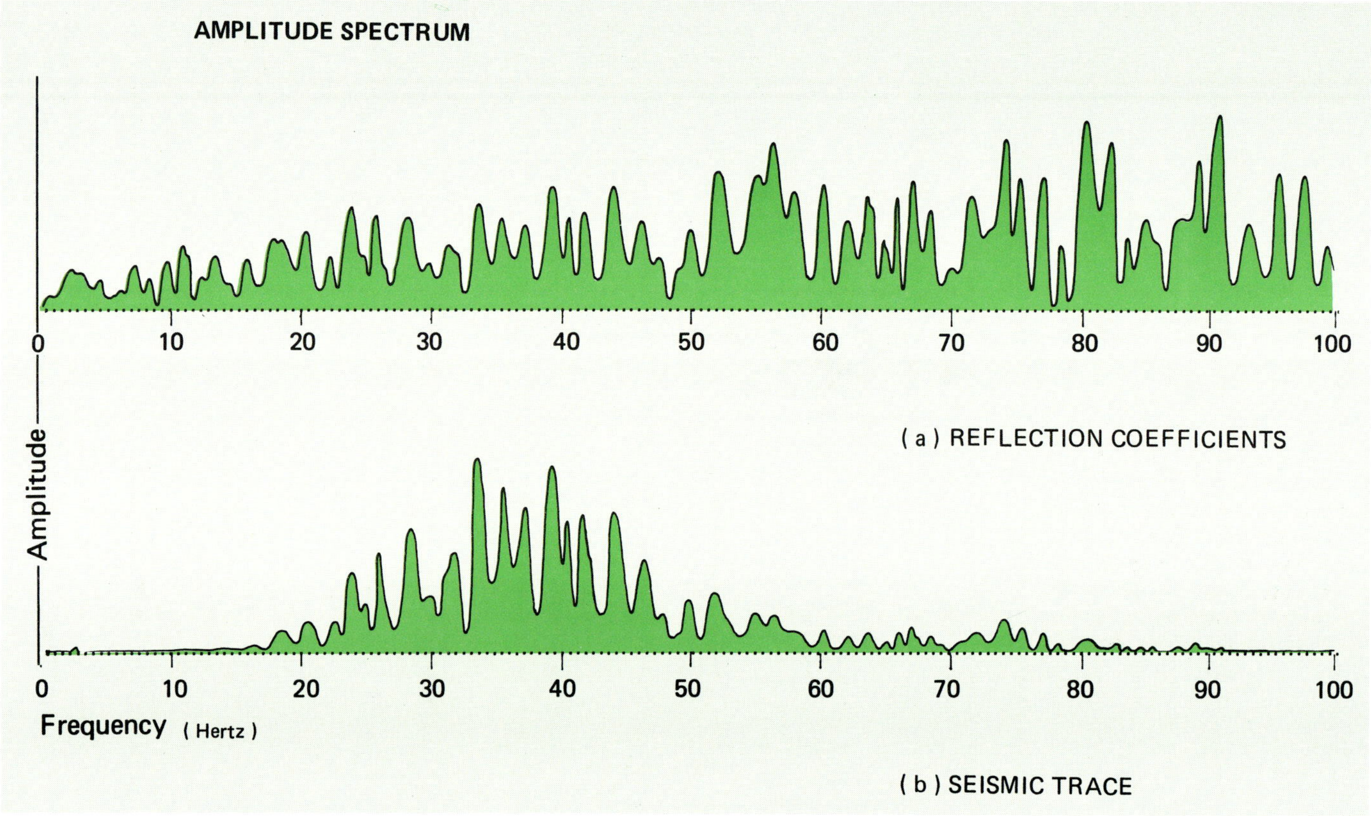

FIGURE 2.23

The spectra of components in the earth signal and the synthetic trace, or in fact, a real field trace, can be compared directly to obtain a measurement of the differences that may exist.

3 DATA ACQUISITION

Seismic Data Acquisition

The basic seismic reflection survey process has remained unchanged in concept since the recording of the first seismic reflection over 50 years ago. A quantum of energy from a controlled source is transmitted into the ground at or near the surface. The energy travels through the sedimentary section gathering information from each reflecting horizon, and returns to the surface. It is captured by a geophone as a short, continuous signal and shunted through amplifiers to a system for permanent recording.

The reflection process consumes the input energy at a rapid rate. The initial seismic return may have a million times greater amplitude than that existing at the end of its short life. Wide dynamic range is a dominant characteristic of the seismic signal, and it is perhaps the major engineering consideration in the design of seismic data acquisition equipment. For the purpose of this paper, dynamic range may be defined as the ratio between the mimimum and maximum signal strength over a range for which signal distortion through a given system can be held below acceptable levels. A system which can record and recover a signal of one millivott with acceptable distortion, and continue to perform satisfactorily up to a level of one volt has a dynamic range of 1,000:1.

A logarithmic scaling unit, the decibel, is commonly used to measure dynamic range. Expression of the ratio in decibels is obtained by multiplying the $\log_{10}$ of the numerical ratio by 20; thus a ratio of 1,000:1 becomes 20 x 3 = 60 db.

In the decimal system of notation a shift of the decimal (or radix point) one position indicates a change by a factor 10, a gain of 20 db. Modern digital recording equipment uses a base 2 binary numbering system instead of the base 10 system used for decimal notation. A shift of the radix point one position in the binary system is a change by a factor of 2, which is approximately equal to a gain change of 6 db. A rough but quick rule-of-thumb calculation of the gain range for binary equipment is to subtract 1 (for the sign bit) from the total number of binary bits available for recording and then to multiply the remainder by 6 to obtain the appropriate dynamic range capability. Thus, a 16 bit digital word can record a dynamic range of (16 - 1) x 6 = 90 db. Perhaps it should be pointed out that, by definition, amplitude of a sinusoid is considered to be the peak absolute value, e.g.,displacement from zero. Numerically, the dynamic range of 16 bits is double that used here, and indeed, in terms of information theory the addition of the sign bit doubles the information-carrying capacity of the binary word.

The dynamic range measures not an absolute amplitude value but a ratio determined by the limits at each end of the scale. To be useful, the limits must fall in the range of significant information. A 60 db system ranging from a threshold of 1mv to 1 volt would be of little use to record a 60 db signal ranging from 1 μv to 1 mv. Often, it is possible to apply a linear scaling factor to the input to provide a match to the system, which is of course the role of the amplifier in the seismic recording system.

At this point it might be interesting to compare a few examples of dynamic range on different scales.

Approximate Dynamic Range Relationships

	Db	Decimal Relationship	Binary Bits (add 1 for sign)
Strong Seismic Field Signal	120	1,000,000:1	20
Digital Magnetic Tape			
21-track (one word)	96	64,000:1	16
Digital Magnetic Tape			
9-track (two bytes)	90	32,000:1	15
Digital Magnetic Tape			
7-track (two bytes)	78	8,000:1	13
Analog Recording FM	60	1,000:1	10
Analog Recording AM	54	500:1	9
Paper Record or Record Section	30	30:1	5

The chart illustrates one of the basic problems of seismic data recording; our seismic signals often have a dynamic range far greater than the capability of the recording system to capture the signals without distortion. When a seismic signal exceeds the dynamic limits of the system, it is distorted, usually by clipping off the top and bottom of the waveform. When it falls below the lower limit, it is very often buried in the background of the noise inherent in the system.

Control of Seismic Energy

One of the basic assumptions of signal processing theory is that each reflection contained in the complex composite signal, that is the seismic trace, is a reproduction of the unit-impulse response. It should equal the shape and frequency content of the unit-impulse response in all respects, and one should vary from another only in amplitude. The amplitude should be a direct function of the amplitude of the reflection coefficient. When this occurs, the solution of the simultaneous equations to determine an inverse filter for deconvolution is good and a very reliable operator is obtained.

Prior to the introduction of digital processing, the emphasis was placed on the mere detection of seismic reflections as an interpretive procedure. With the advent of digital processing it became much more important to define the response with precision.

In addition to the additive effects of noise, the amplitude of the seismic signal arriving at a geophone is affected by a number of somewhat related and, to a certain extent, predictable mechanisms. The most important of these is the change in amplitude which occurs when the seismic energy crosses a lithologic boundary. A predictable amount of the energy is reflected back from the interface to the surface and the remainder continues through. This action produces a seismic reflection. (Under ideal conditions the amplitude of the reflection after processing should be a direct function of the velocity-density contrast encountered at the boundary, which is valuable information for the interpreter.)

On the field trace, the effect of any reflection is to reduce the amplitude of all subsequent events since the input portion of the signal becomes weaker each time part of it is returned to the surface as a reflection. This is only one of many effects which reduce the amplitude of the seismic signal over time (See also **Fig. 2.4**).

The rapid energy loss which causes the great dynamic range of the seismic trace may have two basic components; the direct attenuation of signal amplitude and the loss of frequency components. Of these, the second in the more serious loss. The first is primarily a scaling factor, but the second is a direct loss of information.

Of the many mechanisms contributing to the decay in amplitude of the seismic signal, the four main are as follows:

1) Partial Reflection or Scattering;

This occurs at every acoustic discontinuity in the path of the pulse and can be approximated by a factor $(1 - r)^{bt}$, where b = number of reflectors, and r = average reflection coefficient, encountered between time 0 and time t;

2) Spherical Divergence;

The loss in energy per unit volume caused by the expansion of the wavefront with time. The seismic disturbance spreads out like the surface of an expanding bubble and the amplitude decay factor is roughly $1/t$;

3) Attenuation Loss;

The factor commonly used for the normal "friction" or "heat loss" suffered in passing through homogenous rock, is e^{-aft}, where 'e' is the base of the natural logarithms, 'a' is a constant associated with rock type (fairly uniform for most of the sedimentary rocks), 'f' is acoustic frequency and 't' is elapsed time. Notice here that this factor is dependent on frequency, and

4) Amplitude Distortion Through Frequency Loss;

It is well known that high frequencies are lost much more rapidly in the seismic signal than are the lower frequencies. **(Fig. 3.1)**. There are many reasons for this and one has already been defined. Dispersion is also frequently listed as a cause, but greater losses are probably due to internal reflections. Strong destructive interfer-

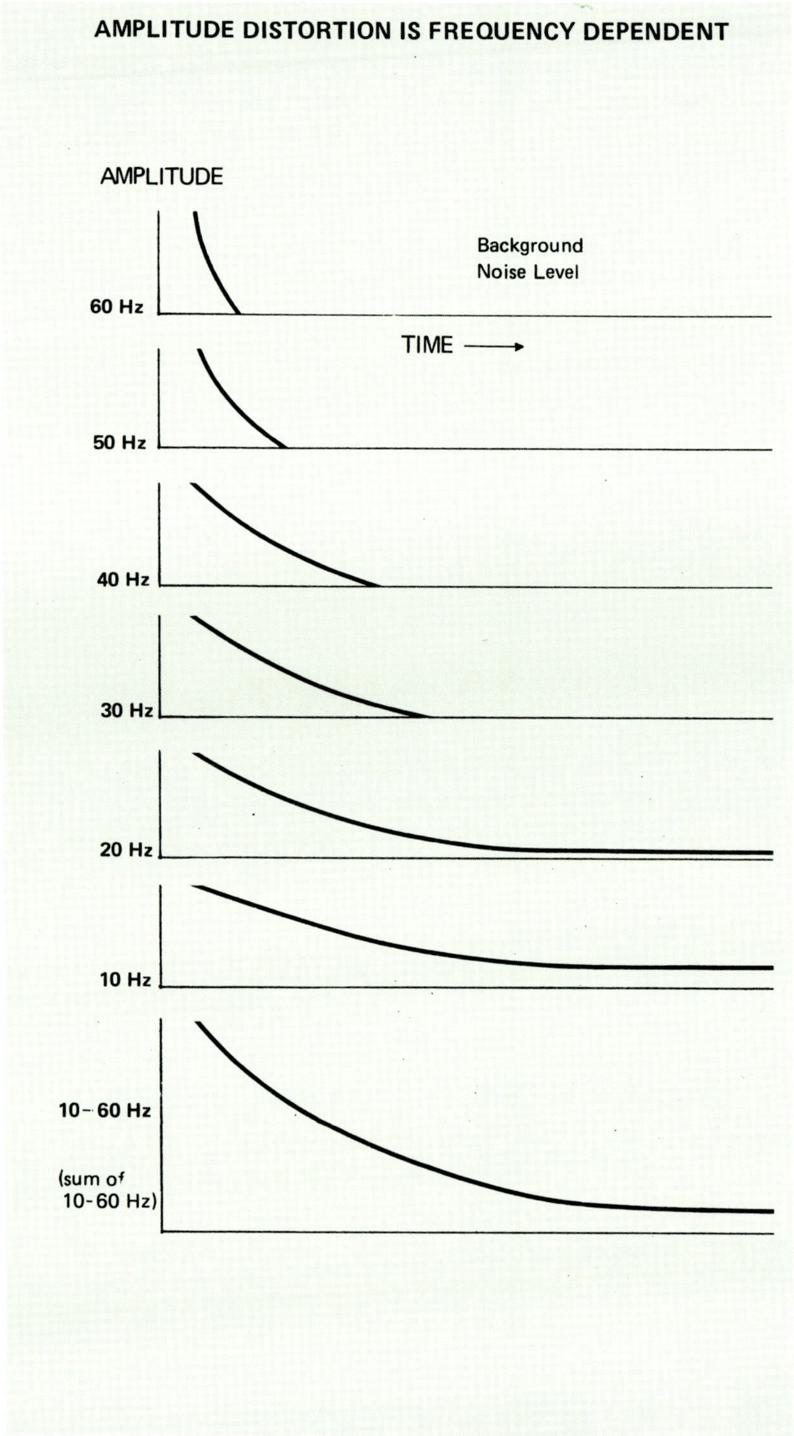

FIGURE 3.1

It is well known that high frequencies are lost much more rapidly in the seismic signal than are the lower frequencies.

ence can occur in layered sequences when reflecting interfaces are exactly some multiple (depending upon polarity) of one-quarter wavelength.

Fortunately, all of the components which contribute to the decay of the trace amplitude have somewhat similar decay functions (**Fig. 3.2**). The several functions occur simultaneously so the effect is additive and results in a final net decay curve which is exponential and approximates

$$B\,(1/t)\,e^{-Ct},$$

where B and C are empirical constants which vary with almost every shot. The result of all this is the extremely rapid decay in amplitude which characterizes the signal received by the geophone.

If the signal can be recorded, or at least subsequently recovered, in a manner which closely approximates the geophone output, the problem becomes the very important one of compensating the recorded trace for the effects of the several decay functions in such a manner that the remaining amplitude variations will be a true indication of the reflection coefficient at each and every boundary within the earth.

Recording Equipment

Seismic instrument design engineers have always grappled with the problem of how to record seismic signals ranging from very strong to very weak with equal fidelity, often over a very short time interval. Many ingenious solutions to the problem have been tried, but they have generally been limited by the dynamic range of the three basic recording media; paper, analog magnetic tape and digital magnetic tape.

Paper as a recording medium has an extremely limited dynamic range, and, in spite of the feedback controls that were designed into early instruments, clipping of signals was a frequent problem. Fitting the enormous range of seismic signal amplitude into the limited dynamic range of the paper recording was the principal motive for the early development of automatic voltage control, (AVC), expander circuits, automatic gain control, (AGC) and programmed gain control.

Each of these systems has its own particular advantages and disadvantages. The AGC amplifier uses a feedback circuit to control the amplitude. The amplitude of the amplifier output is continually monitored and compared to a standard range. If the level is outside the range, a signal feedback causes variable gain to be imposed to maintain the signal level within the dynamic range of the recording medium. The continuous modification of the gain causes continuous modification of the signal amplitude. This, in turn, continuously modifies the amplitude of each unit-impulse response recorded on the seismic trace. The reaction of the system to the feedback control is subject to lag, which may create further distortion of the response. It also may excessively attenuate information arriving a short time after a strong reflection, since the amplifiers are still responding to the high-amplitude signal. Signals distorted in this manner no longer meet the basic assumption that all unit-impulse responses recorded on the trace are identical in shape, and vary only in amplitude and polarity. Nevertheless, the process is well adapted to analog systems and is the principal control system in analog recorders.

The principal disadvantage is that no record was made of the amplification applied to the signal and therefore it has never been possible to reconstruct the signal to its true input amplitude.

Programmed gain and expander systems control the trace under a pre-designed program of amplitude manipulation, based on the expected amplitude variation of the incoming signal.

The advantage here is two-fold. Firstly, the gain program is matched to the decay of signal strength, and does not respond to local bursts of energy. Hence, sudden changes in amplitude, such as a strong reflection followed by a weak one, cause no reaction of the control system and therefore relative amplitudes are recorded in approximately their true relationship. Also, since the control is by program, it is, at least in theory, possible to reconstruct the true amplitude of the incoming signal.

Unfortunately, noise and local variations in strength of the input, and the transmission characteristics of the near-surface layer often cause the signal amplitude to be quite different than expected by the gain program. As a result, several shots are often necessary in order to secure satisfactory records. This action is complicated by the fact that the energy patterns of two successive shots are rarely identical.

Analog magnetic tape recording provided a substantial improvement over recording on paper. But even the best analog recording systems have a limited dynamic range when compared to the seismic signal and must use some form of automatic gain control to hold the signal level within the dynamic range of the recording medium.

The introduction of digital recording equipment finally provided the means to overcome the limited dynamic range of analog recording. It would seem that the digital recorder would provide an instantaneous solution, for a digital measurement of amplitude is limited only by the number of digits available, and any number of digits can

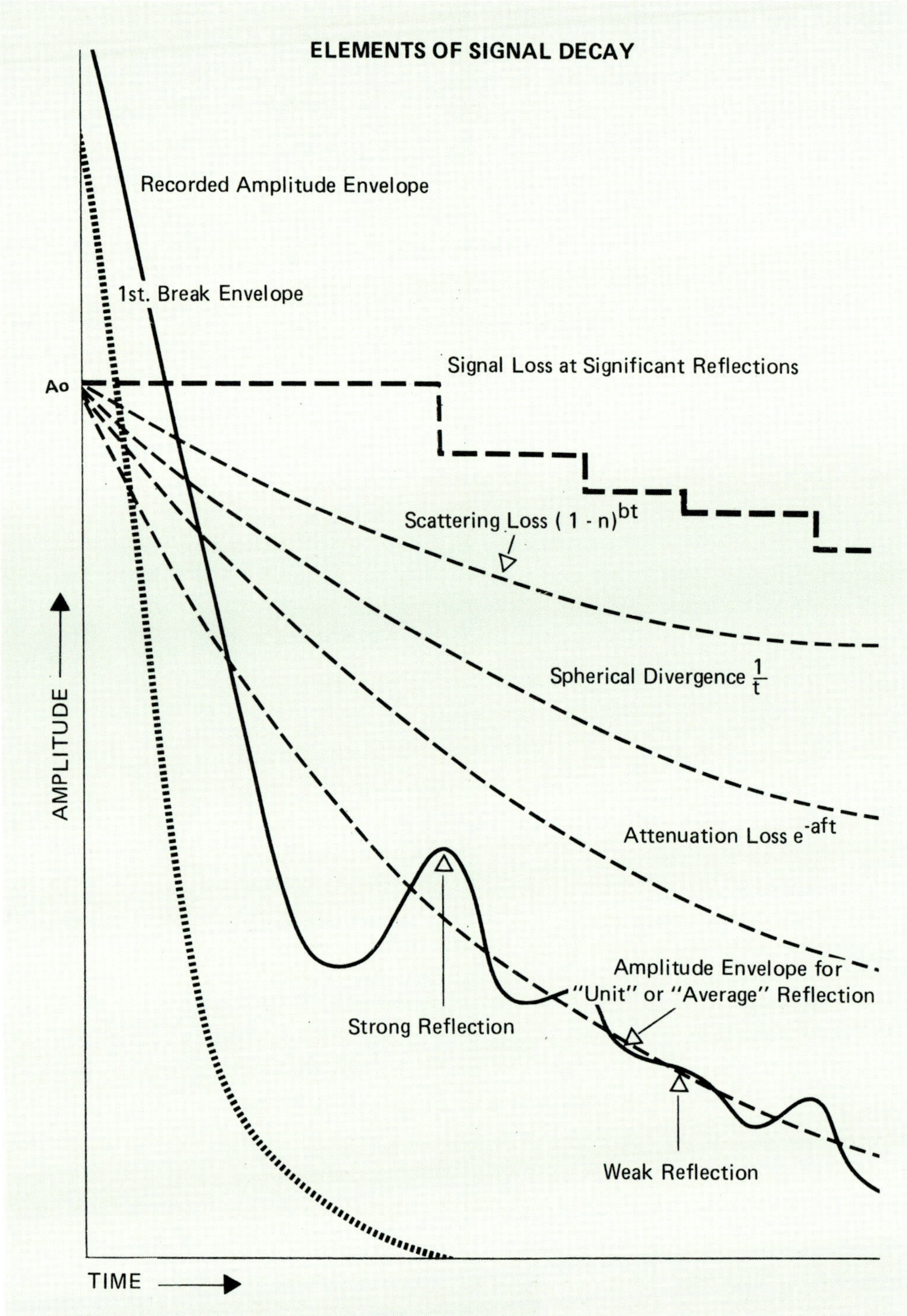

FIGURE 3.2

Most of the components which contribute to the decay of the trace amplitude have somewhat similar decay functions. The several functions occur simultaneously so the effect is additive and results in a final net decay curve which is exponential and approximates B (1/t) e^{Ct}, where B and C are empirical constants which vary with almost every shot. The result of all this is the extremely rapid decay in amplitude which characterizes the signal received by the geophone.

be assigned for the purpose. While this is true in theory, in practice it is limited by the rate at which numbers can be recorded.

The dynamic range of seismic signals requires a binary word of 20 bits plus one bit for the sign. Early tape transports recorded digital data at 556 words per inch at a tape speed of 60 inches per second. Each data word contained 6 bits of information. Thus the recording rate was 556 x 60 x 6 or about 200 kilobits per second.

A standard 24-channel seismic recording of that era, sampling at one ms intervals produced data at the rate of 24 x 1000 x 20 = 480 kilobits/sec. Thus, in order to reduce the rate of data flow to match the capability of the recorder, it still became necessary to modify the dynamic range of the signal before recording. The other solution was to increase the recording rate. Doubling the width of the magnetic tape was one solution, but not all of the increased capacity could be used to store information; some was required for data control. A satisfactory solution was provided with the introduction of the binary-gain amplifier, which suddenly provided recording capabilities closely meeting the demands of the processing.

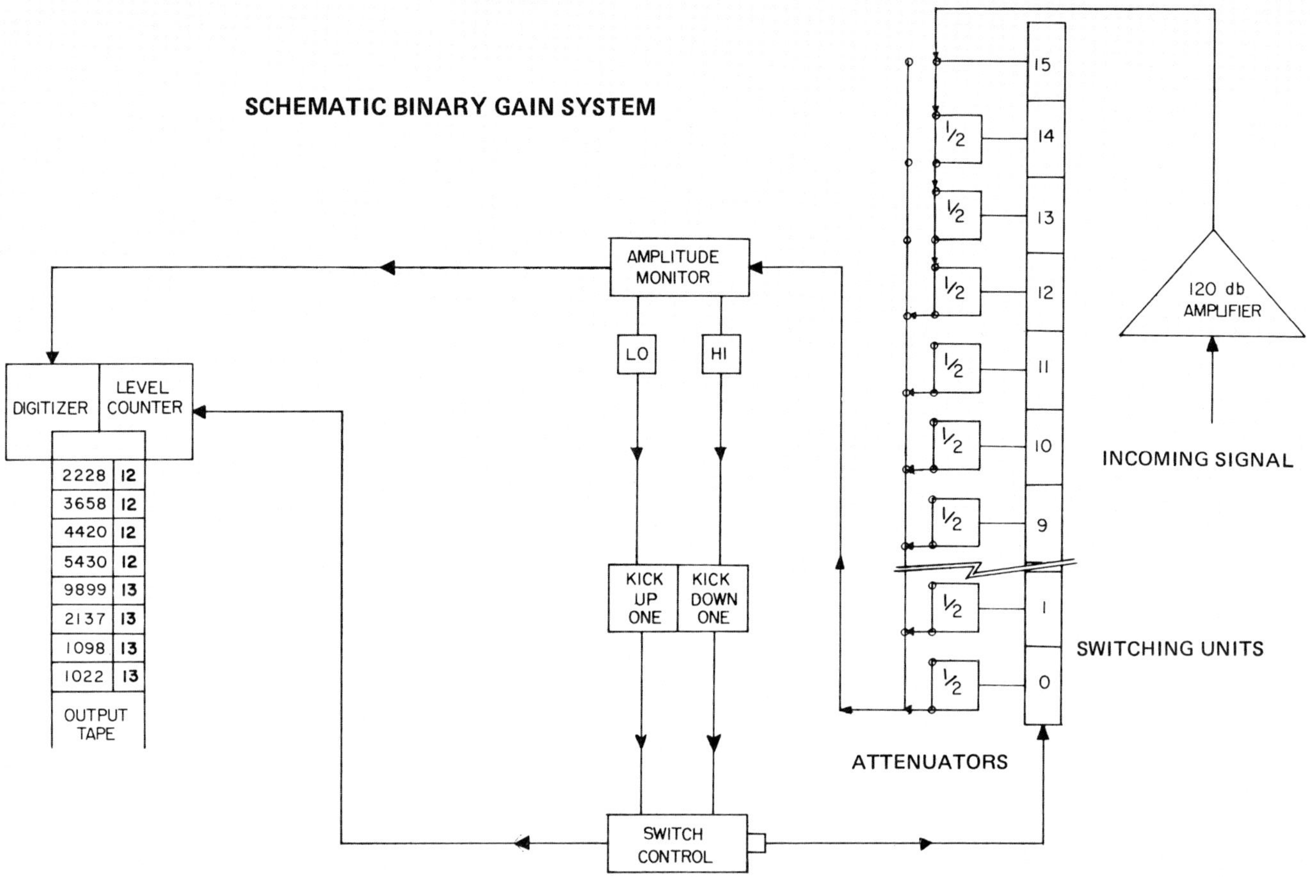

FIGURE 3.3

The principle of the binary-gain recorder may be represented schematically be a series of switches and attenuators Each attenuator will reduce the amplitude of an incoming signal by a factor of two. The amplitude of the output signal is constantly monitored by a metering device which, by a feedback process, allows the signal to neither exceed a certain maximum or fall below a certain minimum level.

Binary Gain Control

The principle of the binary-gain recorder may be represented schematically by a series of switches and attenuators (Fig. 3.3). Each attenuator will reduce the amplitude of an incoming signal by a factor of two. The amplitude of the output signal is constantly monitored by a metering device which, by a feedback process, allows the signal to neither exceed a certain maximum nor fall below a certain minimum level. Apparently the action of the binary gain system is no different than the feedback AGC or AVC systems. The advantage lies in the manner of recording of the signal.

An incoming signal reaching the system is first passed through an amplifier capable of applying a fixed amount of gain over the 120 db range. The amplified signal is then fed through the attenuator system. If the amplitude of the output signal is too high one attenuator will be switched into the circuit, halving the amplitude of the signal. If the signal is still too high, another attenuator will be turned on, and so on until the signal reaches an acceptable level. In a similar manner, should the signal passing through the monitor not reach a minimum level, one of the attenuators will be switched off to double the previous value of the output signal. The amplitude of a signal recorded on the final magnetic tape output can thus always be held within well defined bounds of dynamic range.

The key to the system and the major difference between binary-gain and AGC systems, is that a count of the number of attenuators through which the signal has passed is recorded on the tape in addition to the sample value.

FIGURE 3.4

The actual signal value is carried on the tape as a base number and an exponent indicating to what power of the two the base number must be raised. In practice the system counts the number of switches turned off.

BINARY GAIN MATHEMATICS

POWER OF 2	BIT SETTING	GAIN SETTING	MULTIPLIER
2^0	1111	15	1
2^1	1110	14	2
2^2	1101	13	4
2^3	1100	12	8
2^4	1011	11	16
2^5	1010	10	32
2^6	1001	9	64
2^7	1000	8	128
2^8	0111	7	256
2^9	0110	6	512
2^{10}	0101	5	1024
2^{11}	0100	4	2048
2^{12}	0011	3	4096
2^{13}	0010	2	8192
2^{14}	0001	1	16384
2^{15}	0000	0	32768

AMPLITUDE RANGE $\pm$ 8192

TOTAL RANGE $\pm$ 8192 x 32768 = $\pm$ 268,535,456

Each attenuator divides the incoming level by a factor of two. The attenuator count recorded represents the number of times the signal value which is recorded must be multiplied by two to obtain the original value. Thus, the actual signal value is carried on the tape as a base number and an exponent indicating to or by what power of the two the base number must be raised (**Fig. 3.4**). Later, the true level of the signal may be recovered by merely raising it to the value indicated on the exponent. The reconstructed seismic signal should duplicate closely the signal recorded at the geophone, reproducing with fidelity all of the amplitude variations in the signal which arrived at the ground surface. How well the geophone reproduces the signal arriving from the earth is another story for the coupling at this point remains one of the weakest components of the system. The numerical recovery of the binary gain recording is readily done by a binary digital computer system. Multiplying a binary number by some power of two is done by merely adding zeros to the right-hand side of the number in exactly the same way that a number in the decimal system may be multiplied by a power of ten be adding the proper

FIGURE 3.5

The numerical recovery of the binary gain recording is readily done by a binary digital computer system. Multiplying a a binary number by some power of two is done by merely adding zeros to the right-hand side of the number in exactly the same way that a number in the decimal system may be multiplied by a power of ten by adding the proper number of zeros to the right-hand end of the number. The effect is merely one of shifting digits to the right in recording; to the left in recovery.

PRINCIPAL OF BINARY GAIN RECORDERS

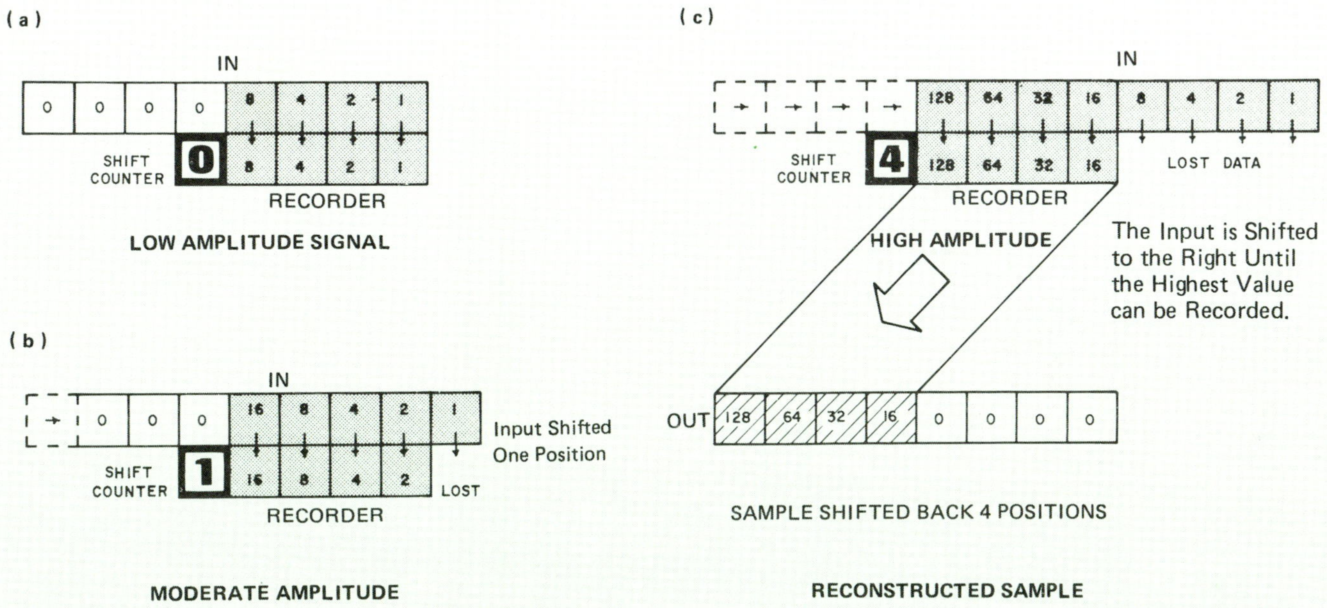

Design Considerations

number of zeros to the right-hand end of the number. The effect is merely one of shifting digits to the right in recording; (Fig. 3.5). Note that the recovered signal is not an exact reproduction of the input. Some loss of the low-order digits is not considered to be serious.

It should be noted that the need for the binary gain switching to step down, i.e., to attenuate the signal, is always much greater than the need to amplify the signal. A drop in the amplitude of a signal requires the instruments to step up the amplification but if they did not it would not be serious. The amplitude on the tape would probably still be at a level sufficient to permit subsequent recovery of the information by application of the proper gain factor. However, the need to step down is very great. Should a sudden burst of energy hit the instruments and reaction be slow, the high amplification applied to the strong incoming signal would saturate the equipment and result in a distorted or truncated signal. For this reason, binary gain instruments are equipped with the ability to step down much faster than to step up. The ranging ability for an up-step usually occurs once every 30 sampling units. If the sampling unit is 2 ms, a step can occur each 60 ms. The ability to step down is commonly available at the rate of one step for each sample.

It will be appreciated that the ability to step either up or down cannot be made completely flexible, say to step up or down every sample position. If this were the case, a single low-frequency sinusoid signal would cause the gain setting merely to reflect the shape of the pulse. As a sine wave goes through zero the gain would step up

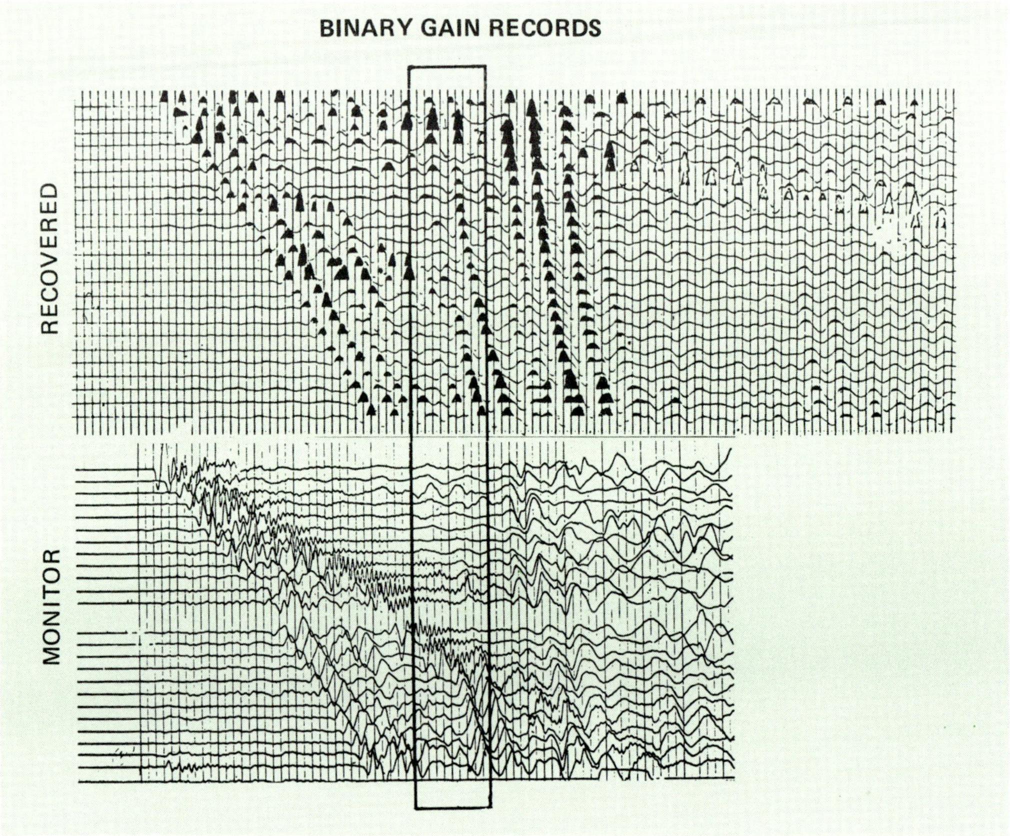

FIGURE 3.6

The binary-gain monitor record which is displayed in the field normally has not been gain recovered and rarely represents the quality of final output.

quickly to amplify the signal and, conversely, as the sinusoid increased in amplitude the instruments would step down. the resulting binary-gain trace would be nothing more than a slave of the sinusoidal wave. In practice, the ability to step upward should not be more frequent than the time period of the lowest frequency which is commonly recorded. For example, the presence of a 20-cycle low-frequency component would suggest that the gain ranging capabilities should not be much faster than every 50 msec.

Note that there is no mechanical or physical basis for this limitation; it is merely to provide a more meaningful record for monitor purposes in the field. If the step facility were freed to move with each sample, the proper trace could be quite readily reconstructed from the recorded monitor signal and its accompanying gain information. However, the monitor values would be recorded as a jagged saw-tooth trace with almost a constant level of amplitude. The binary-gain monitor record which is displayed in the field normally has not been gain recovered and rarely represents the quality of final output (**Fig. 3.6**). The monitor is merely a reproduction of the mantissa value on the tape. Therefore, it always lies between an upper and a lower bound regardless of the actual energy reaching the recorder. Spikes will appear on the monitor trace, corresponding to the binary gain steps. These are quite visible where a gain step has caused a deflection in the smooth output trace.

There are a number of systems for recording the gain level. One system records the actual gain level only once for each 30 samples, but keeps track of the level over the interval by a system of coding. The intervening values can be derived from a two-bit (no pun) code, one bit is set to indicate any change in the level and the other to indicate the direction, whether up or down. The demultiplex routine must then contain a sub-routine to keep track of the gain movements from one fiducial value to the next. The results of the intervening arithmetic must then be checked to make sure it agrees with the actual recorded level. This system of recording, basically the SEG "A" format, is somewhat more economical in consumption of magnetic tape as it uses a smaller number of tape frames to record the sample. It is however somewhat more difficult to handle from a programming standpoint. Field recorded tapes are rarely received in perfect condition. Occasional parity errors or missing bit information may cause one or more tape frames to be lost. When this occurs, the gain level may be lost. A search ahead to the next recorded gain position must then be made and the value carried backwards.

The other common system records the gain level for each sample, which leaves no doubt about the actual gain level, but uses somewhat greater quantities of tape, as a frame must be used for each gain value.

The broad dynamic range of the binary gain amplifier reduces the chance of noise saturation. Hence, there is little or no need to do any substantial amount of filtering in the field. A broad-band unfiltered signal is also best from a processing standpoint, a further reason to perform little or no band-pass filtering in the field. A high-cut anti aliasing filter must be incorporated but otherwise a binary gain crew may operate with no filters in the circuit at all. The instruments can thus provide a more linear response and thereby minimize their contribution to signal distortion.

The instruments do not provide a completely automatic gain process. A manual estimate of the initial gain setting must be applied for control until the arrival of the first breaks. The reason for this is very similar to the pre-suppression on expander or AGC type instruments. If the gain were allowed to range to its natural level, the relatively low amplitude background noise would cause the gain to step up quickly to some level near the maximum. The subsequent arrival of the strong burst of energy from the shot would then be amplified to a level higher than the dynamic range of the instruments, causing saturation and distortion.

It is preferable to have the initial gain setting close to that level at which the gain should be when the initial energy comes in, but the equipment is tolerant of observer misjudgments in the field and, if the initial gain is set too high or too low, the gain will quickly step to the proper level. The level of the gain for each sample may be printed out at the processing center to observe the action of the gain control.

Floating Point Recording

Shortly after the introduction of the binary gain recorder, digital magnetic tape recording rates improved, and another type of encoding of the digital seismic data was introduced, the Instantaneous Floating Point recorder.

This equipment has been hailed by some of its proponents as another major breakthrough in digital techniques. This enthusiasm is hardly warranted, since binary gain recorders do an adequate job, but there are some positive features in the new system.

The floating point system represents an improvement in certain characteristics, primarily in ease of operation in the field. Binary gain requires some estimation of the levels of the incoming primary energy in order to make initial settings. Floating point equipment operates almost completely independently of the operator. Both provide the capability of producing recorded data which meets the demands of the processing systems.

The few minor difficulties encountered with binary gain equipment are almost completely eliminated in the floating point system. Like the binary gain system, the floating point system records the data sample as a two part number consisting of a base number and an exponent indicating the amount by which the base must be multiplied to produce the real value. The format of the floating point number is very similar to the internal format used by some digital computers for engineering calculations. If the exponent is some power of two, the values are similar to binary gain notation.

The design concept of a floating point system can be illustrated by a simple diagram (**Fig. 3.7**). Like binary gain, actual field equipment may vary considerably from the schematic shown here. In practice, commercial binary gain and floating point systems may include many related elements.

The amplifier forms an integral part of the gain control in the floating point system. Each amplifier has a gain

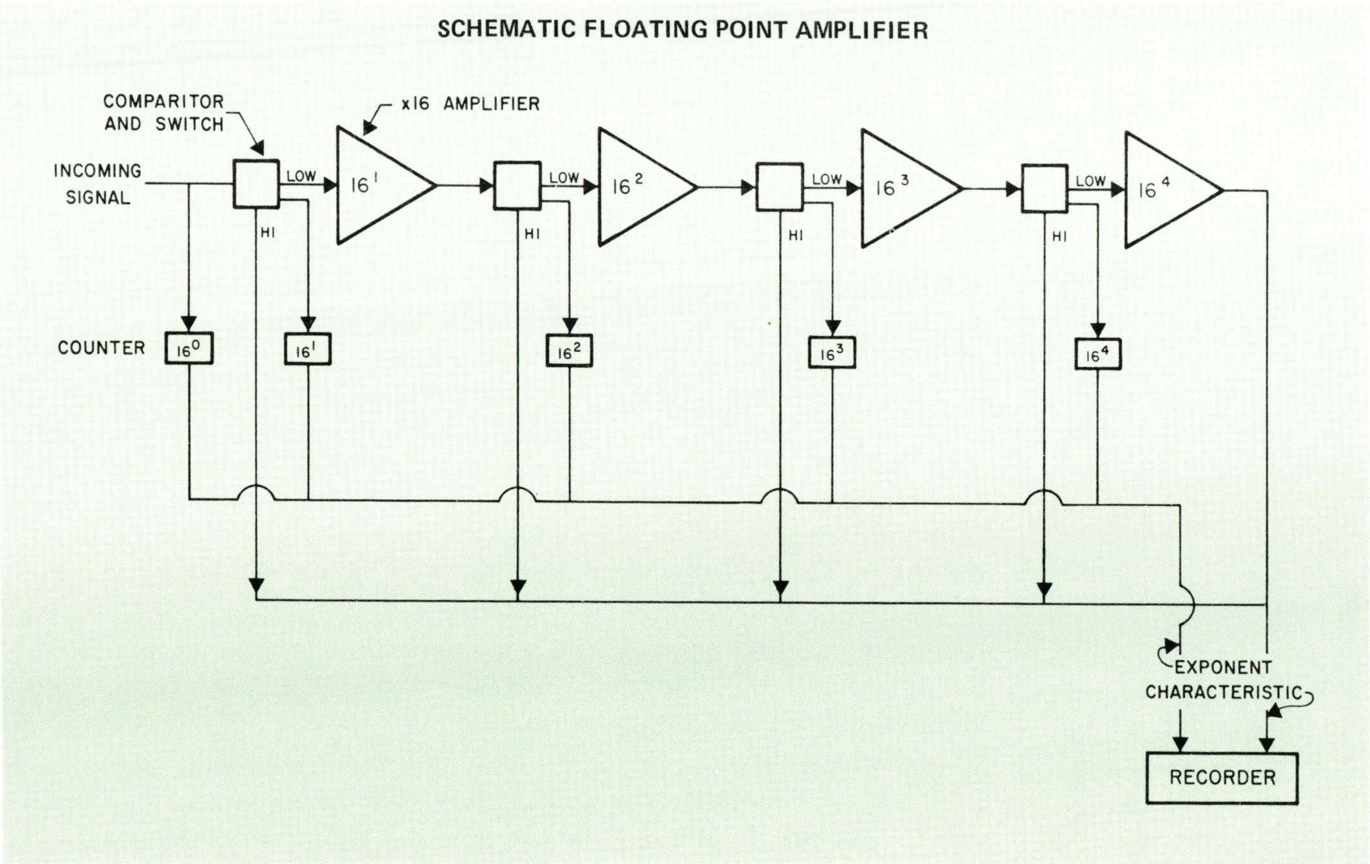

FIGURE 3.7

The design concept of a floating-point system can be illustrated by a simple diagram.

factor which is some fixed integer power of two. In this example, amplifiers having a factor of sixteen will be considered. The amplifiers are switched in series into the next amplifier or to an output line. The output level is monitored by a gain control unit which determines whether or not the amplitude falls below a specified level, in this case one sixteenth of the maximum level permitted by the recording medium. If the signal is below this level the control unit will shunt the signal to the next amplifier increasing the amplitude by a gain factor of sixteen. If the signal level is still below the base level a second amplifier is connected and the process repeated until the desired level is obtained. Conversely, should the amplitude rise to a point approaching the maximum capacity of the recording medium, a control pulse will cause one amplifier to be disconnected from the series, shunting the signal to the output and reducing the amplification by a factor of sixteen. This operation occurs with sufficient speed to ensure that when the sample is taken the output signal lies within the desired bounds. The sample value is then recorded on magnetic tape as a base number and an exponent, the exponent indicating, in this case, the number of times the input has been multiplied by sixteen in order to obtain the base number.

It should be evident that if the floating point system were constructed of amplifiers having a gain factor of two, the net effect of the system would be almost identical to that of binary gain systems, with the recorded signal always representing the input modified by a factor of some power of two. This is indeed true, and the distinction between the types of amplifiers used in the several available commercial systems can be rather blurred.

The other main difference between the two systems lies mainly in the sample and record procedures. Floating point systems resulted from improved digital tape recording technology, which allowed a numerically larger sample to be recorded at an adequate rate. A constraint of the binary gain systems was that only two bytes (tape frames) (16 bits) on standard 9-track magnetic tape were available for recording. Floating point requires as many as four bytes (32 bits) to record a sample in the same format used for scientific notation by many computer systems.

The standard seismic floating point recording format recommended by the SEG Committee on Digital Recording Standards is a 32-bit binary word consisting of three components: Qs for the sign, Qc for the characteristic or exponent Qf for the fraction or base number.

Qs is a single bit which is used to indicate the polarity of the signal. It will remain zero for positive numbers and be set to 1 for negative numbers.

Qc is made up of 7 bits, used to indicate some integer power of 16. The 7 binary bits normally permit an integer range from zero to 127. The actual value is recorded in what is termed 'excess 64 binary notation', which means that all of the values recorded in the Qc word have had 64 added to their true value. Both positive and negative exponents appear as positive numbers. Any recorded value less than 64 indicates a negative exponent.

Qf is a 24-bit binary number. Since it is always less than 1 it may be considered to have a radix point to the left of the most significant digit. Expressed in binary notation the value of the number in positional notation, from the most significant to the least significant digit would be $2^{-1} + 2^{-2} + 2^{-3} \ldots\ldots\ldots + 2^{-24}$.

The recorded value represented by this floating point number may be interpreted as:

$$Qs \times (16^{(Qc - 64)} \times Qf).$$

The main advantage of this system is that the amplitude measurement ranges continuously, and there is no need to have any gain ranging control such as that used for binary gain equipment. Associated with this is the ease of handling. There is no need to manipulate and interpret the series of gain bits as is necessary with one type of binary gain recording format commonly in use. The field operator need not have any concern over gain range rates.

The larger number of bits permits a wide dynamic range. Modern seismic floating point systems effectively have no gain range restrictions, and therefore can record the true amplitude of each sample.

Occasionally through equipment malfunction, bad tape or related problems, one or more of the gain bits in binary gain recording may be lost and it may not then be possible to recover the gain satisfactorily since the gain bits indicate the change in gain from the previous sample. The floating point notation normally contains all amplitude information of each sample, although this is also possible with some binary gain recorders.

These constitute the advantages of the floating point systems, the method that will likely eventually become fairly standard throughout the industry. However, the nature of the improvement is such that it does not obsolete binary gain equipment, but if a choice exists between the two types then the floating point equipment will likely be preferred.

The dynamic range of the floating point recording far exceeds that of the data. This is best illustrated by the fact that although the theoretical dynamic range of the floating point recording equipment is 10^{+75} (several hundred db) the dynamic range of normal seismic field amplifier equipment is limited to 15 bits (90 db). Therefore at least 7 of the significant bits in the floating point word are unused and must be filled with zeros.

Some of the advantages of the improved dynamic range and the ease of data handling inherent in the standard digital floating point exchange format of 32 bits, or four bytes, to record each value may be offset by other considerations. While the 32 bits matches one standard internal computer format, it is somewhat wasteful since the lower 7 bits are normally not produced in the field. This format requires the magnetic tape to handle double the rate of information input for binary gain. It is possible only because the data can be packed on magnetic tape at double the density possible a few years ago. A straight binary notation in a 32-bit word would provide a dynamic range of approximately 180 db, which would appear to be adequate for direct recording over the range of seismic signal. Certain difficulties arise in the implementation and design of direct 32-bit recording, particularly in the analog-to-digital conversion process. For the present, it is likely that both binary gain and floating point equipment will remain in use.

The merits of the floating point system should be reviewed primarily from the standpoint of a field recording system. Subsequent processing procedures applied to the recorded data will be identical to that used for binary gain-recorded data. The first and one of the most important processing steps is to obtain the best possible recovery of the true signal amplitude followed by normalization of the data, to compensate for transmission losses. The need to compensate for the natural energy decay of the recorded signal remains a problem common to both types of data processing.

One of the claims made for floating point recording is that data processing may be done in the floating point format throughout, eliminating any arbitrary scaling decisions which may be required by the more limited dynamic range of fixed point processing. The claim regarding scaling has only limited merit, assuming the comparison is made to seismic data processing commonly in use today. All field data must be subject to some type of scaling in the gain recovery and normalization processes. Also, some scaling, to raise or lower the RMS level of the total trace is desirable to compensate for differences in ground coupling effects in the field. The normal stacking process for common-depth-point surveys is nearly always improved by balancing the RMS level of traces prior to stack, or, in some cases, even applying special scaling factors to the traces prior to stacking. The important point is not that scaling be eliminated but rather that scaling decisions need not be imposed by equipment constraints. In the past, a limited computer word size often forced a truncation of the data with a corresponding loss of information. Scaling now is normally imposed only as required by the process routine to meet the standards of the processing model. The exception to this is found in the process routines of some mini-computers with 16-bit word size. The need for speed in some of these has resulted in truncation or scaling of the data word to a size well below the dynamic range of the input. It is reasonable to state the 16-bit 'mini' is now obsolete for seismic data processing, being replaced by 32-bit machines.

One of the reasons for the popularity of the SEG "A" binary gain tape recording format (gain bit indication) over the SEG "B" (recorded gain level) is that the "A" format uses only two tape frames per sample on 9-track tape and is therefore quite economical of magnetic tape usage. The floating point recording requires four tape frames per data sample, doubling the tape requirements. This may be a particular problem in marine work, where a new reel of tape must be mounted every few minutes. The increased tape usage also requires higher tape speed, or greater packing density or a combination of both, in order to handle the increased amount of data recorded in the same time, and is therefore more expensive.

It is probable that the advent of floating point recording will provide the impetus for further development in the field of high-dnesity recording. The original standard 800 (bits/inch) recording was replaced by 1600 bpi operating at higher tape speeds. Recorders are now available which feature 6250 bpi packing densities. The adaptation of video recorder equipment to seismic recording may be a major advance, packing the equivalent of 100 or more tape reels on a single video tape.

Offsetting the gains in tape transport capacity are the increasing number of channels which are recorded simultaneously. The 24-channel recorder has been replaced by 48-channel equipment, and 96-channel recorders may soon be commonplace. The increasing popularity of 3-D surveying imposes pressures for even more channels.

At the present time, almost all geophysical field systems intercept the seismic reflection signal with a geophone or hydrophone, which converts the seismic velocity or pressure wave to an electrical signal. The analog signal is transmitted through a pair of wires of the recording cable to the instrument system. In the instruments, the continuous analog signal is broken up at some point into a series of discrete digital amplitude measurements. From then on it is handled as a set of separate numerical values, or samples. Each digital sample value, in turn, is represented by a number of discrete bits, each of which may be considered an individual unit of information.

One advantage of having information in discrete samples is that they may be sorted into any sequence, or interleaved, as in a multiplexed record. This feature is used to advantage in a system which uses only two conductors to transmit the signal from the geophone to the amplifiers. The geophone output is digitized near the geophone group location rather than within the recording instruments. The sequence of digital values can be cut into the stream travelling through the pair of conductors to the truck or boat. In other words, the geophone outputs are multiplexed prior to entry into the cable. In the instruments the incoming steady stream of bits is merely segmented into data words containing the corresponding bits for each sample for each trace.

In this manner, very large numbers of channels can be recorded without the need to have a pair of wires for each channel. This promises to facilitate the introduction of multichannel field recording techniques not now feasible. Increasingly, all field instruments are built around a programmable digital processor.

Gain Compensation

Binary gain and floating point have, as their principal advantage, a notation system which provides the ability to reconstruct the signal to a close approximation of the relative signal amplitude as it was received by the geophone. It is for this reason that these two methods are now the commonly accepted means of recording. A relatively simple computer program will reconstruct the signals to their full dynamic range, or true amplitude, as received. This is quite different from the frequent misnomer of true amplitude commonly used to describe final seismic sections. Such sections have had some form of transmission loss correction applied, designed to produce a trace in which the reflection amplitude is directly proportional to the amplitude of the reflection coefficients. But such processes are not always successful.

Reconstruction of the field-trace amplitude is relatively simple; but this true-amplitude recovery does not complete the signal recovery process. At this stage the data processor faces the same problem as the instrument designer; how to compensate for the great change in amplitude over the length of the trace. The critical part of the operation is the determination and application of the inverse gain program to restore the level of the trace to a function of the reflection coefficients, e.g. an approximation to the loss-free synthetic seismogram. Several methods of calculating and applying the inverse gain may be used, but some of these serve exactly to negate the value of using the binary gain instruments.

Perhaps the most popular method in use today is some variation of the programmed spherical divergence curve. A fixed decay curve is designed, usually expressed as a quadratic function of time and velocity. The inverse of this

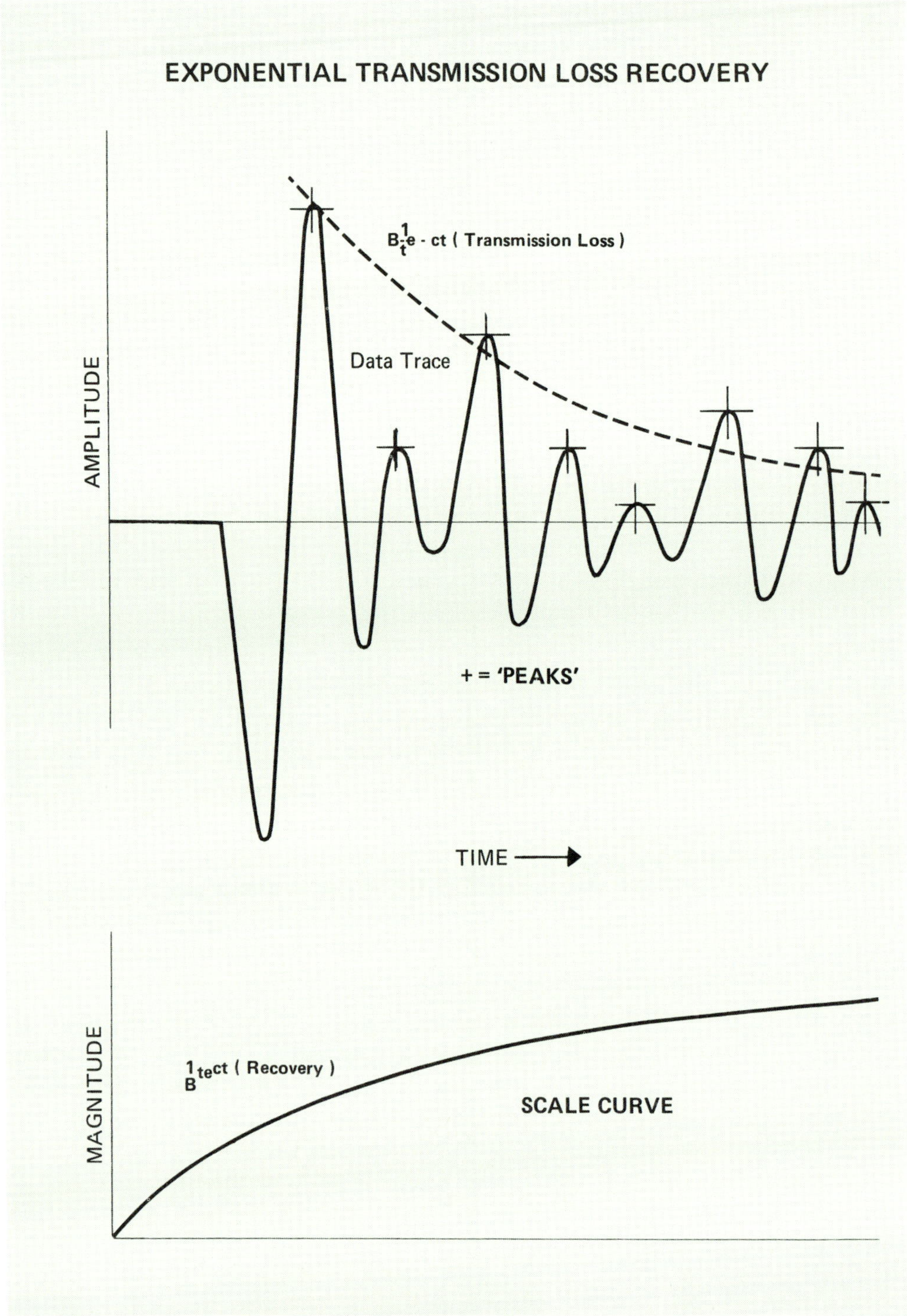

FIGURE 3.8

A popular expression of the decay curve has been the single exponential function $B\frac{1}{t}e^{-ct}$, where B and C are empirical constants. The recovery curve (scale curve) is just the reciprocal function $\frac{1}{B}te^{ct}$.

curve is then applied to all data. While the basic method is the same in all cases, the application varies widely in practice. The same curve may be applied to all traces of a record, but obviously this is in error since the geometry of the spread causes the far traces to receive their first reflections long after the near traces. Distance from the shot has a considerable effect on gain curves. The geometrical distortion associated with normal-move-out compresses the decay curve into a shorter time span on far traces than on near traces, increasing the apparent rate of decay. A different curve is sometimes used for each four or six traces, but this is only a less imperfect method. Correctly a curve should be designed for each trace. **(Fig. 3.8)**.

The entire recovery procedure is illustrated in **Fig. 3.9**. The signal, recorded on the digital magnetic tape, is similar to that plotted in **Fig. 3.9a**. Sharp spikes are present on this trace, corresponding to switching one of the gain attenuators in or out the circuit, changing the signal amplitude by a factor of 2. The gain-count trace is recorded in **Fig. 3.9b**. To recover the true amplitude of the trace, each sample is multiplied by the power of 2

FIGURE 3.9

FIGURE (a)

The signal, recorded on the digital magnetic tape, is similar to that plotted in Fig. 3.9a. Sharp spikes are present on this trace, corresponding to switching one of the gain attenuators in or out the circuit, changing the signal amplitude by a factor of 2.

FIGURE (b)

The gaincount trace is recorded in Fig. 3.9b.

FIGURE (c)

To recover the true amplitude of the trace, each sample is multiplied by the power of 2 corresponding to the gain level. The result is Fig. 3.9c, the fully recovered true-amplitude trace.

FIGURE (d)

An emponential curve can represent the rate of decay of the signal. The inverse of the curve (Fig. 3.9d) (note that it is evidently only a smoothed version of the gain trace in Fig. 3.9b) may then be used as a gain curve to compensate for the exponential decay of the trace and restore it to uniform relative amplitude over its length.

FIGURE (e)

The resulting trace now has no spikes, and each reflection should be a relatively faithful representation of the corresponding reflection coefficient.

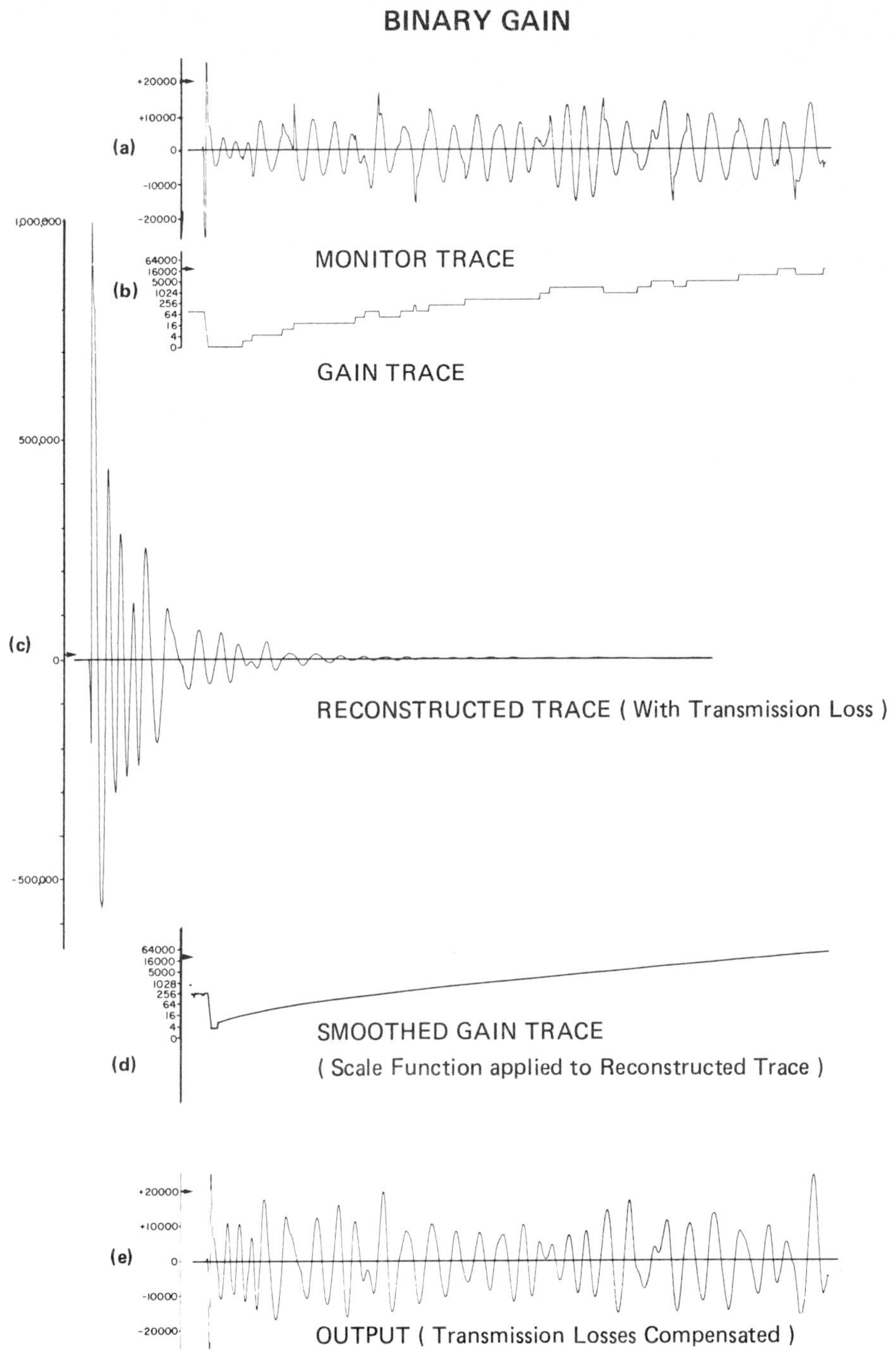

corresponding to the gain level. The result is **Fig. 3.9c**, the fully recovered true-amplitude trace.

An exponential curve can represent the rate of decay of the signal. The inverse of this curve **(Fig. 3.9d)** (note that it is evidently only a smoothed version of the gain trace in **Fig. 3.9b)** may then be used as a gain curve to compensate for the exponential decay of the trace and restore it to uniform relative amplitude over its length **(Fig. 3.9e).** The resulting trace now has no spikes, and each reflection should be a relatively faithful representation of the corresponding reflection coefficient.

The relatively simple gain response curve represents the fundamental difference between analog and digital recording. At first glance, the difference appears to be merely one of procedure, for if a continuous recording of the feedback gain control of an analog amplifier were made, it would also be possible to reconstruct the trace in a manner similar to binary gain. However, this was not done in the past due to limitations of the recording equipment. At best a single gain trace was recorded to represent the average gain of several amplifiers. Early digital recorders, which merely digitized the signal output from conventional analog amplifiers provided no substantial advantage over

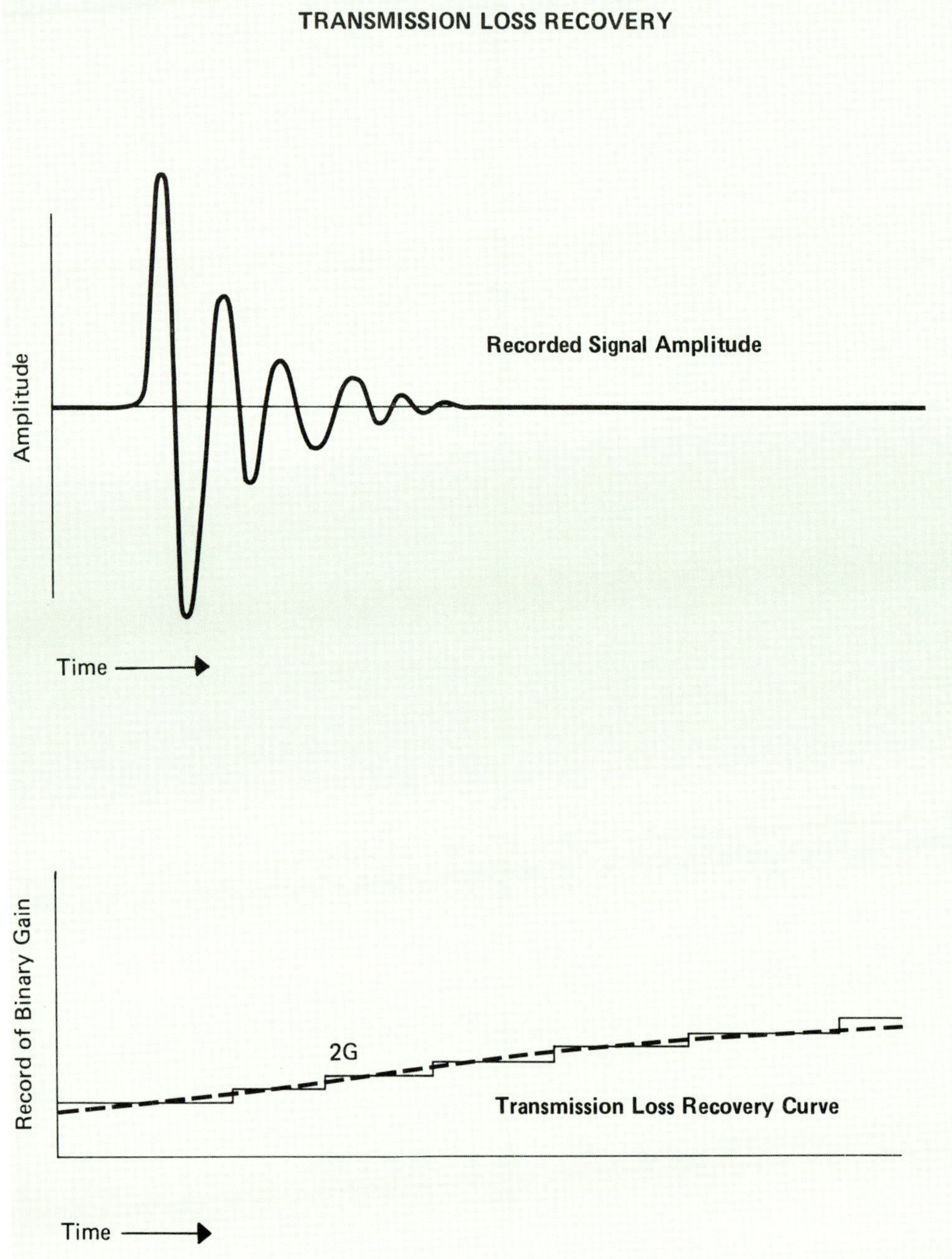

FIGURE 3.10

A smoothed version of the stairstep function, 2G, is used to correct the recorded signal amplitude for transmission loss.

good analog magnetic tape recording. Some wider dynamic range was available, particularly with one-inch tape, and more advanced systems recorded the average gain to provide some measure of recovery. But proper gain recovery was not made before the introduction of binary gain recording.

This method of exponential transmission-loss compensation is a considerable improvement over AGC control, but, like all seismic processing, it makes certain assumptions, principally that the decay is smoothly exponential and does not vary over the line. This is rarely true, since any facies change in the section produces some change in reflectivity.

In the absence of irregular noise patterns and assuming the true amplitude of the recorded trace has been recovered, there are a number of other effective ways of determining from the data itself what the compensation for the transmission losses should be, and which yield an output trace approximating the synthetic trace calculated from reflection coefficients.

Least-squares mathematical techniques can be used to determine the best-fit exponential function from the recorded trace amplitude (**Fig. 3.8**). The inverse of this function may then be applied to the trace as a smoothly-varying scale factor

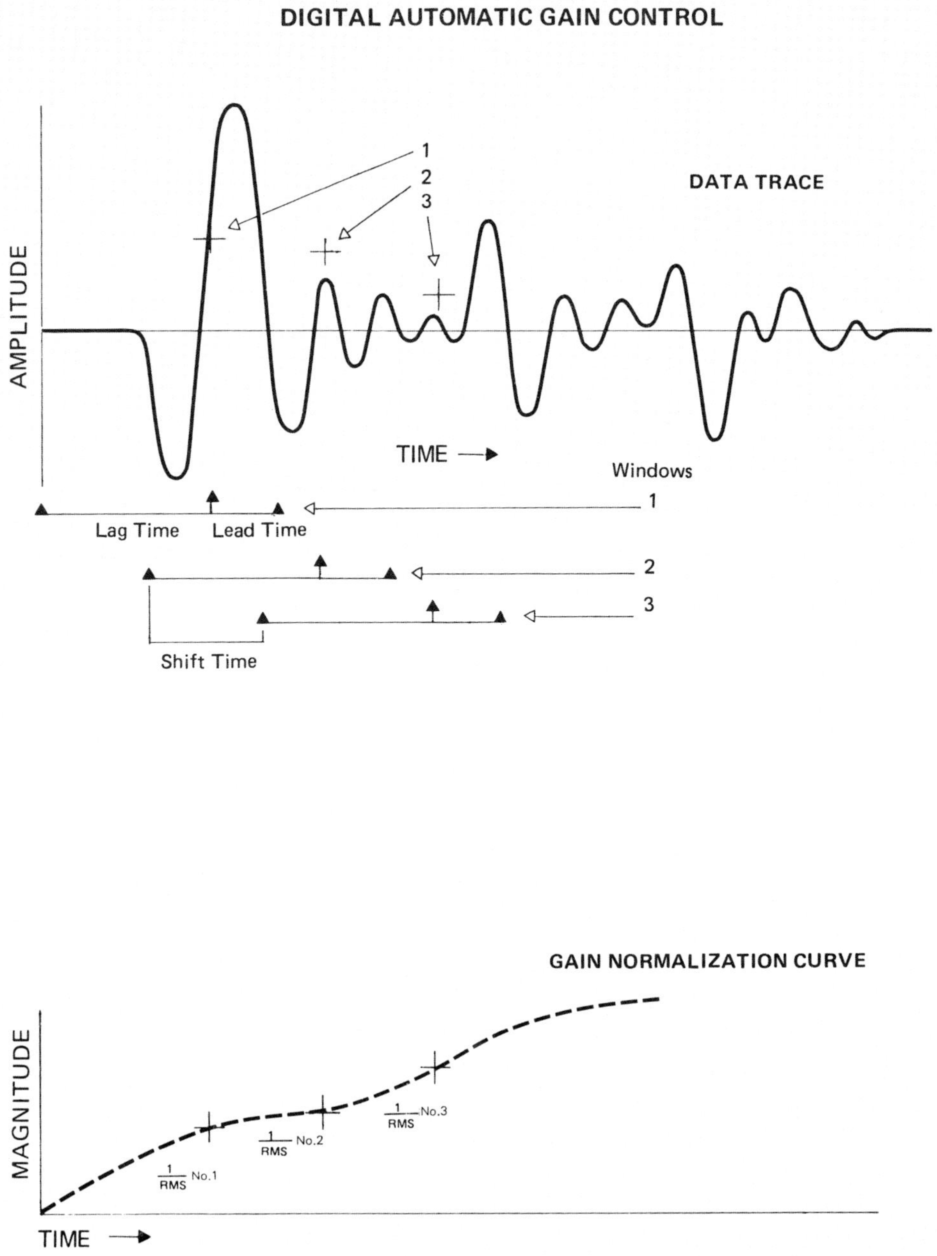

FIGURE 3.11

A modified AGC method uses the data to develop a type of program-gain control. A relatively long sample-averaging window is moved down the trace in large jumps, usually equal to one-half the length of the window. The average amplitude within the window at each point is measured and a smooth curve passed through the several points.

to compensate for the exponential losses. Local variations in the decay curve corresponding to the arrival of reflections will rise above the general smooth background level in direct proportion to their relative amplitude at any particular time.

A related curve calculation is especially well adapted to binary gain recording. A line drawn through a plot of the binary gain steps, will often be nearly a straight line. This is because each step actually represents the power of two, so the line is actually a logarithmic plot to the base two of the gain control curve (**Fig. 3.10**).

Therefore, if the exponential function method is to be used for the trace, it can be done quite easily by merely fitting a straight line through the binary gain steps. The inverse exponential curve is then applied to the data as before.

The signal may not always decay in a smooth exponential manner. A major change in age of the sediments usually changes the reflectivity sequence. A related form of the smoothed binary gain curve may be computed. The model used for seismic reflections suggests that the amplitude of the returning signal should drop after each reflection, i.e., require an increased boost in the gain. The reverse situation should not occur.

FAULTY GAIN PROCESSING

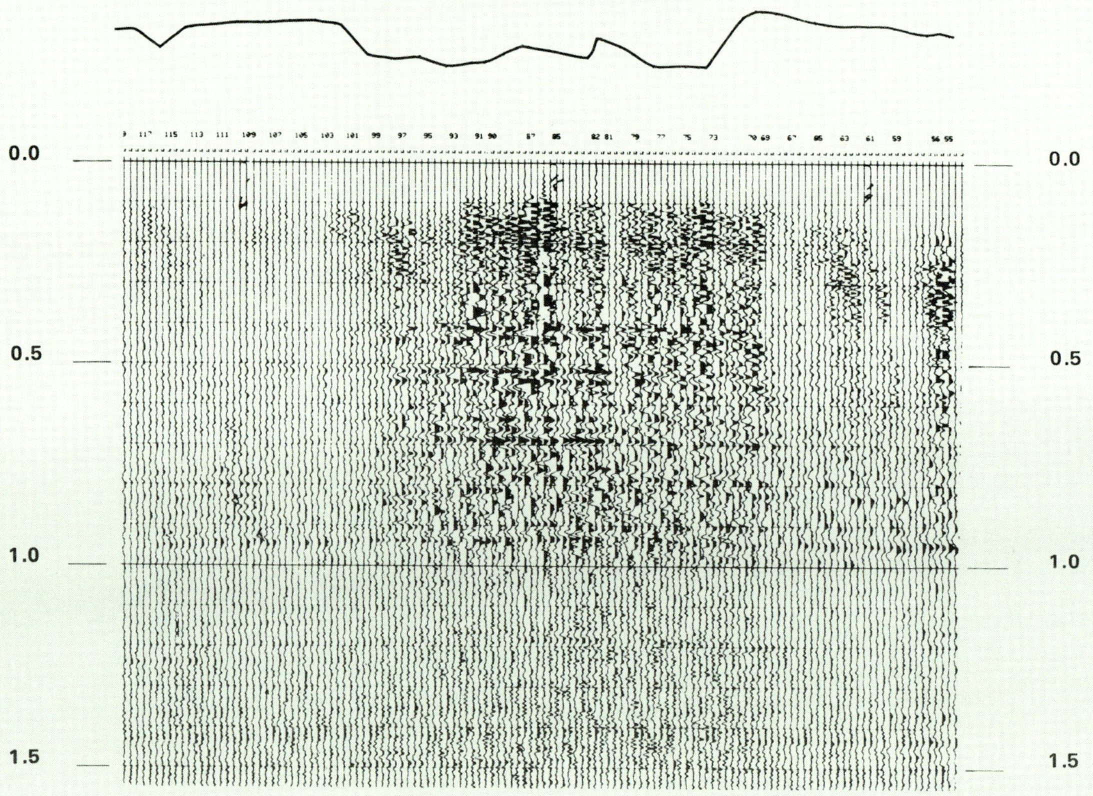

FIGURE 3.12

A common failing of the fixed curve is that it may not take into consideration near-surface changes. In the bottom of the valley a substantial thickness of sediments, present on both sides, is missing. This will obviously affect the transmission-loss characteristics. If the same gain curve is used throughout the results will be as illustrated.

Therefore, if a smooth line, rather than the straight line as above, is drawn through each up-step of gain, ignoring all down-steps, the curve should represent the inverse of the rate of decay. This curve can be used to compensate for transmission losses. The only difference between this and an AGC system is that AGC recognizes all steps up and down.

It is also a simple matter to develop a computer program which exactly duplicates the operation of the analog feedback device. In the same manner as the AGC circuit, the amplitude of the trace is constantly monitored and a variable scale factor is then applied to bring it to some desired level. It should be evident that each sample cannot be examined and treated individually, or the output of the entire trace would consist of a series of square waves, all of uniform amplitude. In practice, digital AGC uses the average absolute amplitude within a sliding window several samples long, and from this average determines a scale factor which is applied to one sample, usually the central sample of the group. The window is then shifted one sample down the trace and the process repeated.

Such a system has as its principal advantage the fact that it is a continuous and entirely automatic feedback process, and never fails to produce an output trace in which the amplitude lies within some specified tolerance. Unfortunately, this process exactly duplicates all of the undesirable effects of using AGC in field amplifiers.

The pulse shape is altered by the tapering effect as the window moves in and out of a single event. As the window slides into the event, the amplitude suppression will gradually increase to the mid-point followed by the reverse effect as the window slides off the peak. This act alters the actual pulse shape. Thus, the effort and expense expended to use binary gain or floating point recording equipment in the field will be completely negated in the data center and all of the advantages of the modern field equipment lost. Lengthening the time window will reduce the distortion and is most often the reason advanced to justify use of the AGC technique. The longer window may mitigate the AGC effect but does not remove it. Lengthening the window has the effect of slowing the AGC response, spreading the effects over a longer time. This appears to be satisfactory but in fact causes the recovery to be influenced by events at some distance from the center of the window. The fallacy may be illustrated by making the window length equal to the length of the seismic trace. This would produce output essentially unchanged from the input except perhaps for a constant scaling factor.

A modified AGC method uses the data to develop a type of program-gain control. A relatively long sample-averaging window is moved down the trace in large jumps, usually equal to one-half the length of the window. The average amplitude within the window at each point is measured and a smooth curve passed through the several points (**Fig. 3.11**). The resulting somewhat irregular curve represents the average amplitude over the length of the trace. The inverse of the curve may then be applied to the trace to perform the normalization function. It requires only some skill in determining the optimum length of the window. This method also has the advantage of being very rapid.

The several recovery methods used may be seen to fall generally into two classes: the data-dependent curve-fitting routines, and the arbitrary fixed-curve-gain procedures. Neither approach is strictly correct and no method is universally effective.

The arbitrary fixed-curve programs assume a constant smoothly-decaying condition, fixed vertically and often horizontally within the section. In practice, this is not true for very many cases.

Along the length of the seismic line, faulting or intrusions can cause the nature of the section to change abruptly laterally, with a corresponding sudden change in the reflectivity characteristics. In some cases, some effort may be made to locate different gain functions at key locations, and interpolate linearly between the control points. The data-dependent curve-fitting routines compensate for this effect automatically.

Unconformities and other major reflection horizons in the earth section often produce very high reflection coefficients. When this occurs a large part of the incident energy is suddenly reflected back to the surface resulting in a very abrupt drop in amplitude. The fixed-program recovery methods fail to recognize the changed loss rates that occur when a strong reflection terminates suddenly, due to a pinchout or a sudden change in the fluid content. A strong reflection will be followed by an immediate drop in signal amplitude, but no such change will occur where the strong reflector is not present. Thus, with a fixed curve, a uniform horizon immediately below will appear to have different response below the strong reflector than elsewhere.

Another common failing of the fixed curve is that it may not take into consideration near-surface changes. **Fig. 3.12** is a short section of seismic line which crosses a deep valley. In the bottom of the valley a substantial thickness of sediments, present on both sides, is missing. This will obviously affect the transmission-loss characteristics. If the same gain curve is used throughout the results will be as illustrated. Some centers seek to compensate for this by applying the gain curve from a constant datum, but this too is in error.

At any given moment in time, the amplitude of the raw seismic trace may be the sum of several different acoustic signals in addition to the desired seismic reflection sequence. The decay rates of the extraneous signals may bear little or no resemblance to that of the seismic signal. Some of these are illustrated in **Fig. 3.13** and may be classified as noise.

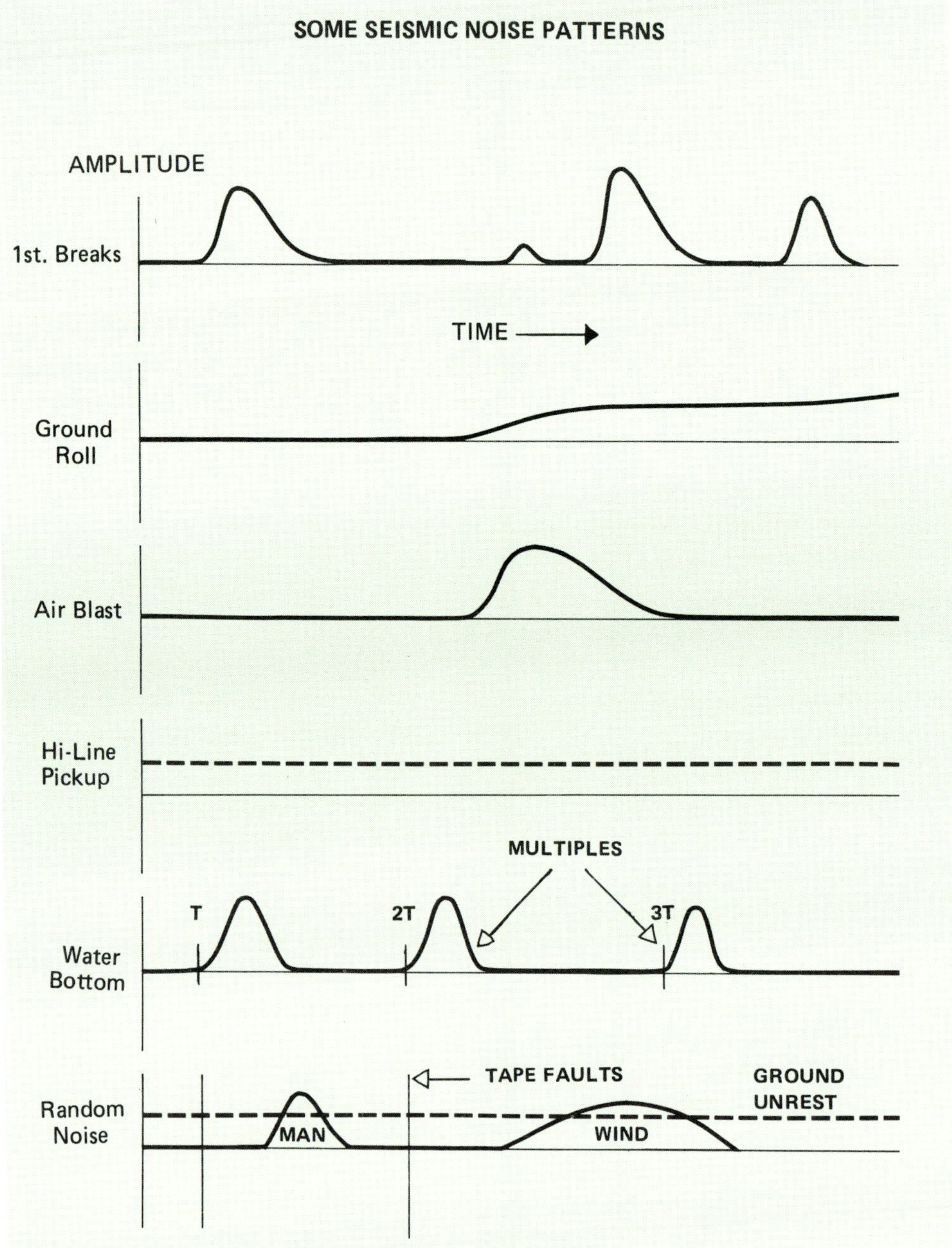

FIGURE 3.13

A any given moment in time, the amplitude of the raw seismic trace may be the sum of several different acoustic signals in addition to the desired seismic reflection sequence. The decay rates of the extraneous signals may bear little or no resemblance to that of the seismic signal.

The decay function of the noise-free signal must basically decrease monotonically with time. Yet, a large amount of noise may cause the recorded amplitude to increase with time. The least-squares method will always produce the best fit in the mathematical sense of a given function type to the information, but the fit may not be best in the physical sense. For instance, in the case of noise as suggested above, the calculated curve will decay at a much slower rate than the seismic signal, and the inverse of this calculated curve will apply scaling factors that are too small, causing loss of data.

The defect in methods which calculate the gain-recovery curves from a measurement of the raw data is that measured raw signal amplitude is always the sum of signal plus noise, whereas the gain recovery should be based on a measurement of the signal only. The ability to recognize such differences is very difficult to design into a process routine. One compromise is to apply some programmed gain such as spherical divergence first, then process to eliminate some part of the noise. The original gain compensating may then be reversed to restore the trace to its original form. A more reliable estimate of true signal strength can then be made by calculation of the gain curve by some

COMPARISON OF BINARY GAIN AND FAST AGC

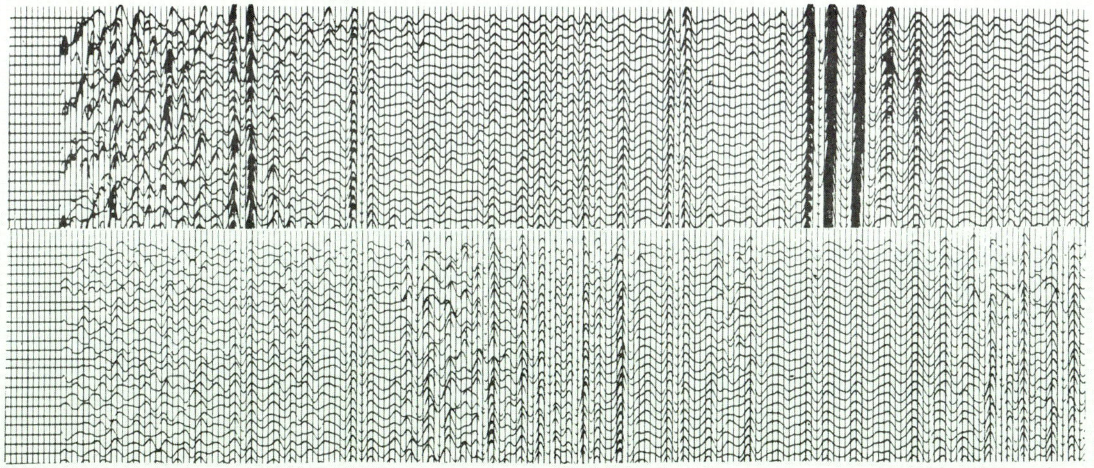

FIGURE 3.14

Binary gain or IFP recorded sections tend to have much better signal-to-noise ratio than AGC-recorded sections before any processing subsequent to trace recovery is done.

selected feedback processing.

Geological Considerations

The success of digital recordings in approaching the basic processing assumptions much more closely is usually attributed to amplitudes of obvious reflections. However, there is another very substantial consideration. In many parts of the world the geologic section is characterized by relatively thick, discrete layers of shale or carbonate which contain few reflecting horizons. Major reflections are obtained only at the boundaries of these units. In such cases there is very little reflection energy returned to the geophone from within the thick units. The corresponding low level of energy passing through AGC amplifiers causes the gain to expand and build up the background energy which may consist primarily of noise. Thus, on the recorded trace, the noise level is built up and reducing the contrast between the reflecting events and the intervening nonreflecting zones (Fig. 3.14). Therefore, basing the gain curve on the major reflections has a sound geologic basis and does lead to improved data quality.

The result of this is that an AGC-recorded section, before processing, tends to be much noisier than one which has been recorded and recovered with the binary-gain system. The noise on the AGC output tape can never be completely suppressed later and causes serious detrimental effects on subsequent processing. Thus, binary gain or IFP recorded sections tend to have much better signal-to-noise ratio than AGC-recorded sections before any processing subsequent to trace recovery is done.

Two records, one recovered with a digital AGC program, the other with a gain-fitting program, are shown in **Fig. 3.15**. At first glance, the AGC record appears to be better. It is less noisy and of more uniform amplitude. The other record has a very strong low-frequency ground roll cutting across the record. However, the noisy traces represent the sum of signal and noise. Hopefully, at some stage in the processing, the noise will be removed, leaving the signal at its proper level. On the other hand, the AGC record has had the sum of noise and signal reduced to the design level of the amplifiers. When the noise is removed the remaining signal will be very weak.

The action of the recovery programs is illustrated on **Fig. 3.16** for the low-noise-level trace 1 of the seismic record. Note the AGC recovery curve dips quite sharply to accomodate a high-amplitude reflection near 1.3 sec. The recovered trace has a low-amplitude reflection at this point. The more correct method ignores the amplitude change, with the result that the recovered trace has a strong reflection at this point. In this case, the amplitude of the event is a more nearly correct recovery of the reflection amplitude.

FIGURE 3.15

Two records, one recovered with a digital AGC program, the other with a gain-fitting program, are shown in Fig. 3.15. At first glance, the AGC record appears to be better. It is less noisy and of more uniform amplitude. The other record has a very strong low-frequency ground roll cutting across the record. However, the noisy traces represent the sum of signal and noise. Hopefully, at some stage in the processing, the noise will be removed, leaving the signal at its proper level. On the other hand, the AGC record has had the sum of noise and signal reduced to the design level of the amplifiers. When the noise is removed the remaining signal will be very weak.

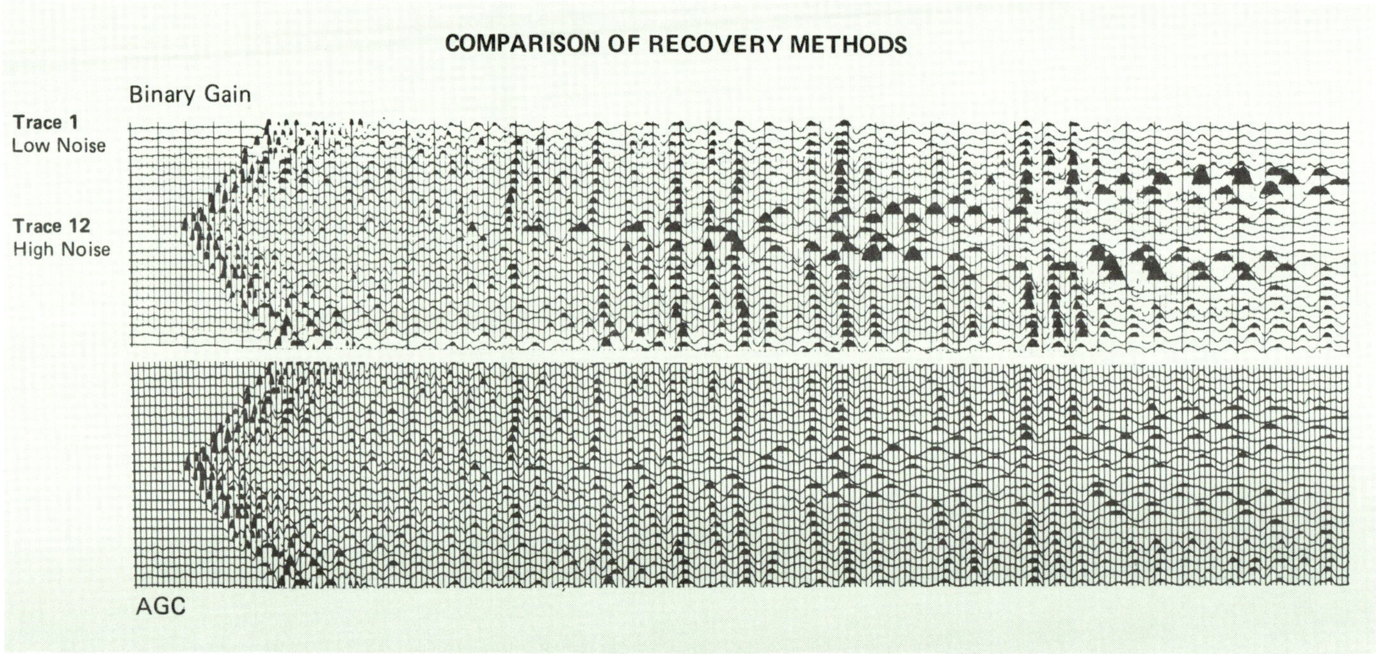

The effect on the noisy trace 12 (**Fig. 3.17**) is more pronounced. The improper gain curve sags from 0.5 to 1.5 sec, holding the amplitude of the recovered trace to a near-constant level. The programmed recovery curve is flat over much of the length from 0.6 to 1.3 sec, and the recovered events are of high amplitude, representing the sum of signal and noise over this portion of the record. After processing to eliminate noise, the signal level will remain close to that of trace 1. Although improved, the programmed-curve recovery is not quite right either, since the curve of trace 1 suggests the recovery curve should rise slightly, through the noisy zone, whereas the curve used was flat over a considerable length.

Conclusions

The conclusion that can be drawn from this is that gain recovery and transmission-loss compensation is a somewhat inexact science. The true-amplitude label is freely applied by processors to the output from a wide variety processes which may not always produce true-amplitude output, assuming the specifications of the term can be defined. The exact nature of the decay curve is rarely understood.

Unfortunately, the important influence that proper gain-recovery procedures exert on subsequent processing is not always appreciated. Also, the effects of processing tend to be cumulative. The quality of output of any one operation directly affects all subsequent processes. Therefore, it is extremely important that the early processes be done properly.

The relatively crude methods which are still in widespread use for gain recovery are symptomatic of the lack of ability to define adequately and to cope with all of the factors necessary to model the transmission system. The is exemplified, and not widely recognized, by the lack of knowledge of the true amplitude characteristics of the recovered trace.

Usually, gain compensation programs are designed so that the final mean amplitude value is uniform over the time length of the trace. This model of the amplitude response is probably incorrect. Velocity increases with depth, and usually the velocity function can be approximated by a simple time function.

FIGURE 3.16

The action of the recovery programs is illustrated on Fig. 3.16 for the low-noise-level trace 1 of the seismic record. Note the AGC recovery curve dips quite sharply to accomodate a high-amplitude reflection near 1.3 sec. The recovered trace has a low-amplitude reflection at this point. The more correct method ignores the amplitude change, with the result that the recovered trace has a strong reflection at this point. In this case, the amplitude of the event is a more nearly correct recovery of the reflection amplitude.

FIGURE 3.17

The effect on the noisy trace 12 is more pronounced. The improper gain curve sags from 0.5 to 1.5 sec, holding the amplitude of the recovered trace to a near-constant level. The programmed recovery curve is flat over much of the length from 0.6 to 1.3 sec, and the recovered events are of high amplitude representing the sum of signal and noise over this portion of the record. After processing to eliminate noise, the signal level will remain close to that of trace 1. Although improved, the programmed-curve recovery is not quite right either, since the curve of trace 1 should rise slightly through the noisy zone, whereas the curve used was flat over a considerable length.

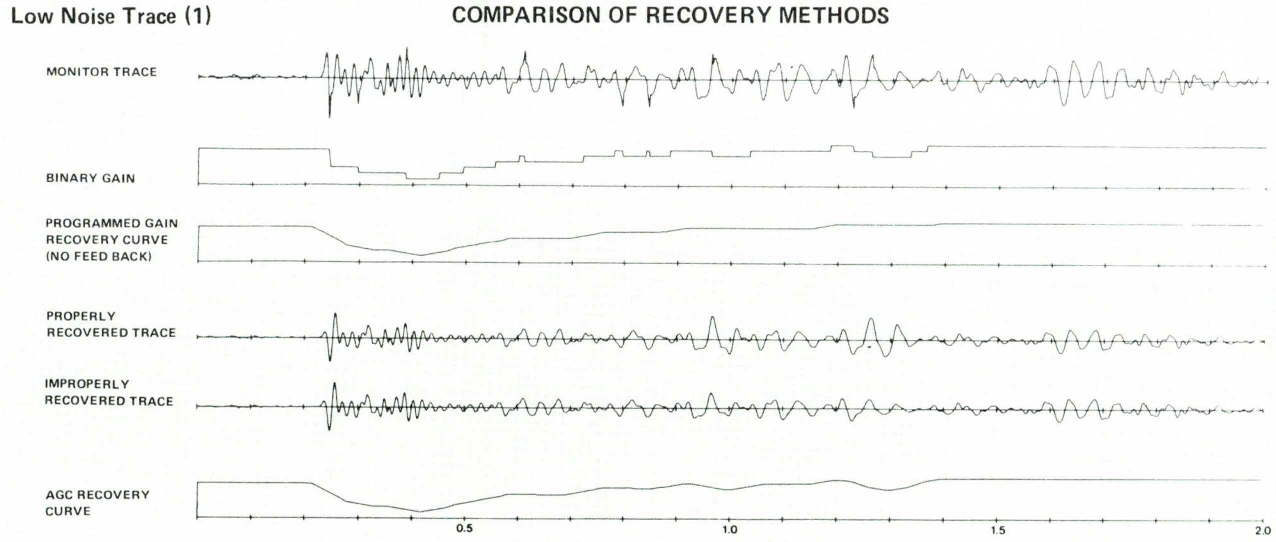

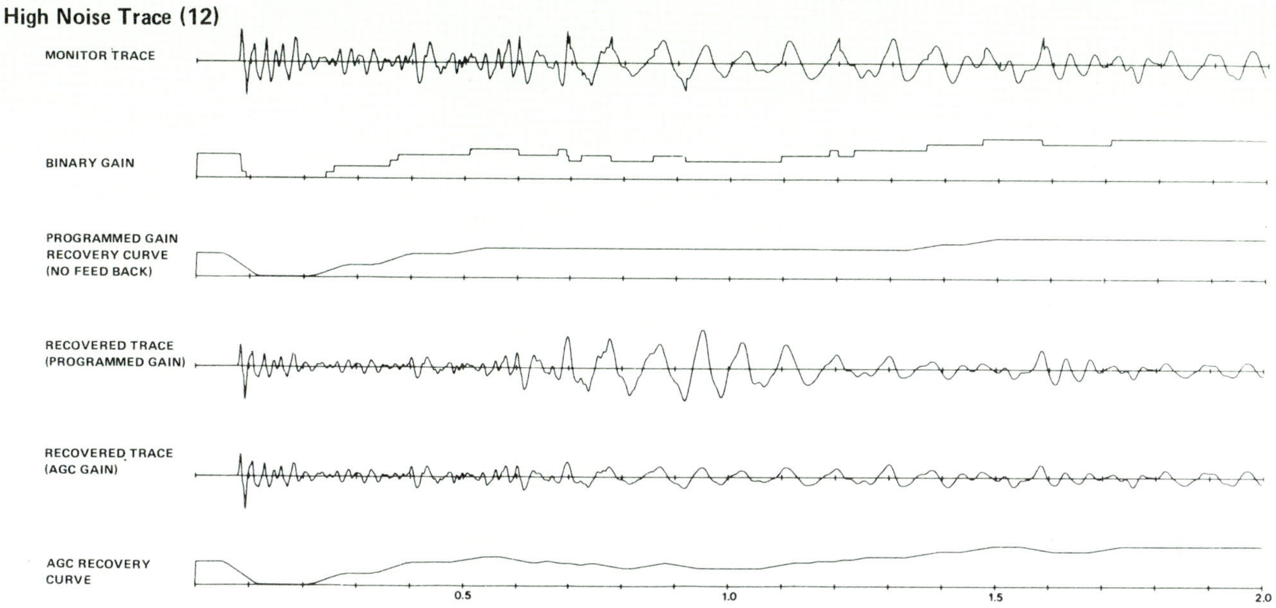

For example, $V_i = V_o + 2kt$ where: V_i is instantaneous velocity at time t

V_o is the datum velocity (at t = 0) and

k is a constant for the area.

For two samples at spacing Δt, $V_1 = V_o + 2k$ and $V_2 = V_o + 2k(t + \Delta t)$.

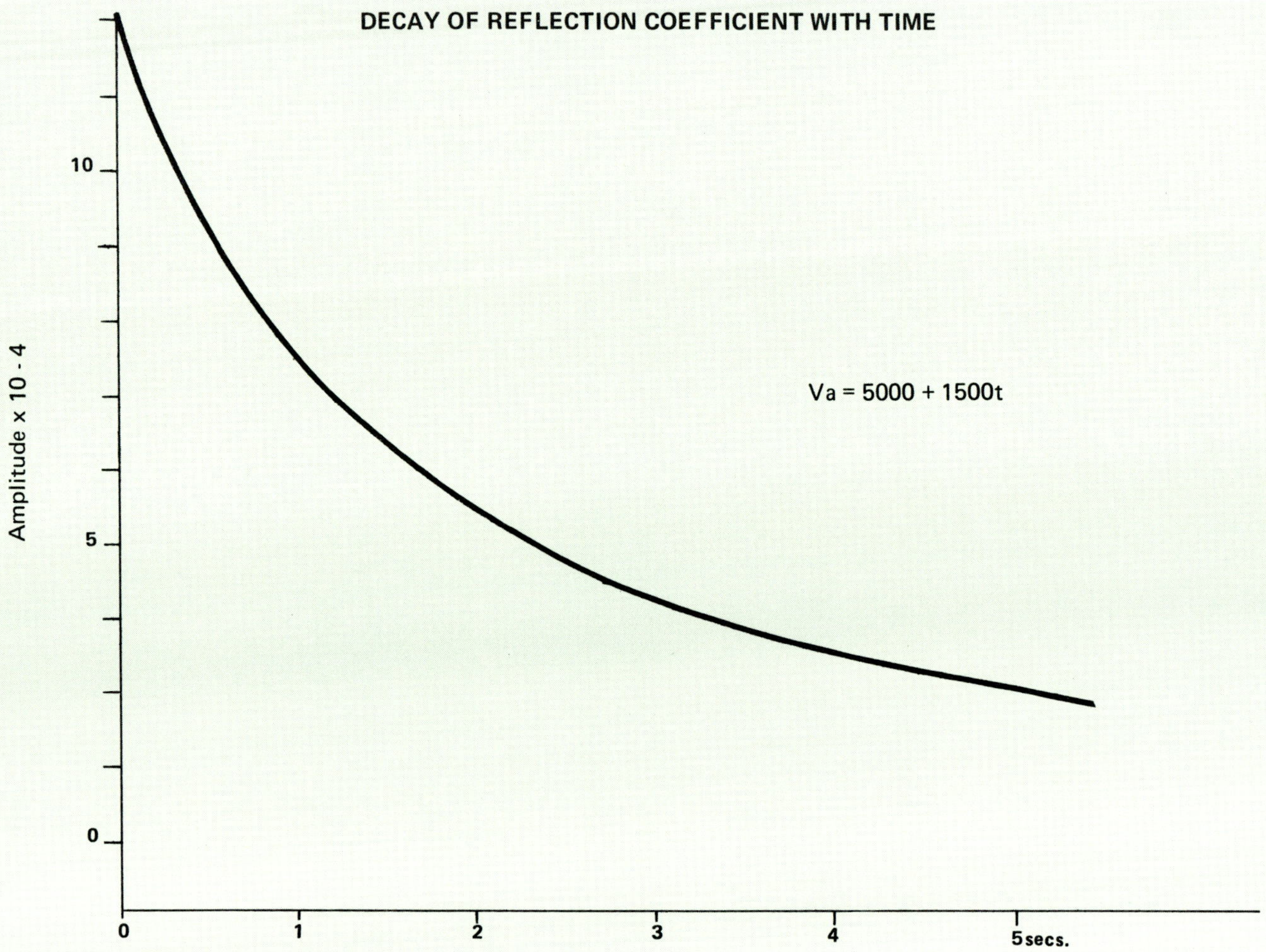

FIGURE 3.18

The average reflection coefficient decreases in amplitude with increasing reflection time.

The reflection coefficient calculation,

$$R_2 = \frac{V_2 - V_1}{V_2 + V_1}$$

becomes, by substitution,

$$R_2 = \frac{2k\Delta t}{2(V_o + 2kt) + 2k\Delta t} = \frac{k\Delta t}{V_o + 2kt + k\Delta t}$$

With increasing time, $k\Delta t$ remains constant, but the denominator increases. Thus, the average reflection coefficient decreases in amplitude with increasing reflection time (**Fig. 3.18**). Deep reflections are weak not only because of transmission losses, but also on the average because of increasing velocity.

4 STACKING

Common-Depth-Point Stacking

Of all the advances in the reflection seismic technique, surely none is more important nor has had greater impact than the invention of common-depth-point coverage, or simply, stacking.

In concept, the procedure is very simple. The distance between seismic source and receiver is varied symetrically to survey the same subsurface point several times. The observations, corrected for geometrical distortion and adjusted to datum, are then summed to obtain the average.

The reason for its outstanding success is that it provides the most powerful method for separating noise from signals of the same frequency. With one exception, all other seismic processes either eliminate both signal and the noise, or merely modify the amplitude without changing the proportion. The exception is multichannel filtering, which is also able to separate signal from noise of the same frequency, but it is effective only in areas of relatively homogeneous dip and is of little value in areas of complex structure. Under proper conditions, stacking will remove several types of noise, but it is most effective against random noise.

Vector Operations

Any signal, be it useful information or noise, can be analyzed by the Fourier transform into its component frequency elements. The contribution of each frequency can be expressed in terms of a vector having a length equal to the amplitude, directed at the angle of phase. Random white noise has a uniform amplitude for all frequencies, but the phase is totally random. Seismic traces represent the sum of signal and noise. Each frequency component is the vector sum of a random noise component and a signal component. For a twelve-fold stack, if all corrections have been properly applied, all twelve signal components of each frequency can be considered vectors having the same amplitude and phase. The twelve noise vectors will also all have the same amplitude, perhaps greater than the signal vectors, but the phase distribution will be random (Fig. 4.1).

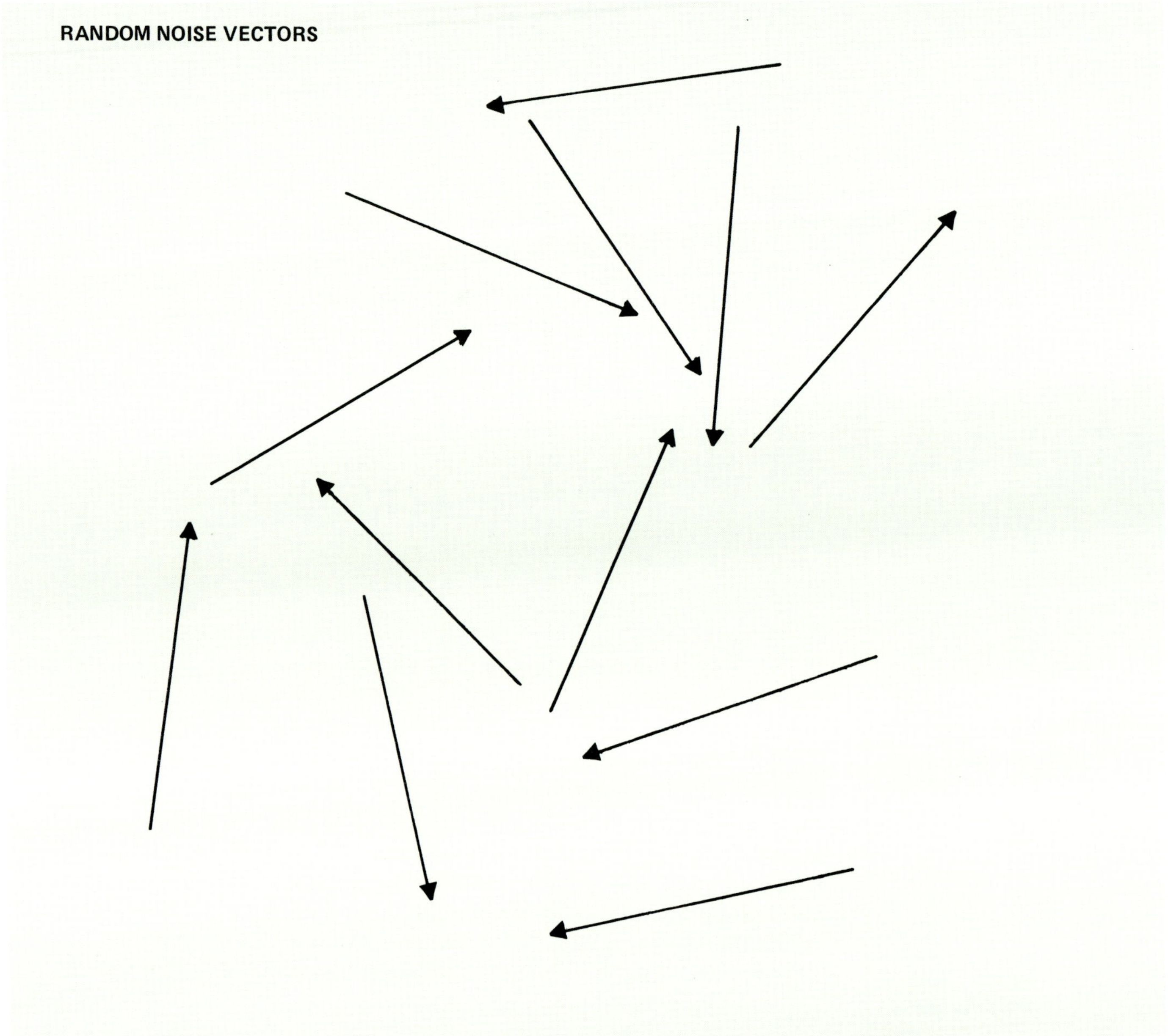

FIGURE 4.1

Twelve noise vectors all with the same amplitude and random phase distribution.

If the ratio of signal-to-noise is 1:4, the signal vectors would sum with the noise, as in **Fig. 4.2** to yield the results indicated by the heavy lines.

In the stacking process, the resultant vector of each trace is summed in turn by vector addition to produce the total resultant. In the addition, random noise components tend to sum to zero, while the signal components sum in phase to produce a resultant nearly twelve times the individual signal amplitude component, directed approximately at the signal phase angle (**Fig. 4.3**).

ADDITION OF NOISE TO SIGNAL VECTORS

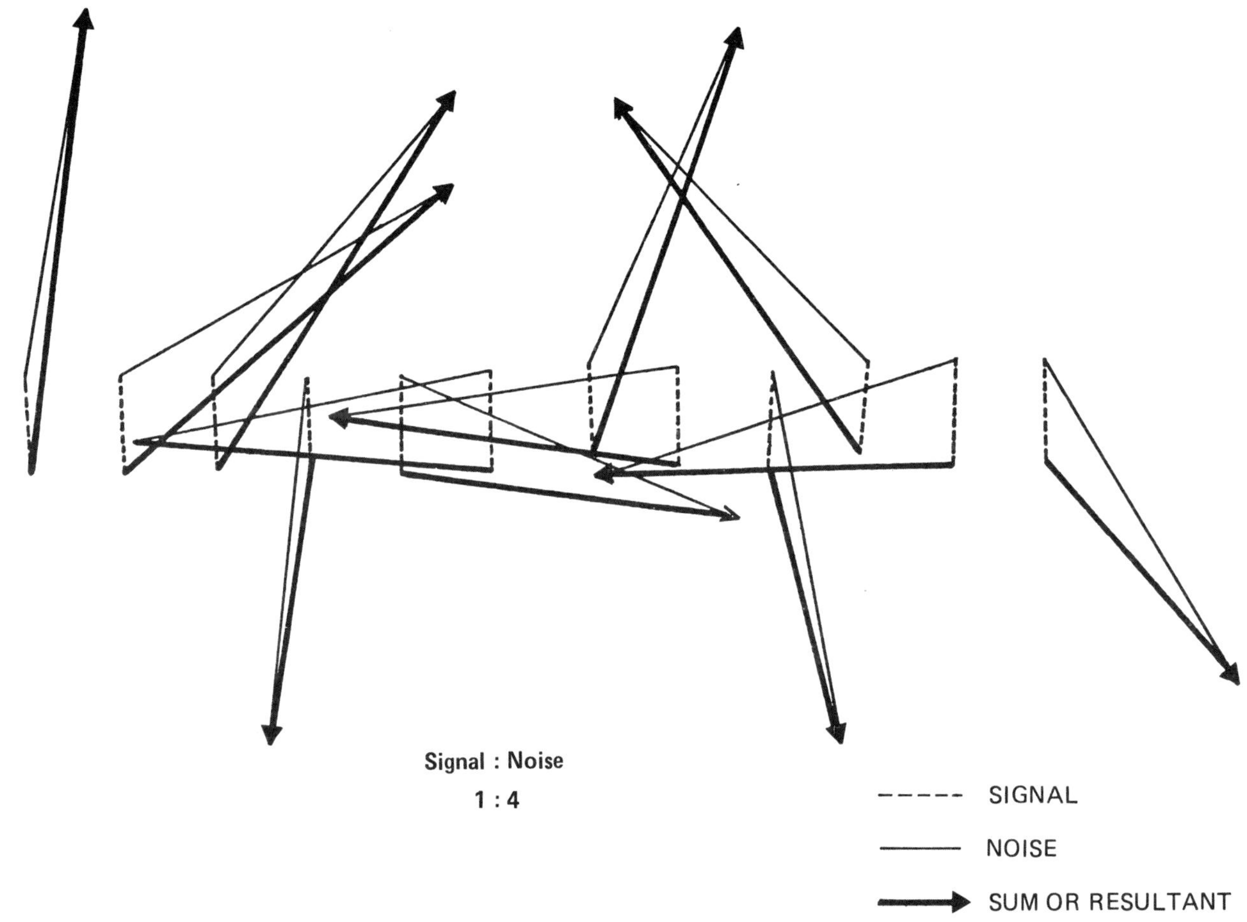

FIGURE 4.2

Signal vectors summed with the noise yields the results indicated by the heavy lines.

The method uses the power of statistics to produce relatively noise-free output. Naturally, if the noise has coherent components these too will sum in phase and be present in the output.

The assurance that the normal moveout corrections and static corrections have been applied in an optimum manner is essential, otherwise the signal components will sum out of phase. Seismic field supervisors have long recognized the damage that one poorly planted geophone can do to the total array response. In the same manner, experienced data processors recognize the disproportionate damage a single bad trace can cause to a stack, and use their experience to mute or discard poor traces in the gather group.

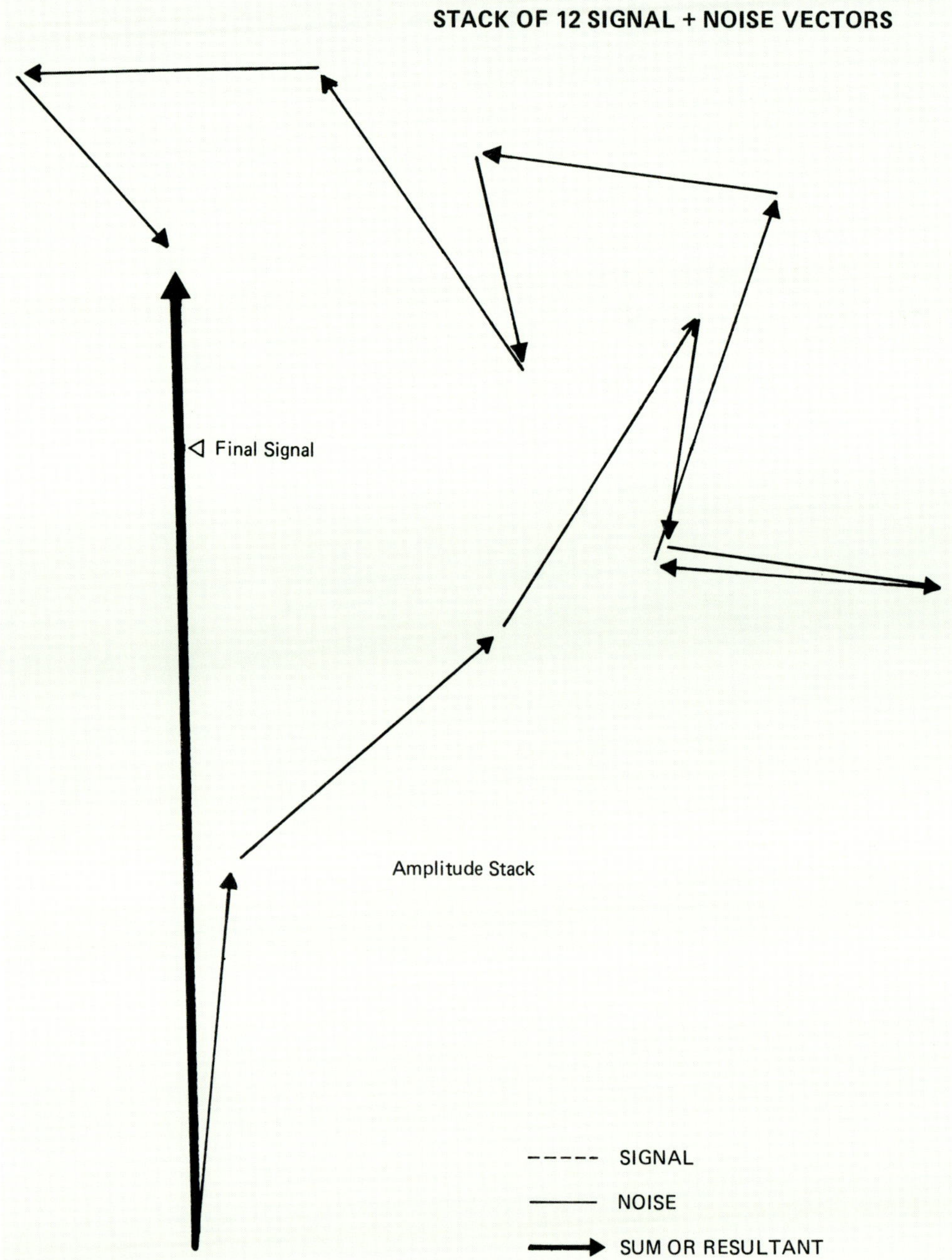

FIGURE 4.3

In the stacking process, random noise components tend to sum to zero, while the signal components sum in phase to produce a resultant nearly twelve times the individual signal amplitude component, directed approximately at the signal phase angle.

This effect can also be demonstrated by vector summation. Assume one trace of the twelve has a moderate noise burst, say 2.5 times the normal noise amplitude. This seems like a small addition to the total noise present on the twelve traces, an increase of about 12% (**Fig. 4.4**). Now compare the result of the stack which includes the noise burst to the previous stack with random noise (**Fig. 4.5**). The phase angle of the signal is distorted substantially and the amplitude is reduced to about 60% of the level obtained when the noise was purely random.

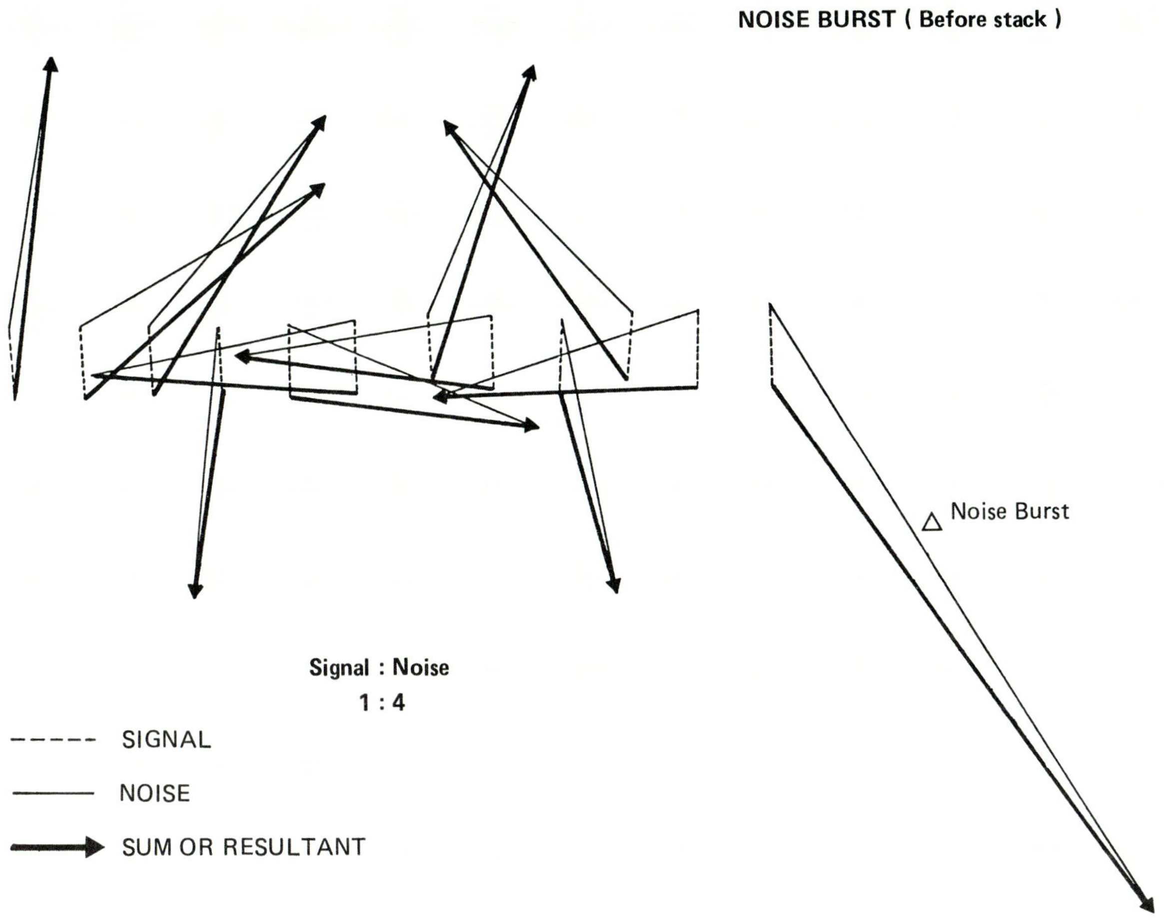

FIGURE 4.4

One trace of the twelve has a moderate noise burst, say 2.5 times the normal noise amplitude. This seems like a small addition to the total noise present on the twelve traces, an increase of about 12%.

This also suggests the reason why the number of stacking elements must be increased to a larger number to gain a noticeable improvement in quality of the output.

Sign-Bit Recording

Over the years stacking has increased from 3-fold to the current 24-and 48-fold. Probably still greater coverage would be used if it were possible to increase readily the number of recording channels. But this is limited by the data rate to be recorded. One solution which is simply to record the sign bit only. Reducing the number of bits per sample from the current 16 to 32 to a single bit allows the number of data channels to increase from 48 to 768 or 1536.

A proposal to record only a single bit, which has a maximum dynamic range of 6db, seems totally incompatible with the enormous industry effort made to achieve wide dynamic range. Nevertheless, modelling confirms the proposal to be totally feasible. In addition to the mechanical and instrumental advantages, important inprovements in signal quality can be obtained, provided the basic assumptions used for the model are met in the field. The model used previously, to demonstrate stacking, summed the resultant vectors from the sums of signal and noise.

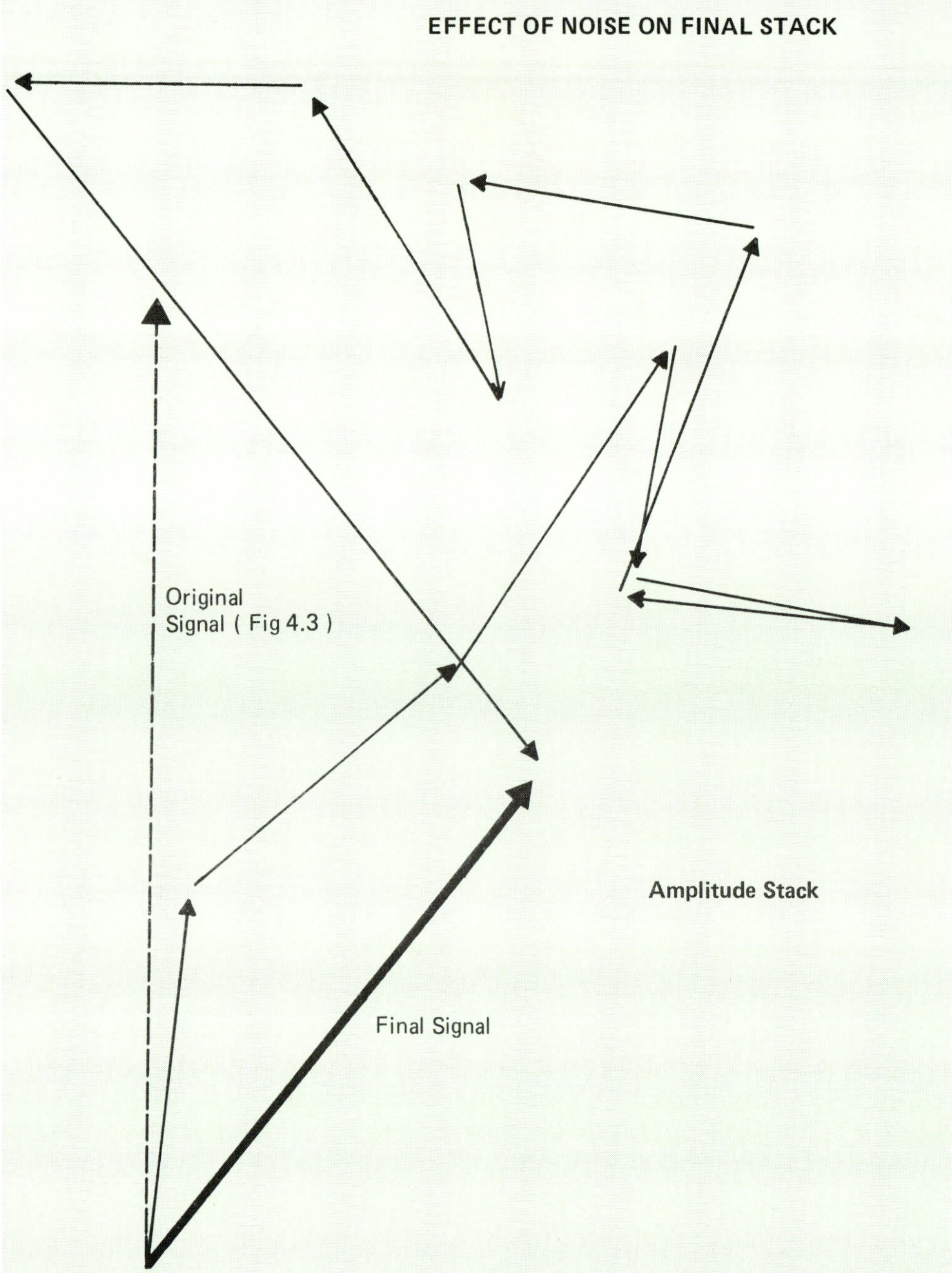

FIGURE 4.5

Compare the result of the stack which includes the noise burst to the previous stack with random noise. The phase angle of the signal is distorted substantially and the amplitude is reduced to about 60% of the level obtained when the noise was purely random.

The phase of a given frequency component determines the position on the time axis of all the zero crossings of a given frequency. Clipping the amplitude vector to the sign bit preserves the zero crossing and hence preserves the phase. The clipping removes any amplitude variation of the components, leaving all vectors with the same unit amplitude. The bit vectors corresponding to the stacking example of **Fig. 4.2** are now included in **Fig. 4.6**. The phase of each bit vector remains the same as its full amplitude counterpart, therefore both the single bit and the full amplitude vectors carry the same proportion of amplitude and phase components. Hence, when the bit vectors are stacked, as in **Fig. 4.7**, the resultant is a vector having the proper direction of the signal component.

In the single bit process, the relative amplitude contribution of signal and noise to each component is effectively converted to a phase angle. **Fig. 4.8** illustrates another aspect of the same model. This time, the amplitude of the signal is twice the previous example, but by definition the amplitude of the noise remains the same. In this case, the resultant of the vector sum of signal and noise is weighted slightly more heavily by the signal, with the result that the bit vectors, although substantially similar

BIT VECTORS FOR SIGNAL + NOISE RESULTANTS
(1 : 4 Signal - to - Noise Ratio)

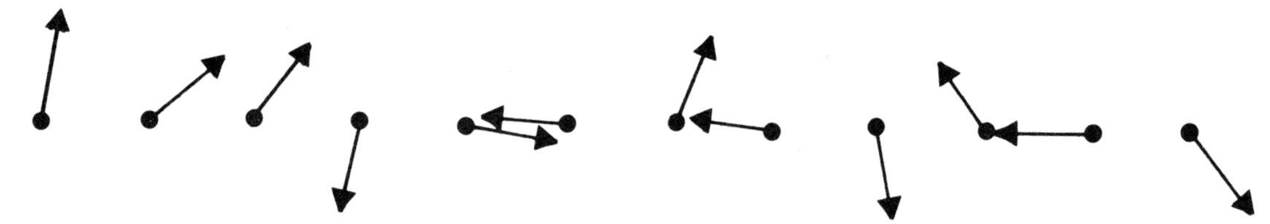

BIT VECTORS (Normalized Resultants)

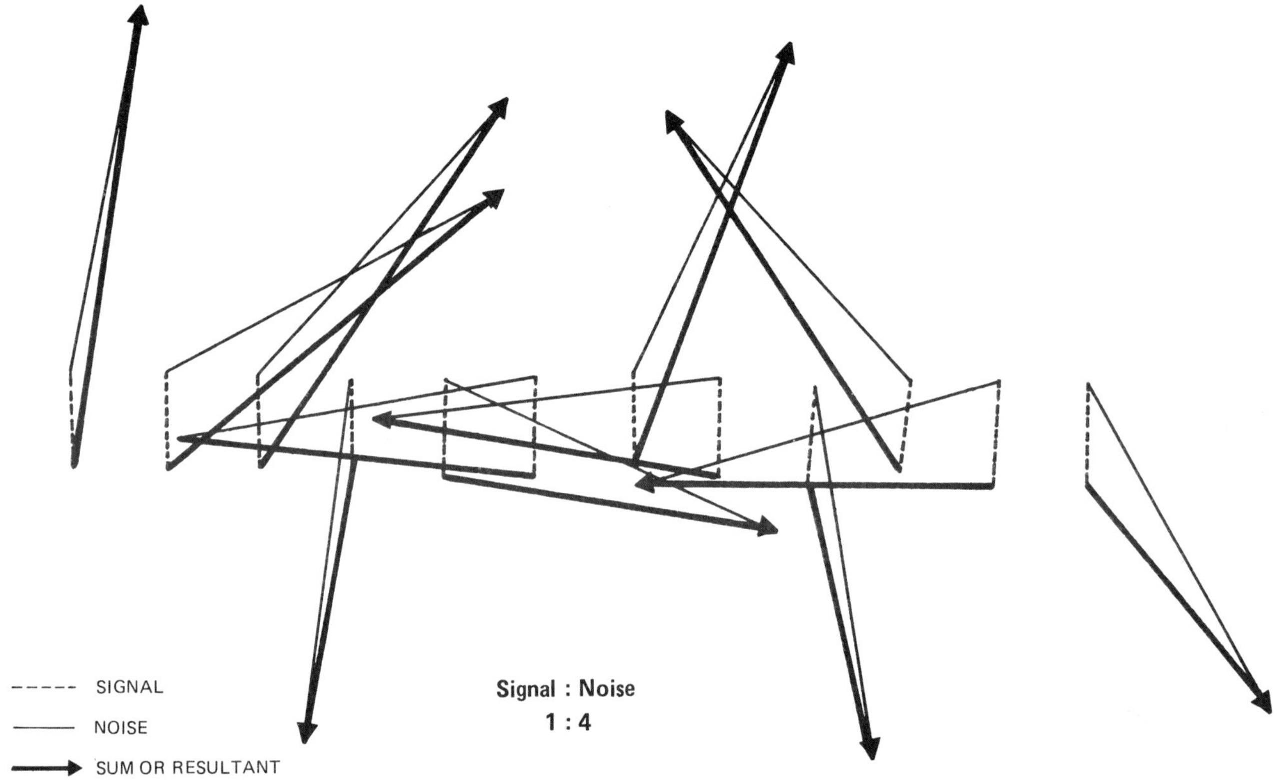

FIGURE 4.6

Bit vectors corresponding to the stacking example of Fig. 4.2 are now included in Fig. 4.6. The phase of each bit vector remains the same as its full amplitude counterpart, therefore both the single bit and the full amplitude vectors carry the same proportion of amplitude and phase components. When the bit vectors are stacked, as in Fig. 4.7 the resultant is a vector having the proper direction of the signal component.

to the preceding example, are directed to a slightly different angle. Summation of these vectors, in **Fig. 4.7b**, produces a resultant which is almost twice the previous amplitude, very nearly maintaining the correct amplitude relationships in the output. Effectively, under the conditions of the model, the amplitude component is converted to phase.

Now return to the example of the noise burst, resulting in the bit vectors of **Fig. 4.9**. The result of stacking these is very nearly the same as before, with only a small deflection of phase and slightly diminished amplitude. (**Fig. 4.10**).

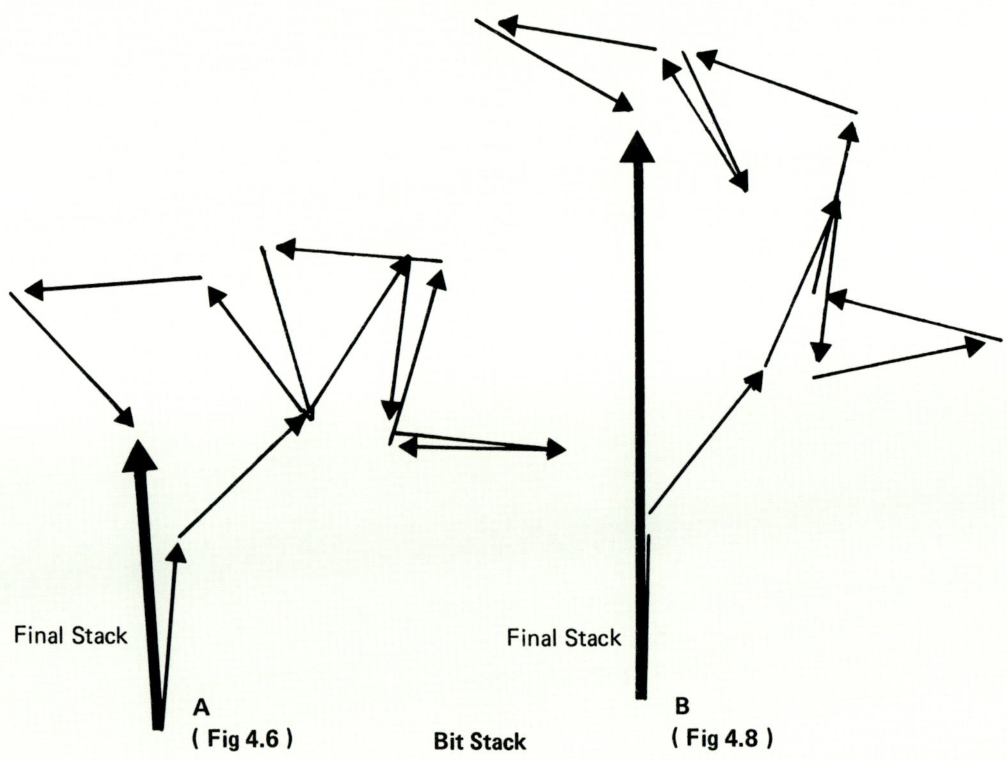

FIGURE 4.7

Bit stacks of Fig. 4.6 (signal-to-noise ratio 1:4) and 4.8 (signal-to-noise ratio 1:2) demonstrate that amplitude relationships in different signal inputs are maintained very nearly correctly.

Vibroseis Application

The system is well adapted to the Vibroseis signal source. Fig. 4.11 illustrates a computer model of the process. A number of Vibroseis sweeps of varying amplitudes are combined to produce a noise free signal. Random noise, including some large bursts, is generated, scaled to substantially greater amplitude than the signal, and the two are summed. The model of the sign-bit trace preserves the zero crossings only. The process was repeated, each time with different random noise, to simulate the several traces of a trace-gather group.

The results compare the conventional and the sign-bit recordings to the noise-free signal for increasing stacking folds. Notice that with only 4 fold, the sign-bit recordings have a superior signal-to-noise ratio. In fact, even the single fold of the sign-bit signal appears to be of better quality than the conventional full range recording. This, and the high quality of the other examples, holds true only for Vibroseis work. The results would be quite different for an inpulse-energy source, such as dynamite.

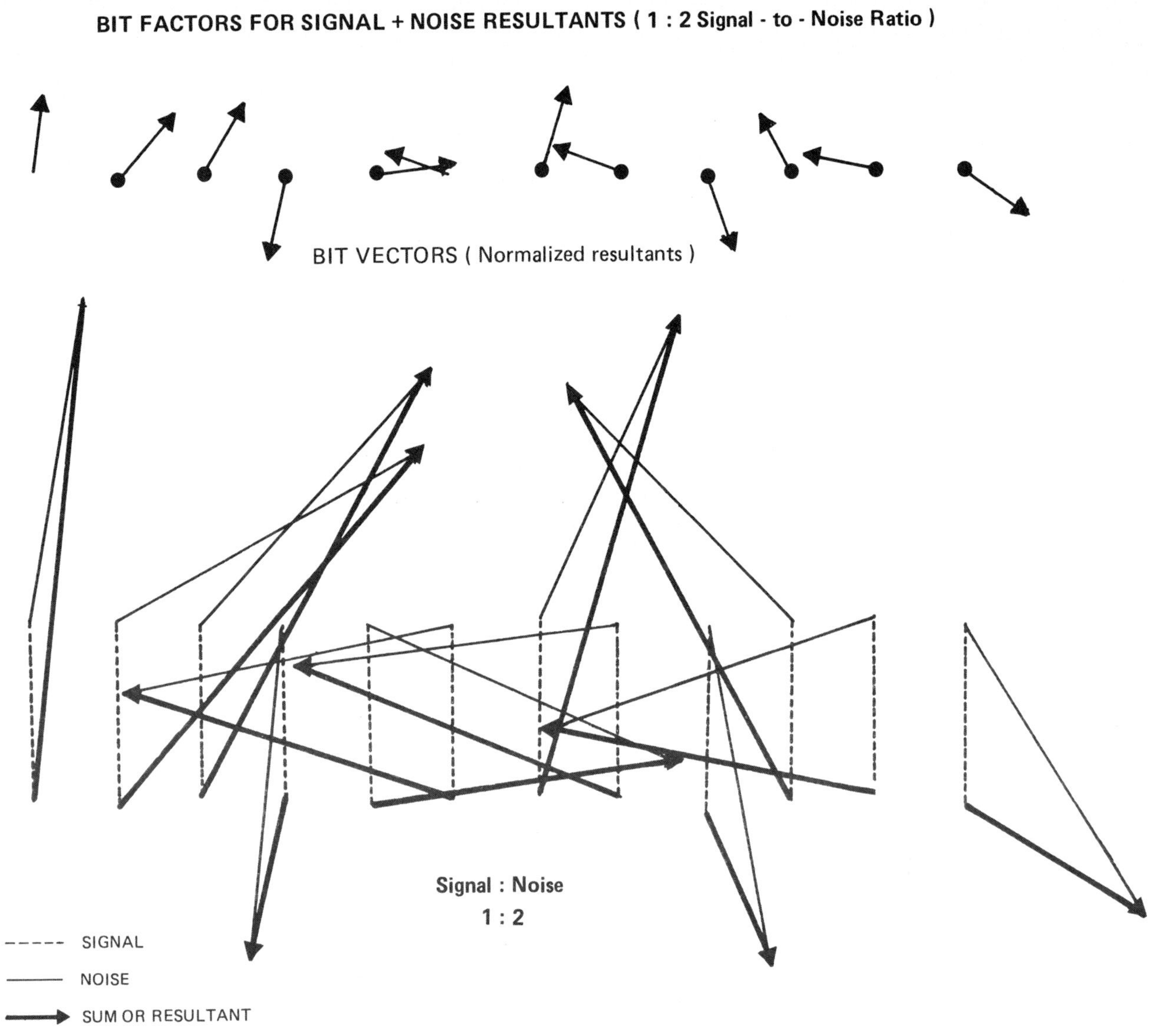

FIGURE 4.8

Another aspect of the same model. The amplitude of the signal is twice the previous example. In this case, the resultant of the vector sum of signal and noise is weighted slightly more heavily by the signal, with the result that the bit vectors, although substantially similar to the preceding example, are directed to a slightly different angle.

Cross-correlation of the signal (the sweep) with itself (the reflected sweep) produces a zero-phase auto-correlation output proportional to the amplitude of the reflected sweep. Correlation of the signal (the sweep) with a spike (a random noise burst) reproduces the sweep at the amplitude of the noise spike. On the conventional recording, noise after cross-correlation is proportional to the sum of all the spike-noise-burst amplitudes, which, in the model, occur at substantially higher amplitudes than the signal. Reducing the traces to sign bits prior to cross-correlation reduces all noise-burst amplitudes to the same amplitude as the legitimate reflections, greatly eliminating the disparity in amplitudes between signal and noise.

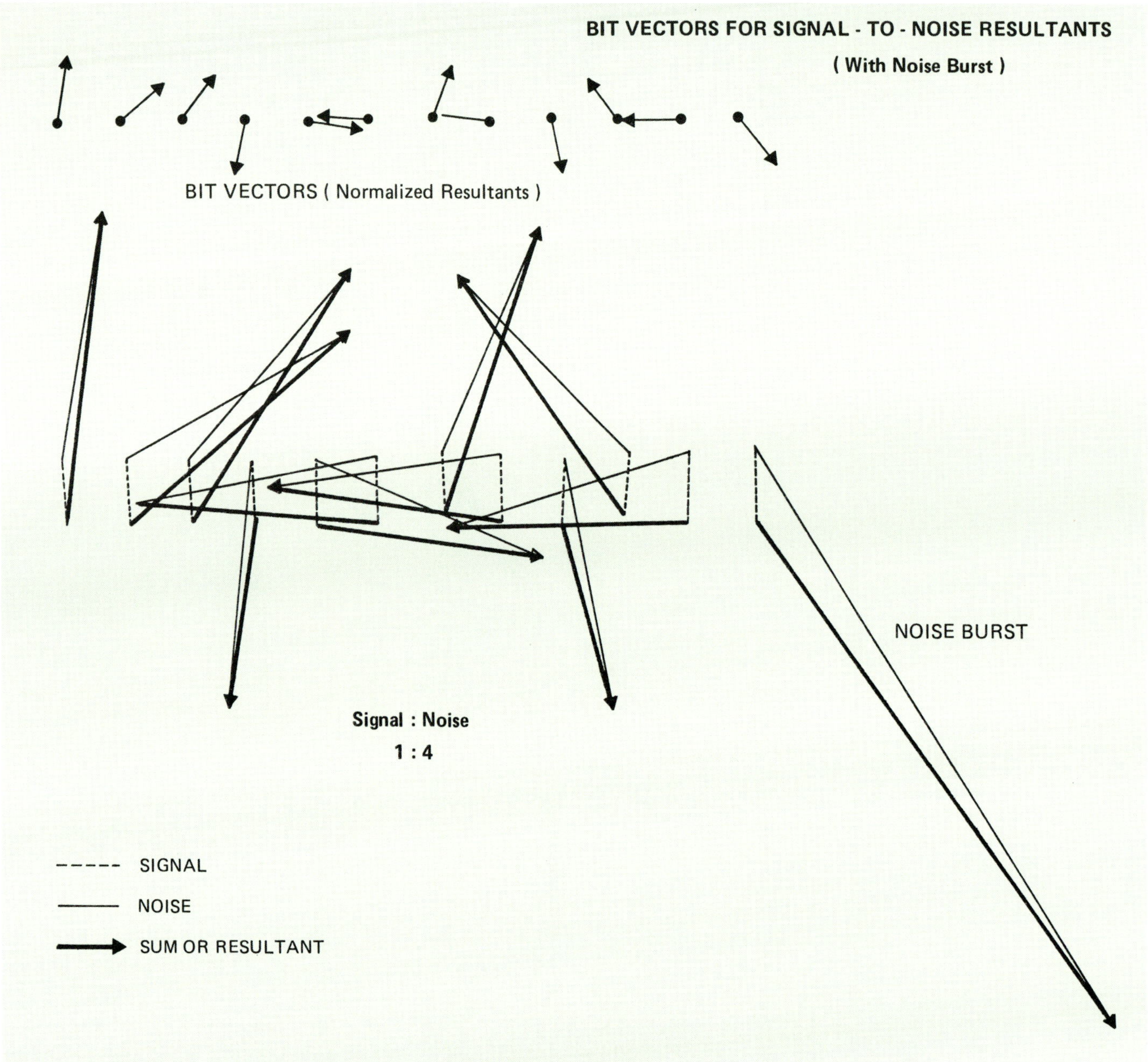

FIGURE 4.9

Bit vectors with noise burst.

The relative signal amplitudes are carried in the phase of each frequency in the Vibroseis sweep. Cross-correlation removes the phase of the sweep from each, leaving only the signal and noise vectors for each frequency, which "sum" in the cross-correlation with the same net result as stacking. Stacking the cross-correlated signals provides further noise cancellation, roughly the same improvement as that obtained from stacking of the same number of traces from a conventional energy source.

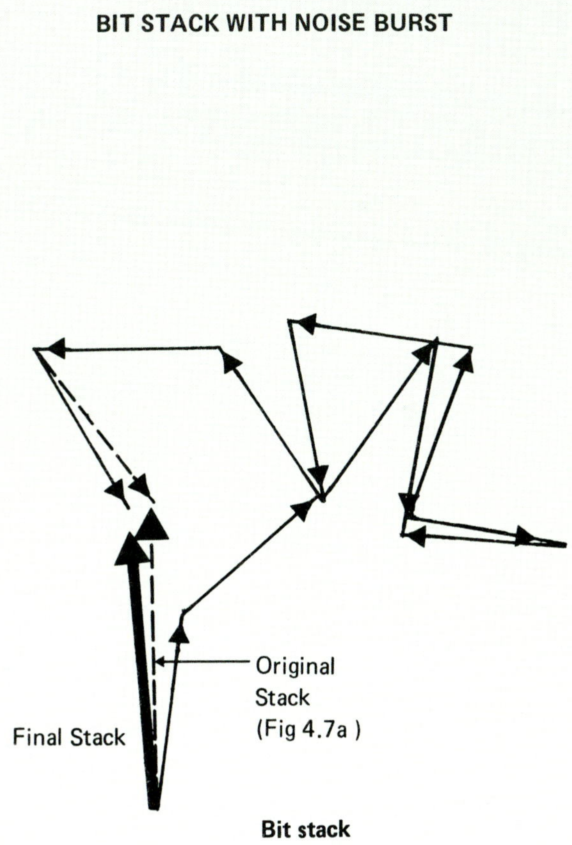

FIGURE 4.10

The effect of the noise burst in vastly reduced by bit- stacking.

The difference is maintained through all comparisons up to 32-fold and clearly, the relative amplitudes have been preserved reasonably well. These experimental conclusions are supported by rigid mathematical proof, in a fine paper by O'Brien et al, 1979.

With the obvious superiority of these results, it may be wondered why the method is not used universally. One of the reasons is the nature of the noise which, in practice, may not necessarily be random. Also, if there is no noise the system will fail to maintain proper amplitude relationships. It is once again a matter of a trade-off of improvement in one area for possible deterioration in others. For this system to work, random white noise must be present, and, if necessary, it can be generated and deliberately added to the field signal in a noise-free area.

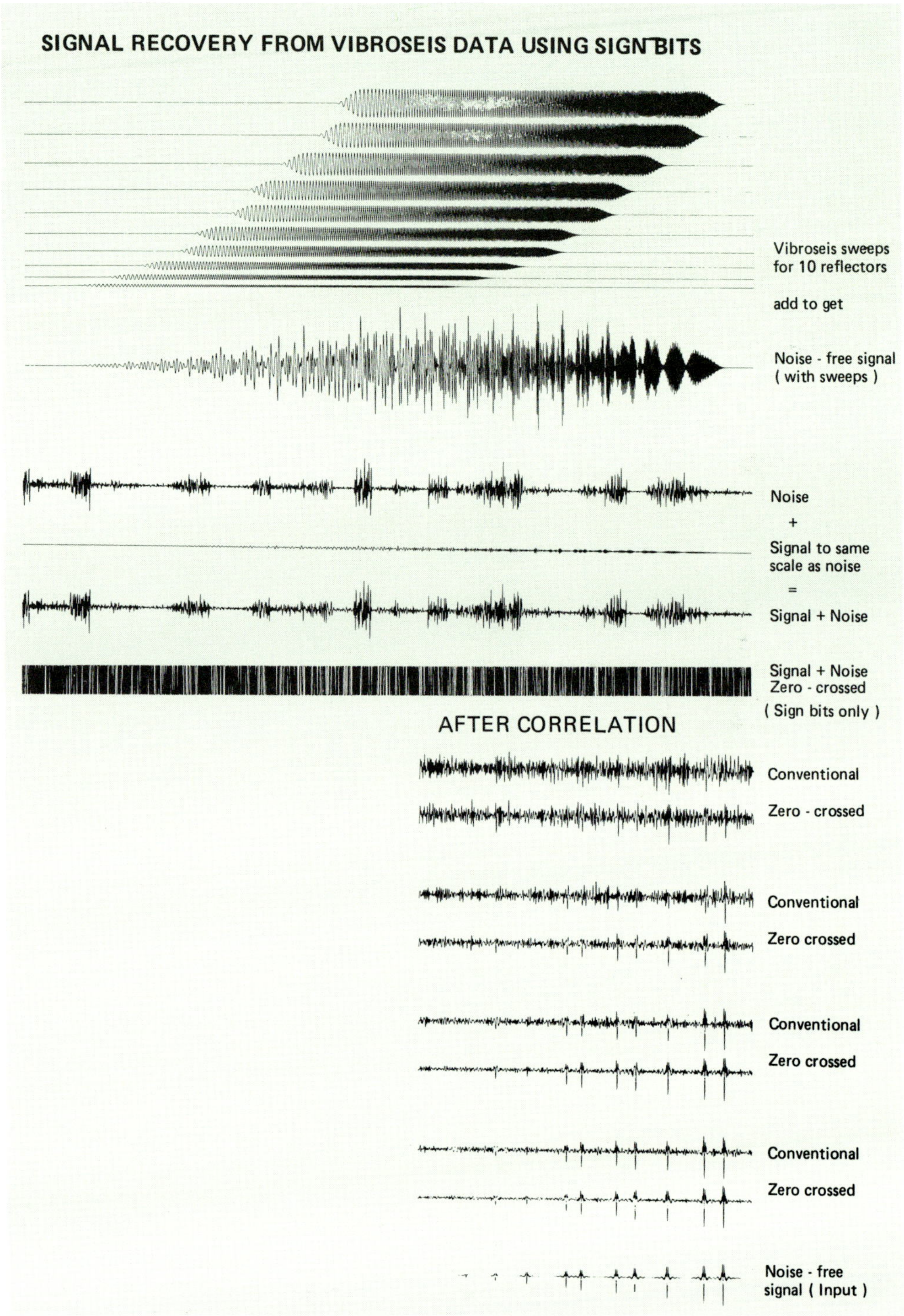

FIGURE 4.11

Fig. 4.11 illustrates a computer model. A number of Vibroseis sweeps of varying amplitudes are combined to produce a noise-free signal. Random noise, including some large bursts, is generated, scaled to substantially greater amplitude than the signal, and the two are summed. The model of the sign-bit trace preserves the zero crossings only The process was repeated, each time with different random noise, to simulate the several traces of a trace-gather group. The results compare the conventional and the sign-bit recordings to the noise-free signal for increasing stacking folds.

It will be interesting to observe future developments of the one-bit or sign-bit system. The ability to increase channel capacity makes other potential improvements possible. One of these is to record very close trace spacings, in effect, each geophone position rather than each group position. This greatly increases the potential for application of multichannel filters, perhaps the next most powerful tool after stacking for removing noise of the same frequency as the signal. It also may provide the solution to obtaining the large number of channels required for three-dimensional work, which has the potential for greatly improved reliability of information in complex structural areas.

5 FREQUENCY FILTERS

Content of Seismic Signals

The capacity of any signal to carry information can be measured in a manner analogous to the volume of a container. Just as volume is the product of height times width times length, information capacity of seismic signals is related to a product of amplitude, frequency bandwidth, and length of the signal. (Fig. 5.1). This is more formally described in the readable book "Symbols, Signals and Noise" by Pierce. A dynamite energy source pulse produces an input signal having considerable amplitude, (height), and bandwidth, but very short length. A Vibroseis [R] energy source compensates for its limited amplitude by extending the input sweep signal over a relatively long length.

Reducing the bandwidth of a signal, as may be done with a band-pass filter, also reduces its capacity to carry information. A reduction in bandwidth not only limits the amount of information but also, to some degree, determines the type of information that can be carried. Yet, in the years prior to the advent of stacking and digital processing, application of a band-pass filter was practically the only procedure available to eliminate noise from seismic data.

Band Pass Filter

The natural seismic response of the earth and of many seismic instruments, Fig. 2.11, is a bell shaped curve peaking about 35 Hz on the amplitude spectrum, remaining at a fairly high level for a half octave or so on either side of the peak, then dropping rapidly. With the knowledge in hand, it could be assumed that recorded signals having frequencies outside this range would be largely composed of noise. Hence, it became an almost automatic procedure to limit the bandwidth of the recorded signal to about 15 to 50 Hz.

This practice carried essentially without change into the era of digital processing. Digital band-pass filters are fairly easy to design and apply. A digital band-pass operator is, in essence, nothing more than the zero phase sum of a collection of frequencies extending over the bandwidth which is to be retained. Cross-correlation of any single frequency with a signal will extract that frequency and that alone from the signal, since the output from a cross-correlation operation can have no frequencies which are not common to both the signal and the operator. Cross-

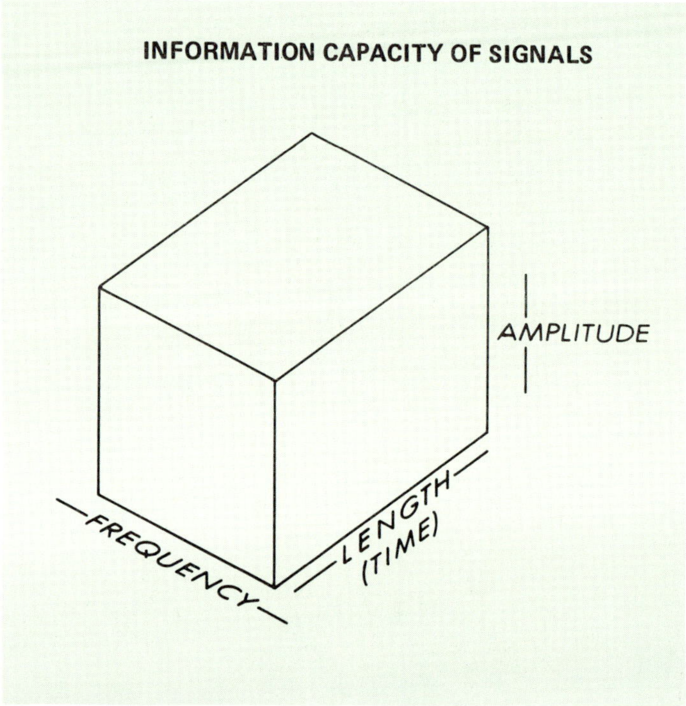

FIGURE 5.1

The capacity of any signal to carry information can be measured in a manner analogous to the volume of a container. Just as volume is the product of height times width times length, information capacity of seismic signals is related to a product of amplitude, frequency bandwidth, and length of the signal.

correlation (or convolution, since the operator has zero phase) of several summed frequencies, with the signal, will extract those frequencies and reject the rest. The several available variations of band-pass operator design exist primarily to obtain slightly improved response characteristics, particularly when short operators are used, and to reduce side lobes.

Fig. 5.2 is an example of the use of band-pass filters to clean up a record. The field record has an obvious interference pattern from low frequency ground roll. Not quite so visible, but apparent from examination, is a band of relatively low frequency multiple energy from .6 to .9 sec.

Application of a time gated filter having a pass of 25 to 55 Hz in the early part of the record, opening to 15-55Hz after 1.0 sec, eliminates the ground roll and apparently creates a clean record. Narrowing the early part of the filter to 30-55 Hz eliminates the evidence of the multiples, which fortuitously have a slightly lower peak frequency than the reflection.

Although this output has been digitally processed, it represents the state of the art of data processing before stacking and digital processing.

Recursive techniques are those in which the operator operates on some of the previous output, as well as the input to obtain a new output point. These filters are popular because the number of arithmetic operations often can be greatly reduced compared to a regular convolution operator.

Band-pass filtering in the frequency domain involves simply modifying the amplitude spectrum to eliminate those frequencies which are not desired. The operation is to multiply all the Fourier coefficients within the pass band by 1.0, and all the others by zero. Obviously, in execution this means merely setting all unwanted values to zero.

A frequency domain operator of this type is termed a boxcar. If desired, to avoid side lobes on the output, a tapered side slope is applied to the ends of the boxcar. Side lobes are an annoying side effect of filters with sharp frequency cutoffs, but they become a minor problem with broad-band signals.

Time Varying Properties

It is a common observation that in many areas, and in particular in fairly deep sections, the frequency content of the seismic trace may diminsh with time. Usually there is increasing loss of the higher frequencies with depth. The reasons for this were outlined in a previous section. The loss violates a basic assumption of most of the applications for seismic processing which have been developed from signal and/or communicaiton theory, namely that the signal is stationary, e.g. a sample taken from one portion of the seismic trace will have essentially the same properties as that taken from another portion of the trace.

Any seismic trace can be Fourier transformed over its entire length. The resulting transform will adequately describe the trace in terms of frequency and phase. The loss of the high frequency at depth will be seen to be related

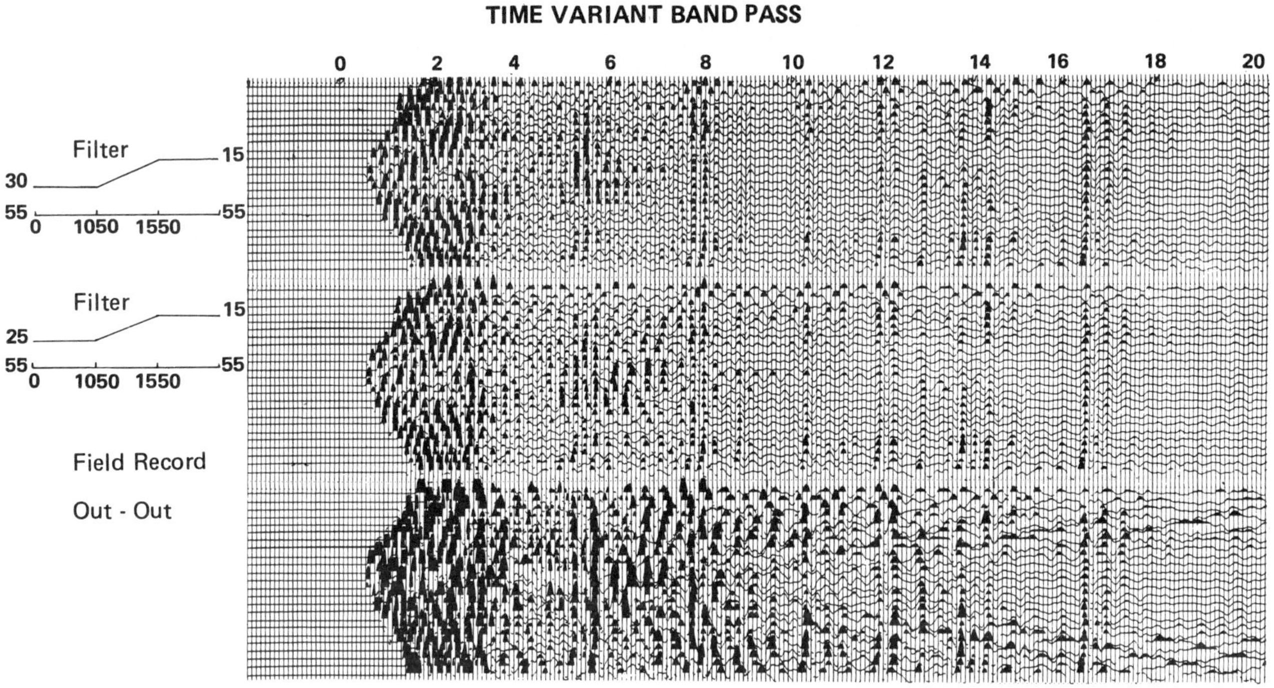

FIGURE 5.2

Time-variant Band Pass filters represent the state of the art of data processing before stacking and digital processing became available.

to the phase characteristics. The higher frequencies will sum in phase to reinforce amplitude over the shallow portion of the section, but they will gradually fall out of phase toward the deeper portion and hence cancel out when summed over that portion. **(See Fig 1.8)**.

If the trace is segmented into two or more pieces, and each disassembled by the Fourier transform, the resulting spectra may be quite different in appearance. In such cases, the high frequency components of the deep section will have lower amplitude than those of the shallow section. The phase spectra will also differ.

This seeming paradox of the pieces not having the same spectrum as the whole is not inconsistant with Fourier transform theory. It should be remembered that the discrete Fourier transform of any signal provides a unique solution of the frequency content of that signal within the constraints of sample rate and length. Changing the parameters in any way, including the length, will also change the solution provided by the transform.
Only when the signal is truly stationary, will two equivalent length segments of the same signal have the same amplitude spectrum, but, even then the phase spectrum will most likely be different. The transforms will be identical only when the signal is both stationary and periodic, and then segmented at some multiple of the periodic interval.

It is fairly common practice to compensate for time variance by applying filter operators which vary with time. Since it is usually difficult to compute and apply an operator which varies continuously with time, common practice is to segment the trace into a number of distinct windows or gates, and treat each gate as a separate signal. The results of the processing are subsequently linked with some smoothing technique at the joint to provide a smooth transition from one segment to the other. This is the basis of most so-called time varying filters. As popularly used, the term is a misnomer. They should be termed gated filters.

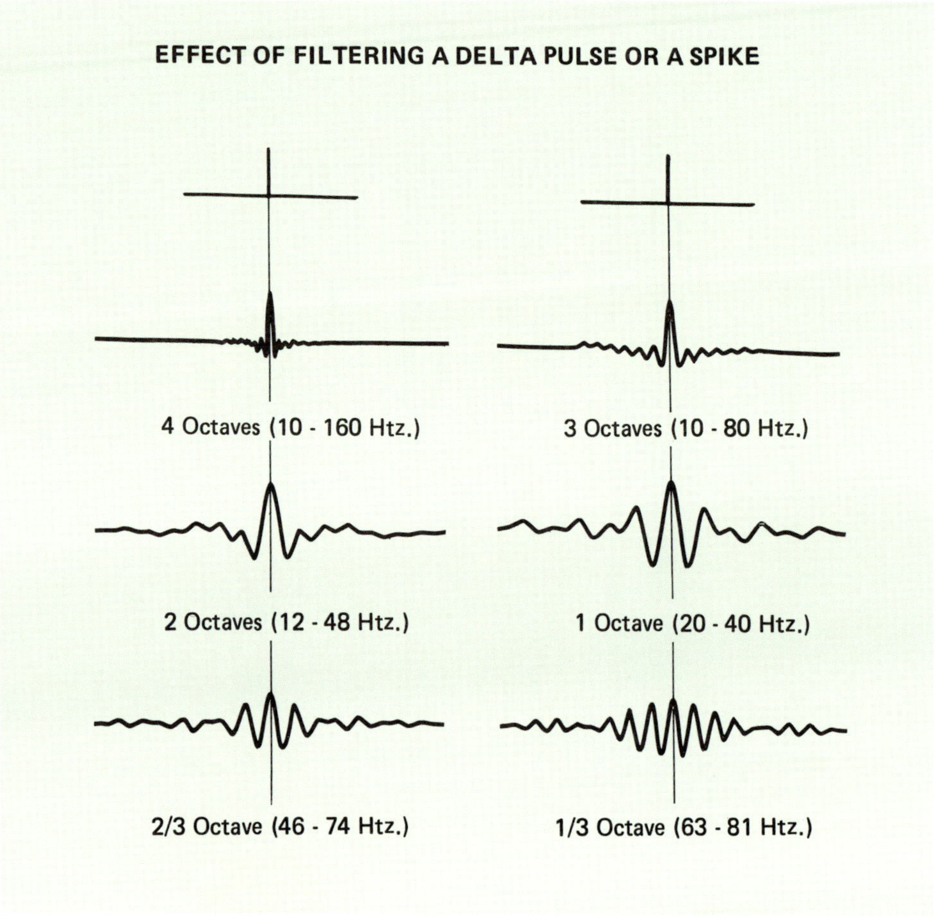

FIGURE 5.3

A zero-phase band-pass filter applied to the coefficient spike will broaden the spike, but it will in no way affect the precision of measurement provided by the method, because regardless of the bandwidth, the position of the peak (or trough) located at the original isolated spike remains exactly at the geologic boundary. What is lost by filtering is resolution; the ability to separate two closely spaced reflecting horizons.

The practice is somewhat suspect for deconvolution. Each deconvolution operator is designed with a minimum phase spectrum. The minimum phase spectrum varies with the amplitude spectrum. As frequency amplitude varies, so does the phase spectrum. A phase shift is a time shift. Hence, it is possible for each window to have a different time shift, which may cause isochron differences. Under the current trend to avoid band-pass filters, there is less emphasis on time varying techniques. Nevertheless, long experience with band-pass filtered data, and the difficulty of interpreting all of the high frequency information of broad-band signals, has established firm impetus for continued use of the practice. However, it should be remembered that the information carrying capacity of any signal is directly related to its bandwidth. Discarding a portion of the frequency spectrum, the result of applying a band-limiting filter, irrevocably discards a portion of information also.

The seismograph method has been, and continues to be, primarily a method for mapping structure. It does so by measuring precisely the reflection times to geologic boundaries within the earth. Band-pass filtering does not materially affect this ability. If a formation contact has a velocity and/or density contrast, it will develop a reflection coefficient. This may be modelled as a spike, having amplitude proportional to the coefficient of reflection. A zero phase band-pass filter applied to the coefficient spike will broaden the spike, but it will in no way affect the precision of measurement provided by the method, because regardless of the bandwidth, the position of the peak (or trough) located at the original isolated spike remains exactly at the geologic boundary. Even very narrow bandwidth signals can provide exact measurements of the location of an isolated reflecting horizon. (**Fig 5.3**) It might seem that nothing is lost by applying band-limiting filters.

What is lost by filtering is resolution; the ability to separate two closely spaced reflecting horizons. Pinch-outs, thin sand lenses, feather edge erosion at unconformities and all features which require detailed resolution, require broad-band spectra if they are to be adequately resolved.

RESOLUTION OF SEISMIC WAVELETS

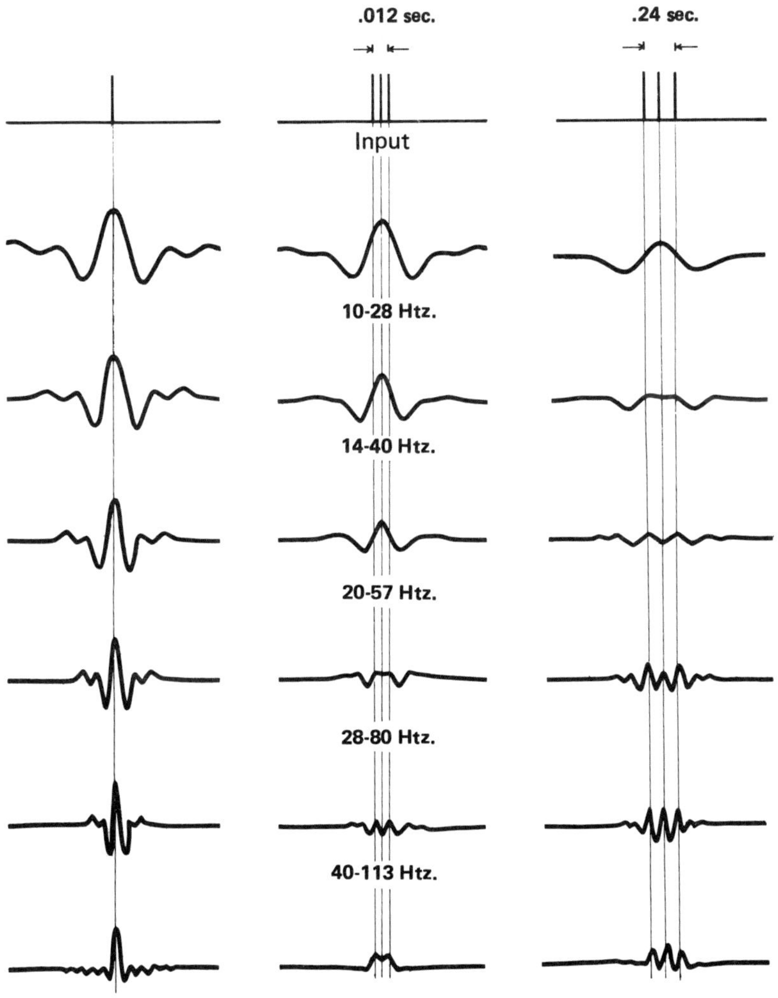

FIGURE 5.4

Pinch-outs, thin sand lenses, feather edge erosion at unconformities and all features which require detailed resolution, all require broad-band spectra if they are to be adequately resolved.

Fig 5.4 demonstrates the importance of frequency content upon the resolving powers of the final signal. A comparison is made between mapping of a single contact and mapping two adjoining thin layers within the section. The top of the diagram illustrates the reflection coefficients of the model; first a single contact, then two sets of two thin layers, the first having a time thickness of .006 sec per each layer, and the second .012 sec. The spatial thickness of the layers will vary with velocity, but may be on the order of 30 to 60 ft for the first example and double that for the second. Such thicknesses are certainly of economic importance.

The three examples have been filtered with several band-pass filters to demonstrate the possibility of mapping these. In all cases, the single contact is mapped with ease. The .006 sec beds are never completely defined, and the .012 sec beds are somewhat ambiguously defined through the low frequency ranges. Use of the mild range filter, 20 to 57 Hz, matches a common filter setting of conventional processing, yet it fails to resolve correctly all but the single contact.

Low frequencies as well as high frequencies are of importance in defining detail. Note that the entire series has the same number of octaves (about 1.5) except for the last. This was done to ensure comparable wave shapes. Compare the last two sets of the series. The 40

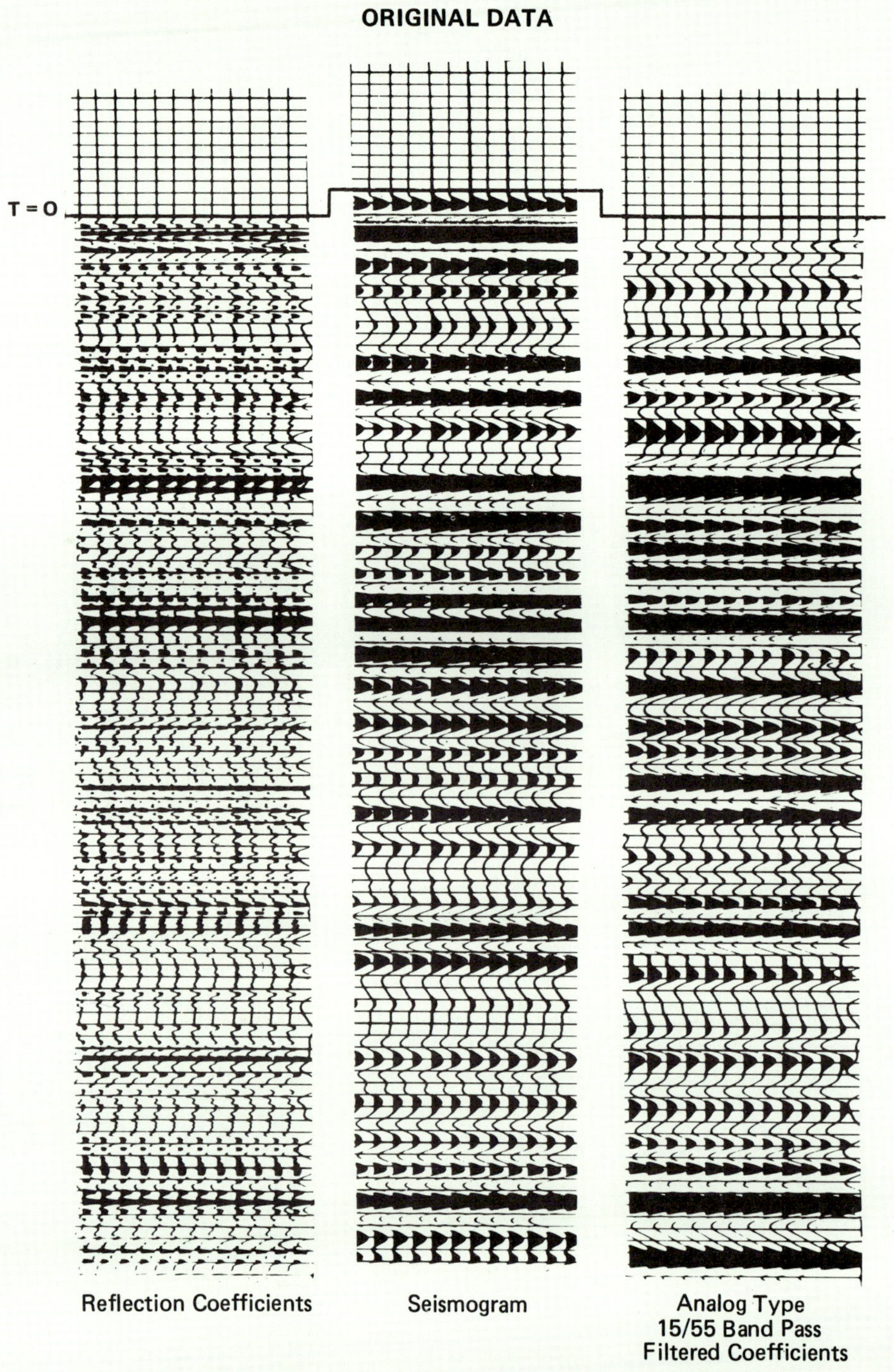

FIGURE 5.5

The difference in appearance between a field seismogram and reflection coefficients, (the ideal seismogram) is due partly to phase shifting and party to band-limiting. A band-pass filtered set of reflection coefficients suggests the major effect is due to frequency filtering.

to 113 Hz provides relatively good resolution, but oscillates, with peaks and troughs of nearly equal amplitude. The addition of the low frequency components to the last set provides a bias which helps to define the correct polarity of the response.

The difference in appearance between a field seismogram and reflection coefficients, (the ideal seismogram) is due partly to phase shifting and partly to band-limiting. A band-pass filtered set of reflection coefficients suggests the major effect is due to frequency filtering. (**Fig. 5.5**) A zero phase analog type filter having broadly sloping sides and a nominal passband of 15 to 55 Hz has been applied to the reflection coefficients, to produce results not unlike the seismogram.

Of particular interest is the development of apparent major reflections on the filtered output where only weak events are visible on the reflection coefficients. The effect is accentuated by autocorrelation of all three signals. (**Fig 5.6**) The somewhat greater definition of the filtered coefficients, compared to the seismic record, in spite of the similar filter bandwidth applied, is probably due to the difference between the relatively sharp slope of the band-pass limits. Even though the filtered output contains much less information than the reflection coefficients, the output appears to contain stronger, better defined reflections.

This is typical of filters, and is one

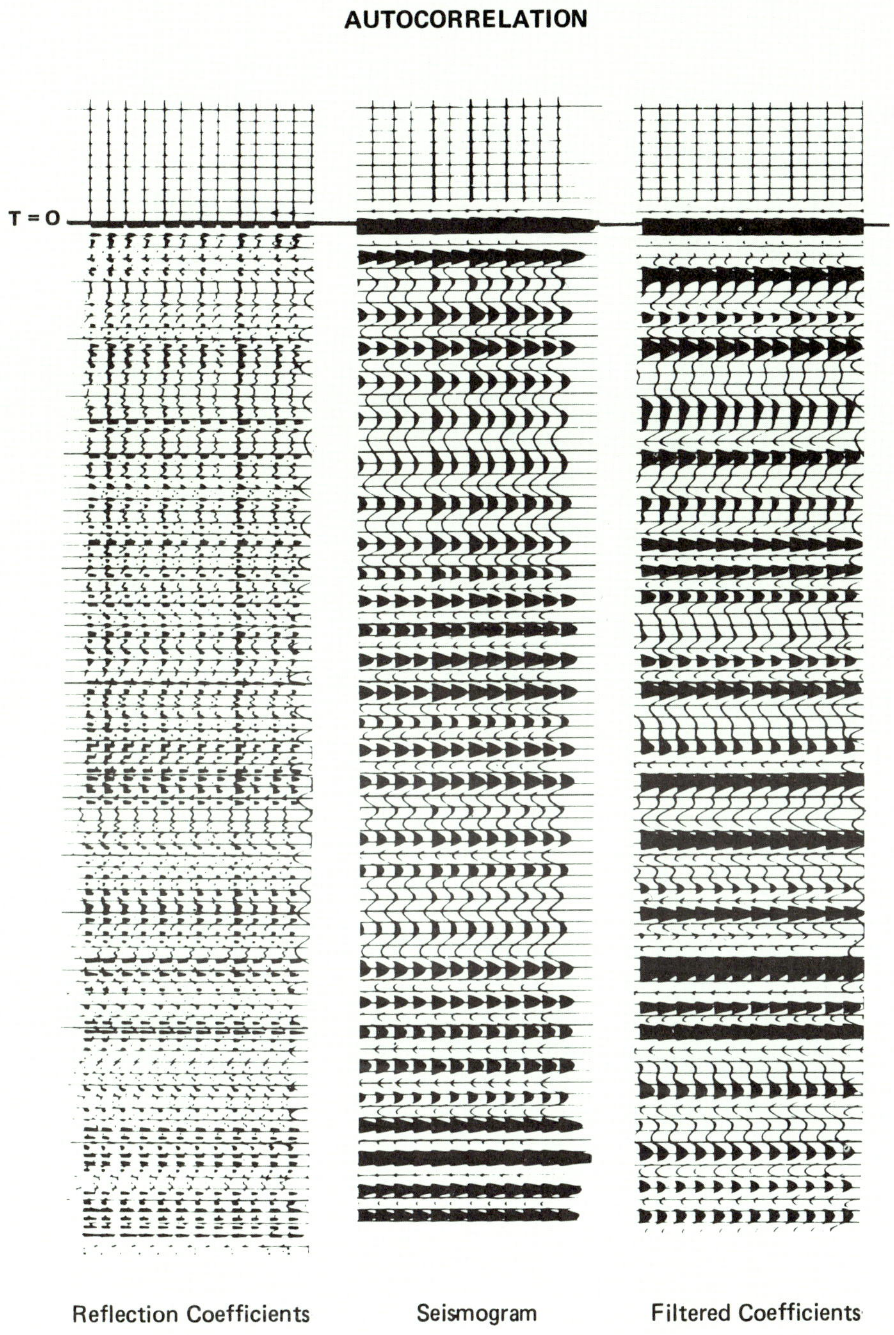

FIGURE 5.6

Band pass filters distort the amplitude relationships of reflectors. Autocorrelation eliminates the influence of phase leaving only the band width effects.

reason why deconvolution output, which contains a much broader effective spectrum than the input, appears to be noisier. Deconvolution can restore weak frequencies to their approximate original level and have the effect of increasing the bandwidth. In fact, deconvolution can only restore weak components, it cannot recover those lost or removed from the spectrum.

Any frequency components which are eliminated by a band-pass are lost forever and can never be recovered from the filtered signal. From a technical standpoint, the best procedure is to apply no filtering to seismic data. Broad-band records may often appear to be noisy and lacking the continuity of filtered sections. Assuming the high frequencies to be information and not noise, the real problem, not often recognized, is that there is so much additional information present that conventional procedures of interpretation become unable to digest the tremendous amount of additional data.

To examine the problem in a different manner, it has been stated that the seismic trace is merely a type of transform of the information contained in an acoustic log. Therefore, the band-pass filter can be applied directly to an acoustic log for the purpose of observing the effects. (**Fig 5.7**) is a sonic log which has been selectively filtered. The original log has been plotted on the left, a high-cut filtered version in the center, and a low-cut filtered output on the right.

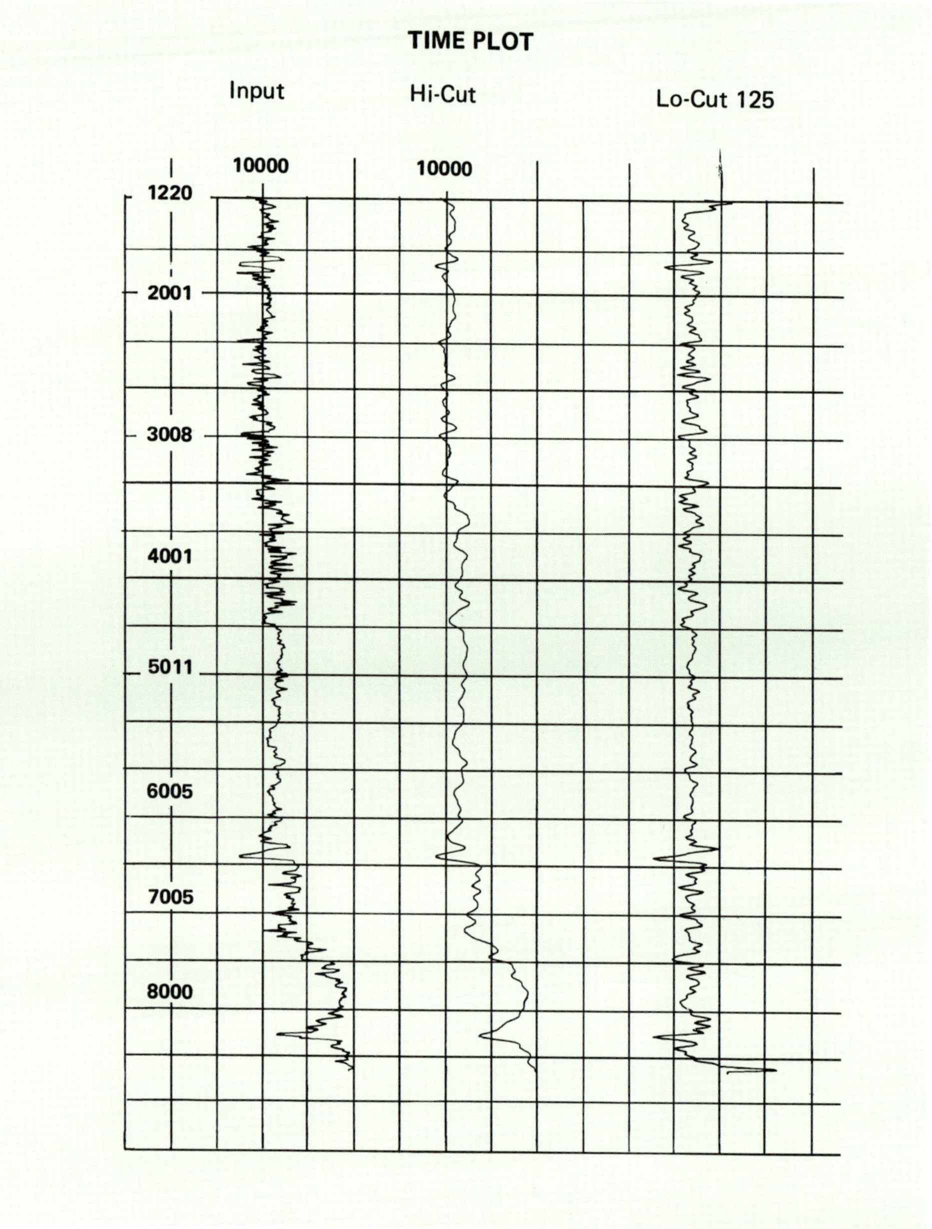

FIGURE 5.7

The seismic trace is merely a transform of the information contained in an acoustic log. Therefore the band-pass filter can be applied directly to an acoustic log for the purpose of observing the effects.

(See also **Fig 12.3**)

It should be pointed out that the term original log is a misnomer. The original log was measured in depth, with an effective sample rate of about 6 inches per sample. To reproduce the same log in seismic time units, the original log had to be integrated over steps of one ms, or about 10 ft, on the log shown. Hence, the so-called 'original' log has already lost considerable fine detail.

Note the great difference in detail revealed on the original log over the filtered log.

In addition to the obvious loss of detail on the high-cut log, there is some loss of amplitude from some of the larger velocity shifts on the log, yet the 0-80 Hz filter applied is relatively broadband. The result of application of the low-cut filter results in output that has lost most of the basic velocity structure of the log.

In general, the effects of a band-pass filter may be segregated into two parts. The cut of the high frequencies diminishes resolution, while removal of the low frequencies eliminates the basic velocity information.

There is a direct equivalence between the information lost by filtering an integrated acoustic well log and a seismic trace. As so often occurs in seismic processing, a trade-off situation exists between the ease of mapping major events on relatively narrow band filtered sections, and the greater detail but more difficult mapping of broad-band signals, which also may be noisier. A very useful test for any problem is to filter a nearby sonic log, to the bandwidth which it is believed possible to obtain the proper effect. If the anomaly can still be detected following this process, it can very likely be detected with the seismic survey, but if not, there is little point in attempting to map it by the seismic method. The final decision should be based on a knowledge of the geologic problem and experimentation with the data to match the filter to the problem, rather than a routine data processing decision.

Although broad-band signals usually contain more information, they are not always critical to the play. For example, in new offshore areas, a very large feature must be found to support the high cost of development and to ensure economic payout. A single clean reflection may be all that is needed to map the required structure, so heavy filtering may be applied with impunity, recognizing that considerable stratigraphic information will be lost. At the other extreme, drilling for a thin productive sand lens in a well developed area may be a sound economic venture only if the seismic data posesses the resolution to map the thin interval.

Today, better techniques than band-limiting exists for the purpose of elimination of noise. For example, modern stacking techniques are very efficient methods at suppressing random noise. Deconvolution can then restore the attenuated portions of the spectrum to full amplitude.

6 DECONVOLUTION

Introduction

Careful stacking of seismic data can remove most of the noise from the recorded earth signals. The signals may have been recorded and processed without filtering of the signal frequencies, nevertheless, the signals will have been modified by the natural filtering from transmission through the earth. Therefore, the spectrum of the recorded signal will most likely suffer considerable attenuation of high, and some low, frequencies (Fig. 6.1). Amplitude of the frequency components is important. A trace may have a very broad frequency band of reliable signal elements but those at low amplitudes can carry very little significant information. The frequency component must be restored to their original level if they are to provide the proper contribution of information to the seismic trace.

The process by which the attenuated elements are restored is another filter operation termed deconvolution. The theory and action of deconvolution can be explained by the use of a simple system model.

Basic Model

Three components make up the basic model used to study the seismic reflection process. They are: a unit impulse input; a system model simulating the combined effects of the earth, the geophones and the instruments on the input pulse; and the output (Fig. 6.2).

The total system effect is simulated by a single transfer function which, when convolved with the input, transforms it into a reasonable approximation of the signal received by a field recorder.

The operation can be reversed, since the convolution operation obeys the same commutative and associative laws as multiplication. Thus, if any two elements of the system are known, the third can be determined.

Deconvolution is designed to reverse the effects of the passage of the input pulse through the system. Unfortunately, the generic term, deconvolution, has been applied to a multitude of processes which may be related, but attack quite different problems.

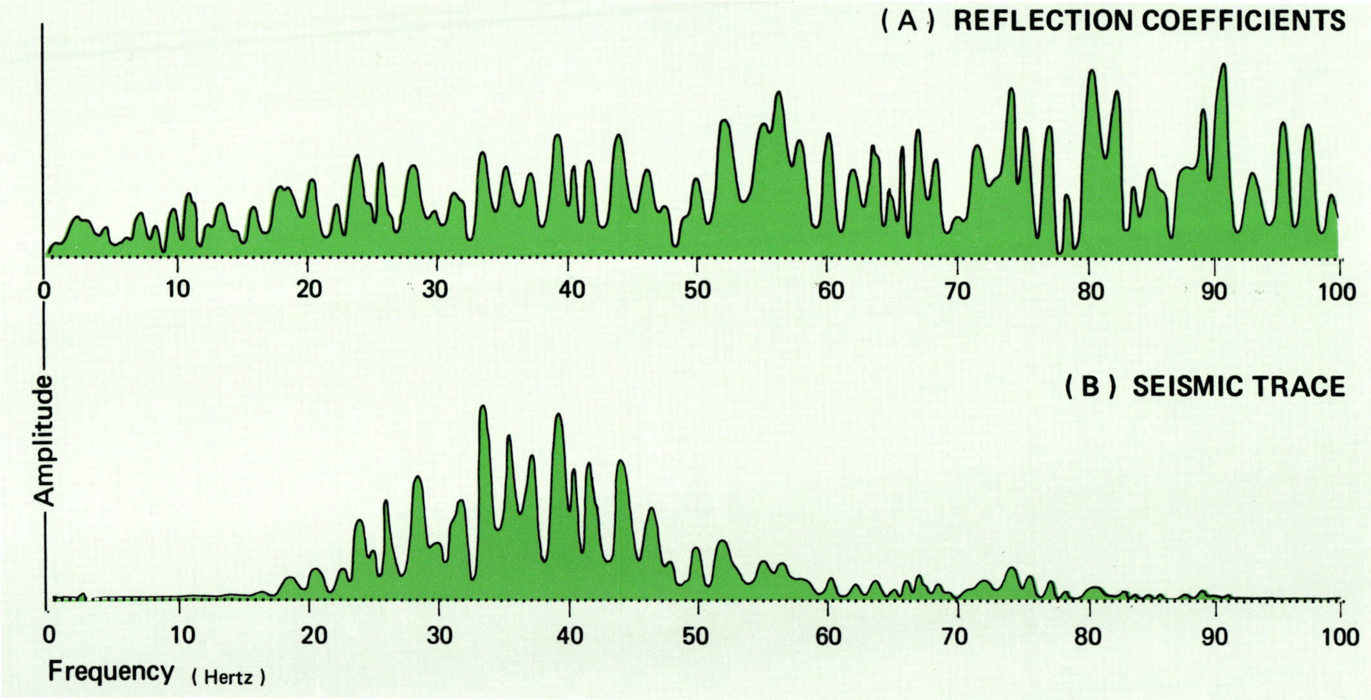

FIGURE 6.1

Transmission of the seismic waves through the earth and instruments attenuates both low-frequency and high-frequency components. Note that the local relative amplitude relationships from frequency to frequency are unchanged, the major difference being a slowly varying change with frequency.

The input convolved with the transfer function yields the output, and conversely the output deconvolved with the transfer function or convolved with its inverse, yields the input. The rub lies in the definition of the transfer function, but under the conditions defined for the basic model this can be relatively simple.

The input of the basic model is a delta function (a time series in which all values are zero except a single value 1.0) which represents the unit impulse. In this case the transfer function is simply convolved by 1.0 to produce the output. Therefore, the output duplicates the transfer function.

The basic equation of the process model is Input * Transfer Function = Output, where * denotes the convolution operation. But for a unit impulse of 1.0, 1.0 * Transfer Function = Unit Impulse Response. Therefore, Unit Impulse Response = Transfer Function.

This operation can also be demonstrated quite simply in the frequency domain. Convolution in the frequency domain is accomplished by cross-multiplication of the amplitude spectrum of the input signal by the amplitude spectrum of the system, or transfer, function. The corresponding phase spectra are added. All frequencies of a unit impulse have unit amplitude so cross-multiplication of two spectra yields a duplicate of the system spectrum as the output spectrum.

If a pure impulse could be input to a seismic system the output would equal the transfer function. It is relatively easy to determine the transfer function of the recording equipment but this is only part of the total system. The shot, its coupling to the ground, the transmission characteristics of the earth, and the coupling of the geophones to the ground are all part of the active system. Unfortunately, noise is invariably added at some stage of the seismic field operation so the output is not the simple response of the input.

Although convolution follows the same commutative rule as multiplication, convolution is not

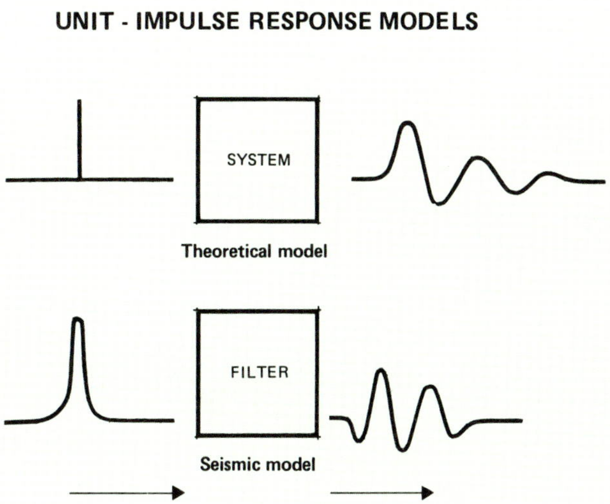

FIGURE 6.2

A suitable model of the seismic recording process lumps all components into one system function which yields a good approximation of the response of the real system to the input signal.

multiplication. Neither is deconvolution division. A form of inverse convolution can be used to obtain the input function when the system transfer function is known, but this operation frequently fails in the presence of additive noise. A more practical process is to convolve the output signal (the seismic trace with the inverse of the transfer function to determine the input signal),

$$\text{Output} * \frac{1.0}{\text{Transfer Function}} = \text{Input}$$

While the operation $\frac{1.0}{\text{Transfer Function}}$

is not defined, i.e., there is no step-by-step equivalent of division for handling functions or ordered arrays of numbers, the desired result can be obtained by computing another function which, when convolved with the Transfer Function will yield 1.0, the delta function, or as close an approximation to it as required. This new function, when convolved with the output will "de-convolve" it to yield the Input.

The single pulse input of the basic model is a symmetrical, instantaneous, zero-phase signal—all frequencies at zero phase, The output is a symmetrical and hence not a zero-phase. The response recorded is spread over several milliseconds of record time, smearing details and interfering with the resolving power of the signal. The reflection process, and hence the system model, is a real-time system, and nothing can come out before something goes in. The input spike has no time width, therefore any time shift of frequency components must be in the nature of a time delay. A time shift is also a phase shift so any phase shift in the system is a phase lag.

Complete deconvolution must, therefore, restore not only the amplitude spectrum, but, to be completely effective, it must also reverse the shift of the phase spectrum. Unfortunately this is often not completely realizable.

Deconvolution Model

The principal problem of deconvolution is one of determining the required parameters, i.e., of defining the transfer function of the system. This is further complicated by the fact that some of the frequency components may be lost in traveling through the system. When this happens the phase component of the missing frequency is also lost. The phase of each component must be defined if the signal is to be described properly. It is not possible to reconstruct an input signal by replacing a missing amplitude vector unless the phase of that vector is also known.

All deconvolution procedures may be related to some model of the process which yields the recorded seismic trace. A synthetic seismogram is an example of the type of modeling process used.

Basic assumptions are used to develop the model, and the success of the deconvolution process is dependent upon the validity of these assumptions under real conditions. For example, the reliability of a process may be degraded by the presence of noise if a noise-free model was the basis of the development. Noise may be added to the model to compensate for this problem, but then the quality of deconvolution will depend upon whether or not the noise recorded in the field is indeed the same type as that used in the model.

Theoretical models, upon which the majority of deconvolution theory depends, assume several, if not all, of the following conditions:

1, the input signal is known,
2, the transfer function is known,
3, the unit impulse response is minimum phase,
4, the output is broad-band,
5, the noise is random white noise,
6, the spectrum of the earth signal (reflection coefficients) is white,
7, the output is stationary, i.e., does not change with time.

In reality the following conditions probably exist:

1, the input signal is not defined,
2, the transfer function is not known,
3, the unit impulse response is mixed phase,
4, the output is narrow-band,
5, the noise may be periodic,
6, the spectrum of the earth signal is not white,
7, the output is time variant.

The fact that such a large number of discrepancies exist has not deterred the pioneers in the field of signal processing. On the contrary, many ingenious strategems have been developed to handle some of the problems, and the assumptions used fit well enough to achieve considerable success. The amazing thing is that some of the processes work so well, given the poor definition of the system. One must admire the great original work done by the developers of seismic signal processing.

Many of the conditions vary, and, unless the interpreter can recognize the situation existing for a particular area, and knows what processes are available, what assumptions were used in their formulation, and their effects on the data, he may not achieve the desired output. The difference between desired and actual output usually represents the effects of errors in the basic assumptions. Differences between theory and practice will nearly always be present to some degree because, as a rule, the basic assumptions are never quite correct.

The system model for the most common deconvolution method assumes the earth signal from a sedimentary section to be a series of random unrelated events exhibiting a completely white spectrum. (See section on Seismic Noise.) A seismic shot initiates a pulse which degenerates almost immediately into a damped wavelet. This travels through the geologic section and convolves with the earth signal, replacing the reflection coefficient at each boundary interface with the wavelet. The resulting recorded signal is the sum of the multitude of small overlapping wavelets resulting from convolution of the input wavelet with the earth signal.

The shape of the unit impulse response, and hence of the transfer function, is unknown, since as soon as the impulse leaves the energy source, coupling of the energy source impulse to the earth, a partially elastic medium, distorts the impulse to a wavelet of unknown shape. The input is no longer precisely defined. Furthermore, the wavelet may be further modified as it travels through the earth, through the geophones and through the recording instruments. Hence, the single most difficult problem in deconvolution is defining the response of the total system.

But all is not lost. A good example of the ingenuity which characterized the early development of seismic signal processing, is the use of the commutative properties of convolution aided by a reasonable assumption to obtain a workable solution.

Due to the commutative properties, convolution of signal "A" with operator "B" yields exactly the same result as convolving "B" with "A". In any series of convolutions the order of operations is not important; it can be changed or combined and the results will be the same. Convolving the input with each separate component in turn yields the same result as convolving the input and all the components in any other sequence. If only the output from convolution of several components is available, and that is convolved with any other remaining elements, the results are the same. It is for this reason that a single transfer function can be used to model the combined effects of the entire system. Instead of making the energy source the input to the system model, if its characteristic response is unknown it can be lumped with the other components which combine to yield the transfer function of the entire system. Then the input to the model can be replaced with a single reflection coefficient which is also a spike, so the original model is still valid.

An extension of the model can be made to fit the conditions of many areas where the geologic column can be assumed to be the result of a random depositional sequence. If such is the case, the earth signal will be a randomly spaced series of reflection coefficients of random amplitude having a completely white frequency spectrum, provided the number of reflection coefficients is large. If the noise is white or absent, the basic model will still be valid.

The input to the basic model was a unit impulse (a single reflection coefficient) having a completely white spectrum. Now the input becomes the series of reflection coefficients making up the earth signal. Under the amplitude assumptions made this will also have a white spectrum.

The output of this system then meets the specifications of the basic model; a white spectrum multiplied by the spectrum of a single transfer function combining all the system components yields an output amplitude spectrum equal to that of the total transfer function. The only difference between the single spike model and the random series of reflection coefficients is in the phase spectrum.

In the time domain, the two models apparently produce quite different results, but this is the effect of the phase differences. Each reflection coefficient has been convolved with, and hence replaced in the output by, the system response. The phase spectrum of the earth signal can be made equivalent to that of the unit impulse response by setting both to zero. The autocorrelation operation will accomplish this by consolidating all of the response wavelets into one zero-phase wavelet representing the average of all the impulse response wavelets on the recorded trace.

White signals yield a single spike at the zero shift point (zero time) of the autocorrelation. If the earth signal and the noise components are both truly white and not band-limited the autocorrelation output will be limited to one point only located at the zero shift point of the autocorrelation. Since the autocorrelation funcion is symmetrical this zero point will be the central maximum of the function.

In reality, both are band-limited, although the noise may have a broader band than the signal. The noise considered here is additive, rather than convolutive. Therefore, the amplitude spectrum of the recorded trace will be the sum of two components; a flat platform-like line representing the random noise with the narrower band of the earth-signal spectrum, somewhat bellshaped due to the effects of the system response, superimposed on it. (This includes, among other things, the effects of frequency-dependent attenuation and scattering of this signal wavelet

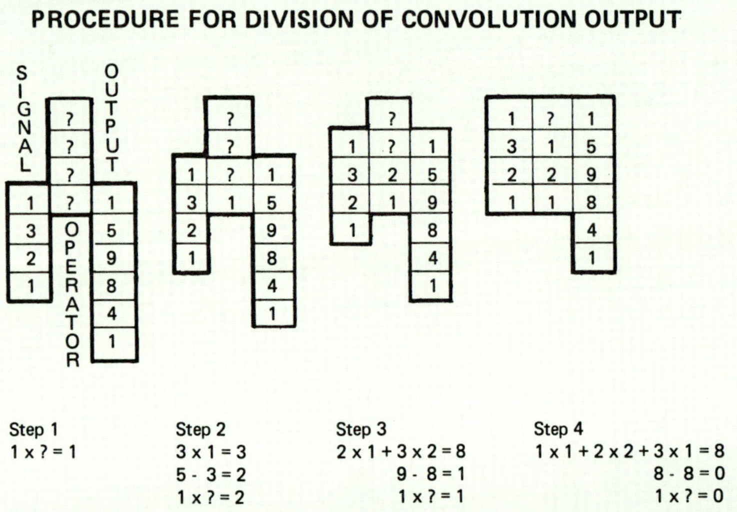

FIGURE 6.3

A form of deconvolution based on the knowledge that the output of convolution represents the sum of cross-multiplications between signal and operator.

as it passes through the earth.) Under these basic assumptions, autocorrelation of the recorded signal will yield a wavelet having the same response as autocorrelation of the system transfer function, weighted by a noise spike at the central peak, but all phase information will have been lost. Since conventional deconvolution is concerned primarily with restoration of the amplitude spectrum, the amplitude spectrum of the system response, for all practical purposes, can be determined by auto correlation of the recorded signal.

The next step is the calculation of the inverse operator. Some mathematical rules of the convolution process are similar to multiplication, so as the operation in the basic model mimics the mathematical expression of a simple product, $A \cdot B = C$, then any one missing element can be determined from two known elements. In multiplication this is always possible provided neither A nor B are zero. In the convolution operation a form of division, equivalent in concept to $C/B = A$ is also possible, as is the equivalent of $1/B$. Therefore the convolution operation $C * 1/B = A$ can be carried out.

The earth-signal input spectrum is assumed to be completely white, with the only difference between it and the single spike input being a different phase spectrum. Hence, autocorrelation of the output of both types of input yields identical amplitude spectra. If the autocorrelation of the output from either input is divided into 1.0 the result will then yield the inverse of the system.

This can be done by a form of polynomial division (**Fig. 6.3**) and indeed the identical procedure was followed in some of the early deconvolution routines. Unfortunately, the division operation has an annoying habit of "blowing up". That is, the division does not converge meaningfully but yields a non-terminating series of increasingly larger numbers. The reader may easily verify this by performing the operation on different data sets. As a rule, the division of a short numerical series into 1.0 always yields a very lengthy result, whereas division of a very long series is required to obtain a short inverse operator.

The problem is solved by the use of a least-squares optimization approach; generally, the well known Weiner-Levinson algorithm, which produces an optimum least-squares approximation of the inverse operator, is employed.

Whatever method is used, be it the least-squares algorithm, polynomial division of the autocorrelation, or some other method, such as factorization of the spectrum, it will yield an inverse operator, which, when convolved with the autocorrelation will cause the autocorrelation to collapse to approximate a spike. If the same operator is convolved with the recorded trace, each of the wavelets which replace each reflection coefficient will be convolved in turn, converting each of these into an approximation of a spike, improving the resolution of the reflection events. The amplitude of each output spike will be proportional to the wavelet amplitude which in turn should be directly

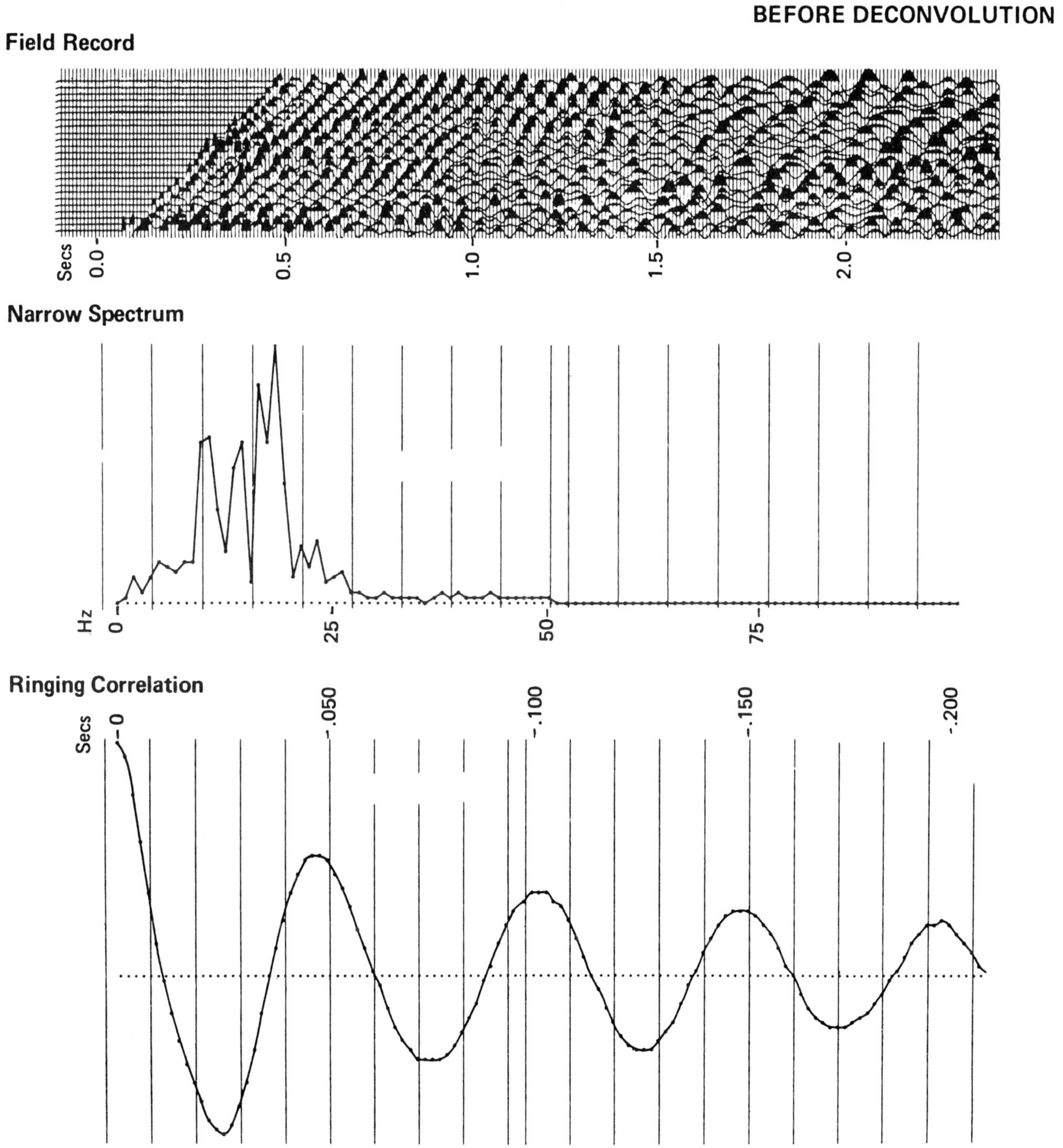

FIGURE 6.4

The field record is dominated by head waves, or 1st breaks, and low-frequency ground roll. The amplitude spectrum shows most of the signal to be concentrated between 8 and 20 Hz. The autocorrelation confirms the narrow-band signal by a slowly decaying uniform oscillation at about 20 Hz. No reflections are evident. Until digital processing made deconvolution possible, very little could be done to improve the record quality.

AFTER DECONVOLUTION

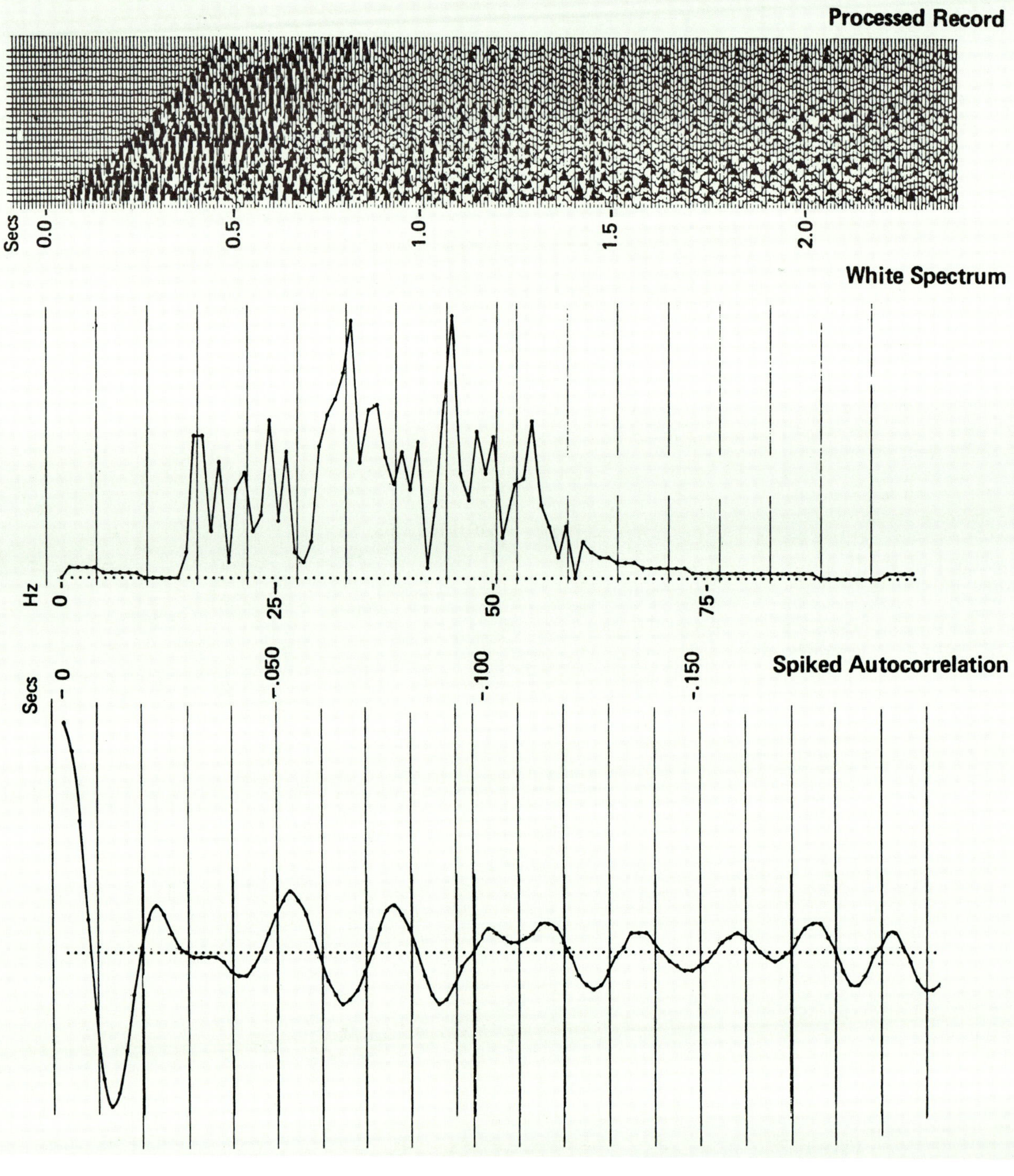

FIGURE 6.6

A basic minimum-phase deconvolution process has restored the amplitude of the signal components from 20 to 55 Hz. Low frequencies no longer dominate the spectrum, and several reflections are now clearly visible. The autocorrelation has a sharp initial peak and decays quickly to an irregular series of small oscillations indicative of a broadband signal. In this case, the deconvolution worked well because the low-amplitude components in the field record were relatively clean signal. More commonly, these would be dominated by noise and deconvolution would merely increase the amount of noise.

FIGURE 6.5

Prior to deconvolution the only data processing technique that could be applied to the record of Figure 6.4 was bandpass filtering which, in this case, is relatively ineffective at improving data quality.

SIMPLE BANDPASS

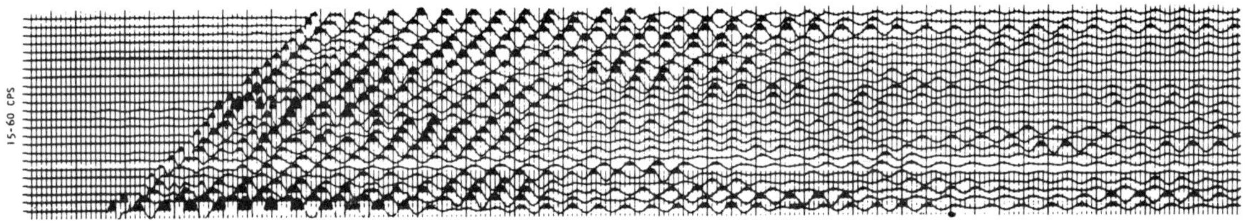

proportional to the reflection coefficient.

If the assumptions apply reasonably well to the real situation, the application of deconvolution as described above will do a good job of reversing the effects of the system and yield a reasonable approximation of the earth signal input.

The sharpness of the output will be a function of the frequency bandwidth. Any frequency components completely lost in data acquisition cannot be recovered by this process and will represent a loss in resolution. The spike can be no sharper than the best approximation possible with the available bandwidth.

From the frequency domain point of view, the field survey process is multiplication of the white input spectrum of the earth signal by the spectrum of the system. The deconvolution process first obtains the spectrum of the output through trace autocorrelation, then computes the inverse and multiplies the spectrum of the trace by this inverse to yield a white spectrum approximating that of the input. Phase effects are ignored, and any frequency components lost in the process are absent from the result.

In spite of a rather elegant theoretical development deconvolution in practice is a messy process, lacking in rigor, definition and often poorly understood by the user. The problem is partly explained by the large number of assumptions that must be made, the frequent mismatch between the data and the assumptions and the lack of analytical procedures for quantitative measurement of the several factors involved.

Furthermore, the term deconvolution is often loosely applied to a wide variety of problems which may be quite different in nature. For example, an operation to reverse the impulse response distortion, and one to eliminate long-period multiples, are both called deconvolution, and indeed may use modified versions of the same computer program, but the problems are quite different.

The seismic field record of Figure 6.4 has no obvious reflections. Rather, it is dominated by low-frequency noise patterns across the front end and ground-roll of still lower frequency over the later part of the record.

The amplitude spectrum confirms the low-frquency content of the record, with three dominant frequencies of 10, 13, and 18 Hz. Other components of the spectrum extend up to 50 Hz but they are apparently too weak to make any significant contribution to the signal.

Amplitude of the autocorrelation decays slowly, a definitive diagnostic of a very narrow-band signal. The period of oscillation is about 54 ms, corresponding to the peak frequency of 18 Hz.

Prior to digital processing and deconvolution, the only data-processing tool that could be applied to a record like this was bandpass filtering. In an effort to eliminate the ground roll from the record, a 15 Hz low cut filter would probably have been applied, giving the results shown in (**Fig 6.5**). Although the ground roll is now gone, no

effective improvement has resulted from applying a 15-60 Hz bandpass filter. The record appears to be cleaner and smoother but there are no reflections evident. In fact, this record now has a very narrow spectrum, since everything below 15 Hz has been removed leaving only a single peak frequency of 16-18 Hz. **(Fig. 6.4)** .
The results of a basic minimum-phase deconvolution, after applying an inverse operator derived by the Weiner-Levinson method from the autocorrelation of **(Fig. 6.4)**, are shown in **(Fig. 6.6)** . The spectrum is now relatively broad-band and white with the frequencies from 19 to 60 Hz making an important contribution to the signal. The effect is immediately apparent on the seismic record. Several reflections are evident, containing useful structural information. The autocorrelation also shows a greatly improved response, with a high initial amplitude decaying rapidly with time.

The example illustrates the basic concept of deconvolution very well. The deconvolved record contains a number of reflections which were not visible on the input. The process worked well because the input record, which was a raw field record before stacking, contained very little noise, particularly in the frequency range above 20 Hz. Although the original record appeared to be noisy, most of the noise was confined to ground roll below 12 Hz, while the rest of the record demonstrated the near-continuous oscillation typical of a very narrow bandwidth signal. This was accentuated even more dramatically in the output from the bandpass filter which, because it eliminated more low frequencies from an already very narrow band, left only one major frequency component.

It is very fortunate that the field record contains so little noise. Had the very weak frequency components in the range from 20 to 50 Hz consisted largely of noise, the deconvolution would have replaced the input, dominated by low frequency noise, with output dominated by a broad band of noise, the result of boosting the weak higher frequency noise components in the deconvolution operation.

Such effects were not well understood during the early days of deconvolution. Application of identical processing to apparently similar data yielded a wide variation in results. Many geophysicists were confused, and data processors were at a loss to explain why deconvolution worked well with some records but not with others. There are several reasons, but a very common one is a high percentage of noise in the relatively low-amplitude portion of the input spectrum.

Before stacking was invented, about the only available remedy for noise was to band-limit the data to the relatively narrow range of frequencies where the signal was stronger than the noise, i.e. the signal-to-noise ratio was greater than one.

If a bandpass filter is applied after deconvolution, it may have the effect of negating the results of a relatively expensive process. The reason for this is that often attenuation is minimal in the relatively narrow band of natural-response frequencies in which signal dominates noise in field data. Filtering, to band-limit the data, was a common operation. Unfortunately this tended to establish criteria for the character of seismic data which became the accepted standard. Hence, there is a tendency to continue the practice.

Today, stacking the data eliminates most of the random noise. The amplitude spectra of noisy field records and the relatively noise-free stacked output may appear almost identical in shape, but stacking will have cancelled the noise components, leaving some valid but very weak frequency components which will contribute valuable information once they are restored by deconvolution. If they are then bandpass filtered, following the practice established in earlier times, much of the benefit of deconvolution will be lost.

Another reason for poor results is a faulty assumption in the basic deconvolution model. A number of deconvolution programs assume the earth spectrum, and therefore the spectrum of the deconvolution output, should be

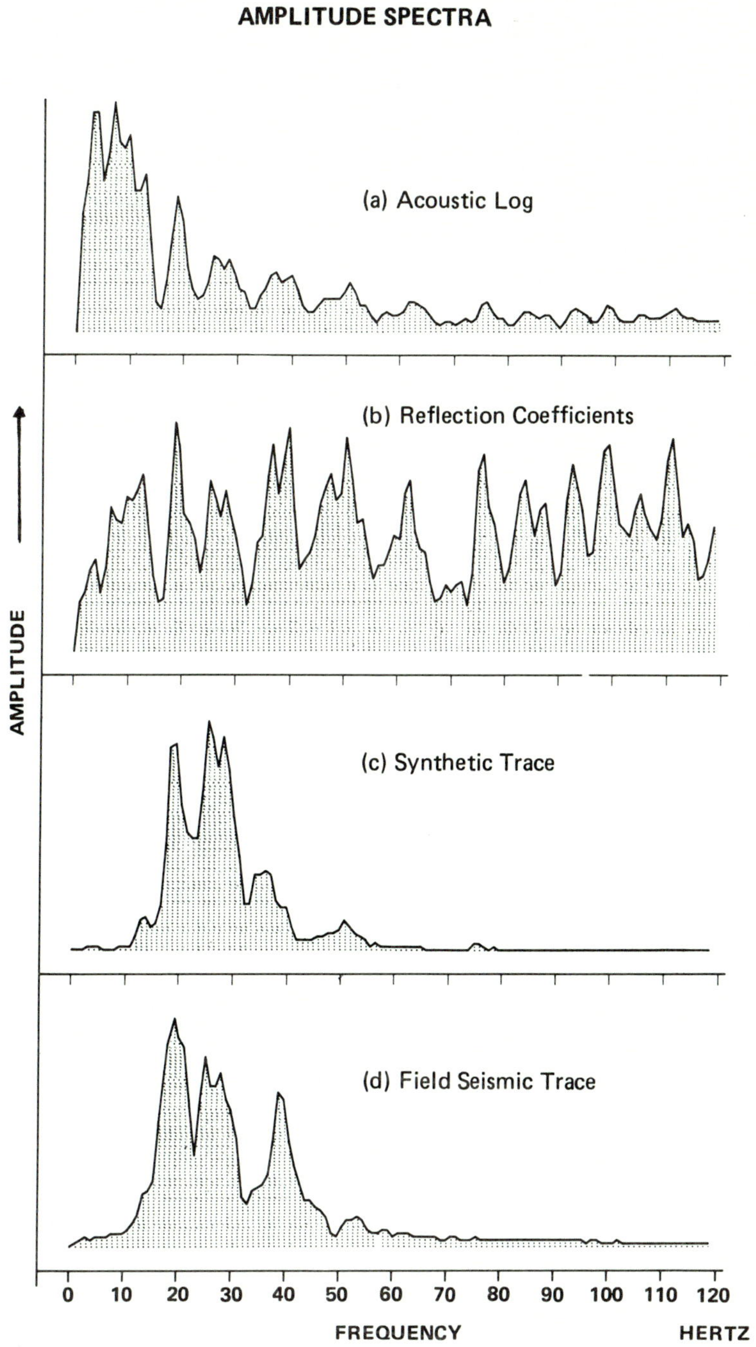

FIGURE 6.7

The sonic log and the reflection coefficients contain the same relative detailed variation in the frequency components but the proportional distribution from low to high-frequency components is quite different. The normal distribution is substantially modified on the seismic trace where high-frequency components are severely attenuated.

completely white. This does not generally agree with facts. Very often the sedimentary section has an amplitude spectrum which is not completely white. The depositional history of a basin often includes cyclic sections. These produce definite periodic elements in the earth signal, and the spectrum contains dominant frequency components generally related to the cyclic nature of the deposition. This produces characteristic irregularities in the spectrum, not unlike the amplitude and character contrasts of reflection horizons in time sections. The character and frequency content of the amplitude spectra for different geologic sections vary widely, validating the assumption of a white signal in some areas and completely condemning it in others.

A thick sequence of deltaic clastics may come very close to meeting the assumption of a white spectrum. On the other hand a section of uniform, well defined lithologic units and sharp velocity contrasts may have strong periodic elements. This, of course, causes both the amplitude spectrum and the autocorrelation to have distinct character. In these cases, a very harsh deconvolution operation, needed to produce an almost totally white output spectrum, will wipe out most of the character of the earth signal and do more harm than good. A

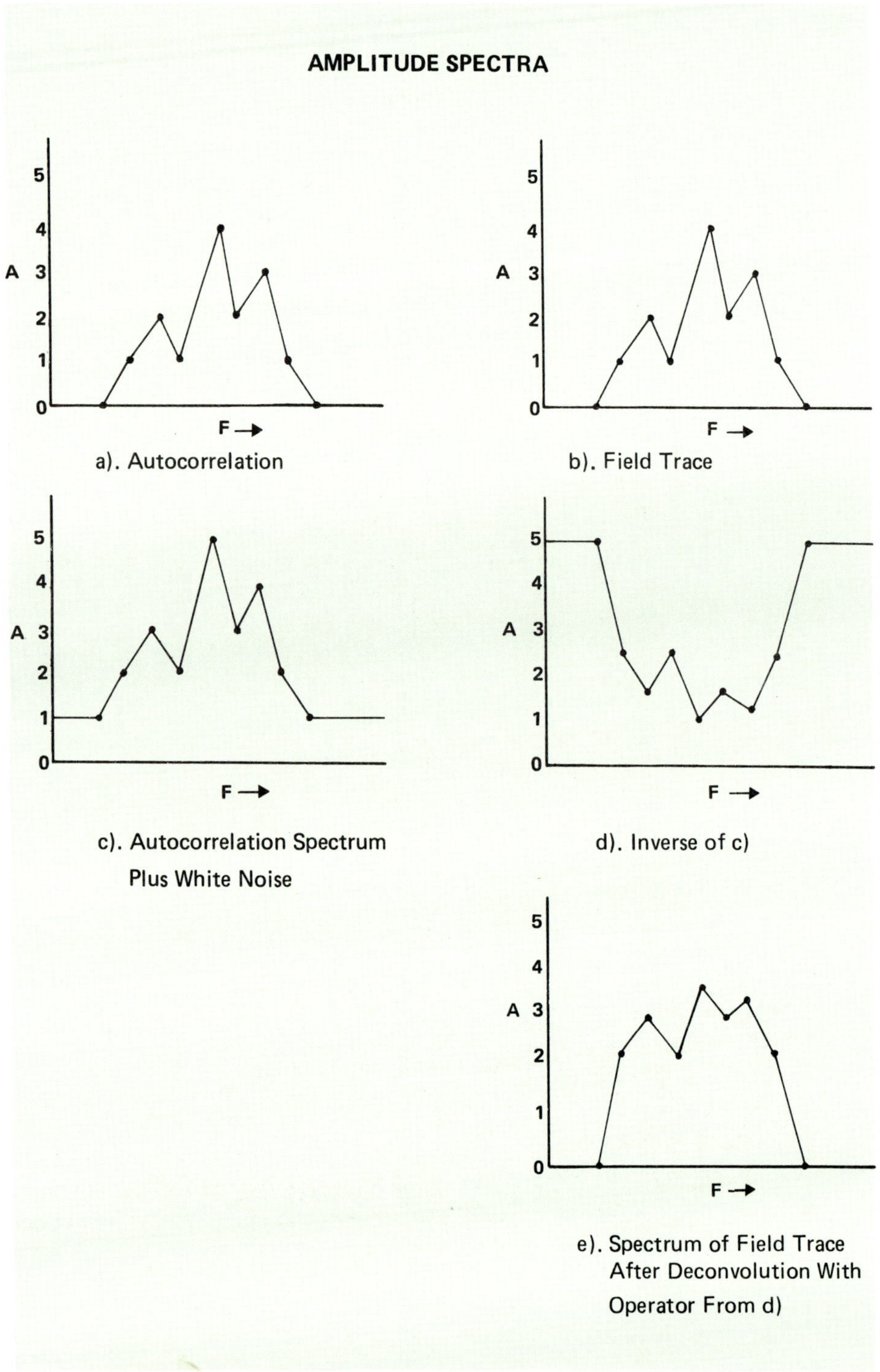

FIGURE 6.8

A normalized autocorrelation has the same amplitude spectrum as the seismic signal but the phase spectrum (not shown) will be quite different. Prewhitening changes the relative amplitude values of the autocorrelation so the inverse is no longer an exact inverse of the spectrum of the autocorrelation and when it is applied to the signal it fails to achieve a complete whitening of the signal spectrum

completely white broad-band signal is often jittery, low amplitude and very difficult to correlate trace to trace.

One of the easiest ways of checking the hypothesis of the white spectrum is to examine the amplitude spectrum of the series of reflection coefficients developed from a sonic log in the area (**Fig 6.1**). Note that, in this example, the spectrum is not completely white but contains significant variations in frequency amplitude. In fact, both the reflection coefficients and the trace show similar variations. Note that both have almost identical local frequency component relationships from spectrum to spectrum varying with the overall attenuation of the spectrum of the record. Therefore, the spectrum character is an essential element in the make-up of the field record (**Fig. 6.7**) since this character is also in the log trace.

Strong dominant peaks on the spectrum may represent dominant geological components having an important periodic influence on the record. When such a situation exists the pure white spectrum assumption may not be valid.

During the early period of deconvolution development the assumption of the white spectrum may have been the only

EFFECT OF PRE-WHITENING ON DECONVOLUTION OPERATION

FIGURE 6.9

Increased prewhitening of the autocorrelation decreases the efficiency of the deconvolution operator and the output remains relatively unchanged by the deconvolution process.

practical one. Detailed spectrum analysis was a very difficult procedure.

Interestingly, a remedy for possible errors in the assumption was found more by accident than by design, due to another quite different problem. Whether done in the time domain or frequency domain, calculation of the inverse deconvolution operator designed to yield a white spectrum in the frequency domain is equivalent to dividing each amplitude component of the amplitude spectrum into the maximum amplitude value. The resulting inverse operator, when convolved with a data trace, should restore all frequency components to a uniform level.

Computer programmers found that quite frequently the routines to design inverse operators became unstable and could not produce a satisfactory operator. A number of causes may produce this but one arises when the amplitude spectrum contains a frequency component with zero amplitude, resulting in an attempt to divide by zero. The solution to this problem is simply to add a small amount of white noise to the spectrum. In the time domain this is accomplished by adding some small value to the central point of the autocorrelation used to design the operator. In the frequency domain this adds a small constant to all frequencies. There are now no zero amplitudes and therefore no zero divisions. The device was called "prewhitening".

Unexpectedly, the results also pleased the geophysical interpreter, who found that prewhitening of the autocorrelation function produced more acceptable output. Understanding why this is so may be facilitated by referring to the diagram of **(Fig. 6.8)**.

A deconvolution operator will have the inverse amplitude spectrum of the signal. In the frequency domain the design of the inverse operator may be depicted as the calculation of an amplitude spectrum which, when multiplied by the corresponding signal spectrum, will yield an output spectrum in which all frequency components have a common amplitude. The usual procedure for this type of normalization is to divide all points into the maximim amplitude value, say 4.0. Thus if the signal amplitude is 2.0, the corresponding operator value will be 2.0. Another frequency component might have an amplitude of 3.0, in which case the inverse operator would have a value of 1.33.

The addition of white noise to the autocorrelation raises all values on the amplitude spectrum a uniform amount. Therefore if white noise of the value of 1.0 is added, the maximum would be raised from 4.0 to 5.0. The point which was 2.0 would now be 3.0 and the previous 3.0 would be 4.0. The inverse operator of the two points would now have values of 1.67 and 1.25.

Convolving the resulting inverse operator with the original trace has the effect in the frequency domain of multiplying the first value of 2.0, on the original trace spectrum, by 1.67 giving 3.33. The value of 3.0 would be multiplied by 1.24, yielding a value of 3.75. Thus, two points, which on the input were 50% and 75% of the maximum, would be at approximately 83% and 94% on the output. Without the addition of white noise they would have been 100%. Thus, the relative differences in amplitude have been reduced but not completely eliminated. The operation does not completely whiten the output spectrum. It merely reduces the amplitude ratio between the signal frequency components.

Fig. 6.9 illustrates the effect. The data are from a noise-free synthetic seismogram. The original reflection coefficients are on the right side of the illustration, the band-limited synthetic seismogram on the left. Very heavy prewhitening of the autocorrelation (about 100% of the spectrum amplitude) produces very little effect. Very light prewhitening (about 5%) produces output very close in appearance to no prewhitening at all; which in turn is a fair approximation of the reflection coefficients. Note that there is some phase shift between the deconvolved output and the reflection coefficients.

Deconvolution with no whitening produces the best approximation to the reflection coefficients within the available bandwidth. The synthetic seismogram has a bandwidth from approximately 15 to 90 Hz, whereas the reflection coefficient frequencies extend from 0 to 250 Hz. The missing frequency components cannot be replaced by deconvolution, therefore, although the output may approach the reflection coefficients it can never equal that series.

Phase Considerations

At first glance, deconvolution of seismic exploration data appears impossible because the majority of the required information is missing. The development of a satisfactory model, and procedure, based on admittedly weak assumptions, provides a very good solution within the constraints of the information available. But unfortunately it ignores the phase components. Phase control is extremely important because the phase defines the time relationships of the events mapped on the seismic trace. Relatively large modifications to the amplitude spectrum have less effect on the output signal appearance than do small changes of phase. Yet, for the most part, traditional

CONSTANT - AMPLITUDE - SPECTRUM WAVELETS

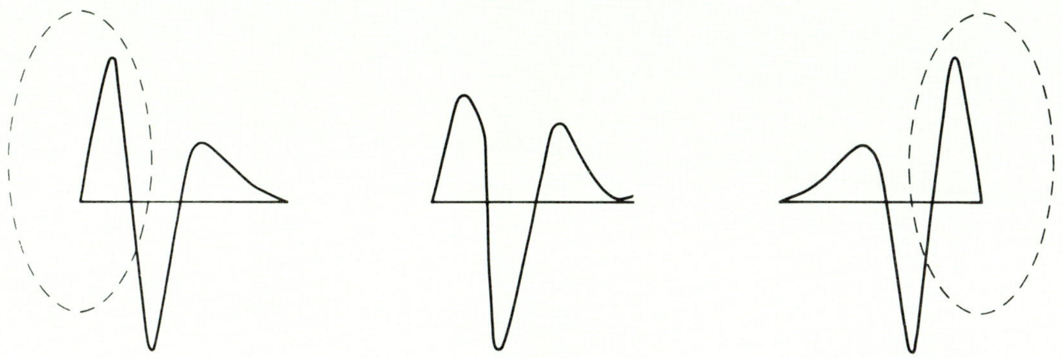

FIGURE 6.10

Energy in the minimum delay wavelet is concentrated in the initial part of the operator. There is only one minimum and one maximum-delay wavelet for each amplitude spectrum, but there are innumerable mixed-delay wavelets.

deconvolution procedures emphasize modification of the amplitude spectrum. This is because the amplitude spectrum can be defined with much more certainty than the phase spectrum. There is no unique definition, in harmonic analysis, of phase shifts greater than 180°. In what might be considered a form of phase aliasing a phase shift of 181° is identical to a shift of -179°

Ideally, deconvolution requires the phase shift applied by the unit impulse response to be reversed, in order to be completely effective. This in turn requires knowledge of the phase of the unit impulse response. For the most part, practice has been merely to try to minimize any phase modification by deconvolution.

Autocorrelation provides the amplitude information from the recorded trace, but in doing so discards the phase information. The phase of the recorded trace can be determined, but there is no sure way of separating the phase of the earth from the phase of the impulse response. Accordingly, any phase modification has been based entirely or another assumption.

Minimum Phase

The input to the basic system model is an impulse (delta function) represented as a time series of ..... 0, 0, 0, 1, 0, 0, ... The time duration of the input pulse is confined to one sample, and the total energy is concentrated in this pulse. The unit impulse spike has a completely flat broad-band white amplitude spectrum, with all frequency components at zero phase.

The system response to the unit impulse is a damped wavelet, extending over some finite time length.

The output is stretched in time over the number of samples which make up the length of the unit impulse response. Since it is not possible for the output of a real-time system to respond prior to some input, any shift of the input frequency components must be a time delay of an amount sufficient to spread the input over the response time. The time delay of the shifted components becomes a phase lag in the frequency domain.

This phase shift, or time delay, must be reversed to complete the deconvolution.

Unfortunately, using the deconvolution method described earlier, there is no way of defining the phase spectrum of the unit impulse response. All phase components are discarded in the autocorrelation. A given amplitude spectrum may be combined with a great number of different phase spectra yielding an equally numerous, but not infinite, number of wavelets having a finite response time. These can be classified (**Fig. 6.10**) into one of three types.

The classification includes one output wavelet corresponding to the least possible delay of the input frequency components. There will also be a maximum-possible-delay wavelet. All other wavelets will have some intermediate combination of phase lags.

The minimum-delay wavelet has several unique properties. The minimum shift ensures that most of the energy will be concentrated at the front end of the wavelet; it has had the smallest possible disturbance of the input within the constraints of time, length and amplitude spectrum shape. Hence, the wavelet rises quickly to maximum amplitude and decays rapidly with time. The wavelet will be as sharp as possible consistent with the bandwidth.

This type of response is common to many systems, including (at least in theory) the typical seismic system. Therefore, the other assumption made in deconvolution design is that the seismic system response wavelet is a minimum delay signal.

This assumption leads to a number of good things. The inverse of a minimum-phase wavelet is also a minimum-phase wavelet. Minimum-phase wavelets are stable convolution operators, which means the operator will converge quickly so short operators may be used. Mixed-phase or maximum-phase operators do not behave so well. For instance, the inverse of a simple mixed wavelet 1, 2, 1, can be found, by polynomial division into 1, to be 1, -2, 3, -4, 5, -6, ...which diverges and cannot be truncated logicially.

Finally, convolution of a minimum-delay operator with a minimum-delay signal produces minimum-delay output. The output of the convolution of wavelet samples with operator samples is n + m = 1 samples long, so the deconvolution process must stretch the output response over a wavelet of greater length than the input although the amplitude of that output may decay quickly. Convolution produces output which has a phase spectrum equal to the sum of the two signals, so the phase lag of the output will be greater than that of either input signal. But the signal is also longer.

The advantages of the minimum-phase system properties make it very desirable for use in seismic processing wherever the assumptions make a reasonable fit to the real situation. Fortunately, they usually do, so the assumption of a minimum-phase condition rapidly become accepted for most seismic work. Every autocorrelation corresponds to a unique minimum-phase wavelet of specified length so, from it, a stable, minimum-phase inverse deconvolution operator can be designed.

If the minimum-phase assumption is valid, the result of the deconvolution will be a high-amplitude, minimum-phase wavelet; which will be the closest possible minimum-phase approximation to a spike.

If the assumption is not valid, the spectrum of the output will, in any case, be whitened which may improve resolution. But the "minimum" phase of the operator will still be added to the mixed phase of the wavelet to yield another mixed-phase wavelet. Some mixed-phase wavelets have their energy spread uniformly over the length of the wavelet, producing a very poor signal as the output of the operation.

The definition of a minimum phase shift should not be confused with a necessarily small phase shift. The amount of the actual phase shift can be substantial. For example, many minimum phase seismic wavelets closely approximate a dipole function (1, -1) which has a constant 90° phase shift. This is still generally considered by interpreters to be a favorable situation, since the resulting very distinct reflections can be mapped with relative ease.

In summary, the simple inverse of a wavelet can be obtained by merely dividing the operator into 1.0, a unit spike, using polynomial division techniques. This is the basis of the Z-transform. In most cases, the division does not come out even but is a continuing series and, if the terms in the series grow as the series lengthens, the operator will be unstable. For this reason "least-squares" techniques have been developed to yield the best possible approximation of the inverse with an operator of some designated length. The inverse operator convolved with the output will restore each wavelet response to a spike within the frequency capabilities of the operator. By nature, the operator is minimum-phase and if the impulse response is minimum-phase the output will be minimum phase and front loaded; i.e.; the first point of the unit impulse response in the input will be replaced by the output spike in the output of the deconvolution process. As convolution with the inverse operator proceeds, the initial output spike should be followed by a series of near zeros, until the next unit-impulse-response wavelet is encountered. The filter must rely heavily on the assumption that the response wavelet is minimum-phase, because it must output the initial maximum value before it can examine the rest of the wavelet. The operator always works with past information to develop the current output point. Thus, the simple least-squares inverse deconvolution process may be termed a predictive operation. This is also known as a physically realizable operation, because it obeys our concept of time; time moves only in one direction and nothing can come out of the system before something goes in.

Operations with recorded data have the advantage that the entire time series is preserved as a "snapshot" of time so the information need not be acted upon the instant it occurs, i.e. real-time operations. Nor must the examination proceed in only one direction. A captured time series can be examined at leisure, so it is possible for a digital filter to "look ahead" to subsequent time values and take action accordingly. This is of particular value in mixed-delay response. The filter can act on the entire event to determine what the output should be at a particular time.

The process of simple inverse deconvolution, using the basic statistical approach, leans heavily on several assumptions. When the assumptions are valid good results may be expected and poor results probably are symptomatic of an invalid assumption. Rarely it may be possible to select some alternative method yielding better results.

The minimum-phase basis for deconvolution has served the industry well, providing a major advance in data enhancement and resolution not possible to achieve readily by any other means. But minimum-phase operations have their shortcomings. Levin (1971) has shown that zero-phase output can provide improved resolution. Nothing in this industry remains static, and progressive interpreters and data processors, always striving for improvement, chafe at the inherent limitations of minimum-phase processing, and try to find ways to overcome them.

The problem is that minimum-phase, as defined, in no way ensures a small phase shift; it is only the least possible phase shift for any specified amplitude spectrum. Often the amount of phase shift is on the order of 90°. A better deconvolution process would reverse all the phase shift of the system wavelet back to zero, as well as modify the amplitude spectrum. But to do this would require the complete definition, including phase as well as amplitude, of the total system response wavelet.

The phase of the recorded output is the sum of the earth-signal phase and the system phase. There is no known method of separating the component parts without additional information. Spectral analysis would provide the desired information if the unit impulse response could be defined precisely, but it cannot. The phase of the instruments and geophones can be determined, but the effects of coupling to the earth and the response of the earth itself cannot. The Vibroseis ®, Maxipulse®, and Sosie ® methods all define the phase spectrum of the input signal and this shows in the generally better processing results that can be obtained compared to the simple impulse system.

One of the first attempts at a solution in marine work was simply to record the waveform resulting from the source impulse. This can be done wherever the water is deep enough to allow the full length of the source wavelet to be recorded before the first sea-bottom reflection returns. The inverse operator corresponding to the recorded wavelet can be calculated quite readily. However, note that the inverse operator obtained in this manner may not be minimum-phase. Most likely it will be a two-sided mixed-phase operator, one which can be calculated only from recorded data. The reason for this is that the zero-phase measurement is referred to the center of the response wavelet. All zero-phase output from band-limited data processing has an anticipatory period equal in length to the decay. The anticipatory portion is non-physical, i.e., output occurs before input, hence it can be obtained only from recorded data since it is necessary to produce output samples in advance of the zero-phase wavelet reference time.

This principle may appear to be violated by real-time field correlators, which for some operations produce zero-phase output concurrently with recording. In practice, real-time refers only to correlation operations; the output is delayed by an amount equal to the length of the operator.

The operation as described is a relatively simple one, but it is incomplete, for it solves only for the signal input and recording elements of the system in marine work, and ignores the effects of the earth. On land, it is not possible to record the signature without some interference from reflections which are received from shallow beds almost simultaneously with the shot. Such limitations do not stop some data processors, who are nothing if not pragmatic, from attempting some related procedure, often with considerable success.

One solution proposed by some data processors is to select a wavelet from some portion of the seismic trace over which they believe a lone wavelet is recorded. The marine equivalent, where the signature is not recorded, is the water-bottom reflection. Several traces are then stacked to obtain an average and the result is used as a typical wavelet. The procedure may yield some improvement over the minimum-phase approach, but rarely, and only under the most unusual conditions on land, is the wavelet not distorted by near-surface reflections. And, in both land and marine data , rarely is the wavelet shape not distorted by interference from earlier or following events. Hence, a solution via this method is rather imperfect at best.

Homomorphic Deconvolution

The basic assumption that the input spectrum is white, or flat, as used for minimum-phase deconvolution, leads to the conclusion that the irregular spectrum of the recorded seismic trace represents the spectrum of the system amplitude response. Yet experience, the results of experimentation, and the output of other systems, such as Vibroseis, show the amplitude response of a typical seismic system to be a smooth, slightly skewed, bell-shaped curve not unlike the response curves published for each set of seismic instruments. It is also easy to demonstrate, from the spectrum of any series of reflection coefficients, that the amplitude and phase spectra of any sedimentary sequence have a characteristically jagged pattern. See (Fig. 6.11).

Thus, the spectrum of a recorded seismic trace which usually has the appearance of a jagged, bell-shaped curve, can be viewed as having two components, a slowly-varying bell-shaped curve produced by the system response, modulated by an irregular jagged sequence corresponding to the spectral components of the reflection coefficients. If the amplitude spectrum were to be viewed as a time series instead of a frequency series, the reflection coefficient components might be considered as high-frequency signal, while the system response might be

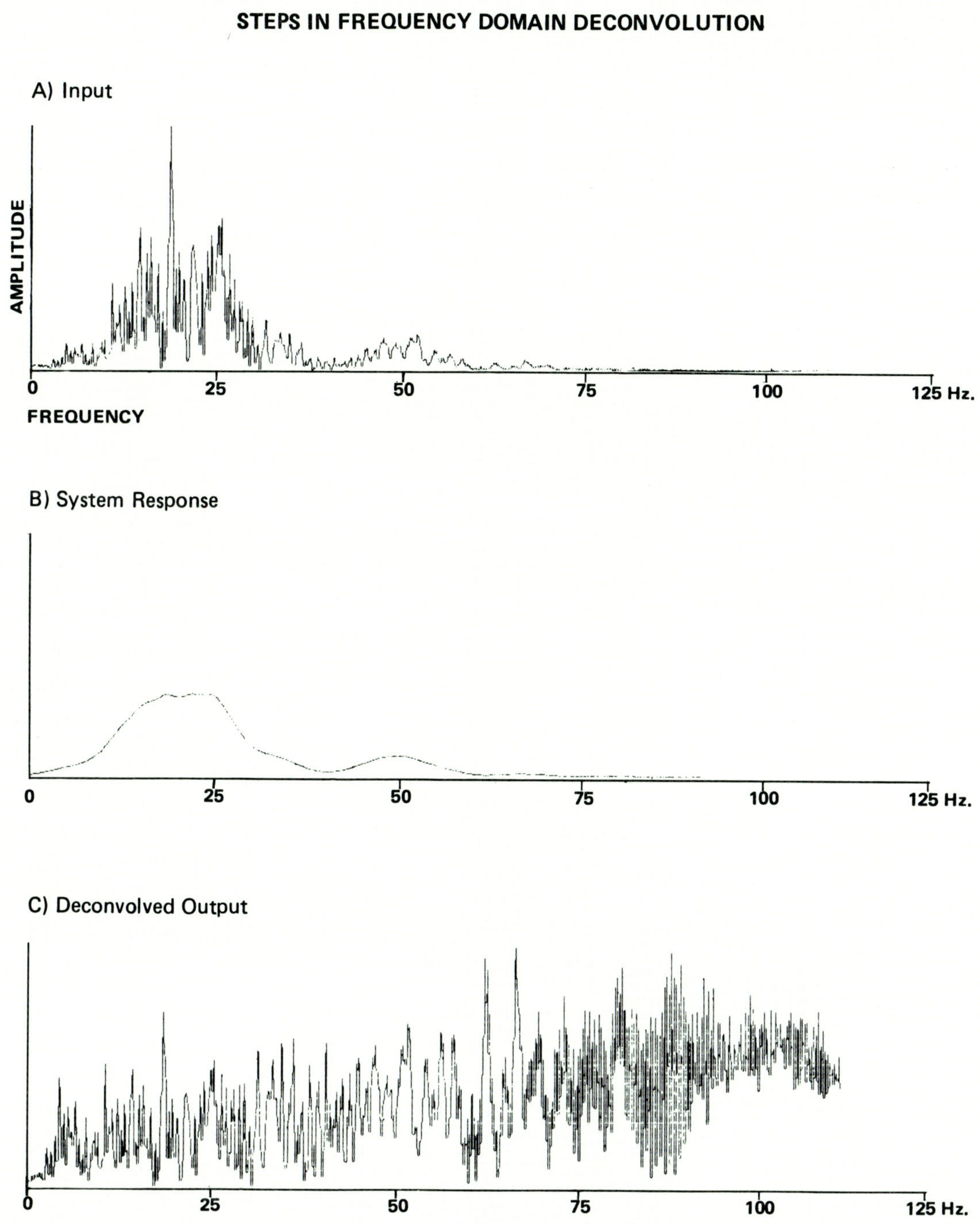

FIGURE 6.11

Frequency domain deconvolution.

considered as low-frequency signal.

This somewhat pragmatic framework forms the basis for an extension of Fourier transform theory to achieve homomorphic deconvolution. The discussion above uses just the amplitude spectrum for illustration but true homomorphic deconvolution is performed on the complex numbers that come out of the Fourier transform and thus phase is also involved.

To avoid confusion, a new set of terms is defined to describe equivalent components in a new domain obtained by a forward transform of the frequency spectrum. For this domain;

Cepstrum	=	Spectrum
Quefrency	=	Frequency
Saphe	=	Phase
Lifter	=	Filter
Zerth	=	Cycles per Nyquist

In the basic model of a time-domain system,the output signal is the result of convolution of an input signal with the system response, In the frequency domain, it is the *Product* of the amplitude spectra and the sum of the phase spectra of the two signals. But the Fourier transform assumes any time-domain signal to be the *Sum* of frequencies over some range. Higher frequencies added to lower frequencies yield the total signal. Hence, a Fourier transform of the *Spectrum* will not meet the basic assumption of summation. However, if the spectrum is converted to logarithms it will be the *Sum* of the logarithms of the two amplitude spectra.

Returning to the amplitude spectrum alone, a Fourier transform of the logarithm of the amplitude spectrum of a recorded signal places it in the Quefrency domain. The amplitude of the several frequency components in the amplitude spectrum will be measured and mapped in the quefrency domain on the amplitude cepstrum as quefrency components. The basic composition of the amplitude spectrum, in which the short-term oscillations corresponding to the reflectivity series are modified by the long-term component of the system response, will yield an amplitude cepstrum split into two major divisions. The low quefrencies are largely derived from the system component, while the high quefrencies correspond to the earth signal. Amplitude deconvolution can therefore be accomplished by applying a low-quefrency lifter, to eliminate the low quefrencies corresponding to the system. This is accomplished by merely zeroing out the unwanted quefrency components. A reverse transform of the remainder, back into the frequency domain, and calculation of the antilogarithms of the spectrum, yields the amplitude spectrum of the earth signal only, the system component having been liftered out.

Notice that the above operation merely modifies the amplitude spectrum. The operation has provided no new information which would permit the phase-lag of the system to be reversed. If the system is minimum phase, its phase can be determined from the amplitude spectrum of the system, since there can be one, and only one, minimum-phase spectrum for any given amplitude spectrum. The next step would be therefore, to low-pass lifter the amplitude cepstrum in order to recover the quefrencies that were rejected in the first stage, take the reverse transform of the low quefrencies, and calculate the antilog of the reverse transform. This should yield the amplitude spectrum of the system response. If the system is minimum phase, the phase spectrum of the system component can be computed, from this amplitude spectrum and subtracted from the phase of the recorded signal. Thus the operation using complex numbers can be broken into two parts, an amplitude deconvolution and a phase deconvolution.

When all the assumptions are met in practice, the homomorphic method is an elegant and effective form of deconcolution. Unfortunately, the procedure does not always work as well in practice as in theory. The system may not be minimum phase, and therefore fails to meet the basic assumption. Additive noise can distort both the amplitude and the phase spectrum, although the method should work well with stacked data, from which much of the noise is eliminated. If the noise is random, determination of the amplitude portion of the deconvolution will still be quite effective since the noise will not modify the shape of the amplitude spectrum. Phase will however, remain distorted.

A substantial advantage of this method is the fact that the system-response-wavelet need not be defined; it is obtained as a by-product of the operation. Practically all other forms of signature deconvolution or wavelet processing require the wavelet to be defined which can be, and usually is, a very difficult task.

Frequency Domain Deconvolution

Just as filtering can be applied equally well in either the time domain or the frequency domain, so can the liftering be done in either the frequency or the quefrency domain. The same routine used to design the time-domain frequency filter can be used to calculate a frequency-domain lifter by merely substituting Zerth for Hertz in the design parameters, and applying the resulting lifter operator to the logarithm of the amplitude spectrum. The determination of the proper quefrency cutoff is a problem but can be determined by examination of the cepstrum. This is a time consuming process but it may be necessary until some experience is obtained in estimating an acceptable low-quefrency cut.

Another less elegant but moderately effective solution in the frequency domain can be fitting a secular curve to the amplitude spectrum to obtain the low quefrency shape. (**Fig 6.11**). It will then be necessary to calculate the inverse of the system estimate and then multiply the original signal spectrum by the inverse to produce the deconvolved amplitude spectrum.

In all cases, the phase can be treated in the same manner as described for the homomorphic method.

In conclusion, an effective method for determining the true signature of the system, and returning the recorded signal to the amplitude and phase of the earth signal, remains one of the most difficult problems in seismic signal processing. Stacking is the most effective procedure to remove the phase distortion caused by random noise. But the properites of systematic or coherent noise, and of the system, at best, can only be approximated.

General Discussion

To the orderly mind of the mathematician or engineer, seismic data processing must sometimes appear to be a chaotic procedure, stumbling incoherently from one unwarranted assumption to another. But having the luck and the unexplained skill of an idiot savant, geophysicists and processors have managed to leave a trail of fortuitous successes based on procedures which can barely be justified, in view of the unreliable data, the almost totally pragmatic assumptions and the unsubstantiated theorems used to develop most process routines.

Justifying much of the practice and proving much of the published theory, is a slippery exercise, perhaps primarily because of the wide variability of conditions. It is probably reasonable to state that somewhere, sometime, every postulate advanced holds true, but probably in a greater number of other cases it can be shown to be not true.

One of the shakiest pillars of process theory is that of the random nature of geological processes. Einstein once reportedly said "God does not play dice", and textbooks refer constantly to the cyclic nature of depositional processes. Yet, the basic developments of deconvolution are based on the fundamental assumption that the earth spectrum is white, and indeed in many, or even most, cases this assumption works reasonably well. Yet obviously this assumption cannot be true in the absolute sense, for if the vertical sequence were truly random, no two vertical sequences would be alike, and then there would be no horizontal continuity within the section. If such were the case, there could be no mapping of subsurface horizons. Truly random series have no correlation, one to the other, but only rarely is there no lateral correlation of one seismic trace to an adjacent one.

A better description might be that supplied by Silvia and Robinson, in their book "Deconvolution of Geophysical Time Series in the Exploration for Oil and Natural Gas" to the effect that any sedimentary sequence is made up like a "prime fraction", or a natural number, which has a unique numerical sequence with no internal correlation but nevertheless is precisely defined and easily repeated. Carried to the same extreme as the random example, above, this would not be quite right either since it would allow no lateral variation in the geologic section; every location would have an identical apparently random sequence. In fact lateral changes do occur and, indeed, for stratigraphic interpretation it is important to understand that the depositional changes which occur vertically with time in any geologic section, may also transpire laterally, and scanning a section either vertically or laterally can involve crossing geological time lines.

Another consideration is that over short segments within any natural or pseudo-random sequence it is quite common for periodic or other numerically interesting sequences to occur.

A single frequency in time is a periodic event which will be represented by a spike on the amplitude spectrum. Any periodic event in time will produce a spike on the amplitude spectrum, at the dominant frequency of the event, and a periodic event on the autocorrelation function. Multiple spikes are symptomatic of several concurrent periodic events in the time series. One periodic event is easy to see in the signal but as the number increases it becomes increasingly difficult for the eye to separate them although they may be clearly defined on the amplitude spectrum.

Frequency filtering, by eliminating some of the interfering periodic components, leaves the rest more visible and any spectrum, whether of a random sequence or not, if filtered to a narrow band (in the extreme case to a single frequency), will be periodic in time.

The only possible spectrum for a purely random signal is absolutely flat. The pattern of spikes on the spectrum of any seismic signal is demonstrative of the periodic elements in the geologic section. Like most geologically related phenomena the condition varies over the face of the earth but rarely is the section, in the main, truly random. Randomness most often accompanies the high frequency components and increasing the bandwidth almost always increases the visual evidence of randomness and probably, almost as often, the measured randomness for the geologic section.

The point of all this is that while the assumption of the geologic section being laid down by a random process may be very nearly correct it is true only in the broad-band sense.

Theory, formulated on the random assumption, may be correct for the full geologic section where the earth system is complete and responds in the broad-band manner.

However, when the appropriate measures or techniques applicable to the full system are applied to only a portion of the information, whether it be a portion of the total sequence in time, or a portion of the total spectrum in frequency, the random-response assumption and/or the action may not be valid. Periodicity may be present within the restricted portion of the total thus destroying the basic assumption.

AUTOCORRELATION FUNCTIONS

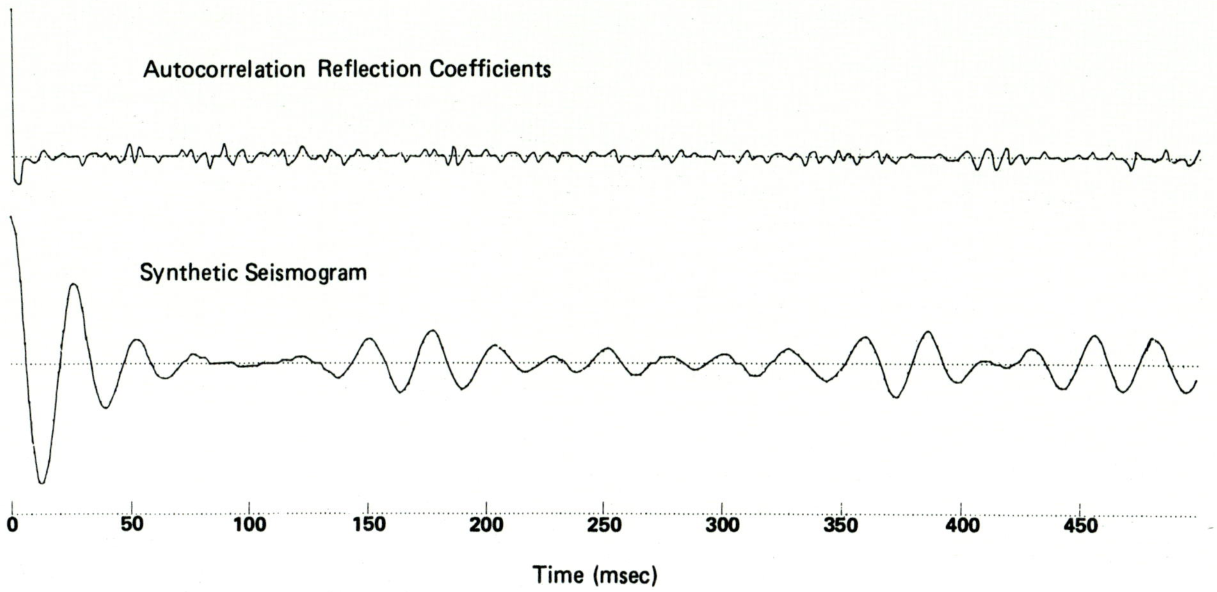

FIGURE 6.12

Peaks near .150 and .350 sec on the autocorrelation might be indicative of multiple reflections on a real field record but in this case the autocorrelation is from a noise-free synthetic which has only primary events. Autocorrelation of the reflection coefficients shows a typical broad-band response.

A good example of the potential for an erroneous assumption is related to the detection, analysis and attenuation of multiples. A multiple, by definition, is periodic. If the period is short, as would normally be the case for reverberations, it may be detected as a spike on the amplitude spectrum. Longer period events are not readily observed on the spectrum and the method most commonly advocated for their detection is the autocorrelation function.

Correlation of a signal with itself removes all phase and consolidates or catalogues all periodic occurrences in regular sequence of increasing period in time. Not only is the autocorrelation often considered to be a definitve indicator of the presence of multiples but the absence of multiple events on the autocorrelation is also believed by some to indicate an absence of periodicity in the corresponding signal.

A very real danger in that assumption is the possibility that uniformly spaced real events may be present on the section and these will correlate to produce apparent multiples on the autocorrelation. (**Fig. 6.12**) shows the autocorrelation of a record with definite periodic events near 150 ms and 350 ms. By the standards defined for multiple detection this is a definite indicator of multiple sequences. Deconvolution by any one of the standard techniques may be used to yield results similar to the upper portion of 6.12. But in this example the autocorrelation of the record is of a synthetic, made directly form the band-limited version of calculated reflection coefficients and, therefore, it can have no multiples. This phenomenon is not uncommon; in many areas the earth signal is not white but shows the effect of some cyclic depositional process particularly when defined by narrow-band signals. The true nature can be determined, as in the example, from the autocorrelation of a simple multiple-free synthetic seismogram in a representative location. If the output contains evidence of periodicity, as does the example, it is very likely the result of coincidental spacing of real geological events within the bandwidth of the signal. Hence if a deconvolution operator were designed to remove multiples over a length which includes these events legitimate information would be destroyed by the process. Furthermore, the inclusion of such an event in the deconvolution design may yield an operator which has an inferior action upon the true unit-impulse response with the result that true events may be seriously distorted or achieve very little improvement in the resolution.

There is some evidence that, here again, serendipity may work in favor of the interpreter and processor. The effects of pre-whitening the autocorrelation of the signal before designing the inverse operator were explained as making it less efficient, and not able to make the deconvolved spectrum completely white, which yielded a more acceptable (if less effective) deconvolution.

Another consideration is the effect of operator length. Sample theory shows that the frequency increment in the spectrum becomes large when a short time signal is used for analysis. Operations can be made faster and cheaper when short operators are used since they take less time to convolve . Unfortunately (or fortunately as it turns out) because of the coarse frequency sampling, the short operator samples the spectrum very coarsely and cannot operate on the spectrum with enough detail to make it completely white. Instead it leaves it looking something like a mildly restored version of what it was before.

Multiple Reflection Suppression

Standard inverse deconvolution is occasionally used, with variable success, to eliminate relatively short-period multiple reflections and/or ghost events. The unwanted event generally lags after the desired reflection events by some short constant time and will be strong enough to be detected in the autocorrelation. Effectively, the procedure to eliminate it is based on extending the minimum phase wavelet assumption to include the multiple event as part of the entire unit-impulse response which is now a mixed-phase event. The assumption is not unreasonable where the system input produces a multiple or ghost response at a constant time delay. Under normal conditions, because the trailing event is a multiple, it will be of lower amplitude than the primary and follow very closely after it. Standard minimum-phase deconvolution should then reverse the effect of both the unit impulse response and the multiple.

PREDICTIVE CONVOLUTION

An excellent paper by Peacock and Treitel in the April, 1969 issue of Geophysics, describes the development of longterm predictive-deconvolution operators for the elimination of multiple events and reverberations, particularly those common to marine work. The predictive-deconvolution operator-design technique they describe endeavours to leave the primary reflection response untouched but makes a deliberate attack upon the multiples. In the routine, an operator is designed to eliminate events which repeat at regular intervals subsequent to the original reflection response. To accomplish this the length of the reflection-response wavelet must be defined as well as the length of the gap between primary and multiple reflections because within this period no action will be taken. The time gap leads to another term " gapped operator " sometimes used to define the procedure.

The assumptions are similar to those for the standard deconvolution but, in addition, specifically assume that; a) the multiples recur periodically, b) they decay in amplitude at a constant rate c) a uniform polarity relation exists (usually alternating with each repitition); and d) that the response wavelet is identical in shape for each event. These are fairly strict conditions for something as ill-defined as seismic data. But the process works fairly well.

As usual, a model is the basis for the development and is perhaps the best way to illustrate the procedure and the

FIGURE 6.13

The deconvolution operator for predictive deconvolution is merely the first two terms of the system model.

FIRST AND SECOND ORDER MARINE REVERBERATIONS

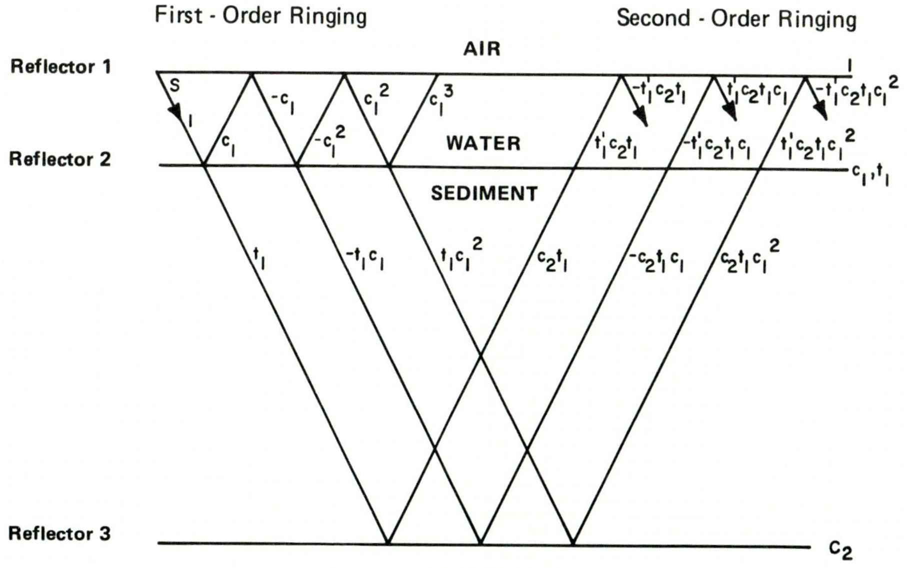

DECONVOLUTION OPERATORS

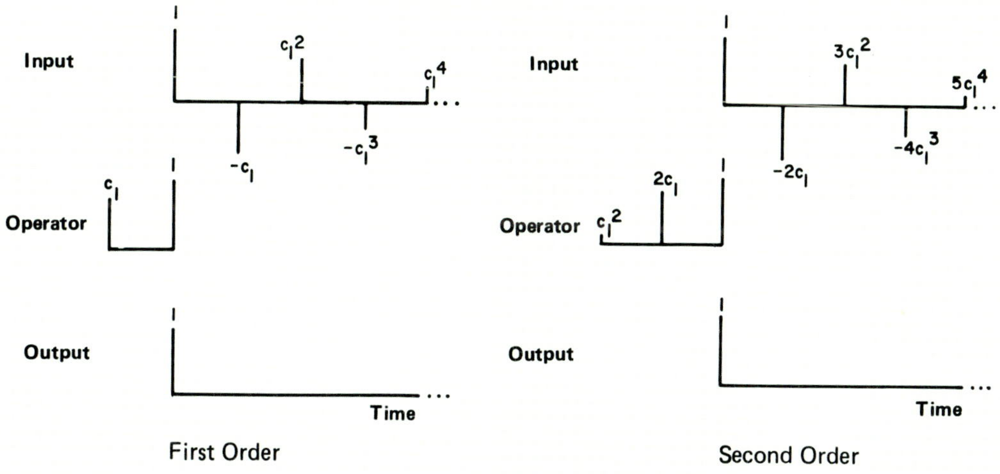

basis for the assumptions. The method is most commonly applied to marine work.

A reflection response to a reflection coefficient R_c (usually the water bottom) is followed at some period of time by a surface multiple. If it is a simple surface multiple, (and this is also a basic assumption) it will be identical in shape to the primary response but will have reversed polarity and amplitude of $-R_c^2$. The procedure assumes a reflection coefficient at the air/water interface of 1.0, which is a reasonable assumption for practical purposes, deriving from the large difference in density of air and water. The polarity reversal is caused by the negative ratio at the surface seen by the upcoming wave. The scaling factor R_c occurs with each bounce from the ocean floor. The simple surface multiples may repeat several times, each time reversing polarity at the surface, and each time scaled by R_c.

The desired deconvolution operator is one which will leave the original reflection response unharmed but destroy all multiple reflections of it. The purpose of the operator is not to convert the reflection response to a spike, as is the case with the simple deconvolution process, but merely to eliminate all multiples of the primary reflection. The sequence of multiples can be modelled by first computing a power series of alternating polarity reflection coefficients spaced at the period of the multiple. If the input (i.e. the energy source) is scaled to 1.0 and located at time zero the initial unit impulse response from the first reflector at the ocean bottom will have an amplitude of R_c. This would be followed by the first multiple, at $-R_c^2$, then the sequence R_c^3, $-R_c^4$, R_c^5, $-R_c^6$ ...., with each term spaced a constant interval in time, t. (**Fig. 6.13**).

The first two terms of the series 1.0, R_c, spaced by a series of zeros over the time interval of the multiple, become the deconvolution operator. The seismic field trace does not record the initial input so the initial event recorded is the water-bottom reflection at a reflection time, t.

Convolution of the first two terms with the field trace will pass the first water bottom reflection unharmed and, if the defined conditions are met, will eliminate all simple multiples from the output. Each bounce of the water-bottom multiple acts as a new source of energy and will be followed by a complete reflected section at the polarity, scaled by the relative amplitude, of the multiple. These multiple reflections should also be eliminated in the process.

Again, in practice, the least-squares approach is used to design the best possible operator. Regardless of the problems a least squares operator can be designed for every case that will, subject to the validity of the assumptions, remove from the total signal all multiple events which lie at the time period specified for the multiple response and the total length of the operator, as specified by the autocorrelation function.

Should normal spiking deconvolution be required, it is a relatively simple matter to perform this operation in a subsequent pass. It should be possible first to convolve the two types of deconvolution operators to accomplish the process in a single pass but, in practice, it seems better to use separate operations.

The above describes the situation for first-order multiples. Second-order multiples are developed from energy reflected from the sub-surface layers striking the surface of the water and acting as a new energy source for a new series of reverberations. The situation here becomes somewhat more complicated and the amplitude value of each reverberation is dependent upon the model. For a simple model, in which energy is returned from a single sub-surface reflector, each first order water reverberation striking the water bottom will reflect some energy and transmit the remainder into the earth to be returned by the reflector. Each second-order reverberation will be reinforced by a reflected subsurface event coming from the corresponding first-order reverberation, and each subsequent even will be further reinforced. The effect of this is to multiply the coefficients of the first-order multiples by a factor equal to the number of the event. For example, after the primary the next following second-order multiple is the second event and will have energy equal to two times that of the equivalent first-order reverberation, the next second-order multiple will be the third event and have energy equal to three times the equivalent first-order event and so on. Again, because of the complexities of the real section, the de-reverberation operator can best be derived by least-square techniques from the autocorrelation function.

Problems in Practice

The predictive doconvolution method is, in theory, ideally suited to the removal of long-period multiple reflections typified by the water-bottom multiples frequently encountered in marine work. In practice, it is often not very effective, as the result of two fundamental and contradictory differences between the theoretical model and the real system. In the first place, the theory requires a precisely constant time interval to be maintained between successive multiples. The predictive deconvolution operator uses information from each impulse encountered on the trace to generate a predicted multiple occurring at an exact time interval later. In real data, this condition nomally is met only on the very near seismic traces. At greater distances, the normal moveout changes the time interval. When this happens, the predictive deconvolution process may add, rather than remove, noise. Correcting for normal moveout before deconvolution may solve the interval problem, but then the response wavelet at each level is stretched in a non-linear manner, making each successive event different and not well defined by a simple predictive operator.

The second problem is related to gain recovery. Theory requires the amplitude of the successive multiples to be in strict geometric ratio. Only in the original true-amplitude signal is this likely to occur. A variable gain compensation will tend to destory this ratio. This may not be a serious problem if a uniform geometric amplitude ratio is maintained between multiples. This might be the case if an exponential compensation gain recovery curve has been applied. Even if the multiples have all been brought to the same amplitude the operation is still valid with 1, 1 as the two point operator.

The problem may not end there and other factors will often foil efforts to obtain satisfactory results. Often a simple water-bottom model is not adquate to describe the real situation. Typically, the water bottom is not a simple reflection interface represented by a single spike reflection coefficient but rather a complex transition zone. Then, instead of the replication of the unit impulse which normally occurs when convolved with a sharp-boundary coefficient, the signal returned to the surface is the convolution product of two waveforms. This, in turn, is then reflected from the surface to be convolved once again with the bottom response. Convolution of two signals, having sample lengths of n and m respectively, produces output n+m-1 points long. Thus, successive multiples quickly degenerate into lengthy, complicated, mixed-phase responses. (**Fig. 6.14**) (Note the example also illustrates still another problem, that of a ghost event due to the collapse of the bubble produced by the explosive.) Multiples from a hard bottom usually do not suffer serious distortion of the waveform.

Deep water is often noted for severe multiple problems. The multiple event is often much stronger than the primary reflections occurring at that time. The reason is the different rate of attenuation. During the time span between the water-bottom reflection and the first multiple, the primary signal travels through the layered sedimentary section, continuously returning reflected information to the surface, and in so doing follows the normal energy decay pattern. Meanwhile, the multiple travels only through water and suffers practically no attenuation in the period between each multiple reflection. The greater the time interval between primary and multiple, the greater the contrast in amplitude that may be expected. When the primary-to-multiple amplitude ratio is high it

FIGURE 6.14

A seismic section showing severe water bottom multiples and a short reflection from a bubble collapse of the input shot.

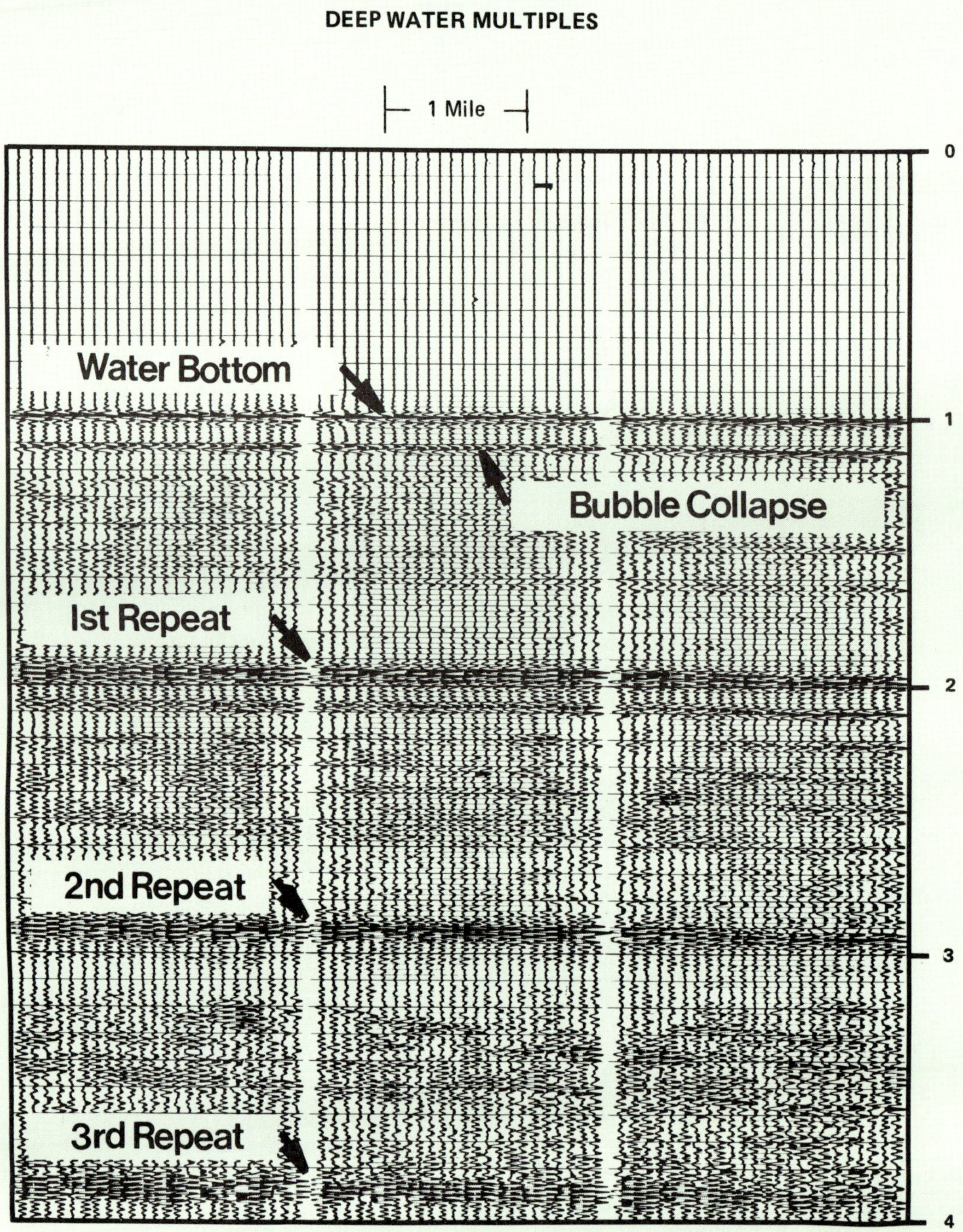

may not be possible to eliminate the evidence of the multiple by stacking.

These several effects usually preclude resolution of this most serious of problems affecting marine work. Stacking is one of the most effective methods, provided there is sufficient velocity contrast between multiple and reflection paths to provide adequate differential moveout between primaries and multiples. Increasing the distance between shot and near geophone can help by increasing the differential moveout between reflection and multiple. Muting of the near traces before stacking may also help somewhat. It should be noted that broad-band data can be helpful, since there is greater separation between multiple and reflector, i.e., events are all shorter in time.

Considerable effort continues to be made toward the development of new techniques, some in the field, some in the processing center. A successful field technique, which requires some special data processing, is the super-long source array (SLAG) which generates a more effective stack.

Unfortunately, some of the data processing solutions offered in the data processing market are merely cosmetic, improving the appearance of the seismic section but possible misleading the user into a false sense of security and resulting in erroneous interpretation. An example of a fairly common technique of this type is simply to mute surgically, or to attenuate severely, the obvious expression of the successive multiples of the water-bottom reflection prior to stacking. The resulting stack is often much approved in appearance, but may be extremely misleading for the following reason. Water bottom multiples, distorted or not, are never limited to the mere repetition of the bottom. On the contrary, each water bottom multiple acts as a new primary energy source each time it bounces back from the water surface, so that not only the water bottom is repeated, but the entire reflection sequence follows each multiple of the water bottom. A marine recording may therefore be considered as the sum of several repetitions of the primary record, alternating in polarity, each spaced at a delay equal to the travel time within the water layer. This model remains admirably suited to predictive deconvolution theory, and subject to the constraints described earlier, can be correctly treated by the method.

A more accurate model would be to convolve the series of multiple coefficients with the entire primary signal sequence.

Stage of Application

Marine reverberations which occur in the water layer, are merely a special case of the multiple problem where the multiple reflecitons are more closely spaced, generally to such an extent that the multiple reflection occurs within the time span of the unit impulse response.

It is interesting at this point to review the stage in processing at which deconvolution should be applied.

Field seismic records are usually distorted by the geometrical effect of normal moveout. Referring to a layered model, reflections recorded from two reflecting horizons directly below the energy source will be separated in time equal to the vertical interval travel time between the two events. At the far trace the same two horizons will both arrive later than the near trace due to the increased length of the travel path. The far reflections travel slightly different paths and, due to the path geometry, the difference in travel time between the two reflections on the far trace will be less than the near traces. The net effect of this is a time delay on the arrivals of all events on the far trace. But the shallow event is delayed more than the deeper event thus compressing the entire section in time on the far traces. A time compression causes an increased in frequency so the spectrum of the earth signal is shifted toward the high end of the spectrum.

In every case the reflection coefficient at the boundary triggers the unit impulse response. The geometric compression of the reflection coefficient spacing has no effect on any given impulse response, since the entire response

FIGURE 6.15

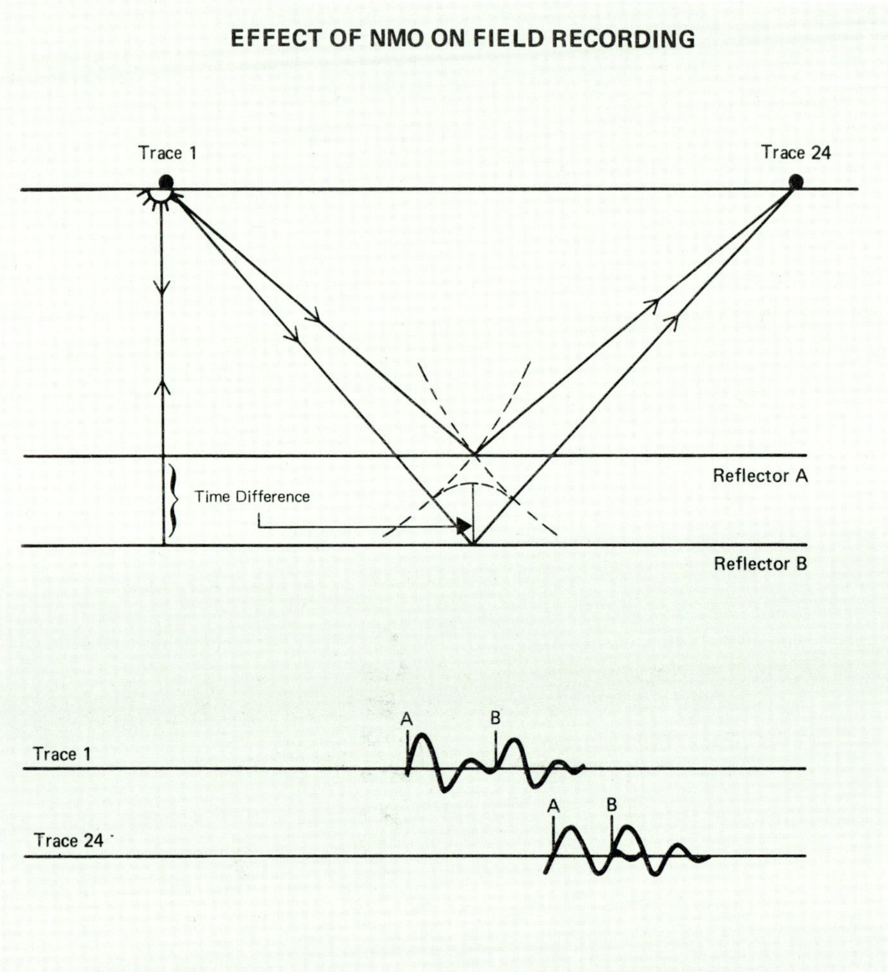

for any reflection travels over the same path. The only effect is that the following deeper reflection will come in slightly closer to the upper reflection than it will on the near trace (**Fig. 6.15**).

The net effect of normal moveout is to shift the spectrum of the earth signal but to leave the spectrum of the unit impulse response untouched.

Application of a proper normal moveout correction should return the earth signal to its proper position but will apply a variable stretch to the unit impulse response on the far trace. From this it may be concluded that optimum results from a deconvolution process designed to return the wavelet to an approximation of the reflection coefficient should be obtained when applied to the trace before normal moveout correction has taken place.

On the other hand, deconvolution before stacking is almost certainly affected to some degree by the presence of noise in the data. Stacking will remove a substantial part of the noise leaving a much more certain definition of the signal components. Errors in normal moveout calculation and/or in statics tend to attenuate some of the components, the effect increasing with higher frequency, where the short period is more susceptable to small time displacement errors. Accordingly, a second deconvolution after stack, to normalize the spectrum, is also good practice.

7 VIBROSEIS

Vibroseis

The Vibroseis seismic energy source system was developed by the Continental Oil Company of Ponca City but may be used by other operators under a license arrangement. It differs from practically all other forms of seismic reflection exploration in that the signal which is sent into the ground is not a pulse but is a carefully controlled sweep frequency. (Fig 7.1) The input signal is long, usually 7 sec or more, and normally ranges smoothly from one end to the other of some defined frequency band, as for example from 17 to 67 Hz.

The signal is impressed into the ground by one or more vibrators driven under control of a sweep signal generator. The signal can be chosen so that the band which is emitted corresponds closely to the natural response of the area. Thus, those frequencies which would in any case be attenuated need not be transmitted. The amplitude spectrum of the particular sweep signal may differ very little in bandwidth from that of the unit impulse response. The important difference in the characteristics of the input lies in the well defined phase components of the sweep signal. This makes a tremendous difference from the processing standpoint because the input signal now has both the amplitude and the phase spectrum well defined. The amplitude spectrum is white (flat) over the sweep bandwidth and the phase spectrum is a smooth acceleration function. In other words, the conditions of the transmitted signal now meet the basic assumptions of classic signal processing theory in that a known signal of defined frequency, amplitude, and phase, has been transmitted, and is subsequently received contaminated by noise. The problem is then one of locating and removing the defined signal from the background noise.

The transfer function for this system is a sweep signal instead of the unit impulse response wavelet. The process may be modelled by a convolution of the sweep signal operator with the reflection coefficients of the layered earth model. In terms of frequency domain operations, the output is the product of the amplitude spectrum of the earth signal multiplied by the amplitude spectrum of the sweep, and the sum of the phase spectrum of the earth signal plus the phase spectrum of the sweep.

FIGURE 7.1

The Vibroseis seismic energy source, developed by the Continental Oil Company, is long, usually 7 sec or more, and normally ranges smoothly from one end to the other of some defined frequency band, as for example from 17 to 67 Hz.

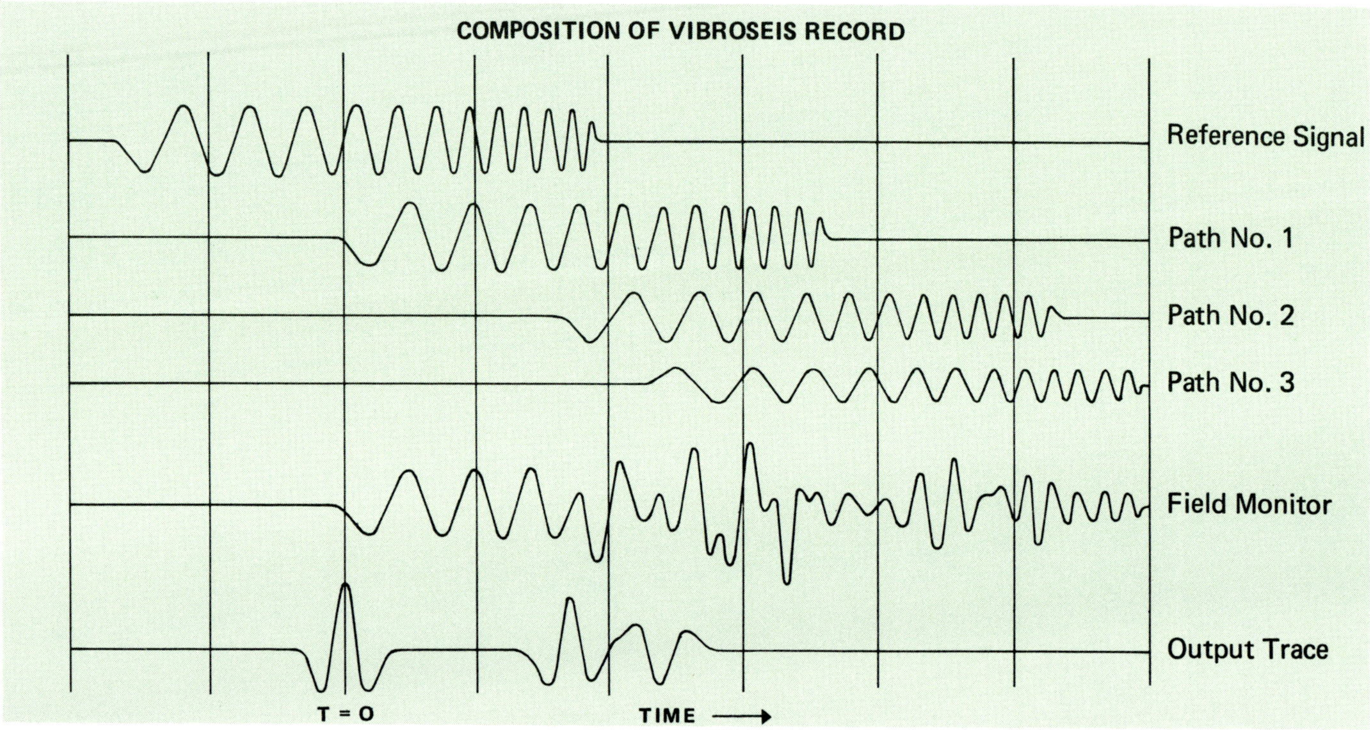

This uncomplicated model responds well to relatively simple treatment. Crosscorrelation has been mentioned as a good method to extract a known signal from noise. The conventional method used by data processors to extract the earth signal from the recorded signal is to crosscorrelate the sweep with the recorded signal. The process is done before applying NMO. Crosscorrelation of the sweep with the record has an effect upon the phase exactly opposite to that of convolution, cancelling completely the phase shift which occurred in the original convolution. In simple terms, the original product of the field operation is:

$$A_e(\text{ earth }) \cdot A_s(\text{ sweep }); \qquad \phi_e + \phi_s.$$

Crosscorrelation of the recorded signal with the same input sweep signal produces:

$$(A_e \cdot A_s) \cdot A_s; \quad (\phi_e + \phi_s) - \phi_s$$

Consolidating and cancelling out terms, the result of the crosscorrelation becomes simply:

$$A_e \cdot A_s^2; \quad \phi_e.$$

All the influence of the phase of the sweep upon the phase of the field record has been eliminated, leaving the individual reflection components of the processed record with the phase of the original earth signal. Each recorded reflection coefficient should now be recognizable as the best approximation of a spike possible with the range of frequencies contained in the sweep signal. This is exactly the effect which we wish to obtain with wavelet deconvolution and which is so difficult to achieve on signals from conventional seismic signal generators. The amplitude of the output is weighted by the product of the amplitude spectrum of the sweep, but, since the sweep spectrum is completely white, the squared spectrum is also white, which merely multiplies the earth spectrum by a constant, which is not serious.

It is interesting to reiterate here that the output of any convolution or crosscorrelation operation contains only those frequencies which are common to both operator and signal. Therefore, the Vibroseis output record or correlogram can contain no frequencies which are not contained in the sweep, and possibly some of those may be missing if any sweep frequencies were absent from the field record. The uncorrelated field Vibroseis record may of course contain frequencies outside the sweep frequencies, but these must be noise external to the sweep, or harmonics, since the noise is additive and is limited in bandwidth only by the bandwidth capacity of the recording instruments.

The reflection events on the processed Vibroseis output should provide the best possible definition of these events within the resolving powers of the frequency content of the sweep. The outut record should be almost exactly the same as the result of passing the reflection coefficients through a band-pass filter having the same bandwidth as the sweep. Each reflection coefficient will be replaced by the zero phase wavelet representing the sum at zero phase of the frequency components of the sweep. From another viewpoint, the reflection coefficients have been convolved with a band-limited zero phase wavelet. The definition of the record is now limited to those layer boundaries which are spaced no closer in the section than the period of the highest frequency component present in the returning sweep signal. Thus, if the sweep signal were 20 to 60 Hz, the minimum bed separation which could be resolved, would be that represented by one-half Hz. At an average velocity of say 10,000 ft per sec, and allowing for two-way time, the minimum bed separation that could be measured would be roughly 83 ft.

It is a mathematical impossibility to try to identify the top and the bottom of a thinner bed, say a 20 ft sand lens, with information containing 60 Hz as the highest frequency. No matter how experienced the interpreter, at least one of the tops, or possibly both, is bound to be wrong. This does not preclude the detection of a much smaller interval, or mapping a very small change in elevation of a single boundary from trace to trace, assuming that the corrections to the adjacent traces have been correctly applied. In this case, the only limiting factor is the ability of the system to reproduce small changes in phase and of the interpreter to detect phase change. The time from one trace to the next is related to the relative phase positions of the two traces. A small change of say 5° in phase at 60 Hz is 5/360 of a period of one wavelength or .015 sec, or roughly 1/5 of a millisecond change.

The correlogram which is the normal output of Vibroseis processing contains the amplitude spectrum of the earth coefficients multiplied by the square amplitude spectrum of the sweep. If the amplitude spectrum of the sweep is not uniform, the influence of the sweep on the earth signal can be detrimental, strongly accentuating some frequencies, and attenuating others. The most common distortion is the result of a sweep which decreases in amplitude at high frequencies. If the same sweep is used in the crosscorrelation process, squaring the sweep boosts the relative amplitude of the low frequencies substantially, and the output is dominated by low frequency content.

In such cases, an operation in the frequency domain can compensate for the effect of the sweep. The Fourier transform of both the recorded signal and of the sweep, define both the amplitude and the phase spectra of both signals, and permit operations on the amplitude or the phase spectrum of either as desired. Hence, it is simple to divide the amplitude spectrum of the sweep into the amplitude spectrum of the signal and thus remove the effect of the sweep completely.

$$\frac{A_e \cdot A_s}{A_s} = A_e \qquad (\phi_e + \phi_s) - \phi_s = \phi_e$$

The phase relationship is handled exactly as in crosscorrelation by subtracting the phase of the sweep from the phase of the trace to cancel it out of the signal. The resulting transform should contain only the amplitude spectrum of the earth signal (band-limited) and the phase of the earth signal, which is exactly the goal of true deconvolution. Thus, this one pass results in the best model of the earth that can be obtained within the limitations imposed on the system by the frequency response, by the noise which has been added, and by the ability to define the sweep signal correctly.

The same operation is done in the time domain by computing a correlation sweep in which the amplitude varies as the inverse of input sweep, and the phase is shifted to the same amount as that of the input. For instance, if the amplitude of the input decreased with increasing frequency, the amplitude of the correlation sweep will decrease with frequency. Crosscorrelation of the inverse sweep with the recorded signal will accomplish the desired results.

$$A_e \cdot A_s \cdot \frac{1}{A_s} = A_e \, , \; (\phi_e + \phi_s) - \phi_s = \phi_e$$

Note the completely contrary approach of the Vibroseis system to that of impulse energy source deconvolution. The general model for deconvolution of an impulse energy source assumes a completely white earth signal, which is convolved with an input wavelet in the seismic recording process. Therefore, since the earth signal spectrum is white, and flat, the shape of seismic field record spectrum is entirely the result of the earth signal spectrum multiplied by the system.

As described here, the Vibroseis system is exactly contrary. It assumes the spectrum of the Vibroseis system input to be completely white, so the shape of the amplitude spectrum of the field recorded signal is the shape of the earth signal.

In fact, neither assumption is completely true. The reason that both methods work reasonably well is that the true situation lies somewhere between the two assumptions.

If Vibroseis is to provide reliable results, it is very important that the amplitude relationships found on the recorded trace be truly representative of the signal received. The sweep is of much longer duration than the reflection time through the sedimentary section so at any point in time the recorded signal will represent the sum of components from a great number of reflected sweeps. The use of AGC or similar gain control devices is to be abhorred from the digital processing standpoint because they alter the amplitude content of the signal in a non-linear fashion. A far better match of equipment to the system is the use of floating point or binary gain. Programmed gain control may also be used with a substantial degree of success because the ratio of the initial final recorded amplitude is not nearly so great on Vibroseis as it is on dynamite shooting. The first arrivals from the dynamite shot carry a massive burst of very broad-band energy to the instruments. The initial energy coming from the Vibroseis sweep is a narrow band signal from a much weaker source than dynamite. The relative energy received from the reflectors in the Vibroseis recording will like recordings from other energy source signals, decay exponentially with time, but each returning sweep is so long it overlaps all following reflectors, maintaining a more uniform amplitude over time than the short impulse response from traditional energy sources. After the elimination of the sweep by crosscorrelation, application of an inverse gain program to the recovered trace should provide an acceptable approximation to a non-attenuated trace.

1?

VIBROSEIS

Although the Vibroseis input is much weaker than that of a dynamite pulse, it is a very good signal for use in seismic surveying. A paper by Tanner & Cook points out that the Vibroseis system trades off signal amplitude for signal length. Just as volumetric measure is the product of length times breadth times height, so is the information capacity of any signal, including a Vibroseis sweep, related to length (time) times breadth (frequency) times height (amplitude). Shannon & Van Vleck have formulated a theory for this, which has been extended and related to computer theory by Pierce.

To a large extent, practice of the Vibroseis method obeys the theory and very good seismic records can be obtained with the Vibroseis system in many areas. The fact that a Vibroseis sweep contains only a fraction of the energy that it is possible to produce with dynamite, results , in some areas, in a weak returning sweep signal being overwhelmed by the natural ambient noise. In spite of the fact that only a narrow range of frequencies is input into the system, the recording truck used to record the Vibroseis operations should contain a band-pass filter to limit the frequency spectrum to approximately that of the Vibroseis sweep. The additive noise coming from exterior sources may have a frequency band broader than the sweep. In order to reduce this noise to the minimum level possible, it should be blocked or band-pass filtered to keep it from being recorded with the signal.

The Vibroseis method presently continues to enjoy increased popularity for many reasons. These include the advantage of the non-destructive energy source, the well defined input signal, and the success of digital processing techniques to improve output compared to the analog method. The long field records are somewhat costlier than conventional records to process in the initial stages. Digital processing costs are almost a direct linear function of the number of samples, and the Vibroseis records are often two to four times the length of normal records.

Offsetting these advantages is the fact that the Vibroseis sweep frequencies were historically limited to an upper limit of about 60 Hz. This was considered an advantage. Processing could be done at a 4 ms sample rate since the highest frequencies used in the sweep fell below the corresponding alias frequency. Some contractors processed Vibroseis data routinely at a 8 ms sample rate to minimize processing costs.

The current interest in broad-band techniques has prompted developments of field equipment and technique to input and recover a broader band of frequencies. This has been successful and signals from 8 to 100 + Hz are routinely recorded by many Vibroseis field crews.

Once the signal is recovered, the physical characteristics of the output records are similar to those of impulse sources with one major exception, and the remainder of the processing is routine. The important exception is that the wavelet representing the reflection is (at least theoretically) a band-limited zero phase wavelet with the center peak at the reflection time. If this is true, little further processing will be required other than stacking to increase the signal level. When the process works well, i.e. the conditions meet the assumptions, it is a one-pass process for both signal extraction and deconvolution.

This may be true in many cases but not all; although the phase of the input has been compensated for, the remainder of the system may impose its own characteristics on the recorded signal. There are still many areas in which the application of the deconvolution process will produce an improvement of the results.

The answer to this seeming paradox lies in a thorough understanding of normal output of two types of systems, of deconvolution and its effect on the seismic record, and to a large extent, the opinion of the interpreter. The last item is of particular importance since the basic output of the two systems may be quite different in appearance. From a signal processing viewpoint, both should ideally eventually be the same.

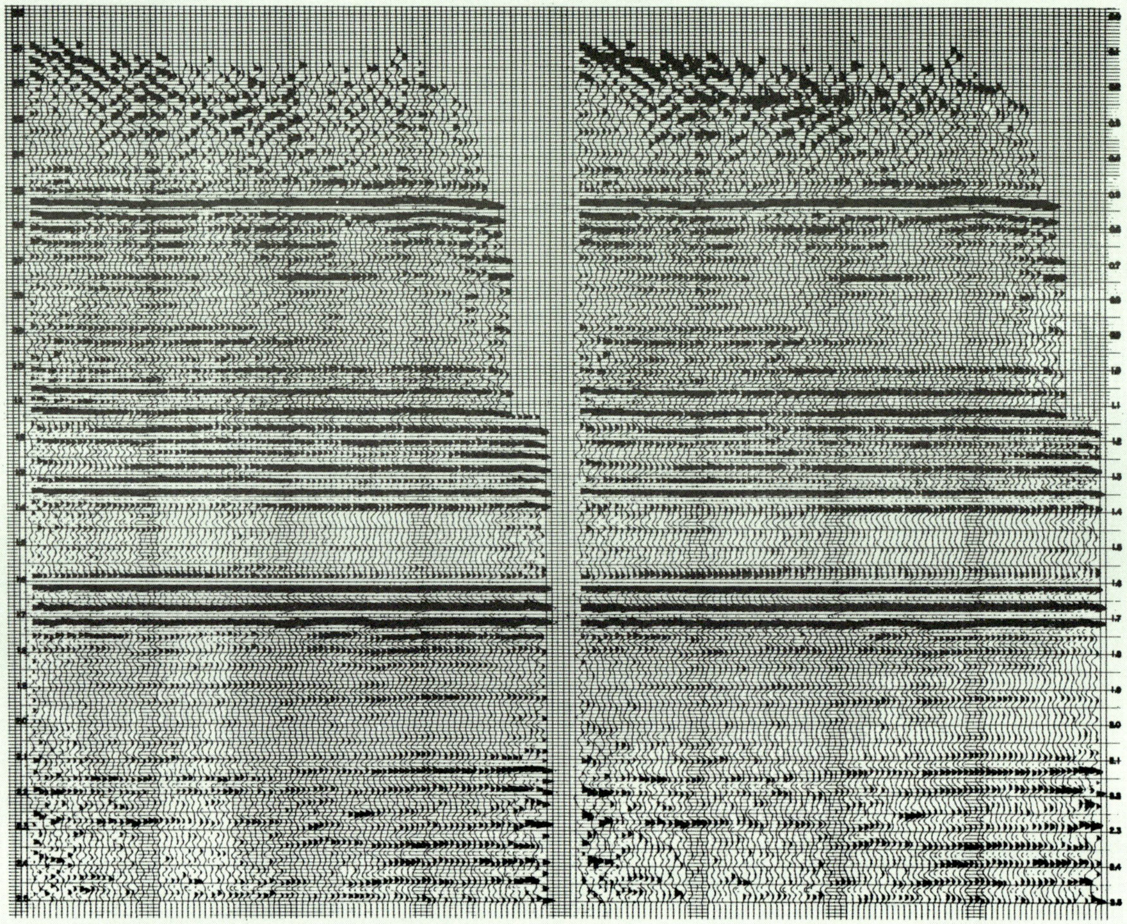

FIGURE 7.2

The phase of the Vibroseis sweep is not subject to the considerable phase lag of impulse input, yet in spite of the basic theory, it may return with a phase modification due to the system. Compare a zero phase deconvolution of a Vibroseis section to the input. Obviously there is little change, confirming the good characteristics of the original output. Some improvement in the reflection below 1.0 sec may be noticed.

The operation of extracting the Vibroseis sweep signal from the recorded signal seeks to obtain the effect of cancelling out all of the phase effects of the sweep and of minimizing or eliminating the effect of the amplitude spectrum of the sweep. As stated, this is very close to ideal deconvolution except for the effects of noise.

The ideal output of any signal recovery process should be identical to the convolution of the earth signal (reflection coefficients) with a zero phase band-pass operator of a bandwidth equal to the limits of the system response. Each reflection will be represented by a zero phase wavelet (the band-limited approximation to a spike) with the peak centered at the reflecting interface. The wavelet is symetrical with side lobes leading and trailing the peak amplitude which represents the horizon.

The output wavelet of the Vibroseis processing should be a close approximation of the spike but, if some attenuation of the spectrum has occurred in transmission of the sweep through the earth, the output wavelet may still need to be sharpened by applying deconvolution. The phase of the Vibroseis sweep is not subject to the considerable phase lag of impulse input, yet in spite of the basic theory, it may return with a phase modification due to the system. This can be assumed to be minimum phase. Generally, frequency domain deconvolution is preferred where the returning sweep has little or no phase lag. **Figure** 7.2 compares a zero phase deconvolution of a Vibroseis section to the input. Obviously there is little change, confirming the good characteristics of the original output. Some improvement in the reflection below 1.0 sec may be noticed.

Another problem, usually less frequent on recent work than on earlier surveys, is that the sweep spectrum may not be generated perfectly due to mechanical problems, and this will affect the spectrum. If this problem is known, say from the recording of the pilot sweep, the crosscorrelation operator should be modified in some way to compensate for the imperfect input sweep signal. Use of the recorded pilot sweep will be effective for phase problems, but an effort to remove irregular amplitude of the input by crosscorrelation with the improper sweep operator accentuates the irregular spectrum. Deconvolution may be a better solution.

The model used for conventional deconvolution assumes the earth signal to be completely white and that any modification of recorded signal spectrum is entirely due to the system response. If this is valid, then the inverse spectrum of the recorded trace may be used to describe the amplitude spectrum of the sweep signal recovery operator for crosscorrelation. The result should be perfect amplitude deconvolution within the band-limit of the sweep. However, if the spectrum of the earth signal is not white, the output trace may be over-deconvolved by this method. In any case, modification of the Vibroseis spectrum is advocated where indicated as it should produce a better output trace.

Under good conditions, the signal extraction process yields a good approximation of deconvolution. The process also acts as a band-limiting filter, rejecting all frequencies outside of the sweep, **(Fig 7.9)**. The relatively narrow band often used for the sweep, makes it imperative to retain all the useful frequency components. Sweep frequency components which return no reflection information should be further rejected, often with a selective time varying filter.

Recorded Sweep

Standard field procedure is to record the sweep signal used to drive the vibrators. This may be considered equal to the transmitted message of standard signal theory. In fact, coupling of the vibrators with the ground may produce a quite different input. The recorded sweep often provides a better crosscorrelation operator that the theoretical sweep. Certain other advantages accrue, as well as a potential disadvantage.

The Vibroseis method provides no sharp impulse by which to time the initiation of energy. The vibrators must be accelerated from rest and there may be some variation in the rate. Timing schemes exist but may not be completely reliable, so that initial time variations of several milliseconds are possible. If the sweep is recorded simultaneously with the record, any variation in the start time will also affect the sweep. If the indicated time zero point on the record is incorrect, both sweep and recorded signal will be delayed the same amount. The delay amounts to a phase shift, which will be the same for both sweep and record, and will be cancelled by crosscorrelation of the recorded pilot sweep with the record process. Thus, the output record will always have a true zero time. A synthetic sweep will not have any inherent delay, and the output of this crosscorrelation operator will still contain the original delay.

Any other phase variations in the pilot sweep, and these occur from time to time, will be reversed if the recorded sweep is used for crosscorrelation. The output wavelet will then always be zero phase, except for any phase shift during transmission. A synthetic sweep, with a perfect phase spectrum, will not compensate for variations in the

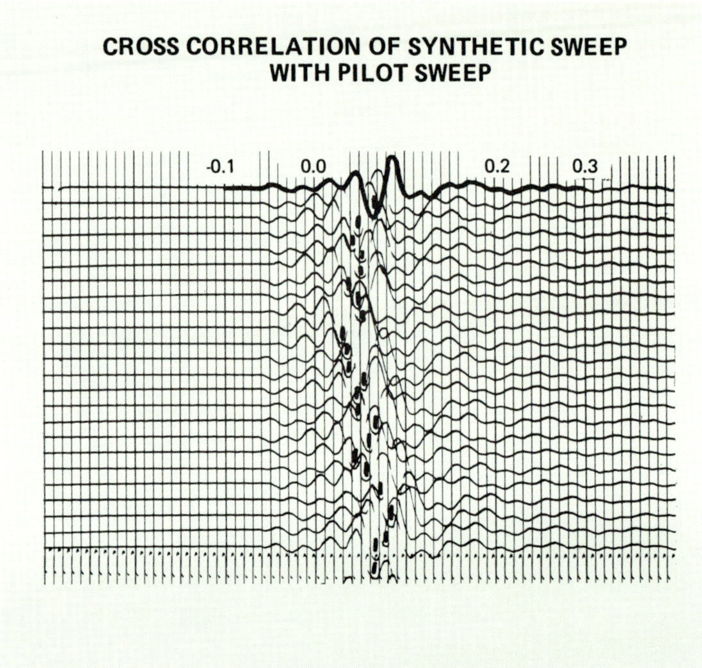

FIGURE 7.3

If the synthetic sweep is cross correlated with the recorded sweep, the output will provide an indication of the quality of the recorded input sweep.

input.

If the model and recorded sweeps are crosscorrelated the output will be the quality of the recorded input sweep. **Figure 7.3** is a collection of crosscorrelation of the theoretical sweep with pilot traces from several early Vibroseis analog field records.

Two things are evident. Severe irregular time delays, in excess of .050 sec, characterized the recordings. The phase spectrum of the pilot was not uniform, as evidenced by the asymmetrical appearance of the output. This example serves well to illustrate the point that the sweep trace should be monitored. Errors of this magnitude are not likely with modern equipment. These problems may be corrected by applying a static shift to the record, and a phase shift to the output. If the pilot sweep is used for the operator, the above two items should self compensate. Note that these procedures are effective only if the pilot sweep represents the sweep that actually went into the ground. Coupling, mechanical problems with vibrators, and imperfect transmission, may all make the recorded sweep quite different than the pilot. Therefore, the autocorrelation of the sweep should be monitored. Drastic frequency changes may be indicative of equipment problems.

Harmonics

Quite frequently the vibrators will produce a noticeable harmonic. This is usually the first harmonic, so a sweep of say 10 to 50 Hz will produce a harmonic of 20 ot 100 Hz. (Fig. 7.4). Some of the harmonic frequencies will fall in the same portion of the spectrum as the sweep, In the example, frequencies of 20 to 50 Hz will be present

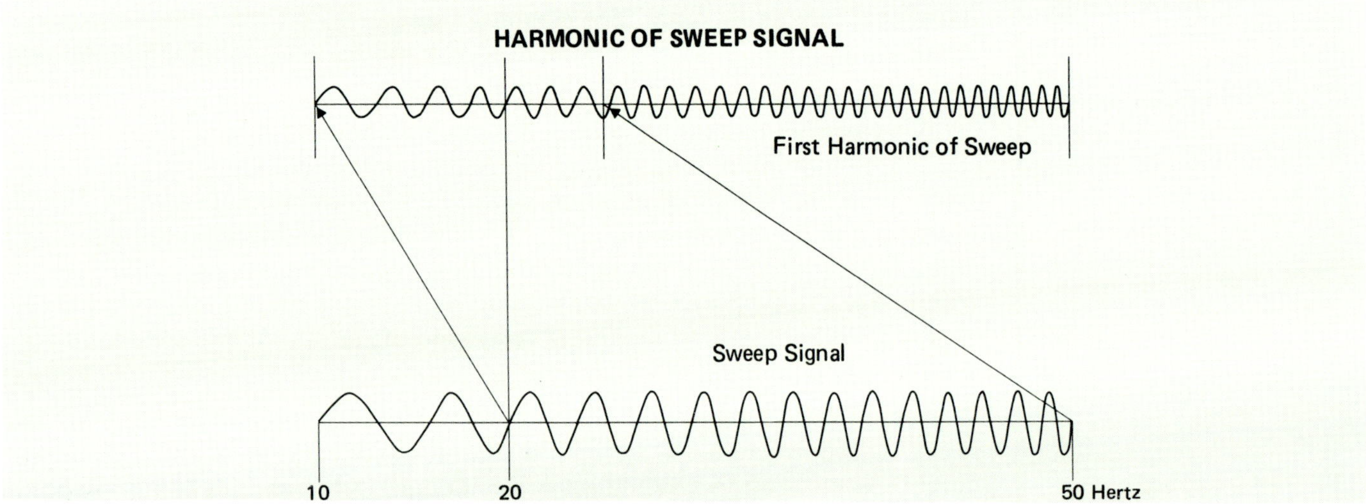

FIGURE 7.4

Quite frequently the vibrators will produce a noticeable harmonic. This is usually the first harmonic, so a sweep of say 10 to 50 Hz will produce a harmonic of 20 to 100 Hz. Some of the harmonic frequencies will fall in the same portion of the spectrum as the sweep.

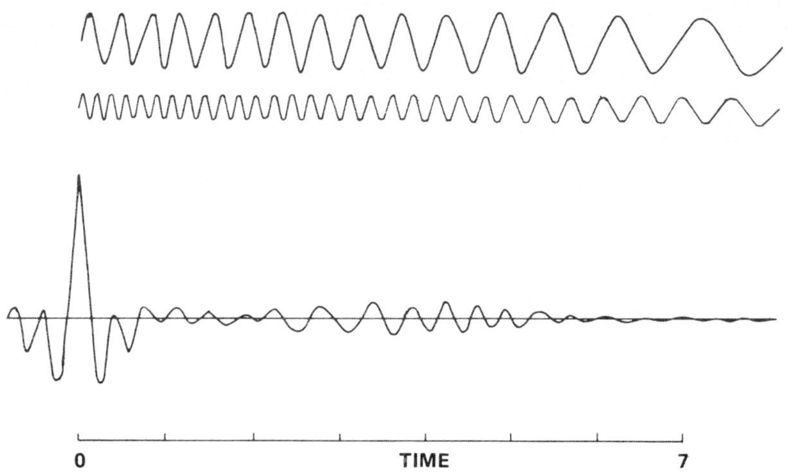

FIGURE 7.5

Correlation of a record with sweep harmonics with a synthetic sweep (no harmonics) will produce the desired correlation spike at the first point of the sweep. The synthetic sweep will also correlate with the harmonic, but since the phase spectra are different, the output for a 7 sec down sweep, from 50 to 10 Hz would extend harmonic noise from 20 to 50 Hz over the record time of 1.75 secs to 4.375 secs.

in both the sweep and harmonic, but will be displaced in time. The 20 Hz energy will be displaced ¼ of the sweep length, and the 50 Hz 3/8 of the sweep length.

Correlation of this record with a synthetic sweep (no harmonics) will produce the desired correlation spike at the first point of the sweep. The synthetic sweep will also correlate with the harmonic, but since the phase spectra are different, the output will be a sweep wave extending from 50 to 20 Hz, over a time of -4.375 sec to -1.75 sec for a 7 sec sweep. A down sweep, from 50 to 10 Hz would reverse the effect, so the harmonic noise would extend from 20 to 50 Hz over the record time of 1.75 sec to 4.375 sec. (Fig 7.5)

If the recorded sweep is used, both of the above harmonics will exist in the output (Fig 7.6)

Vertical Stacking

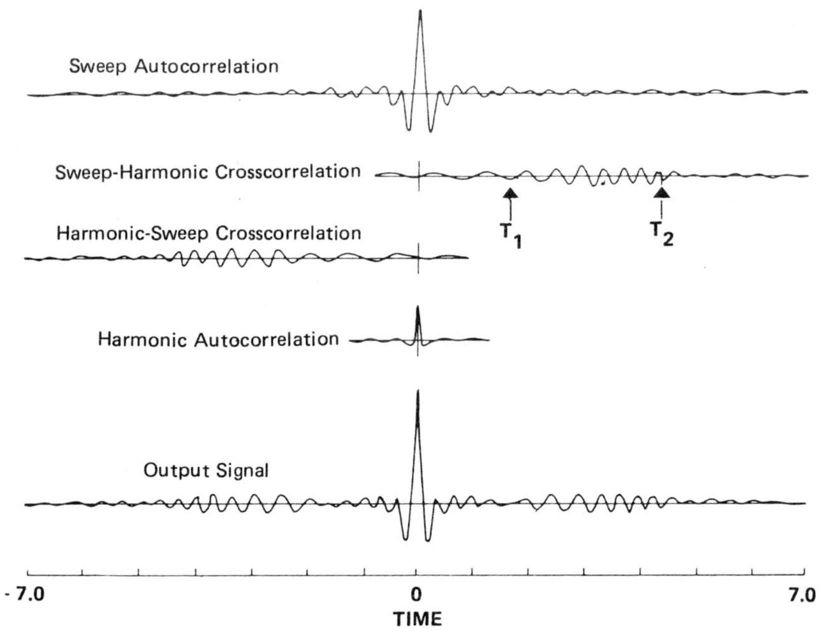

FIGURE 7.6

If the recorded sweep is used, both of the above harmonics will exist in the output.

The rather weak signal of the Vibroseis record in the presence of high ambient noise often requires several individual records from approximately the same location to be stacked into a single record before processing. When this is done, utmost care should be taken to ensure that the initial time of the records be exactly aligned. Static shifts cause frequency components to be vector summed out of phase, attenuating or even cancelling the component. A 50 Hz sweep component will be completely cancelled out if a static error of one-half Hz, or .010 sec exists. The trend toward broader band signals and higher sweep frequencies aggravates the problem, and vertical stacking in the field becomes increasingly less desirable.

It is interesting to analyze a series of field records, each consisting of one more additions to the vertical stack. Quality of the

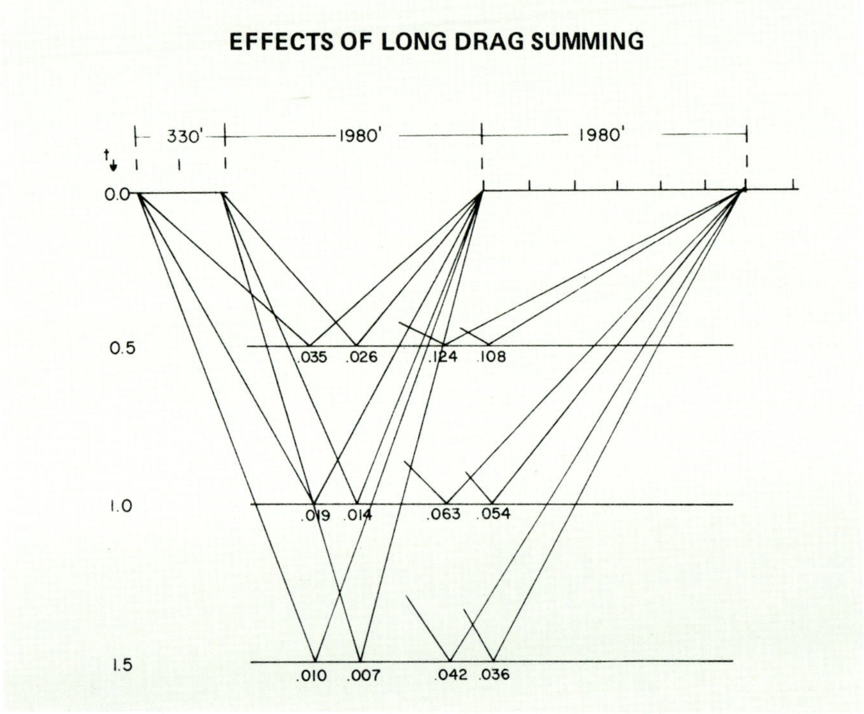

FIGURE 7.7

The cummulative effect of the long offset and the long drag over which the several sweeps are obtained, gives rise to some severe non-linear time shifts associated with the normal moveout.

processed record will improve with each addition to some point, at which it may decrease. The decrease is due largely to misalignment of phase, usually at increasing distance, with consequent cancellation

Considerable surface noise is generated by Vibroseis, with the result that the vibrator may be located a considerable distance from the recording spread, often a full spread length. The multiple records for vertical stacking are obtained by shifting the vibrators a short distance to a new location for each record. The reason for this is to obtain some uniformity in the coupling of the vibrators to the ground which is a problem, as it affects the transmission of the sweep. Poor coupling will produce a distorted sweep. Several points vertically stacked will average out the variations.

Unfortunately, the cumulative effect of the long offset and the long drag over which the several sweeps are obtained, gives rise to some severe non-linear time shifts associated with the normal moveout. As an example, in a typical case, an offset of 1980 ft with a drag of several sweeps over 330 ft produces a .009 sec time differential at a

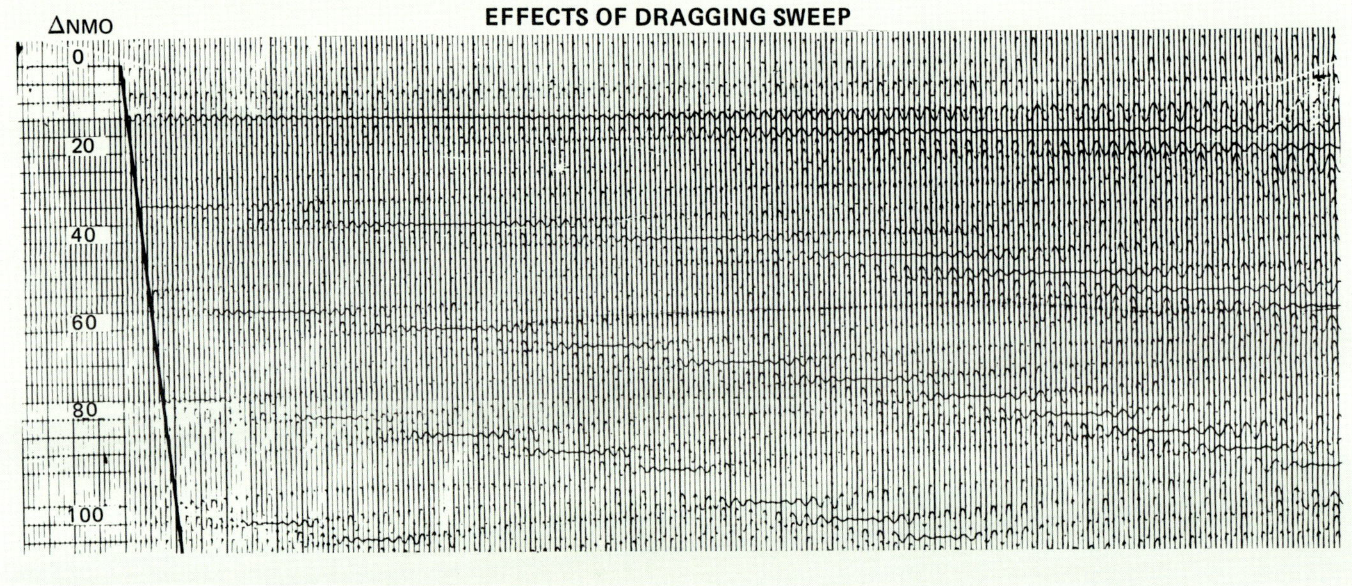

FIGURE 7.8

To illustrate the effect of long offset a sweep at time zero has been summed with a second sweep, at increasing time shifts. Note the distortion of the sweep amplitude, including complete cancellations, at a varying combinations of frequency and time shift.

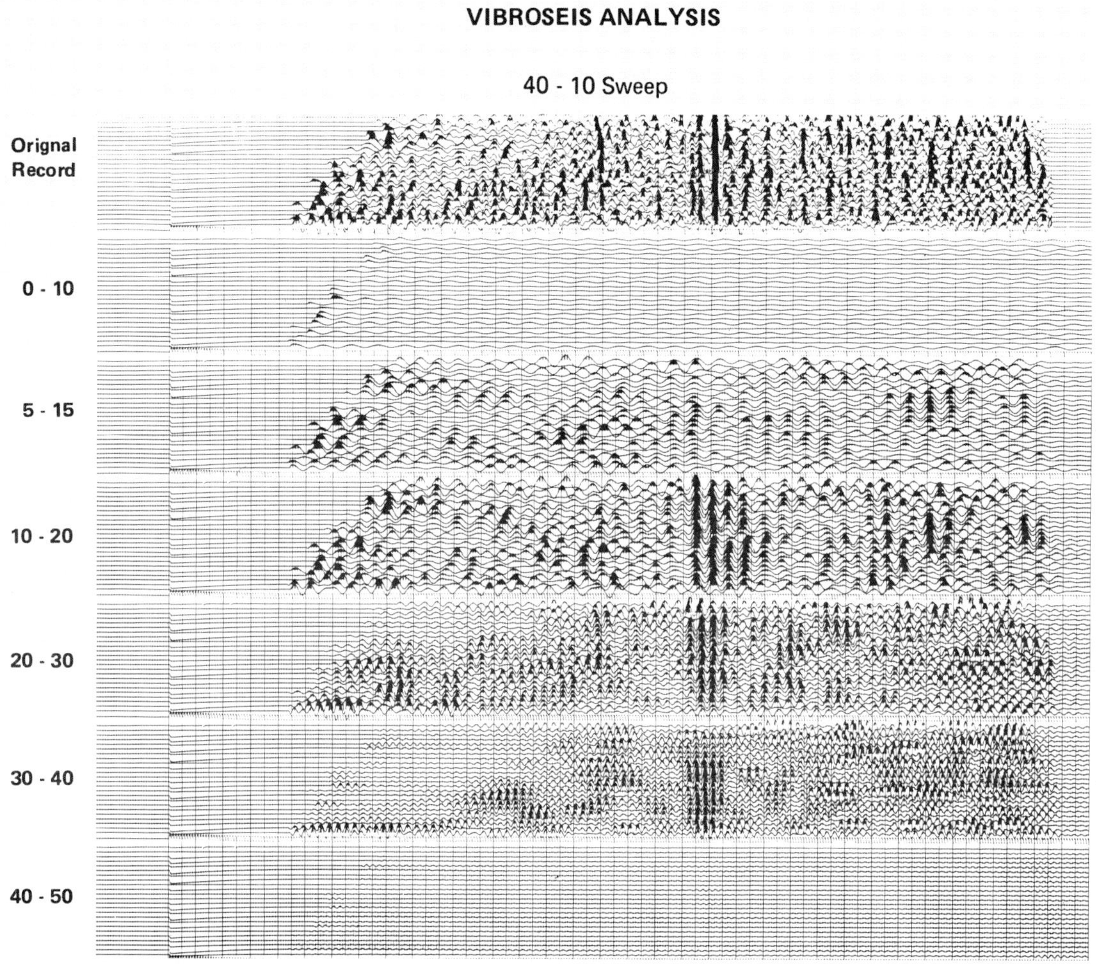

FIGURE 7.9

Under good conditions, the signal extraction process yields a good approximation of deconvolution. The process also acts as a band-limiting filter, rejecting all frequencies outside of the sweep. Contrary to theory, and to the usual results observed, high frequencies are very weak in the early part of the record, but increase in amplitude with time. This is purely the result of moveout differences in the vertically stacked field data traces.

depth of one sec. (**Fig** 7.7) These recordings will be summed, and any high frequency energy will be cancelled.

To illustrate the effect a sweep at time zero has been summed with a second sweep, at increasing time shifts. (**Fig** 7.8) Note the distortion of the sweep amplitude, including complete cancellation, at a varying combination of frequency and time shift. Similar effects can be found on a band-pass display of the processed output record. (**Fig** 7.9) The processed record has been filtered with a series of relatively narrow band-pass filters to indicate the information carried by different frequencies. In this case, there is something very strange about the results. Contrary to theory, and to the usual results observed, high frequencies are very weak in the early part of the record, but increase in amplitude with time. This is purely the result of moveout differences in the vertically stacked field data traces.

This leads to another very important point, the manner in which the recording instruments handle vertical sums in the field. The Vibroseis field record does not appear to have the rapid substantial decay in amplitude common to impulse sources, because of the length of the sweep. The signal from the first reflection is still being recorded when

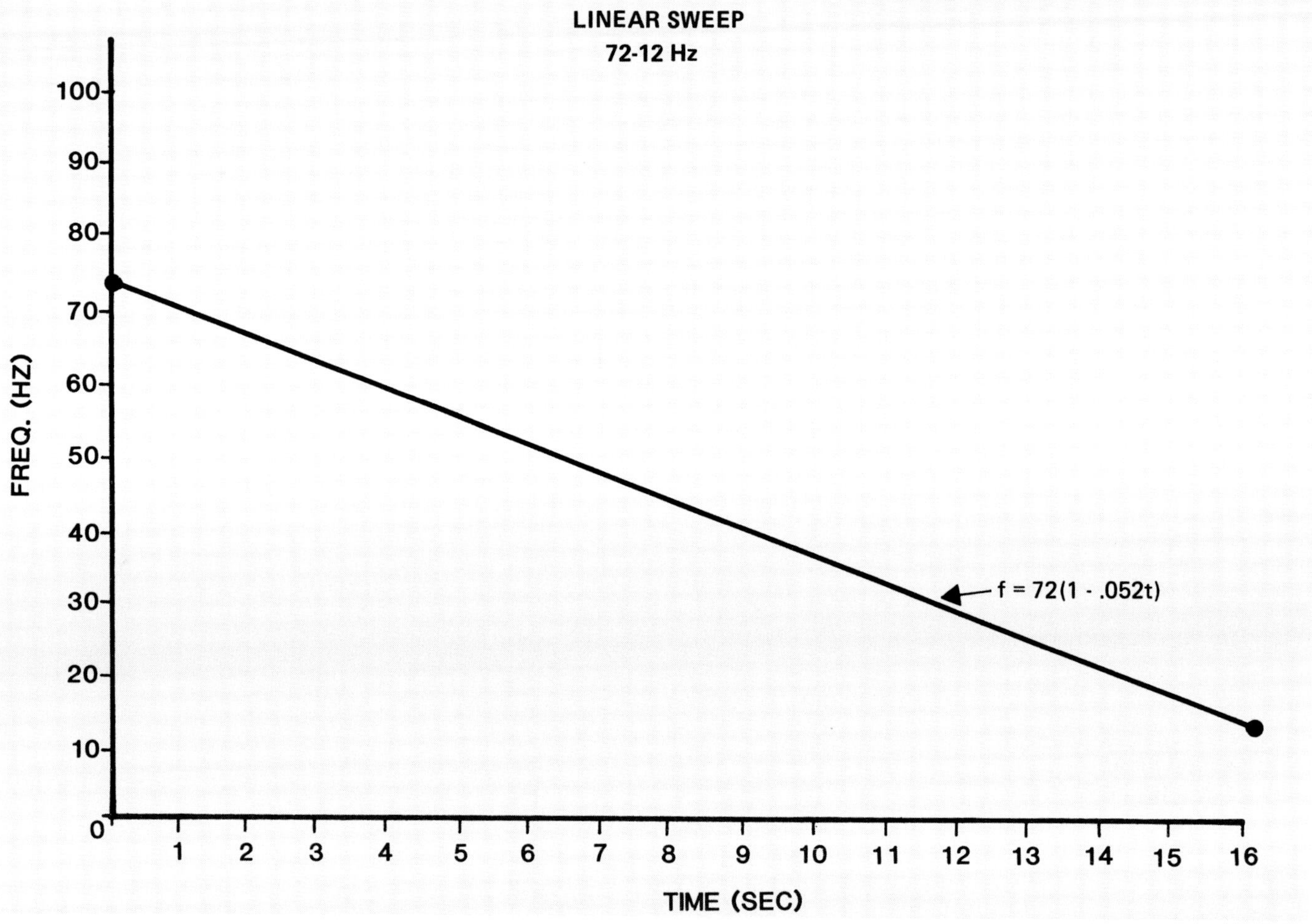

FIGURE 7.10

The linear sweep uses a uniform acceleration function, spending the same amount of time at each frequency.

the deepest reflection arrives. The spreading of energy over the length of the field record, plus the relatively low level of input, allows a Vibroseis record to be made with fixed gain settings, or if binary gain equipment is used, a minimum number of gain steps. When several records are summed in a vertical stack, however, the total amplitude becomes sufficiently large that some gain control becomes necessary. Unfortunately, for reasons of economy or problems with data rates, gain changes may not be properly recorded. For instance, one summing device merely sums the signal portion of binary gain instruments and ignores the gain bits. Another applies a rather slow AGC to the sum. While yet another clips the input to a small number of bits before summing.

While there may be some arguments that such procedures are valid under some conditions, (see the section on one-bit stacking), such procedures are, in fact, totally unacceptable except under very particular circumstances, and should not be used.

With increasing emphasis on expanded frequency bandwidth, and the need for higher resolution, Vibroseis users have had to scramble to extend the upper end of the usable frequency range. The earth is well known to attenuate

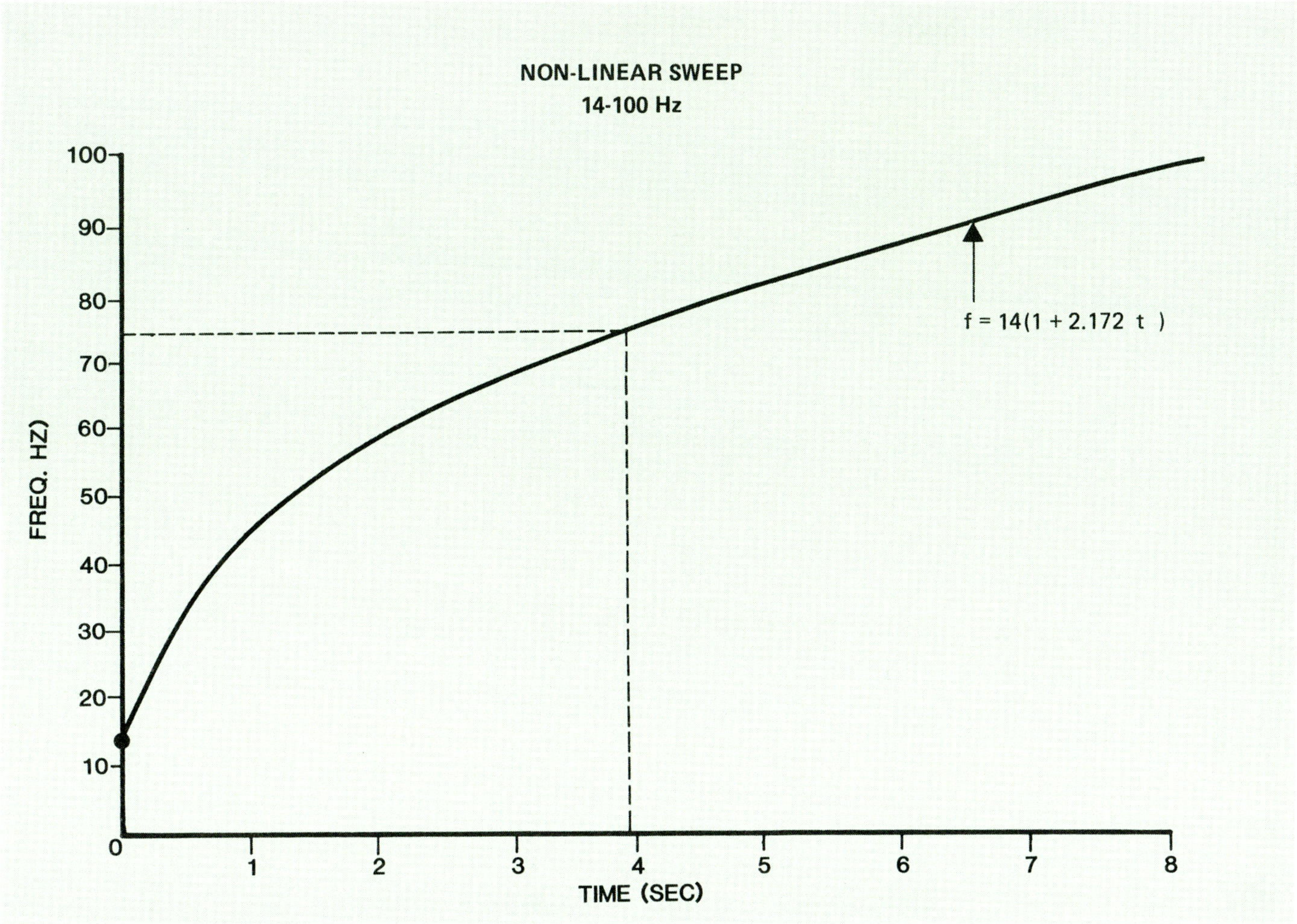

FIGURE 7.11

The rate of acceleration of the non-linear sweep decreases at higher frequencies, allowing more time, and hence concentrating more energy at the higher frequencies.

signals as a function of frequency, while problems of generation of the higher frequency signal, and coupling of it from the vibrator to the earth have all constrained the upper limit of the recorded signal.

One customary practice in such cases has been simply to lengthen the sweep, which slows down the acceleration of the vibrator, **(Fig 7.10)**. However, the relative input amplitude remains constant over the spectrum, while the response will still show the severe attentuation of the higher frequencies.

One solution to the problem is to use a non-linear sweep. For the example shown, **(Fig 7.11)**, the acceleration function is calculated as the square root of sweep time. As a result, the sweep spends little time at the lower frequencies, but lingers much longer over the higher frequencies. For the 14-100 Hz sweep used, approximately ¼ of the sweep time was spent in the lower half of the frequency range and ¾ in the upper half. As a result of the emphasis on high frequencies, it was possible to shorten the sweep length, cutting field production time and subsequently processing time.

The resulting records, **(Fig 7.12)**, clearly show better definition over the high frequency range.

FIGURE 7.12

The comparison of these two surveys is not on an equal basis, since different sweep frequency ranges were used. However, the 72 Hz limit on the original survey was believed to have been the feasible limit at the time the survey was made. The broad band record was obtained with a shorter sweep than the original.

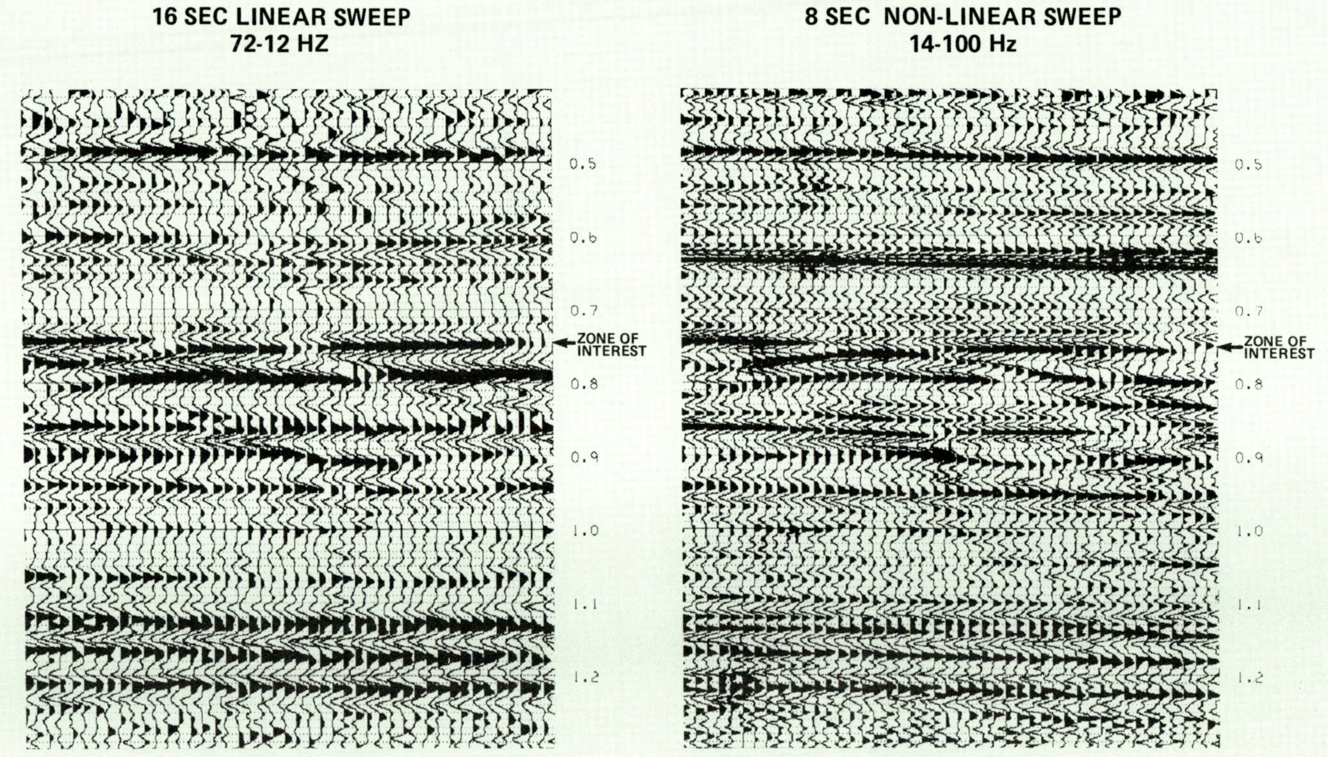

FIGURE 7.12a FIGURE 7.12b

Summary

Early Vibroseis work was characterized by low frequencies and very little character. This was due largely to a series of problems which tended to eliminate the high frequencies. These included vibrators which attenuated high frequencies and multiplied vibrators which were not properly locked in phase, again causing the high frequencies to be attenuated first. These two problems have been solved.

Crosscorrelation with low-frequency dominant-sweeps aggravated matters, as did great numbers of multiple sums over long drag distances. As a result, in many cases narrow band, low frequency records became associated with Vibroseis output.

This has now changed. Vibroseis records can be obtained with a full range of frequencies, and hence good resolution. It requires good equipment properly maintained, and good field procedures based on a knowledge of the effects of some field parameters. If this is not done, field procedures may preclude good broad-band information.

All in all, from the processing standpoint, the Vibroseis method offers an excellent approach to the problem of obtaining and enhancing seismic data.

VIBROSEIS* Registered mark of Continental Oil Company

8 VELOCITY ANALYSIS

Time and Velocity

The objective of the seismic survey method is a reliable measurement of the travel time between a signal source and a receiver. The results of a reflection survey are displayed on a time section, and geophysicists map structure in units of reflection travel time. A geologist, on the other hand, is more interested in the depth of the horizons, and the thickness of the section components measured in units of meters or feet.

The key which links the time domain to the depth domain is velocity, which in itself can be a significant diagnostic tool for the determination of lithology.

Geologists and geophysicists are often guilty of underestimating the influence that velocity exerts in all phases of the seismic process. In processing, the use of improper velocity can completely negate the benefits to be obtained by stacking or migrating data, for example.

Velocity variation causes the conventional time-display section always to have some degree of vertical distortion and if the velocity function changes laterally the geologic dip can be quite different from time dip. A relatively small change in velocity can either manufacture or wipe out a significant anomaly on a time section. Migration, which is very dependent upon precise velocity determination, can greatly enhance the definition of structural features when it is properly applied, but it can also be very damaging to the data if the velocity function is not correct.

The potential danger of velocity distortion of time structure maps has always existed, but prior to the advent of digital processing, the operational difficulty and expense of deriving a reasonable approximation of true velocity and then compensating for its effects was often so great that it was simply ignored. Unfortunately, as is so often the case in seismic data handling, early practice formed habits and established standard operational procedures

VELOCITY CALCULATION

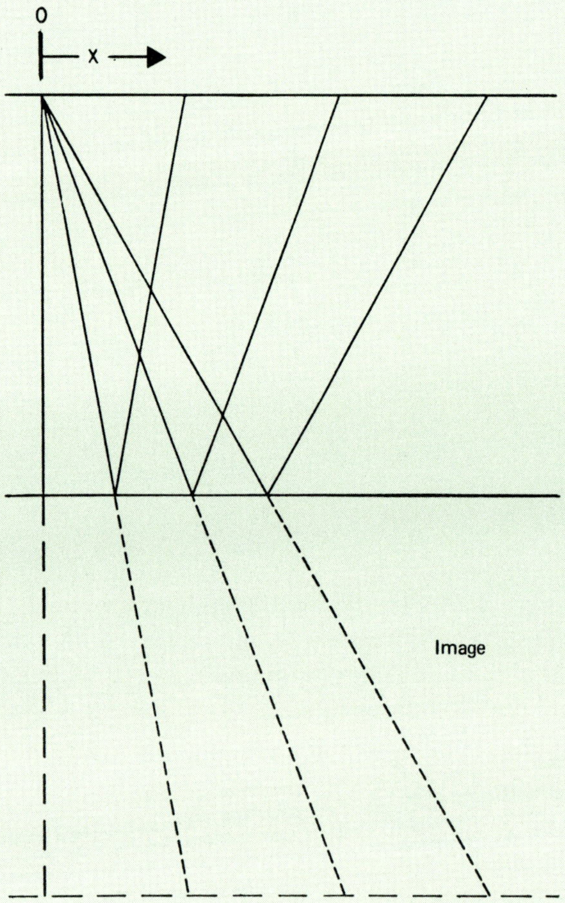

$$X^2 = V^2 T_x^2 - V^2 T_o^2$$

$$V^2 = \frac{X^2}{(T_x^2 - T_o^2)}$$

FIGURE 8.1

Normal recording geometry provides a right triangle relationship between near and far trace recordings of the same reflections, which can be exploited to yield velocity.

that remain the accepted standard long after digital computing makes refined procedures available to compensate for previously ignored conditions.

This topic of velocity determination and its relation to seismic data is, in itself, of sufficient interest, importance and complexity to fill an entire volume. All seismic field records contain the means for determining velocity information to some degree of reliability, but prior to the advent of digital processing an entire seismic survey could be completed without a velocity determination. Interpretation was done directly on original paper field records, continuity of horizons was carried from record to record by mapping near-trace times where dynamic effects are not significant, and the remaining traces were merely used to correlate continuity across the space between. Velocity determinations were used primarily as an aid to depth estimations but often, time structure maps were considered adequate.

Velocities can be obtained by direct borehole measurements (velocity survey) or derived from seismic field records. The normal moveout approximately follows the expansion of geometry of a right triangle (**Fig. 8.1**) so the familiar X^2, T^2 method of plotting trace-time squared against trace-distance squared was a fairly common early method. Occasionally, a rather time consuming expanded field profile was shot for more precise measurement of the velocity function, or direct measurement of velocity was accomplished by shooting a downhole well survey. A geophone lowered in the borehole measures the direct travel time from the surface to several depths.

The advent of magnetic recording and the production of continuous analog NMO-corrected sections required some routine determination of velocity. This initial data processing development to produce a continuous record section was significant because it made it possible to use all of the traces for an interpretation. Normal moveout

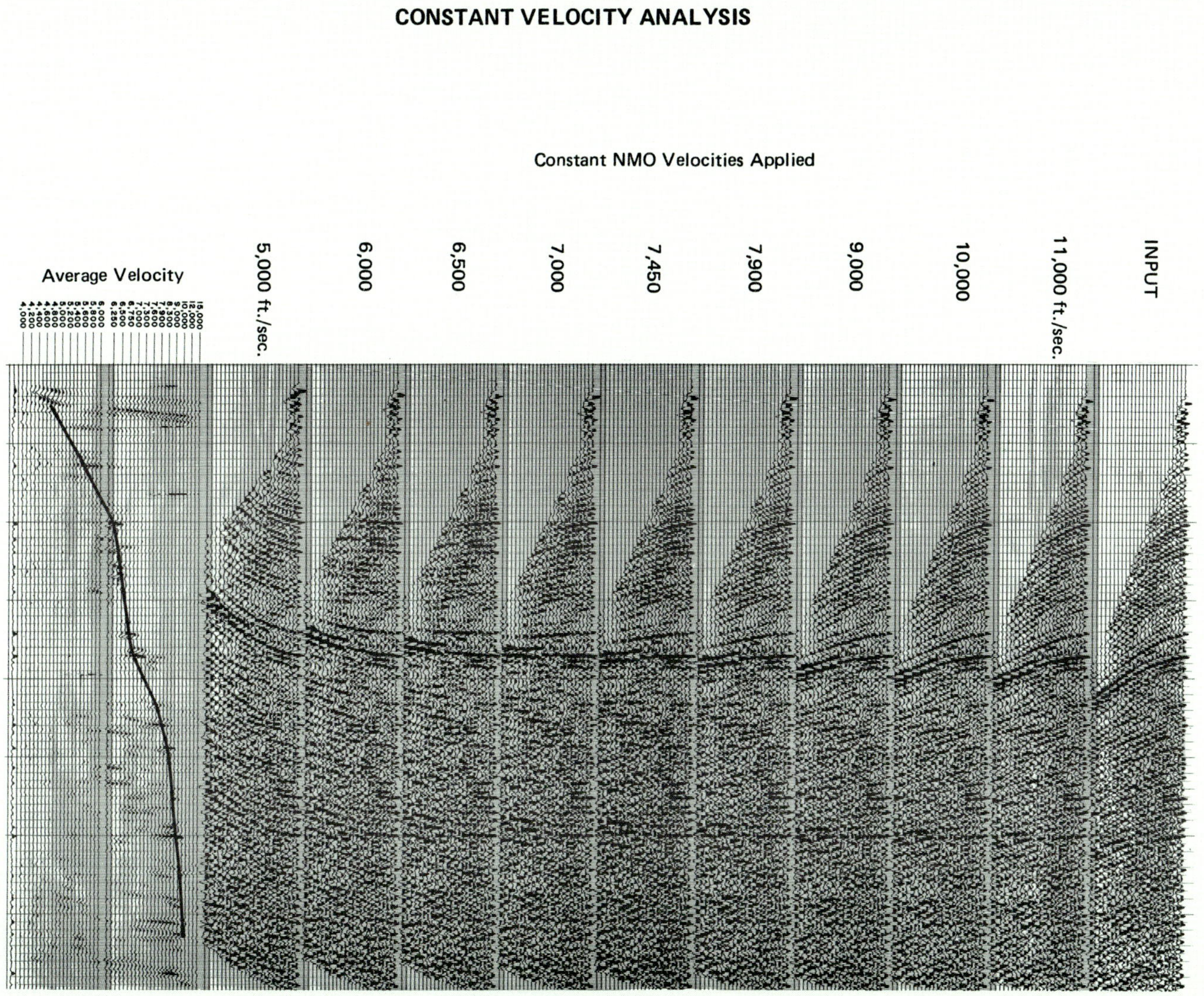

FIGURE 8.2

Constant velocity analysis applies normal moveout corrections to a common depth point trace collection (right) for a constant velocity to yield the corrected traces for each velocity (center). When each set of these are stacked, the results (left) show high amplitude at the levels where the correct velocity was applied to stack the reflection in phase, while the others not in phase cancel out.

corrections were applied with a mechanical cam, which required some effort to design and cut. Small amounts of normal moveout remaining on the 100% coverage then in use were readily detected and visual adjustments could be made to allow for the error, or the section could be re-run with a residual velocity correction. Hence, local velocity variations were often ignored in the processing sequence.

The introduction of multi-fold coverage, and long recording spreads, required more precise calculation of the velocities used for dynamic correction, and recognition of changes in the velocity function, so that the corrections could be adjusted accordingly.

Traditional velocity analysis methods proved inadequate to supply the amount of spatial control required, and were often not economic, particularly when performed in conjunction with the high speed single vessel marine seismic surveys that developed about the same time.

Fortunately, the same multi-coverage stacking techniques which initiated the requirement for automatic velocity analysis techniques also provided the means for a solution.

Constant Velocity Stack

The basic concept of this velocity analysis procedure is simple, varying only in the mechanics of operation. The seismic traces of a common-depth-point gather group are the basis for each velocity analysis. Suitable static corrections, (ideally referred to a datum near ground surface) and data enhancement procedures are first applied. A series of normal moveout corrections, each based on arbitrary constant velocity, are then applied to each trace of the data set. The NMO-corrected traces for each velocity are then stacked to produce a single output trace. This calculation is repeated for each constant velocity, until the range of velocities applied extends from the minimum to the maximum likely to be encountered in the area (Fig 8.2). The velocity increments are not necessarily uniform, but may be rather small for the application of the slower velocities, which yield large normal moveout corrections, and larger for the higher velocities.

For any given constant velocity, the NMO correction will prove to be incorrect over most of the record length. However, at some level within the section, a given velocity may produce the optimum NMO correction which yields the best stacked trace at that level. This is evidently the best stacking velocity to use for that depth at that location. Further examination of the stacked trace output for the entire range of velocities will determine the optimum correction velocity at each reflection level. Interpolation between the selected points yields a continuous stacking velocity curve for the entire sedimentary section. This completes the basic process, but in practice many variations are used.

For example, many display techniques have been developed. The collection of stacked traces may be scanned horizontally by a computer to pick the maximum amplitude at each level. A plotting device can draw horizontal time/amplitude lines to indicate the maximum amplitude picked **(Fig 8.3)**. Another presentation is to contour the amplitude of the output, providing an easier visual selection of the maxima. Some criteria other than amplitude, such as coherence, power, or a simple sum of sign bits, may be used to select the optimum velocity at each level.

The results are interpretive to a certain degree, depending upon the amount of noise and the number of multiple reflections present. Multiple events may have higher energy content than legitimate reflections at a given time level. Multiples are detected on the output plot by observing that the maximum amplitude peaks have the same average velocity as shallower events often at one-half the time. Noise and spurious energy, refractions and other phenonema associated with the reflection process may produce spurious amplitude peaks also.

The several possibilities for the occurrence of spurious maxima preclude totally routine mechanical picking of the stacking velocity curve. The quality of velocity determination usually benefits from intervention by a skilled interpreter. Yet totally manual operation may introduce bias or yield insufficient detail. Manual analysis is often least effective where it is needed most, such as in areas of very poor data, steep dip, or when a poor selection has been made in the range and increment of the constant test velocities applied, or in the number, spacing or location of the velocity determinations. Unless the velocity increments used in the process are sufficiently small, misinterpretation may arise from velocities which straddle the correct velocity at a given horizon and hence never yield the optimum-value point on the velocity chart. There may be a tendency to skimp on the number of velocity

Automated Analysis

VELOCITY ANALYSIS DISPLAY

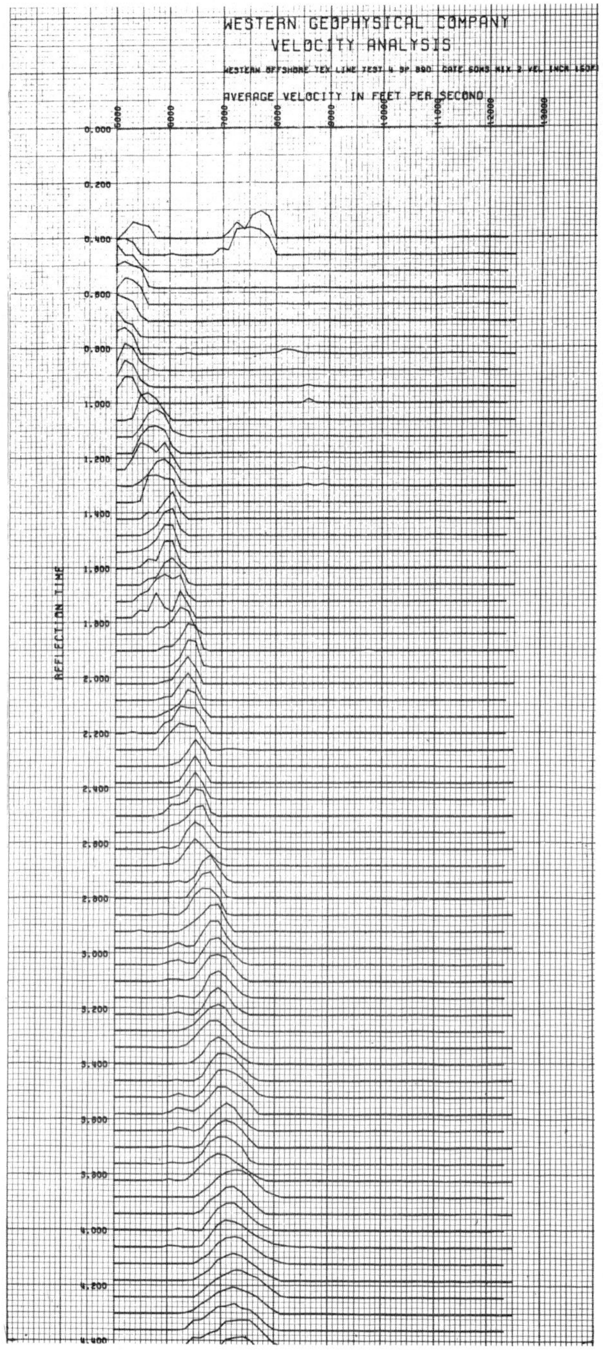

FIGURE 8.3

An early form of display for constant velocity analysis. The output of the analysis is scanned horizontally for the maximum amplitude, which is then output while all others at that level are suppressed.

selections used. Probably no less than 40 constant velocities should be used to obtain sufficient control. The trace-gather groups for the input ideally should map essentially flat dip.

The key to automated analysis is to exploit the power of the digital computer to examine very large amounts of velocity information and then to analyze it with detailed statistical methods. The degree of precision required may be dependent upon the end use. Velocity for interpretive work must be very carefully developed, and even for stacking purposes results will be better if more precision is introduced into the analysis procedure.

A display of the sequence of NMO corrected records (**Fig 8.2**) is perhaps the most valuable aid to interpretation. It is usually not necessary to display the full range of constant-velocity results; 10 to 15 representative velocities may provide adequate quality control for the average stacking velocity analysis from which structural times sections will be produced. Careful examination of these records together with the stacked trace output from all velocities will indicate the relative reliability of the information and possibly detect any situation in which the velocity increments have been overly large.

Variations in the display procedure may be used to improve the final output. The constant velocity stack may be applied to several adjacent trace groups, or even to an entire line. Instead of a simple stack of the traces, the cross-product may be used to amplify the amplitude relationships. The output may be enhanced by further data processing to improve the visible continuity on the display records and thus make it easier to correlate the results.

A popular related display technique is to display the coherence of the semblence function of the gather group rather than the simple stack. There are probably as many display techniques as there are data processing contractors.

A mechanical variation of the constant velocity stack process is to apply a geometric sequence of constant time shifts to the traces rather than calculate normal moveout. Static time shifts are applied to each trace along a proportionate scale which varies as the square of the distance. The output is much the same as before but the parameters, instead of variations of velocity against time, will be variations in Δt or moveout time against vertical time. This method has the advantage of being somewhat faster inasmuch as each trace in its entirety is moved by a fixed static amount. The display may be somewhat more difficult to interpret and involves eventual calculation of variable velocities corresponding to the constant time shifts.

A quite different approach to the determination of the moveout uses the cross correlation process. Each trace is cross-correlated in turn with the adjacent trace and point of maximum correlation noted. The result will be a series of shifts obtained across the record for each of the individual time gates. From this, the amount of Δt at all levels of record can be obtained, providing the necessary information for determining the velocity.

A by-product of this method is the possibility of developing static information from the time shift information. The time shifts over the several gates at increasing time should be a smooth variable amount of dynamic effect and a constant shift for each trace static.

The Geological Dip Problem

The problem of developing reliable velocities analysis for steeply dipping sections is considerable. Where dips are high the common-depth-point information will not represent a true common-depth-point but will smear over some distance. This will contain a component of dip. The problem is accentuated where the quality of data is such that several gather groups must be used in order to obtain sufficient redundancy of data to eliminate some of the noise. The error is greatest for single-ended records. The difficulties may be alleviated, although not always satisfactorily solved, by first displaying a simple 100% section of near traces from multiple coverage records to obtain some idea of the structure across the line. The location of the velocity analyses can then be selected to minimize the influence of dip.

It should be noted that many of these situations will yield velocities outside of the normal physical range which are, nevertheless, velocities which will yield the best stack of the traces. Further processing may be required to obtain a reasonable approximation to true velocities.

Stacking Velocities

Velocities obtained directly from the geometric properties of seismic field recordings are designated stacking velocities (or, less correctly, RMS velocities) to distinguish them from true vertical geologic velocities. It has long been recognized that velocity information derived from seismic field information does not match the vertical velocities measured in boreholes. The explanation for the difference lies in the laminar nature of the sedimentary section and the physical laws which govern the transmission of sound.

If the sedimentary section were homogeneous the vertical reflection time through it would merely be the thickness divided by the velocity, (doubled for two-way time). The reflection time at a receiver separated from the source would be proportional to the increased distance of travel along the hypotenuse calculated for the right-triangle relationship.

When the section is layered, however, the simple relationship no longer holds. Reflections exist because adjacent layers have different velocities. The travel time through each layer will be the thickness of the layer divided by the

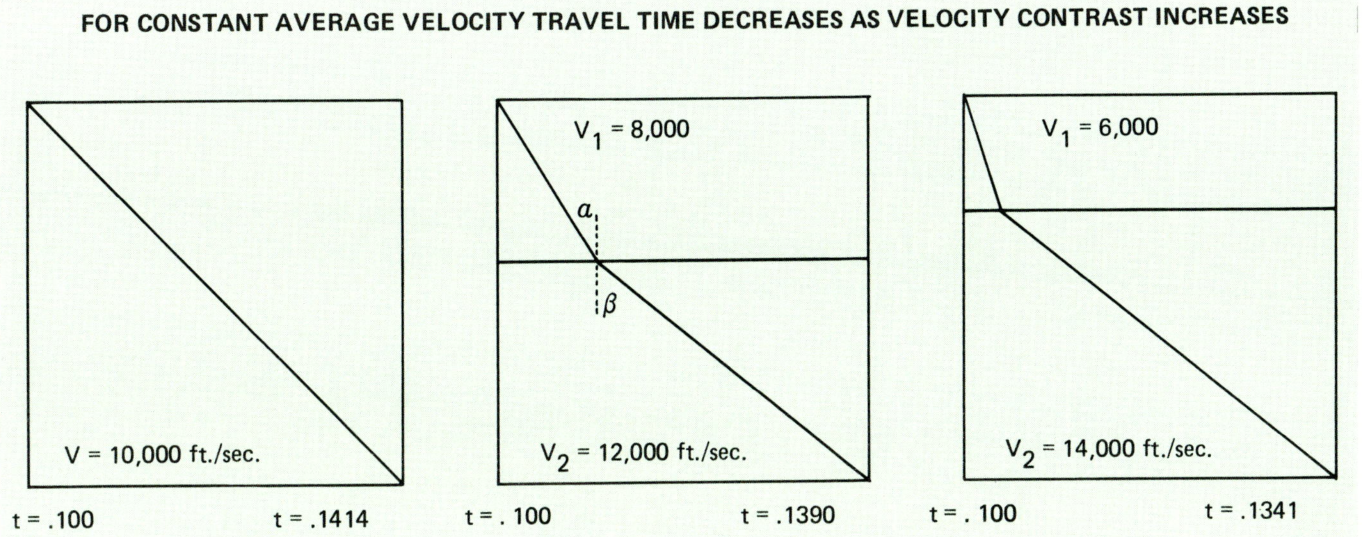

FIGURE 8.4

The vertical travel time through each of the figures is constant, but times to a distant receiver will vary as a function of velocity contrast and thickness in a layered sequence.

corresponding velocity, doubled for two-way travel time. The total travel time of a vertical wave travelling through the section will be equal to the sum of the travel times through all of the layers. But the offset time will no longer be a simple sum of travel times extended along the simple right-triangle relationship. The reason for this is that the travel path will not be a straight line but one dictated by Snell's Law for incident and emergent angles through a layered medium (**Fig. 8.4**).

$$\frac{\sin a}{\sin \beta} = \frac{V_1}{V_2}$$

At each velocity layer interface the travel path will be refracted through the section at a new angle. The ray path will have the shortest possible travel time through the layered section, even through the distance travelled is greater than along the straight line path.

Whereas the average vertical time, and hence vertical average velocity, through a layered section may be exactly the same as through a homogeneous section, the travel time to a given distant geophone will nearly always be shorter through a layered earth than through a homogeneous section. The distance traveled along the ray path will be greater than a straight ray path, but higher percentage of the total travel path will be through the high velocity material so that the total elapsed travel time is actually less than for the straight ray path.

The route travelled by the signal through the layered earth to a distant geophone is unknown to an observer at the surface. If velocity is to be calculated the only readily available distance estimate is the straight ray path, but if this distance is divided by the short travel time it yields an apparent velocity higher than the actual. This difference will increase with distance. Also, if anisotropy is present, the horizontal velocity component will weight the measurements.

At very small offsets, the travel path approaches the vertical and the error becomes negligible. Unfortunately, the geometric ratios at small offsets are too small to yield reliable velocities. Assuming other things to be equal, the mechanical reliability of a velocity determination increases with distance of the geophones from the source, but the difference between the true vertical velocity and the apparent velocity also increases.

For the purpose of common-depth-point stacking, determination of the apparent velocity is desirable because it is the one which will give the best stack. True velocity, if measured without error in a borehole, will be slower than the apparent velocity and result in over-correction of the field data.

Stacking velocities are sometimes referred to as RMS (Root Mean Square) velocities but this is not quite correct. RMS velocity is a mathematical weighting of vertical velocity information to approximate the minimum travel path within a layered velocity sequence.

$$V_{rms} = \sqrt{\frac{\Sigma V_i^2 \cdot t_i}{\Sigma t_i}}$$

The results of this calculation will yield a velocity which will match the stacking velocity at some point of distance. If the only velocity information available for stacking is a borehole survey the measured velocities should be recomputed to RMS velocities, which will more closely approximate, but not necessarily equal, the required stacking velocities.

Once determined, the stacking velocity may be input into the normal moveout routine to compute the dynamic corrections for each trace prior to stacking. Strictly speaking, as the distance increases the stacking velocity should also increase. These differences are of relatively minor importance in stacking, if the stacking velocity has been determined from the data.

Common-reflection-point trace groups, rather than common-shot trace groups, are used for the input to be automated velocity determination programs to avoid the effects of dip on the velocity determination on the assumption that the data are all from a common subsurface depth point. In fact, this is not the case. Dipping beds cause the subsurface reflecting point for the common-depth-point group to shift slightly updip with increasing distance. The result will be a further velocity change but it will still be the best stacking velocity.

Stacking velocities need not fall in the range of physical velocities. Near-surface velocity variations, such as those caused by irregular layers of caliche, limestone, permafrost, or improper time measurements resulting from poorly corrected elevation changes, can cause high or low stacking velocities. Nevertheless, in all cases these are generally the velocities that provide the best stack of the data so they should be used accordingly. (**Fig. 8.5**) shows the result of a near-surface wedge of limestone pinching out along the line. As a result the travel path through the limestone had a shorter time than the other, distorting the apparent velocity.

If the stacking velocities are used in automatic migration processes and/or in the calculation of depth sections, the resulting depths will always be too deep, because the stacking velocity values will always be high.

Automatic section migration can be extremely sensitive to incorrect velocities, and although the stacking velocities may be used, some correction for the spread distance may be necessary when the migration grating is calculated.

Vertical Velocities

STACKING VELOCITIES ARE NOT NECESSARILY REAL VELOCITIES

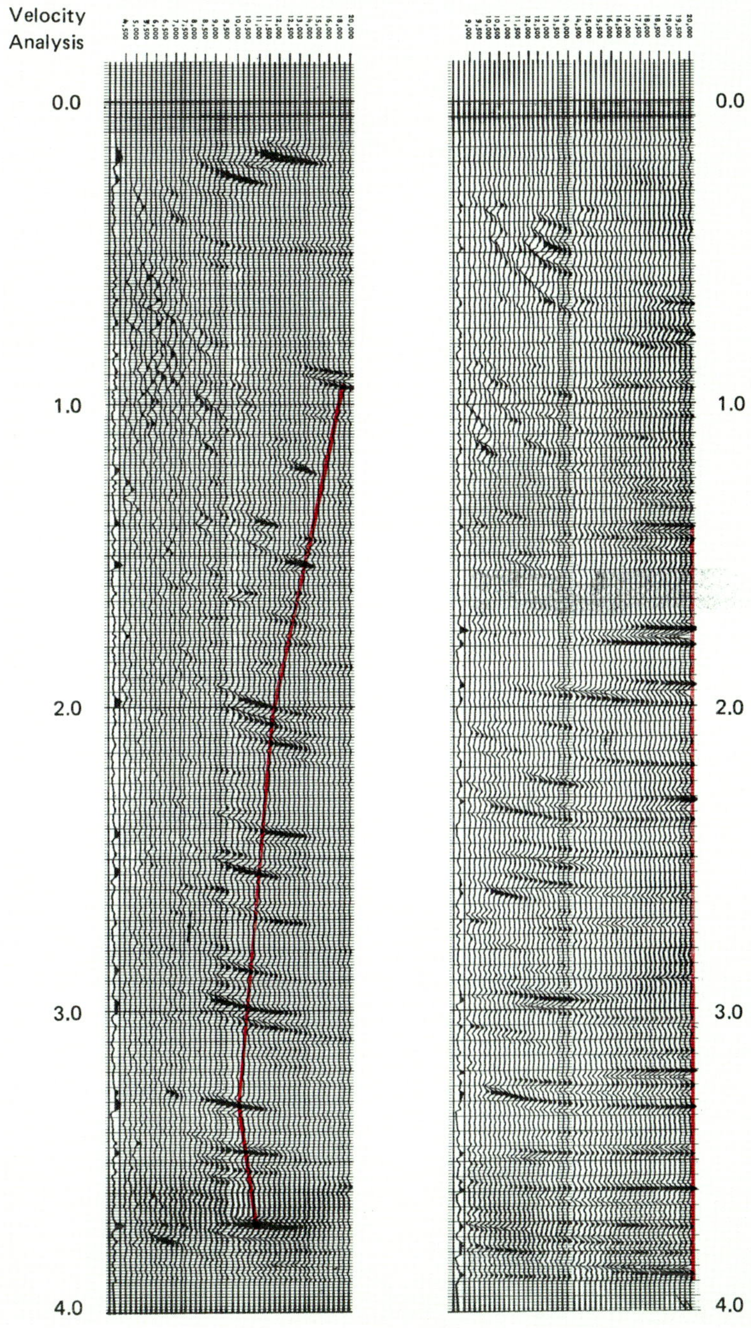

FIGURE 8.5

Two stacking velocity analyses that yield unlikely velocity functions due to near surface conditions. Nevertheless, the velocities shown will yield the best stack of the traces.

Vertical velocities , if destined for use in geological work to determine thickness and depth, or to indicate possible lithologic changes, must have a much higher degree of precision than stacking velocities. Unfortunately if derived from stacking velocities, the velocity source data are not very precise. The desired information is often buried in noise and the measurements are highly leveraged. That is a small error in source data measurements leads to a large error in the output. Hence a large part of the difficulty in extracting vertical velocities from stacking velocities is not so much the conversion, but rather in gathering a statistically valid data set which has adequate detail for the purpose.

Check-Shot Well Surveys

Traditionally, the only reliable source of vertical velocities is the borehole checkshot survey. The advent of sonic logs provided another source of information, but they are not always reliable, due to occasional mechanical or instrument problems and more frequent borehole conditions such as formation alteration around the borehole and large cavities caused by washouts. More importantly, the upper 800 or 900 ft. of hole is often not logged as it usually contains surface casing which precludes reliable readings from the surrounding rock.

The original (and still highly recommended) method of checkshot survey consists of hanging a geophone at discrete levels in a borehole, usually at or near all key geologic tops, and firing a check shot near the surface for each position. The geometry and arrangement of equipment usually requires modest corrections to the travel times, which then

can be used to calculate average velocities. The interval velocity between each depth point is also calculated. This procedure, combined with the relatively few major reflectors mapped, seems to have fixed into place the common model of a sedimentary section which can be divided into a few major divisions having discrete constant interval velocities (and discrete reflection coefficients). In actual fact, as can be seen from any sonic log, the velocity within the section is a continuously varying function, rarely having either the sharp step functions or the uniform interval velocities commonly assumed.

Check shot surveys are the most reliable source of velocity information but are not necessarily without error. Sources of error consist mainly in failure to recognize the first arrival time, due to excessive attenuation of the signal in transmission. This yields a velocity which is slower than the actual value.

Other sources of error may include a refracted break, when the energy source is offset a substantial distance from the well. The signal travels down to a high velocity reflector, then nearly horizontally along it to the borehole. If the energy source is close to the borehole, the first arrival signal may travel down through the surface casing, shortening the travel time. In both cases, the result is higher-than-true velocities.

One of the logging service companies now provides a check shot service simultaneously with borehole logging, primarily for offshore operations. This procedure uses an airgun suspended in the water as an energy source.Instead of simply recording the travel time, a reference geophone is suspended in the water as a zero time reference and travel times are measured as the difference between the arrival at the reference geophone and the borehole geophone. A relatively large number of check shots are obtained for a survey.

The procedure of using the difference between the near-surface reference geophone and the borehole geophone eliminates any instrument lag, while the conventional practice of recording the normal time break and single geophone through seismic instruments leaves the normal instrument time lag in the observations. A comparison between the true travel times measured by the continuous check shot survey, to readings obtained by the conventional method, reveals the substantial time lag common to conventional instruments. True travel times, compared to recorded seismic times, will differ by amounts varying from 25 ms to 100 ms, depending on the instruments, with about 60 ms being the most common amount. This difference should be recognized when comparing check shot surveys to conventional seismic recordings.

The quality of these surveys is usually excellent. The numerous closely spaced shots provide very close control of the vertical velocities at the location.

Sonic Logs

From a theoretical standpoint, the continuous sonic log should be the best and most complete form of velocity information possible to obtain at a given location. The method uses a logging tool, which contains both a rapidly repeating signal source, and one or more receivers. The tool is lowered to the bottom of the borehole and slowly drawn up the length of the column, firing as it goes, and continuously recording the velocity across the small interval between the source and receiver, or between two receivers, to provide a continuous, very detailed velocity profile of the geologic column.

The sonic log was first introduced about 1950, and was one of the first of new generation of borehole logs to complement the then existing electrical logs, by measuring some property of the geologic column other than electrical. The sonic (and, to a lesser degree, the density) log provides a direct link between geology and seismic geophysics. Geologic tops are very commonly picked at the boundary of a major age division or at some pronounced change in lithology, or both. Such changes are commonly also associated with a change in velocity and velocity

THE THREE COMMON CONFIGURATIONS OF SONIC LOGGING TOOLS

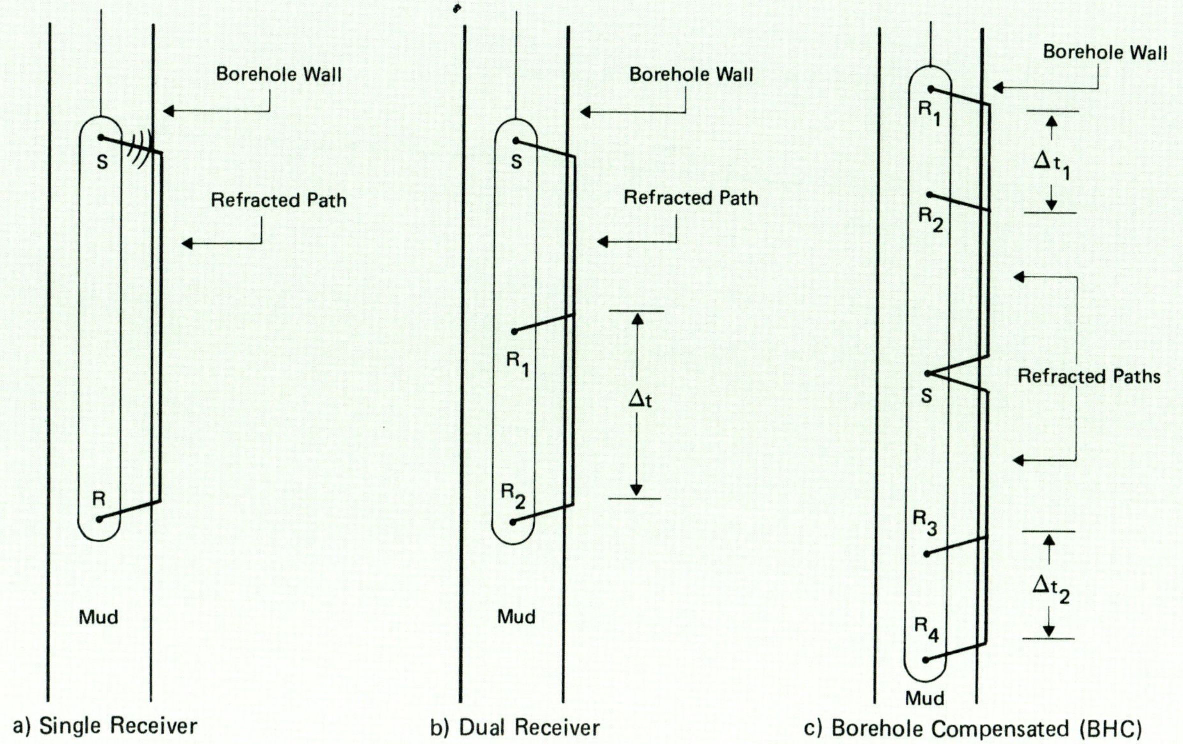

FIGURE 8.6

The borehole sonic log measured the travel time over a short interval of borehole wall. Dual receiver and Bore Hole Compensated configurations reduce errors caused by travel through borehole diameter.

contrasts yield seismic reflections. Faust has shown that velocities in the sedimentary section are a function of lithology, depth, and geologic age. Porosity will also lower the velocity within a given lithology.

Sonic logs measure the transit time (which can be expressed in terms of velocity) over a very small interval of the geologic column repeated at very small increments. Depending on the logging speed and the firing rate of the gun, the increments are probably about 3 inches. For all practical purposes, they yield a profile of instantaneous velocity over the geologic column surveyed. Integration of the transit-time measurements yields a detailed travel-time profile through the sections.

A sonic log measures refracted rather than the reflected travel time measurements of a seismic reflection survey or the direct travel of the check-shot survey. The logging tool is suspended in a borehole filled with drilling fluid (usually drilling mud) having a velocity only slightly faster than water. Sound waves travel from the source in the tool, through the drilling fluid to the formation, where they are refracted along the borehole wall to travel through the higher velocity rock material, and then back through the drilling fluid to the receiver. (Fig 8.6)

The first arrival of a refracted signal always has traversed the shortest-travel-time path, which may not be the shortest distance, and so can have an important influence on the resulting log.

Originally, sonic logging equipment used a single source and receiver **(Fig 8.6a)**. The recorded travel time is the total travel time over the path followed from the source through the drilling fluid to the borehole wall, through an interval of rock and back once again through the fluid to the receiver. A caliper device is used to press the tool against the wall of the borehole and minimize the amount of mud the signal must travel through, although this is not always successful. Travel through the drilling fluid is usually considerably slower than through the rock, so recorded travel times were nearly always excessively long and the recorded velocities slower than rock velocities.

This system was soon improved by the addition of a second receiver, **(Fig 8.6b)** using it to record the difference in transit times between the two receivers. In this manner, the travel times through the drilling fluid largely cancelled out, and a much superior measurement was obtained. However, boreholes occasionally are not true to gauge and have irregular variations in diameter. When this occurs, although the travel path from the source to the borehole wall will be the same for both receivers and cancel out, the length of the travel path back to the receivers may be different when an enlargement in the hole, due to a washout, is opposite only one of the receivers. Then a difference in time through the borehole fluid will occur and the mud times will not cancel. To some degree, compensating equal and opposite errors will arise as the tool crosses the washout, so the total integrated time will not be seriously in error, only the values of a few local measurements will be false.

A further refinement was added with the introduction of the Bore Hole Compensated tool, (**Fig 8.6c**)in which two sets of receivers at opposite ends of the logging tool are used, and the results averaged. This tool reduced the percentage of bore hole errors, but has a slightly smoothing effect on the output curve, although not enough to make an important difference for most applications.

Sonic logs are frequently used to compute synthetic seismograms for velocity information, and the results are usually accepted almost without question by the majority of geophysicists. Yet there are several sources of important errors on sonic logs which must be learned and recognized by any user.

The two key considerations are the nature of transit of any refracted signal path, and the condition of the hole surrounding the tool.

Logging runs are made infrequently, and only after drilling several thousand feet of borehole. In the drilling process, the pumping of drilling mud up the borehole annulus around the drill pipe tends to erode the sidewalls, particularly through the softer formations. Often large washouts can occur; an effect most common in the upper part of the hole where formations tend to be softer and exposed to drilling erosion for the longest period of time. Even though these washout effects may be compensated for with the dual-receiver or the BHC log due to the geometry of refracted paths, almost invariably borehole washouts tend to increase the measured travel time on sonic logs. In the soft, young formations frequently found on offshore continental shelves total integration errors of 200 to 300 msec over a few thousand feet are not uncommon.

Check shots help determine the true travel time and the sonic log can be corrected accordingly. Where check shots are not available an examination of the caliper log, if one was run, provides a reliable guide to potential errors. Velocities across enlargements in hole diameter should always be viewed with skepticism. A reliable means of verifying appreciable errors is simply to correlate the log, or a synthetic produced from it, to the seismic section at the well location. If the seismic section exhibits shorter time intervals than the log, it is most likely correct and the log is wrong.

Logs can be edited to compensate for recognized errors. But there is no guarantee that any of the several ways of accomplishing this are correct. Some operators merely shift the entire curve the amount required to integrate to the correct total time between check shots. This is incorrect because the resistant members will probably have recorded correct velocities while the errors may be confined to the (usually) low-velocity beds in the washed out intervals. In other words the errors vary with geology while the correction assumes a uniform rate of error.

Another common editing practice is simply to draw a straight line across the washed out segment of the log at or near the velocity of the more resistant materials on either side. This is also often incorrect. Washouts are most likely to occur in the softer, lower-velocity formations and most commonly the velocity of the washed out material is lower than the more resistant surrounding material.

The best solution appears to be one which distributes the corrections as a function of apparent velocity, applying the greatest proportion of correction to the lowest apparent velocity.

Other causes of attenuation include, most commonly, cycle skipping which is usually the result of differential attentuation between the near and far receivers. Most commonly the signal arriving at the far receiver is attenuated more than at the near receiver. The timing device may not recognize the first arrival at the far receiver but sees a later cycle, effectively lengthening the measured time increment between near and far receivers.

An opposite effect may occur in very porous formations. Heavy drilling mud is used in the borehole to transport cuttings of the hole, to stabilize the borehole walls by exerting pressure against them to avoid blowouts, and to seal off porous zones which might otherwise drain off the drilling fluid. In the last instance drilling mud leaking into a porous formation will separate; the fluid component leaking into the formation while the solids component, acting as fibrous material, mats across the porous zone to plug off the leaky formation, building up a mud cake until there is no more leakage through it. But transit time in porous formations (which are also the hydrocarbon reservoirs) increases as porosity increases, and the presence of gas greatly increases the transit time. In such cases, travel through the mudcake may be faster than travel through the formation, and the log will exhibit the higher mudcake velocity across the porous zone. This is common, but often not recognized on the log, since the error in transit time is small. If hydrocarbons fill the formation, examination of an electric log, in conjunction with the sonic, may show where such conditions exist.

Similar response may occur across fracture or vuggy porosity. The sonic signal will travel around fractures, through the higher-velocity matrix rock, and not reveal the porosity. Seismic reflection waves, travelling through the formation, may be influenced by the total effect of the fracture porosity, or vugs, and provide a more reliable measure of travel time.

Sonic signals in cased holes will always travel down the casing at the velocity of sound in steel.

These common phenomena can, and frequently do, distort the travel times and velocities obtained from sonic logs. Other sources of error include equipment malfunctions, miscalibration and incorrect depth measurements. All of these can yield incorrect velocities, and when the sonics are used for model studies they yield misleading synthetic seismograms. The net result is that although sonic logs provide the highest and potentially the most accurate velocity resolution, they also include the risk of several types of error, and they should always be examined critically, checked, and, where possible, verified by check shots.

In spite of the many advantages of the sonic logs, it is quite surprising how many logging programs are designed

Vertical Velocities from Stacking Velocities

without sonic logs, or the sonic is logged only a short distance across the zone of interest.

The increasing availability of stacking velocity information has prompted interest in techniques to obtain true vertical velocity information from stacking velocities. Although basic conversion techniques are fairly simple, the major problem in converting stacking velocity information to vertical velocities is the order of precision required, since it is much higher than is normally required for stacking purposes. Inherent sources of error in stacking velocities may preclude the determination of true velocity. The typical stacking-velocity function often requires only a few widely spaced points on the time axis to define a curve commonly accepted as adequate for the purpose. The tolerance for error is fairly large, and even substantial geological changes rarely cause major velocity variations.

When stacking velocities are properly converted to vertical they yield an average velocity curve equivalent to the integration of instantaneous velocity over time. The average velocity curve tends to converge to a constant value with increasing time. Often the required information is the instantaneous, or interval, velocity curve which can be derived by calculating "first differences" of the average velocity curve. However, in the process, small, seemingly insignificant errors in average velocities blossom into unacceptably large interval velocity errors. There are methods for improving the information required, but they require additional processing.

In any case the first step is to determine a method to extract reliable vertical velocity information from stacking velocities. One procedure is to assume that the stacking velocity equals the RMS velocity. Dix proposed the following solution to obtain vertical interval velocities from a series of stacking velocities;

$$V_i^2 = \frac{(\bar{V}_i^2 t_i - \bar{V}_{i-1}^2 t_{i-1})}{(t_i - t_{i-1})}$$

Where: $\bar{V}$ is the stacking velocity at reflection time t and
V is the interval velocity between t_{i-1}, and t_i.

The Dix approximation is very good for uniform sequences of flat dip where the spreads are not overly long. Elsewhere, a better method of determining true vertical velocities, (but still reliable only for relatively flat sections) is by a ray-path-tracing procedure which calculates, in an iterative fashion, the true travel path. There are a number of approaches to this problem but all are based on distance, time and Snell's Law combined with optimization techniques which determine the minimum-time travel path. Commercial routines usually have some degree of logical analysis built into them to ensure that the values picked are physically reasonable. Under favorable conditions they will provide a good estimate of true vertical velocity. But for many reasons, including anomalous surface conditions, stacking velocities often may be a relatively uncertain source of true velocity data.

In the first place, the source of stacking velocities is the normal moveout curve which is a relatively insensitive function of the average, or stacking velocity curve, particularly at high velocities. Average velocity, in turn, is the integration over time of the continuous instantaneous velocities within the sedimentary sequence as would be logged by the borehole sonic, and it is also less sensitive to local velocity changes compared to instantaneous velocity. If a sonic log is integrated to produce a continuous average velocity log, and is then plotted in detail, all of the events on the sonic log will be apparent in the very shallow part of the section on the average velocity curve, but they will show diminishing contrast as time increases (**Fig. 8.7**) . The normal moveout curve calculated from the average velocity curve, when suitably scaled, is very similar in appearance to the average velocity curve, but the broad range of velocities is reduced to a range of only a few milliseconds of normal moveout.

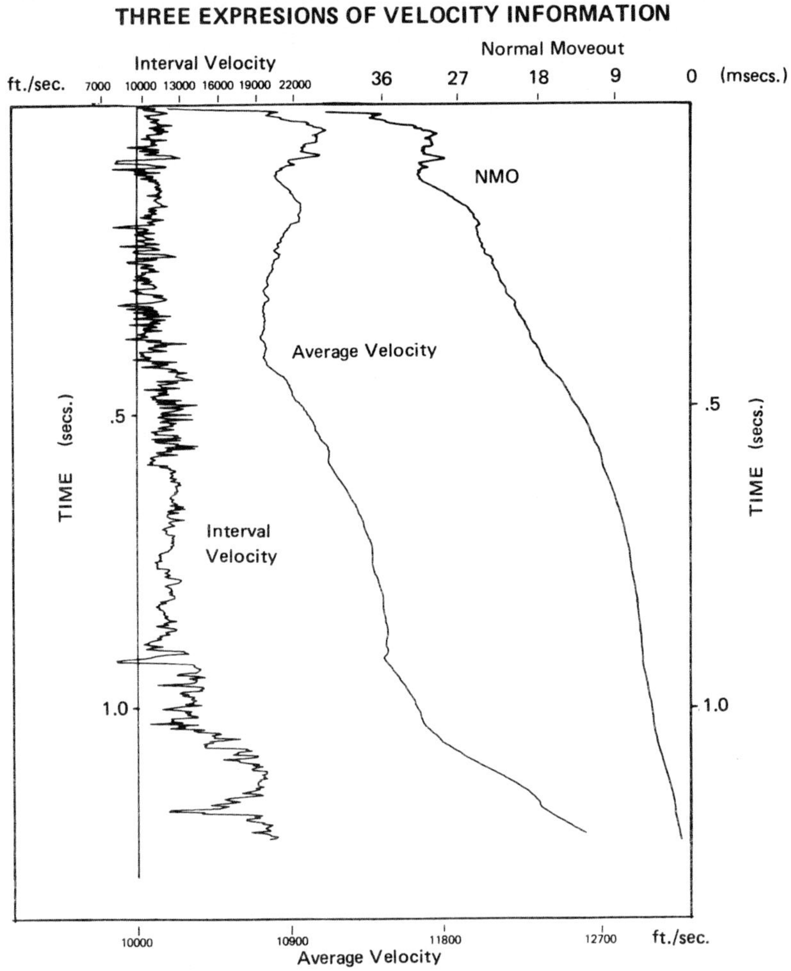

FIGURE 8.7

The sedimentary section does not usually contain large intervals of constant velocity, but rather a continuously varying instantaneous velocity profile (left) of a sonic log, changing as a function of rock lithology. If integrated over total travel time it yields the average velocity profile (center) over the velocity range at the bottom of the figure, as would be measured by check shots in a borehole velocity survey. The computation which combines average velocity with spread geometry yields the normal moveout curve on the right side of the figure. All three curves are merely different representations of the same basic information, but their resolution varies greatly.

The normal moveout curve, although a direct function of average velocity, is further complicated by geometric distortion (**Fig 8.9**). The geometry factor is merely a smooth exponential curve which varies with depth and/or distance for any given constant velocity, but when the average velocity increases with increasing reflection time, as it normally does, the velocity component cuts across the geometric component. The normal moveout data therefore become an expression of the geometry distorted by the velocity spectrum.

Increasing depth and velocity both serve to reduce the amount of the normal moveout curve (**Fig 8.8**).

As a result of the low contrast the output will change very little between adjacent velocity applications and it may be difficult to select the best-fit velocity except on high-amplitude reflections. In the usual case, velocity analysis values tend to be picked at major reflecting events, ignoring all intervening information. This action tends to reinforce the common assumption of a relatively small number of major rather abrupt velocity changes, separated by uniform interval velocities.

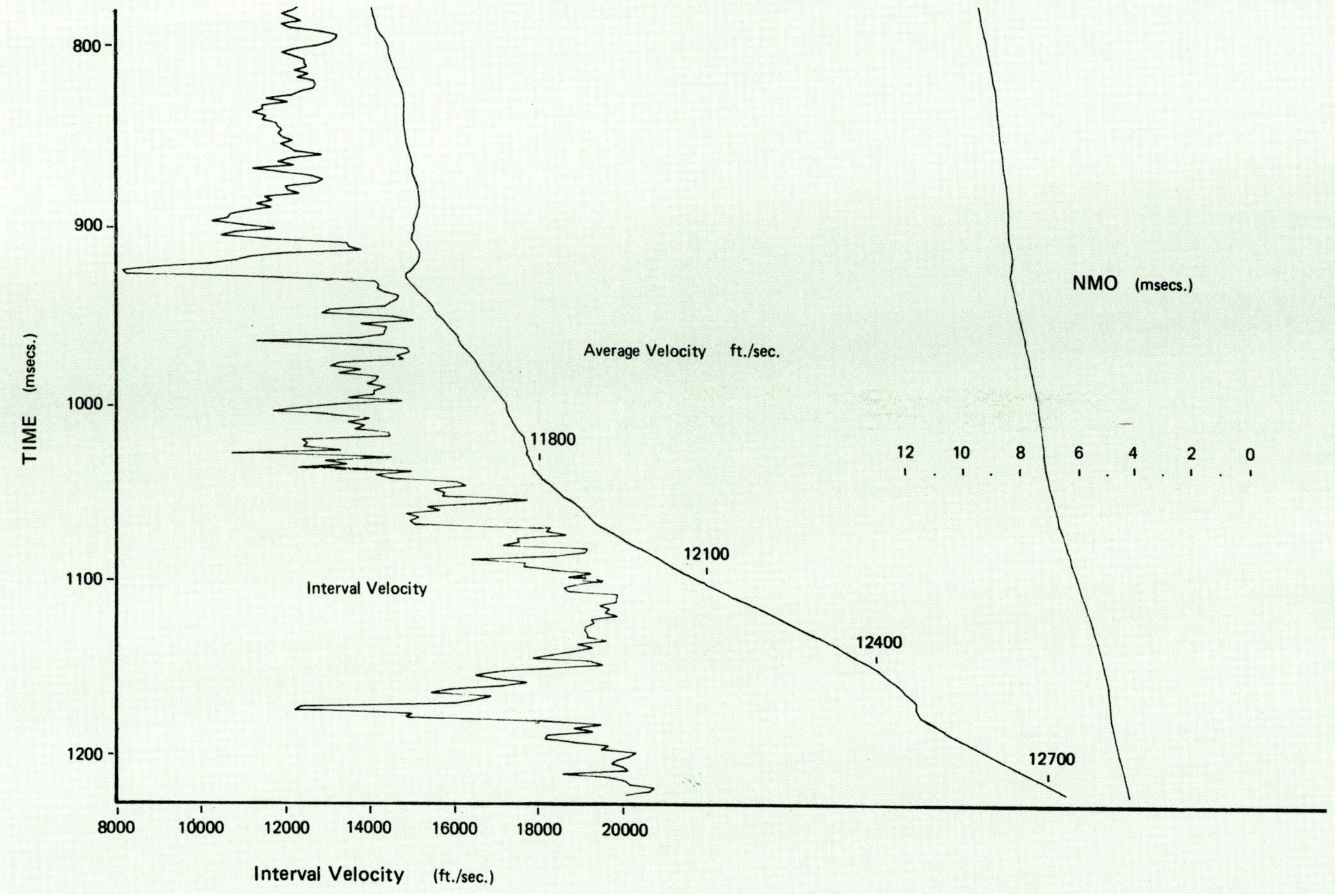

FIGURE 8.8

Expanded scale of the lower part of Figure 8.7 illustrating the very small changes in normal moveout from which data processing endeavours to extract the corresponding interval velocity information.

SAMPLE PROBLEM STACKING VELOCITIES

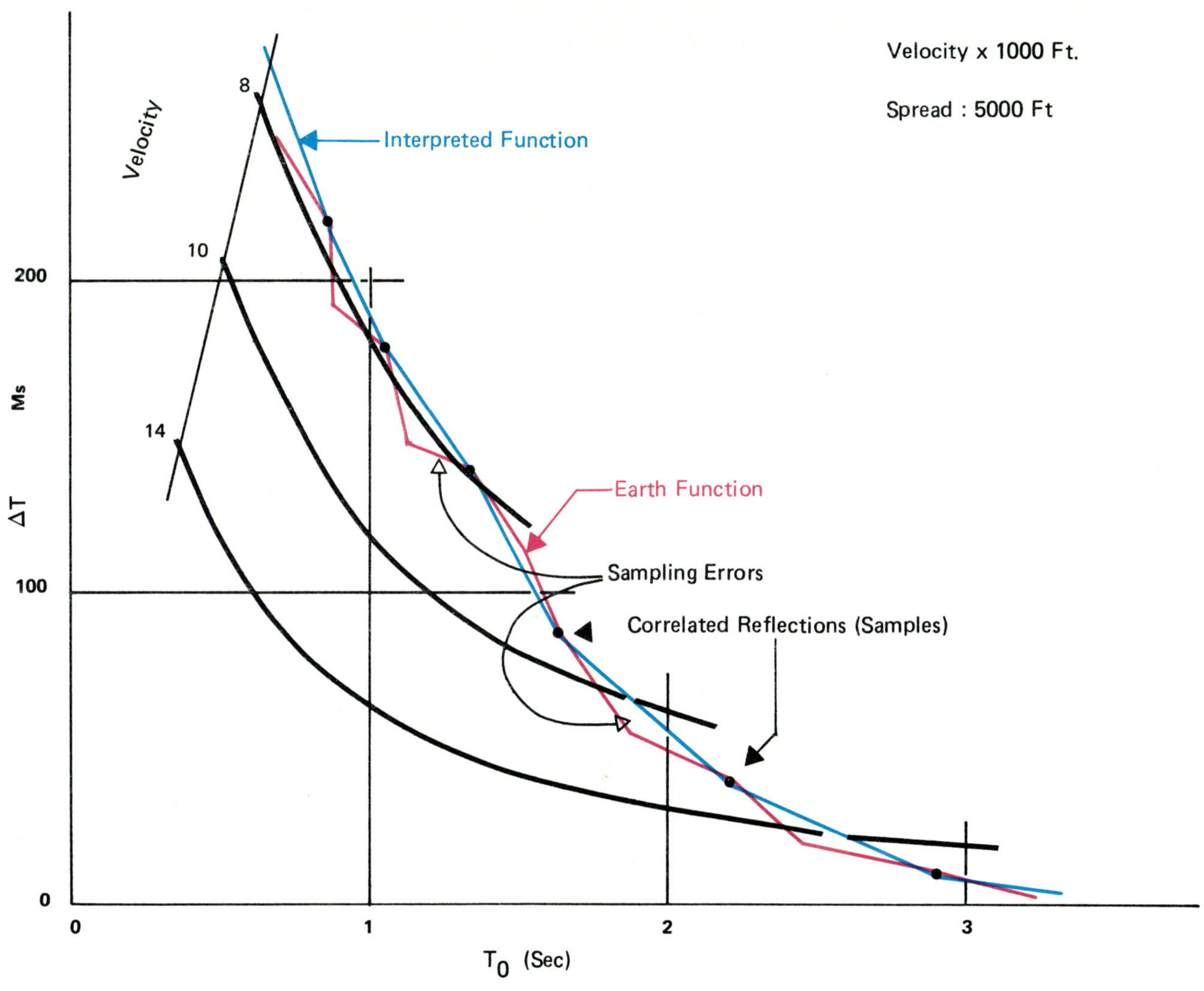

FIGURE 8.9

The smooth curves represent the geometric distortion of the constant velocity. The subsurface velocity function is the coloured curve cutting across the constant velocity curves as velocity increases at greater depths.

A sonic log, even one which has been high-cut filtered to the upper seismic reflection frequency range, will show that this is indeed a mistaken assumption. Many substantial changes can occur in rapid succession due to thin beds in the section. Many of these are alternate contrasts, at the top and bottom of thin beds, but nevertheless they can be substantial.

Also, not all velocity changes occur as sharp step functions. Velocity changes can occur in either of two ways, and only one of these is a major contrast across a boundary, resulting in a strong reflection. Most velocity changes are gradational or occur as a series of much smaller steps in succession, so that over a period of time represented by, say, 100 ms on a seismic record, a major velocity change may have accumulated but at no place does a major shift occur. These gradual changes may not produce strong reflections but account for many major velocity shifts in the geologic section, and are the result of a very common physical situation in the earth, such as a sand grading upward to shale.

The Low-Frequency Component of the Velocity Function

It can be shown very simply that a sonic log which, under good conditions, provides a very faithful representation of the velocity response of the sedimentary section, when integrated to a vertical time scale, has a very broad spectrum of frequencies, the result of rapidly varying lithology and velocity within the sedimentary column. Each change in velocity produces a reflection coefficient. The series of reflection coefficients, which make up the earth signal, has the same frequency spectrum, except that the relative emphasis, or spectrum density distribution is different. Sonic logs are heavily weighted at the low frequency end of the spectrum, while reflection coefficients, representing a form of derivative of the sonic, have their greatest emphasis at the high frequency end of the spectrum. Filtering the sonic log with successive high-cut frequency filters (Fig. 8.10) shows the basic velocity structure of the log to be made up of low frequencies. In fact, the basic shape of the log is contained in the very same range of low frequencies that are normally excluded from a seismic recording.

The illustration confirms a concept introduced earlier; instantaneous velocity is best described by a continuous function rather than by the blocked or stepped interval velocity that is normally used by geophysicists to represent the interval velocity structure of the section.

In fact, as illustrated by **Figures 8.7** and **8.8**, the normal moveout curve, the average velocity curve and the sonic are all merely different expressions of the same continuous curve of velocity information found in the earth whether on the sonic log or in the seismic data. Therefore, any form of the information can and should be treated with the same procedures and constraints as would be used in processing any other given form of the same information.

The use of a sampled data series to represent continuous frequency functions is governed by sampling theory. This includes the definition of the rate at which a given function must be sampled in order to maintain its fidelity, or, conversely, the limits of frequency that can be defined by a given sampling rate. Only when the sampling constraints are properly observed can the data be analyzed statistically and/or filtered correctly in order to remove noise.

The model of a geologic section made up of a few homogeneous layers, having precise interval velocities bounded by abrupt steps at the formation boundaries, is a very gross approximation to the real situation and, in the case of a continuous sequence of deposition, or the pseudo-random sequence which forms the model for many deconvolution processes, it is patently erroneous.

Strictly speaking, even the usual stacking or check-shot velocity samples, in which interpolation between control

LOW FREQUENCY CONTOURS OF A SONIC LOG

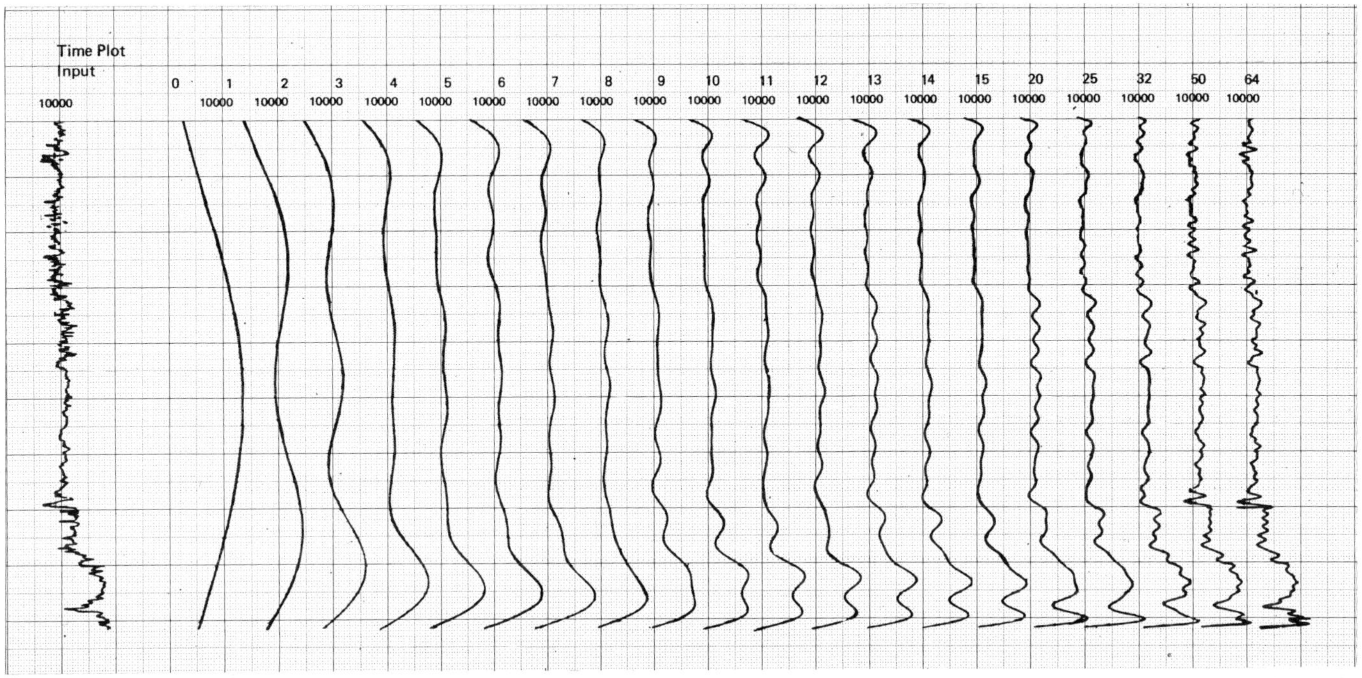

FIGURE 8.10

The integrated sonic log on the left has a frequency range extending from 0 Hz to 500 Hz. The high frequencies have been filtered off to demonstrate the resolution remaining in the low frequency components. Note that the basic velocity structure of the log is contained in the 0 - 15 Hz range, most of which includes frequencies not normally recorded in the seismic signals.

points is used to define an average velocity which increases linearly with time, does not fit the constant-interval-velocity model. Since a constant interval velocity would produce an average velocity curve which would asymptotically approach the interval velocity, a ramp interval velocity function might yield a better model. If derived velocity information is considered as a continuous frequency function it will generally result in a much better fit to sonic log data than any model which purports to define exact interval velocities over large increments.

The frequency with which a velocity analysis or any continuous signal is sampled determines the highest frequencies which can be defined uniquely. Sampling theory requires at least two samples per wavelength of the highest frequency present in any signal. If the normal moveout is picked every 100 msec (probably better than average), then the highest frequency that can be measured is one having a period of 200 ms, or 5 Hz. It follows, then, that the sampling rate must be increased if frequencies higher than the limit imposed by an existing sample rate are to be measured.

Another related problem with deriving velocity information from normal moveout (and probably another argument for fitting smooth functions to velocity samples) is that small errrors in stacking velocity measurement blossom into very large errors in interval velocities. Common practice in constant-velocity analysis is to apply a fixed set of constant velocities to a collection of traces. The selection of the best velocity fit is made by a visual, or sometimes automatic, selection process. But the results are extremely sensitive to small errors in the choice of velocity and, to a lesser extent, to the precision with which the reflection time can be determined. Hence, if the correct velocity for a given reflection time is not included in the applied set, the selected best fit will be larger or smaller than the correct value, and an error will be introduced.

VELOCITY ERRORS RESULTING FROM PICKING ERRORS

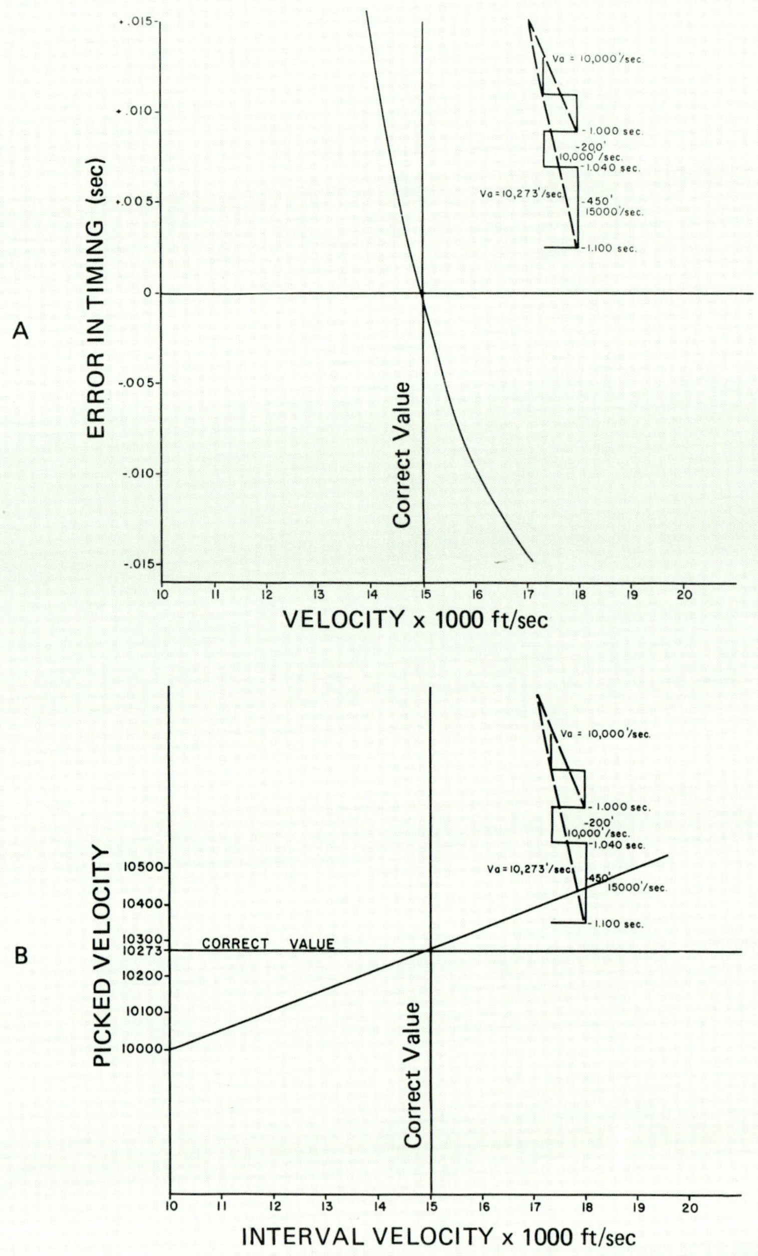

FIGURE 8.11

Correct evaluation of velocities is critically dependent upon precise picking of values. The errors vary with the condition, but the model demonstrates a typical case. If the picked value is located on the left side, a horizontal line extended to intersect the diagonal, and dropped to the bottom, it will provide the apparent values.

To illustrate, an arbitrary model has been developed (**Fig. 8.11**) for a rock unit having an interval velocity of 15,000 ft/sec over an interval of 450 ft. These are rather typical parameters from real data. In the illustration, the exact stacking velocity required is 10,273 ft/sec. The constant-velocity analysis might have used test velocities of 10,000 and 10,500 ft/sec, giving the interpreter the choice between two erroneous values. The chart demonstrates that an error of a few hundred ft/sec in the selected average velocity may cause several thousands of ft/sec error in the derived interval velocity (**Fig. 8.11b**). The problem may be exacerbated by relatively large steps between each constant velocity used for the velocity analysis. A fixed set of velocities used for several analyses may tend to apply a bias toward a nearby, but incorrect, value in the results. A better method is to vary the test velocities on a random basis within specified step sizes.

A smaller but still important error can result from errors in the time picks (**Fig. 8.11a**).

It is extremely unlikely, with the customary procedure of making a relatively small number of velocity picks, with widely spaced time, velocity or space increments, as is commonly used for stacking velocity determinations, that reliable determination of true velocity and the corresponding implied lithology can be obtained.

The problem is analogous to that of obtaining a detailed subsurface map from sparse seismic data which is

VELOCITY ANALYSIS 7 HZ

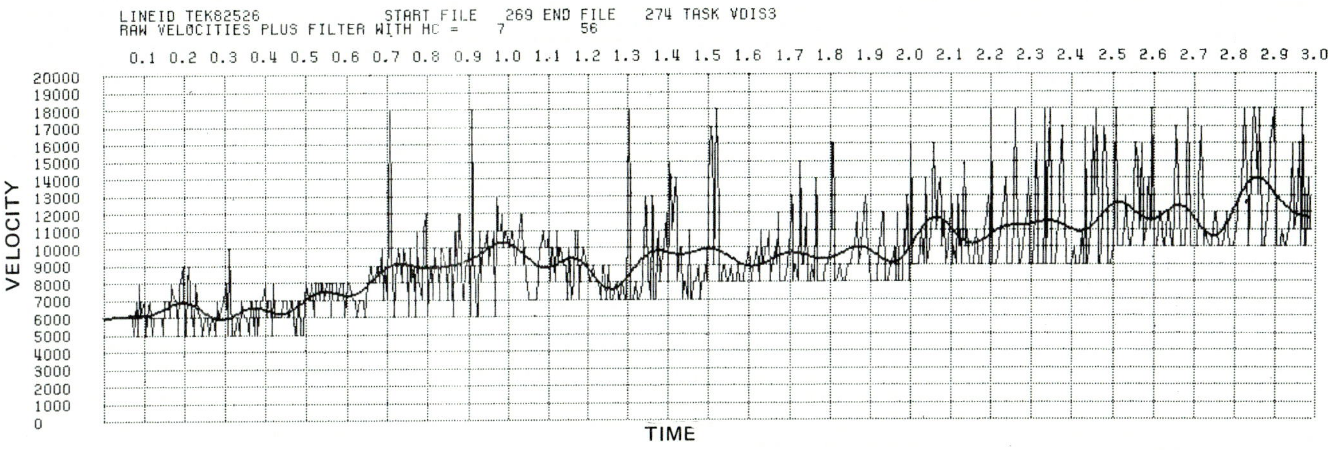

FIGURE 8.12

Under favorable conditions, a filtered function of machine picked velocities will yield a detailed velocity function.

distorted by noise, although the degree of the problem may be different. Like seismic data, the useful information can be enhanced by applying statistical processing techniques, including redundant coverage, proper sampling and filtering.

A major improvement of seismic data quality is obtained by stacking several observations of the same subsurface point. Stacking is a statistical process which uses the average response of several repetitive recordings of the same seismic reflection event. Another way of looking at stacking is to consider it as a type of array, or velocity filtering, which passes those events that are in phase, (flat dipping on the trace-gather collection) and filters out those that have some other distribution. The constant-velocity stack method exploits the filtering properties to maximum advantage.

Manual selection of velocity values can be very tedious. Mechanical methods of picking the optimum stacking velocities are numerous and operate with varying degrees of success. The machine selections are often erratic, particularly in zones of relatively low amplitude reflectors, noise bursts or multiples. But a machine can pick a large number of points which then can be statistically evaluated. The example of (**Fig. 8.12**) illustrates rather unusually good results of a machine-picked constant-velocity analysis producing an output value each 8 ms. The raw results are extremely erratic and apparently of little use, but application of a 7 Hz low-pass filter to the results yields a smooth, continuous, low-frequency velocity curve.

A single velocity analysis curve of this type probably contains residual errors. But several curves, closely spaced, can be filtered or averaged in two-dimensional or three-dimensional space to produce more reliable results.

The technique is illustrated by a series of examples (**Fig. 8.13**). The first display is a series of velocity analyses, one every six shotpoints, used to compute dynamic corrections for marine data. One of the aids toward effective use of velocity information is proper presentation. A useful form is to plot velocity as a function of reflection

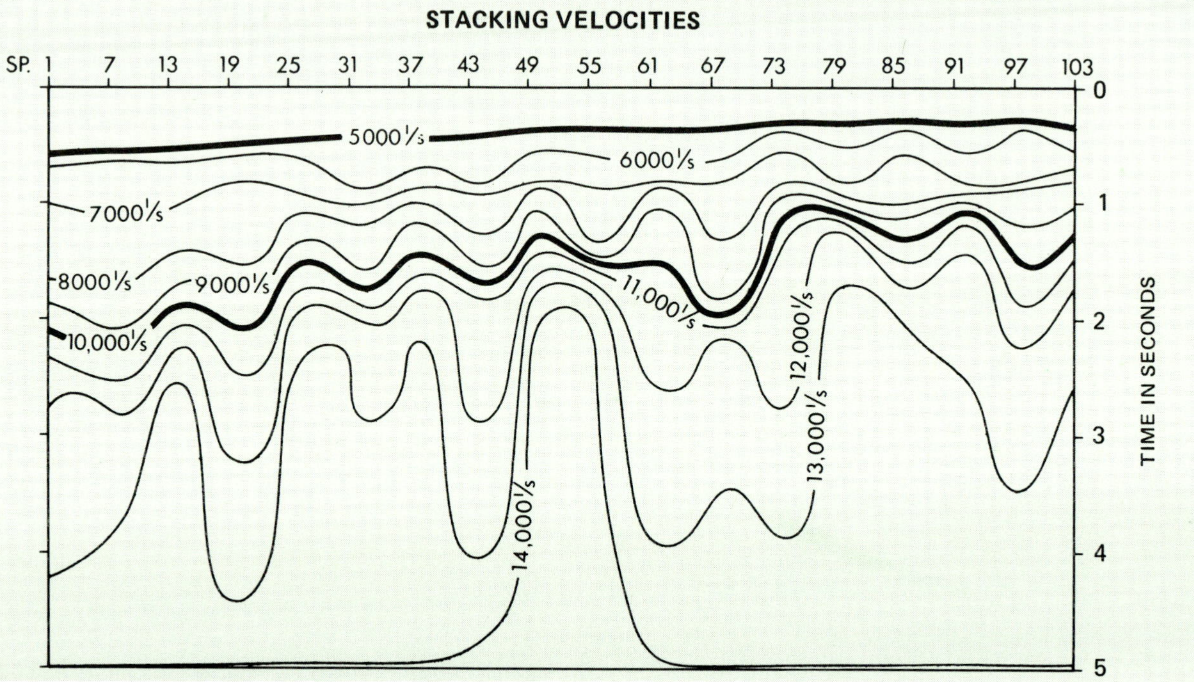

FIGURE 8.13a

Contours of stacking velocity functions obtained every six shotpoints yield a map which is geologically improbable. Seismic data indicates a relatively flat section. The variations are the result of normal errors in stacking velocity analyses.

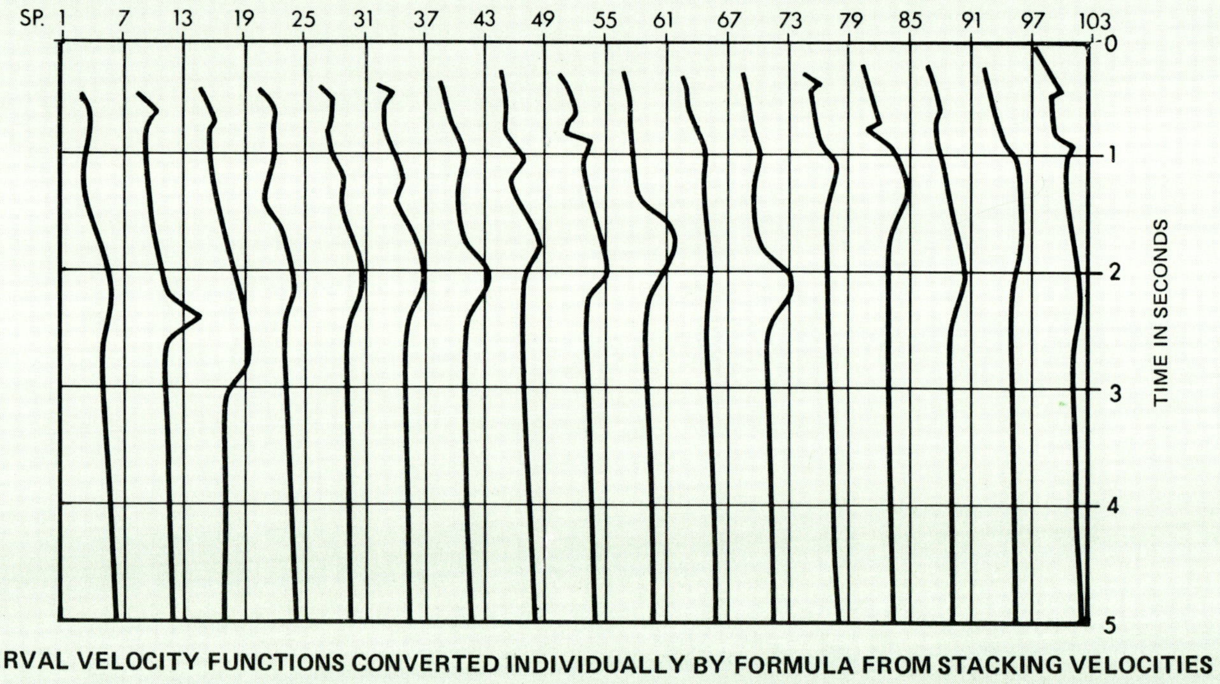

FIGURE 8.13b

Evaluation of the stacking velocities does not yield useable information. Variations in stacking velocities are magnified in the interval velocity calculation.

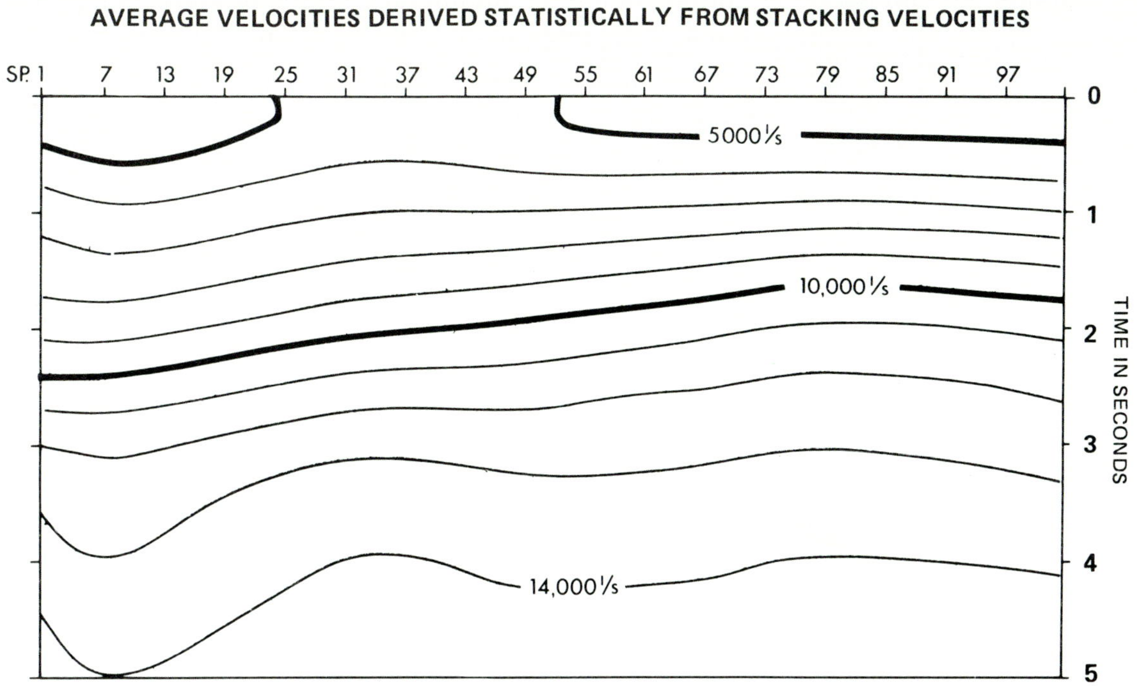

FIGURE 8.13c

Surface filtering techniques applied to the stacking velocities of Fig 8.13a, and ray tracing to produce average velocities, yields this much more plausible result.

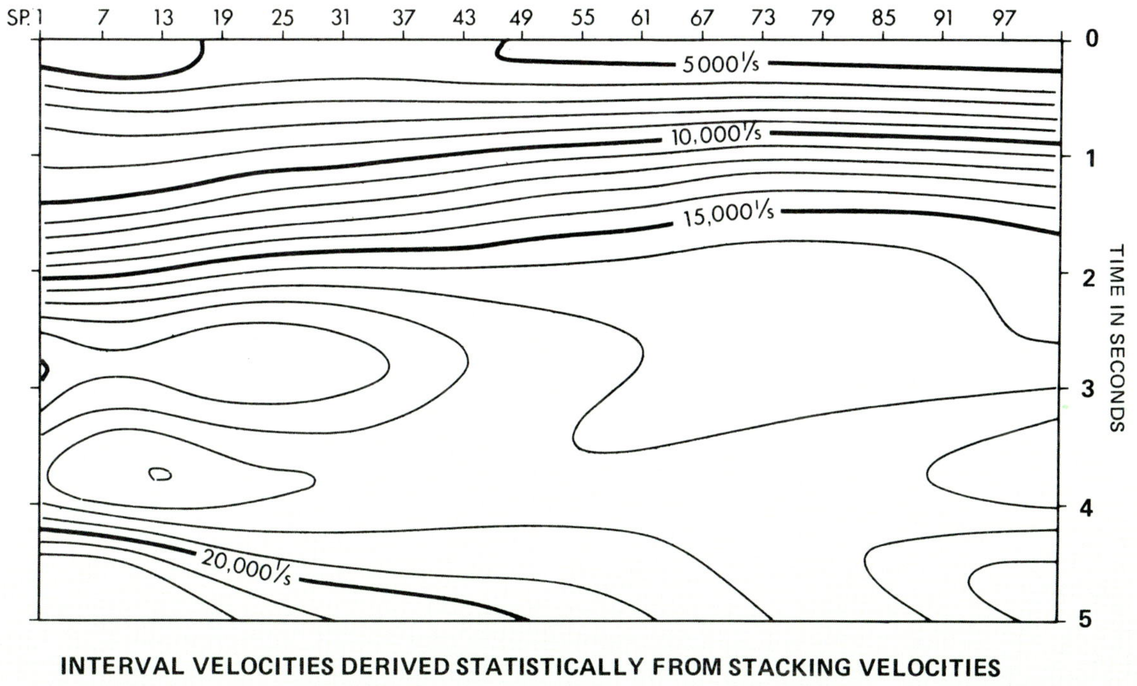

FIGURE 8.13d

Continuous evaluation of the average velocities above yields this instantaneous velocity surface. The change in rate near 15,000 ft/sec marks an unconformity. The high velocity pod on the left side marks a reef in the section. These data were extracted from the input of 8.13a.

time (**Fig. 8.13a**) at scales corresponding to the seismic section display, contour the velocities and use the velocity section as an overlay to show the relationship between velocity and seismic data.

For this example, the horizontal scale of the sample velocity section has been compressed in order to accentuate differences. It can be seen that the stacking velocities are very erratic, and likely do not represent geological conditions. Note at depth the increasing variation resulting from decreasing resolution.

These source data can be converted to instantaneous vertical velocities by interpolating between the stacking velocity control points and applying the Dix method of approximation to obtain vertical velocities. The results of this operation were also plotted (**Fig. 8.13b**) but due to the extremely erratic nature of the results contouring would be even more unreasonable than those of the source data, so it was not done.

The fundamental data were next reprocessed by increasing the frequency of the stacking velocity picks, by applying surface filtering techniques to the output section, and then converting the stacking velocities to vertical velocities by a ray path modelling technique.

The average velocities resulting from this operation (**Fig. 8.13c**) are now smooth and represent a reasonable correlation with geological expectations of subsurface structure. Buried in these improved results is even more significant velocity information which can be obtained by carrying this process one step further.

The instantaneous velocity on the sonic log is a derivative of the average velocity curve. The vertical average velocity data can be evaluated for interval velocity stepwise over the sample increment, say 8 ms, to develop a continuous instantaneous interval velocity curve. This has been done to the data of (**Fig. 8.13c**). The results (**Fig. 8.13d**) are quite revealing. The upper part of the section shows a uniform increase of velocity, consistent with the dip of the section, to a depth of approximately 1.5 sec. At this point, the velocities suddenly change; over the left half of the line very high velocities are noted at depth. The velocity at 3.0 sec at the extreme left hand end in 20,000 ft/sec, a rate commonly associated with dense carbonates. This velocity gradually decreases to the right, then suddenly drops to about 17,000 ft/sec in the center of the line, a value more typical of limey shales or sandstones.

The feature mapped is a carbonate reef with dense backreef facies to the left, a central zone of vuggy porosity and forefront facies of limey shales to the right.

It is to be emphasized that the final velocity interpretation came from the same basic source data. No change in basic data gathering technique is required to obtain this improved quality of information. The velocity data are obtained from the conventional field shooting and can be extracted by applying some of the same basic statistical procedures and sampling rules which apply to all seismic data. Conversely, deterministic evaluations, for example, the effort to predict lithology on the basis of two individual velocity picks, say 100 ms apart, is very prone to error.

Clastic Sections

Clastic sections present some of the most favorable possibilities for evaluations of this type. A typical offshore section features a thick layer of clastics resting on a severely eroded unconformity, often the surface of much older rocks usually with much higher velocity, which represent the acoustic basement. Velocities within the upper clastic section are normally quite stable and vary directly as a function of compaction and/or lithology. Horizons within a section of this type are rarely continuous, so that the standard procedure of mapping continuous horizons is precluded. This type of section is well adapted to statistical velocity analysis.

The average velocity computed for over 2,200 locations of a single mapping horizon has been plotted against

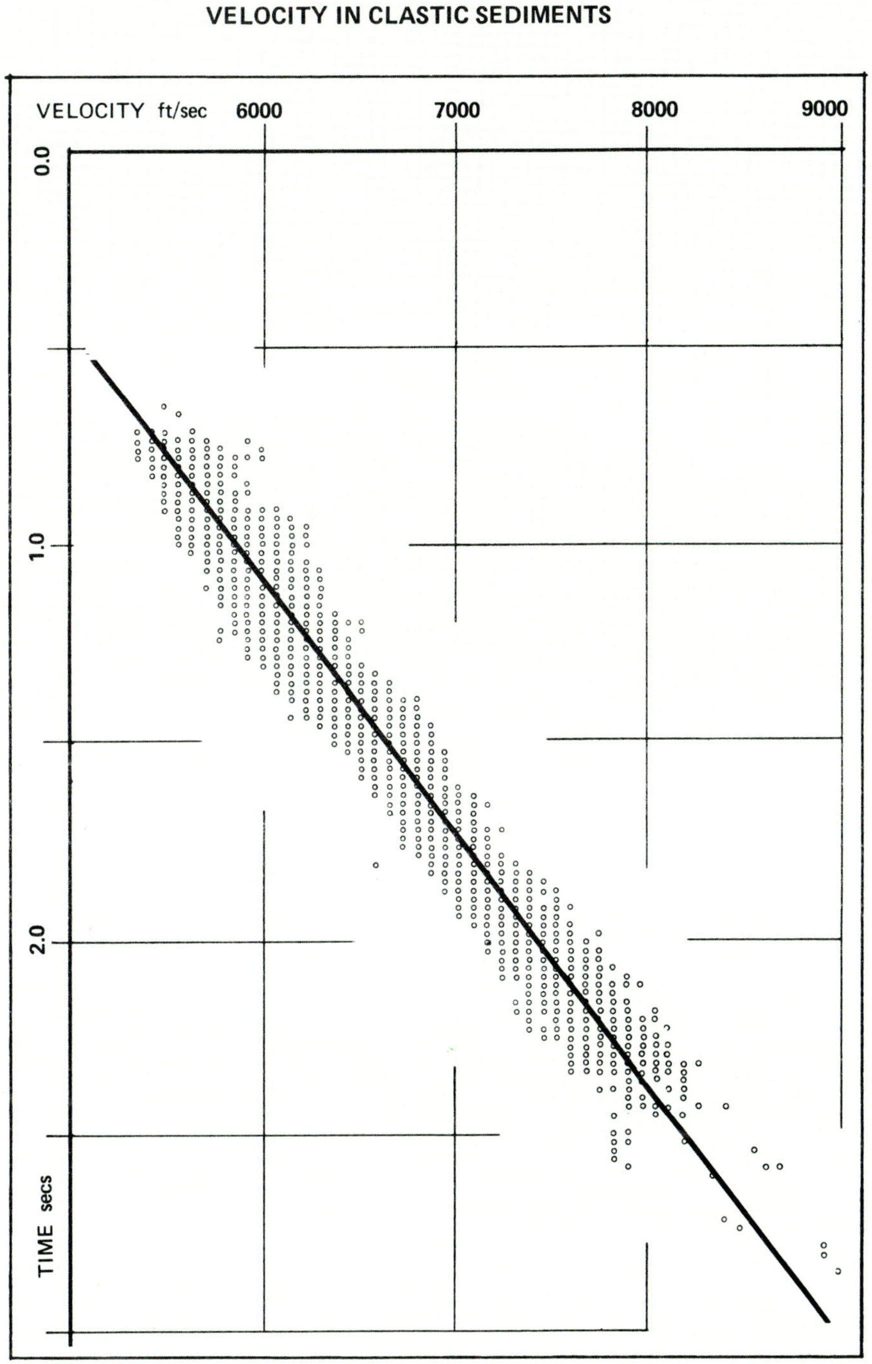

FIGURE 8.14

The heavy line represents the increase in velocity resulting from shale compaction due to loading of the overburden. Values left of the line probably correspond to shaly facies while those on the right side contain more sand.

two-way reflection time (**Fig. 8.14**) for a marine survey. The horizon used in the evaluation varies considerably in depth due to the considerable structural relief in the area. The values conform closely to a straight-line function of average velocity over time. This is a typical response for sections of this type.

Note the close clustering of all the values about the central line. If a profile were drawn horizontally through the data set at any level, it would take the form of a Gaussian curve centered about the trend line. In the area of this example, local velocity deviations from the norm for the area are significant in terms of lithology. In general, sandier facies tend to be of higher velocity and the shalier facies of lower velocity. Overriding the lithology effect is the effect of compaction which can be removed by the simple determination of the normal velocity function from the area as shown, and subtraction of the loading factor from the observed velocities. It will be found that after such an operation, mapping the deviations of residual interval velocity, as a function of location, will map changes in lithology which may be related to depositional trends, such as those corresponding to deltas or old shorelines.

Summary

Velocities derived from normal moveout analysis of seismic records normally fail to match the vertical velocities measured in a borehole. The error increases with increasing separation between the shotpoint and the geophone, a result of anisotropy and the physical laws of acoustic transmission through a layered medium of variable velocity. Anisotropy, the variation between vertical and horizontal transmission velocity in a given bed, is often only a minor factor in the deviation. This is fortunate because, as shown by Levin (1978), only the horizontal component of velocity in anisotropic layers can be determined from surface measurements.

The major velocity variation is due to the erratic path that sound follows in its travel through a layered earth. Fortunately, the stacking velocities obtained from seismic reflection records are optimum for normal moveout calculations to correct seismic records before stacking. Hence the term stacking velocities.

9 SEISMIC TRACE ATTRIBUTES

Introduction

Most of the effort in seismic data processing has been directed toward noise reduction and signal enhancement. Great progress has been made and continues to be made. The high quality signals that are now available have shifted some of the emphasis towards extraction of information, other than reflection times, from seismic data.

For example, velocity information has always been a valuable by-product of seismic reflection surveys. Procedures for its determination preceded the advent of digital computers. Automation now provides more precise velocity information in much greater quantities than was previously possible and, of equal importance, it has prompted a much better understanding of the nature of subsurface velocity and its relation to seismic data analysis.

A number of related properties of the seismic trace, including amplitude and frequency, and their derivatives, are grouped together under the general heading of Seismic Attributes. Although these expressions may or may not be a direct function of geological processes, at times they do demonstrate geological changes rather dramatically, and hence may be useful as techniques to highlight anomalous features in a seismic section.

Many of the special process techniques depend on the digital computer for implementation, but it is a credit to early geophysical workers that most of the so-called new techniques were predicted, and often performed in a crude way, many years ago. Many early attempts to turn theory into practice failed simply because they were premature. The concept was not faulty, but insufficient computer power, or the limited dynamic range of recording instruments, and/or poor signal quality (particularly prior to the development of the stacking technique) all precluded satisfactory results.

It is important to review the earlier attempts from time to time, and perhaps renew experimentation in some of them. Just because some theory did not work at one time, does not mean it never will. Many a research laboratory has declared some process impossible only to have someone later make it work, and many an oilfield has been discovered in an area which had been declared barren after someone drilled a few dry holes.

There are two routes to the discovery of secondary processes. The first is by research and experimentation, based on those properties or effects which can be predicted by known geophysical theory. The second is purely pragmatic; the result of an alert observer recognizing a link between a specific data response and a subsurface condition.

This was a fairly common characteristic of early geophysical papers. Authors reported some observed phenomenon, but sometimes found it difficult to express an equation, or to describe it in meaningful terms of geology or geophysics.

It is important to be able to explain an observed phenomenon, or a new method, in terms of geological and physical processes. Geophysical surveying is a means of gathering geological information. After all, it is geology, not geophysics, that controls the generation, migration and accumulation of hydrocarbons. Geophysics merely provides a means to observe geological conditions. Applications based only on observed geophysical phenomena do not gain widespread acceptance, primarily because of lack of appreciation for the physical causes. Nevertheless, a phenomenon observed should not be ignored, since later discoveries may provide the explanation. With a better understanding of the reflection signal and the earth response, the reasons for many early observed phenomena became clear, and it became possible to describe them in terms of geology.

Amplitude Mapping

An award winning paper by Ben Rummerfield, published in Geophysics, 1954, is a good example of a pragmatic approach explained by theory. Mr. Rummerfield's technique used reflection quality to add another dimension to the interpretation of seismic records. The paper illustrated an area in which sand bars can be located by mapping the quality of a diagnostic reflection. A contour map of reflection quality, graded good, fair, poor, and questionable, outlined a series of shoe-string sandbars. Probably unwittingly, this paper was a forerunner of modern Bright Spot theory - the change in reflection amplitude due to changing fluid content, porosity, or lithology. The changes in reflection "quality" noted by Mr. Rummerfield were most likely changes in amplitude and/or polarity, related to the presence or absence of gas in the formation.

Another example of an idea which was ahead of its time was a development reported several years ago. C. H. Savit, in an interesting paper in Geophysics, 1960, describes an instrument termed the seismic energy summarizer, a device to consolidate and measure the average energy of two or more seismic traces. The continuous measurement of amplitude versus time was plotted on the seismic record as an auxiliary trace. Usually, the signal was taken from the gain control circuits, so it measured the gain level of the amplifiers. Similar traces were popular on many early instruments, since they provided the observer (and interpreter) with an indication of true signal strength.

A case history in the paper demonstrated results of a test line across three wells, over which the summarizer trace closely followed the amplitude relationship predicted for the presence of a thin sand body (**Fig 9.1**). When the sand body disappeared, the corresponding energy on the summarizer trace disappeared also. These amplitude changes were not present, or at least not visible, on the standard seismic reflection trace since the automatic gain control circuits of the time tended to bring all reflected information to a common amplitude level.

Unfortunately, as is often the case, the response could be identified only in a known situation. Many other conditions, as can be observed on **Figure 9.1**, produced greater changes on the summarizer trace than those at the zone of interest. Accordingly, it was not easy to interpret the data in an unknown area, and the device did not prove to be a commercial success.

Today we know both these procedures to be quite justified by theory. Amplitude mapping is exploited by combining the use of a modern digital recorder with careful digital processing and display. The resulting true amplitude sections respond readily to changes in lithologic units and can be useful for mapping pinch-outs, lithology changes, and so on. The main problem remains one of interpretation. Reflection amplitudes are merely ratios of velocity contrasts between two subsurface units and a change in either will initiate a response. Amplitude changes may have their source in so many different subsurface conditions that their interpretation may be ambiguous.

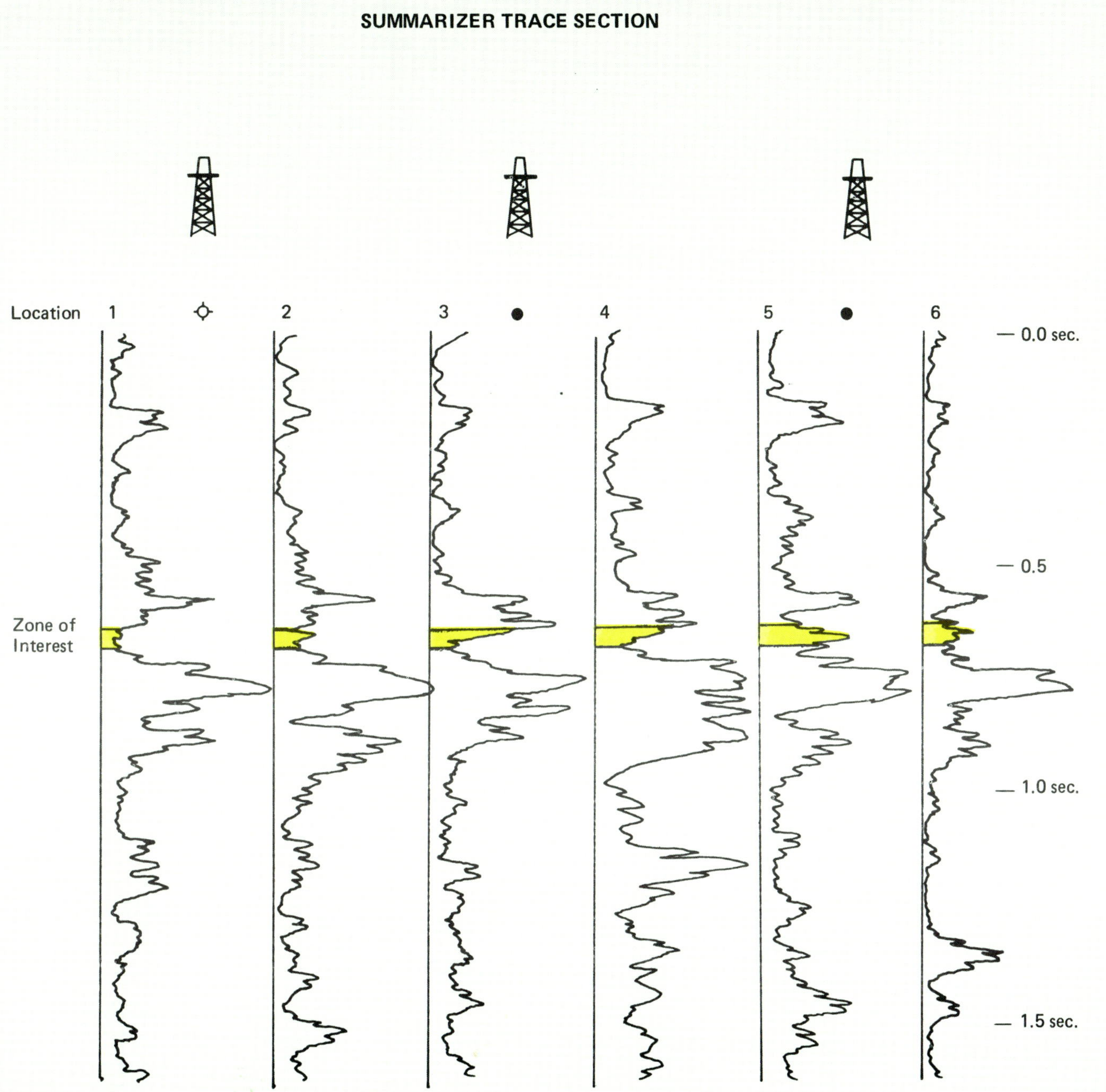

FIGURE 9.1

The summarizer trace was an early attempt to map true reflection amplitudes. The example shows the change across a known petroleum reservoir. Other insignificant conditions, produced equal or greater changes, so the method was not successful.

Bright Spots

The introduction of binary gain and floating point instrument made it possible to record the full amplitude range of signals returning to the geophone. If the true trace amplitude is recovered, and compensated for transmission losses in a manner which recognizes and compensates for such losses as they occur, then the relative amplitude of reflection events on an individual seismic trace should, to a degree determined by bandwidth, (and assuming no multiples), correspond to the reflection coefficients in the ground. Therefore, amplitude changes on the processed trace should map changes of porosity or lithology in the sedimentary section.

The so-called Bright Spot is a sudden increase in reflection amplitude generally associated with the replacement of fluid within a reservoir rock by gas. This source of a Bright Spot is not unique. Similar effects may result from other conditions, such as coal beds, in the section. The change is not always an increase in amplitude, although it often is, and the Dim Spot is a related phenomenon.

A basic understanding of the conditions which produce Bright Spots, and the response that may be expected, can be obtained readily from simple models. The mathematics of Bright Spot occurrences can be reduced to simple approximate equations which are quite adequate to cover many situations. If all possible influences, such as shear and compressional velocities, saturation, pore pressure, etc., are taken into consideration, the calculations become very complex.

The reaction is based on a simple well known principle: that the interval velocity through a porous material represents a weighted average of velocity of the rock matrix and the velocity of the material filling the pore space.

Although a precise definition of the average velocity through a fluid filled porous rock is a complex problem, requiring a lengthy equation, a good approximation in clastic rocks is provided by Wyllie's equation.

The Willie equation defines the approximate transit time through a porous matrix filled with fluid as follows:

$$\frac{1}{V_a} = \frac{\theta}{V_f} + \frac{1-\theta}{V_m}$$

where:

V_a = Average velocity

V_m = Rock matrix velocity

V_f = Fluid velocity

θ = Porosity, expressed as a decimal fraction

The Wyllie equation is a simple formula used frequently in well logging practice. Although primarily designed for sandstones, it will also work reasonably well in carbonates.

The Wyllie equation, plus the simple approximation,

$$R_c = \frac{V_2 - V_1}{V_2 + V_1}$$

is sufficient to model the Bright Spot process.

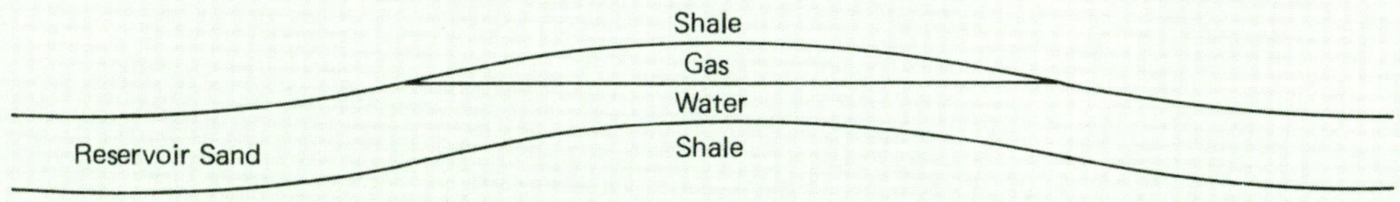

SAND POROSITY	θ	=	10%
SAND VELOCITY	Vm	=	14,000 ft./sec.
WATER VELOCITY	Vw	=	5,000 ft./sec.
GAS VELOCITY	Vg	=	1,000 ft./sec.
SHALE VELOCITY	Vs	=	8,000 ft./sec.

WYLLIE'S EQUATION : $\frac{1}{Va} = \frac{\theta}{Vf} + \frac{1-\theta}{Vm}$

SAND w/WATER	Va = 12,000 ft./sec.	Rcw	$= \frac{12{,}000 - 8000}{12{,}000 + 8000} = \frac{4000}{20{,}000} = 0.2$
SAND w/GAS	Va = 6,000 ft./sec.	Rcg	$= \frac{6000 - 8000}{6000 + 8000} = \frac{-2000}{14{,}000} = -0.14$
GAS/WATER CONTACT		Rc	$= \frac{12{,}000 - 6000}{12{,}000 + 6000} = \frac{6000}{18{,}000} = 0.33$

FIGURE 9.2

Model of the conditions which create a Bright Spot and a Flatspot.

If the pore space in a given reservoir rock is filled with salt water, and at some other location the water in the same rock is displaced by low velocity gas, the interval velocity through the reservoir rock will become appreciably slower. Assuming the velocity within the rock surrounding the reservoir to be constant, a significant change in the reflection coefficient at the gas/water boundary should result from the change of fluid filling the porosity.

Figure 9.2 is a model of the response of a lenticular reservoir with a gas cap, surrounded by a shale. Water-filled porosity in the reservoir rock causes its average velocity to drop, so that a contrast is produced at the shale/reservoir interface. Over the gas cap, gas replaces the fluid, and causes the average velocity of the reservoir rock to be lowered a substantial amount, increasing the contrast with the overlying shale and producing a seismic reflection of reversed polarity.

Assume a porous sandstone, with a matrix velocity as shown, interbedded with shale. The porosity will be filled with fluid. Using the Wyllie's equation for water-filled sand porosity, the average velocity would be:

$$\frac{1}{V_{wp}} = \frac{.1}{5{,}000} + \frac{.9}{14{,}000}$$

$$= .00002 + .0000643$$

$$= 84.3\ \mu sec/ft$$

$$V_{wp} = 11{,}860\ ft/sec$$

Similarly for the gas cap:

$$V_{gp} = 6{,}086 \text{ ft/sec}$$

The reflection coefficient at boundary between shale and wet sand:

$$R_{sw} = \frac{11{,}860 - 8{,}000}{11{,}860 + 8{,}000}$$

$$= .019$$

The reflection coefficient at the gas cap:

$$R_{sg} = -0.14$$

And at the gas/water boundary:

$$R_{gw} = 0.33$$

The relative reflection amplitude, which is a direct function of the reflection coefficient, will be:

water wet sand	=	0.19
gas cap	=	-0.14
g/w boundary	=	0.33

Obviously, the presence of wet porous sandstone should cause a noticeable reflection in the model. The gas cap will show a reversal in polarity over the water filled sand, but not a great difference in reflection amplitude. The most prominent, and perhaps surprising, result is the response at the gas/water interface; a very strong reflection coefficient of 0.33. Since this contact is a flat plane, it produces a flat spot on the seismic section. Many interpreters regard the flat spot **(Fig 9.3)** as a much better criterion for a gas indicator than the Bright Spot.

If oil replaces water in the porous sand, a very modest change takes place. Oil has a velocity of about 4,200 ft/sec.

Oil sand: $V_{op} = 11{,}350$

$R_{so} = 0.17$

Compared to: $R_{sw} = 0.19$

A change in amplitude at the oil interface should be recorded, but it is doubtful if the contrast would be noticed in routine interpretation. A similar comment applies to the flat spot at the oil/water boundary. It should generate a reflection having a relative amplitude of only 0.02.

The Bright Spot method depends largely upon detection of a gas cap through a sudden change in amplitude on the seismic section. It depends on two fundamental conditions:

1. Velocity contrast relationships which will produce the expected change.
2. Data acquisition and processing done in a manner to preserve the amplitude relationships found on the field signal.

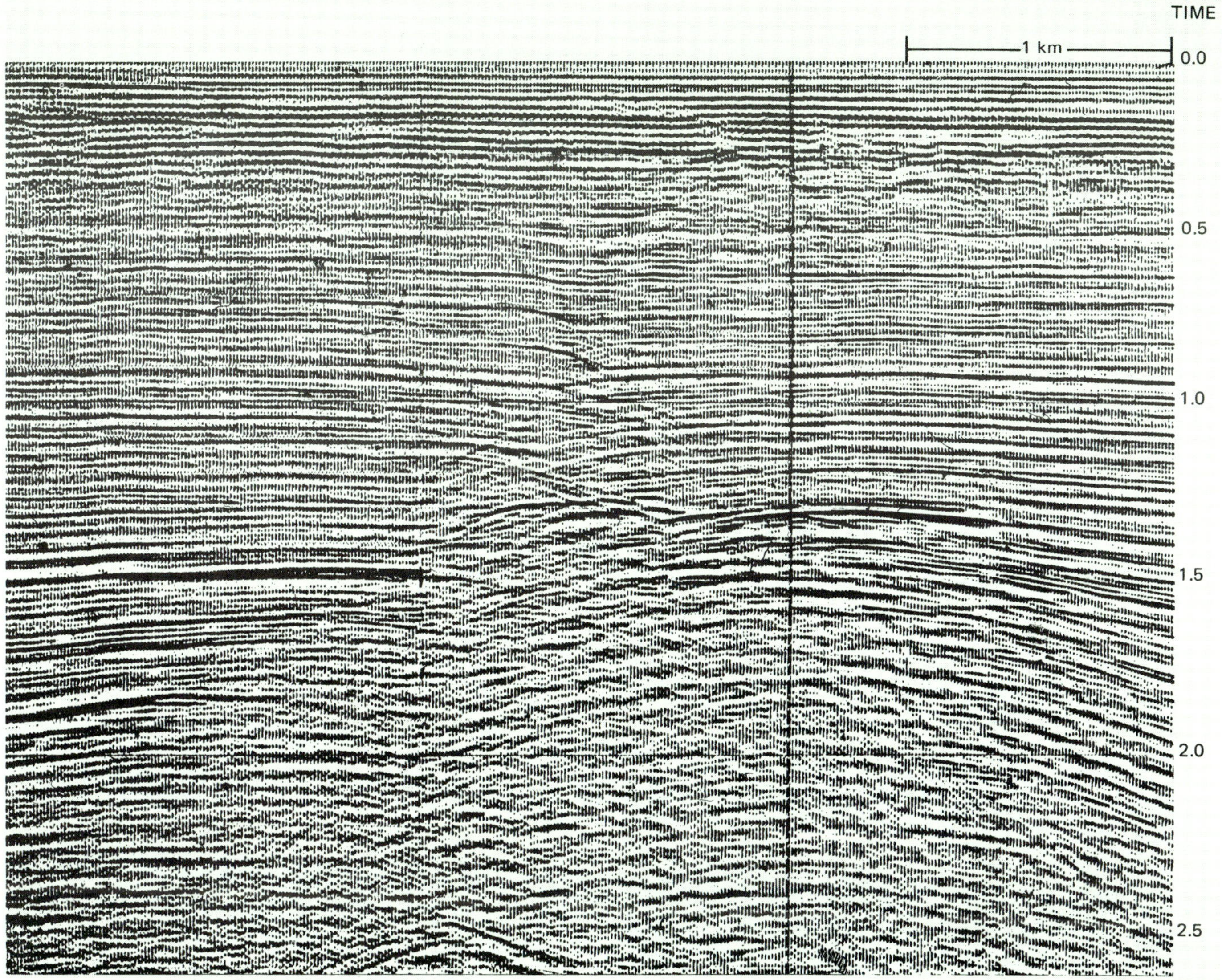

FIGURE 9.3

A least two apparent Bright Spots appear on this seismic section. The one on the left includes a well defined flat spot and is the better location.

Physical Considerations

Formation fluids invariably have lower velocity than the matrix rock. Gas always has lower velocity than either fluids or the matrix rock. Hence, porosity in a given matrix rock will always reduce velocity and if the porosity is filled with gas, it will have a still lower velocity. Any such change in the rock will usually result in a change to the reflection coefficient at the boundary, varying in amplitude as a function of the fluid filling the porosity.

However, the amplitude of the response is strictly a function of the velocity contrast. If a water filled reservoir rock has a higher average velocity than the surrounding material, fluid replacement by gas may lower the contrast such that the reflection disappears. Frequently the velocity relationships are such that the change from fluid fill to gas fill causes the corresponding reflection to reverse polarity.

For these reasons, whenever possible, it is always important to determine the nature of the velocity relationships in order to make a confident interpretation of the situation.

Amplitude is a function of contrast, or ratio, in velocities. If a given porosity causes the velocity of a rock to drop by a fixed amount, say 3,000 ft/sec, the ratio of change will be greater in low velocity material than in high. For this reason Bright Spots are usually more visible in young sediments than in highly consolidated, older sandstones and carbonates. Nevertheless, they are generated in carbonates, as shown by **Figure 9.4**.

A Bright Spot response can be obtained from stratigraphic conditions not related to reservoir fluids. Pinch-outs, erosional surfaces and laminar sections can all generate Bright Spots. In particular, combinations of thin beds, which vary in thickness from one-half to one-quarter a seismic wavelength, can show dramatic amplitude changes due to constructive interference, particularly on narrow band signals. Sudden changes in lithology can cause velocity changes equal in magnitude to those caused by fluid changes. Coal beds produce strong Bright Spots.

The best direct hydrocarbon indicator, at this date, is the flat spot, and it is one that can be readily explained in terms of geology and physical response. It is by far the most reliable indicator since there is no other known phenomenon which will cause a similar response, and any flat spot in the midst of a structural fold on a seismic section can usually be drilled with a great deal of confidence **(Fig 9.3)**.

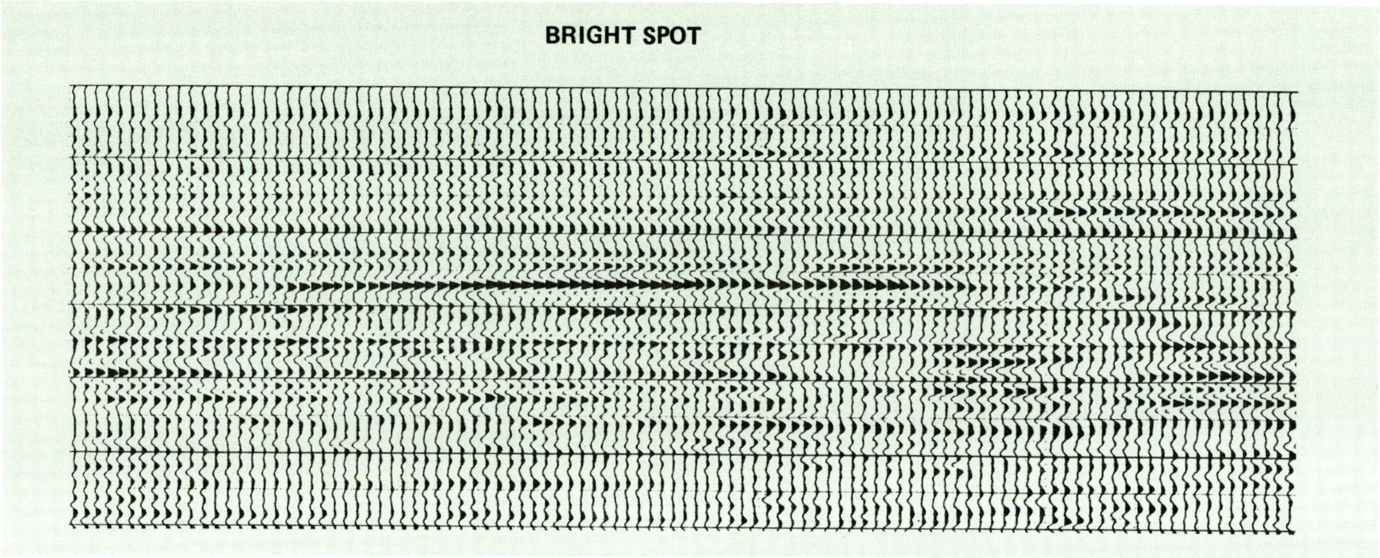

FIGURE 9.4

Bright Spots tend to show less amplitude in carbonate sections due to the proportionately small change in velocity caused by porosity, compared to that in clastic sections.

The simplest of the four main seismic attributes, and the one most closely related to geology, and perhaps the most commonly used, is reflection amplitude. Under favorable geological conditions, and optimum processing, it yields direct hydrocarbon detection, by means of the Bright Spot.

No special transform processing is required to produce the reflection amplitude attribute, it merely requires care and good practice at every stage of the normal process sequence. Important is a proper gain recovery and transmission loss compensation routine. Zero phase wavelets are a more certain indicator of amplitude and polarity than minimum phase wavelets. Above all, no time-varying or automatic gain scaling of any kind should be used.

Data Processing Considerations

The field instruments must be capable of recording true amplitude data over the full range. The data processing must recover and compensate the signal for transmission losses in a manner which will preserve the proper amplitude relationships. This is one of the neglected areas in the data processing application. Many data centre software packages include arbitrary scaling procedures which degrade true amplitude relationships to some greater or lesser degree. As a result, the product sold as a Bright Spot section may be only partially so.

The proper gain curve application is one which compensates for the actual rate of amplitude decay and recognizes decay changes resulting from lithology changes, yet avoids any AGC effect. It must also recognize the step-out time shift and divergence effects on far traces.

If an adaptive system is used, compensation for the amplitude decay should not be made until some initial processing has been done to eliminate noise. In subsequent processing, all time-varying processes should be suspect. Time-varying scaling can be very detrimental, as can be time-varying filters, since many such programs have built-in automatic scaling routines. Band-pass filters can cause amplitude changes particularly when the pass band is narrowed. This does not preclude the use of filters in processing, it merely points out the need to make sure that when a comparison is made between any two points on the seismic section, the processing has been such that the variations be a true representation of the subsurface.

Frequency Mapping

For many years geophysical interpreters have relied on reflection character to supply a third dimension to the time and distance measurements made to describe the subsurface. Character is not easy to describe, and much less so to define in quantative terms. It is the shape in waveform, or of a reflection package, which the experienced interpreter has found to be diagnostic of certain significant conditions in the subsurface. It might be expected that such features could be described and classified in some more significant manner. That another dimension exists is suggested from experiments in signal analysis of sedimentary sequences. It has been found that formation characteristics, as defined in the signal spectrum, are carried through the frequency transforms of well logs.

There is a sound geological basis relating some of these spectral characteristics to the history of a basin. Take, for example, the growth of a pinnacle reef. The first stage is often the development of shelf or platform upon which the reef growth is initiated (**Fig 9.5**). Over time the reef continues to grow and may reach a substantial elevation. Eventually the growing reef is killed and eventually buried, usually by an influx of sediments which are deposited around and eventually over the reef. The reef is structurally competent and supports the relatively thin overlying sediments, but the shaly material surrounding the reef compacts, accentuating the drape resulting from differential disposition of the overburden over and around the reef.

Due to the differential deposition and concurrent compaction, the bedding within the shale section over the reef will be thinner than in the same sequence in the off-reef section. The travel time between boundaries in the off-reef section will therefore be slightly greater and hence the signal series will have apparently lower frequency compared to the same section over the reef. The degree of frequency change will, of course, depend on a number of factors, including the height of the reef, the amount of deposition over the reef, and the degree of differential compaction. In any case, the frequency content through a shale section over a reef should be higher over the reef than in the basin.

A test of this hypothesis has been illustrated by a cross section processed over a known reef **(Fig 9.6)**. An amplitude spectrum is produced for a specified window of each trace in order to detect changes in the frequency content. The corresponding amplitude spectrum may be plotted out at some suitable scale along the same line of section as the seismic data to provide a continuous frequency analysis related to the source data.

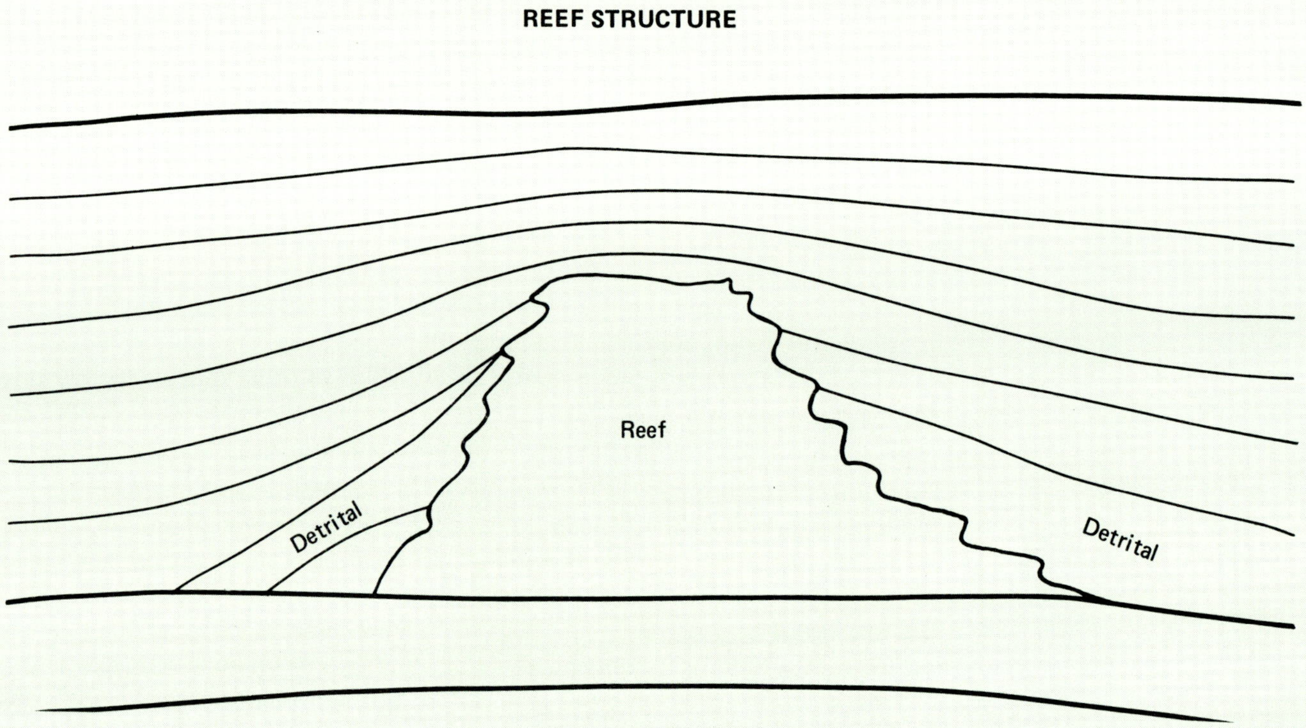

FIGURE 9.5

Compaction and drape over a reef mass causes time lines to thin over the reef, creating a higher frequency response over the crest than away from it.

Figure 9.6 illustrates the application of this process. The eastern extremity of the line used for the illustration crosses the Rainbow "A" pool reef development in Canada. The slight sag in front of the reef edge is due to salt collapse. For the test a window was selected from the shale section immediately overlying the reef. The frequency content of the window is determined by taking the cosine transform for each trace. The results are shown at the bottom of the section. In this case the scale has been plotted so frequency increases upward in order to obtain a frequency shift which moves in the same direction as the structure. Off-reef, a very distinct band of frequencies occurs about 37 Hz. As the line approaches the reef the frequency begins to rise slowly, until over the top of the reef, it reaches approximately 47 Hz, or in other words, a shift of approximately 10 Hz or about 35% of the original frequency.

Similar frequency changes should be associated with other geologic phenomena. For example, under favorable conditions it is possible to detect the presence of a sand lens lying on a boundary which produces a reflection.

The cause of this can be illustrated by a sketch of a particular low velocity sand layer lying in a shale section **(Fig 9.7)**. The contrast between the shale and the sand will provide a velocity coefficient which, for a sand of lower velocity than the shale, would be negative at the top of the sand. The base of the sand will have a similiar amplitude of coefficient but the polarity will be reversed. A frequency analysis of the section will indicate a periodic component of twice the time through the sand, resulting from the opposite but equal impulse response received at the top and the bottom of the sand bed. The result of this should be a well defined frequency component in the spectrum which will vary in frequency as a function of the sand thickness. Within the resolving power of the spectrum, very thin sand beds should appear as high frequency. As the bed thickens and the period between the top and the bottom events lengthens, there should be a very definite decrease in the frequency of the energy component. Thus, not only does the method provide an indicator of the presence of a sand lens but the frequency should be a function of the sand thickness, with the frequency decreasing for increasing amounts of sand thickness. An actual example across a known field is illustrated in **Figure 9.8.**

A problem common to all of this type of frequency analysis is resolution. A related requirement is the ability to detect and interpret a significant change in the results. Large structures, such as the reef example, are usually quite visible, without any need for frequency analysis. The sand bar is thin so there is always a danger that the window

CONTINUOUS FREQUENCY ANALYSIS

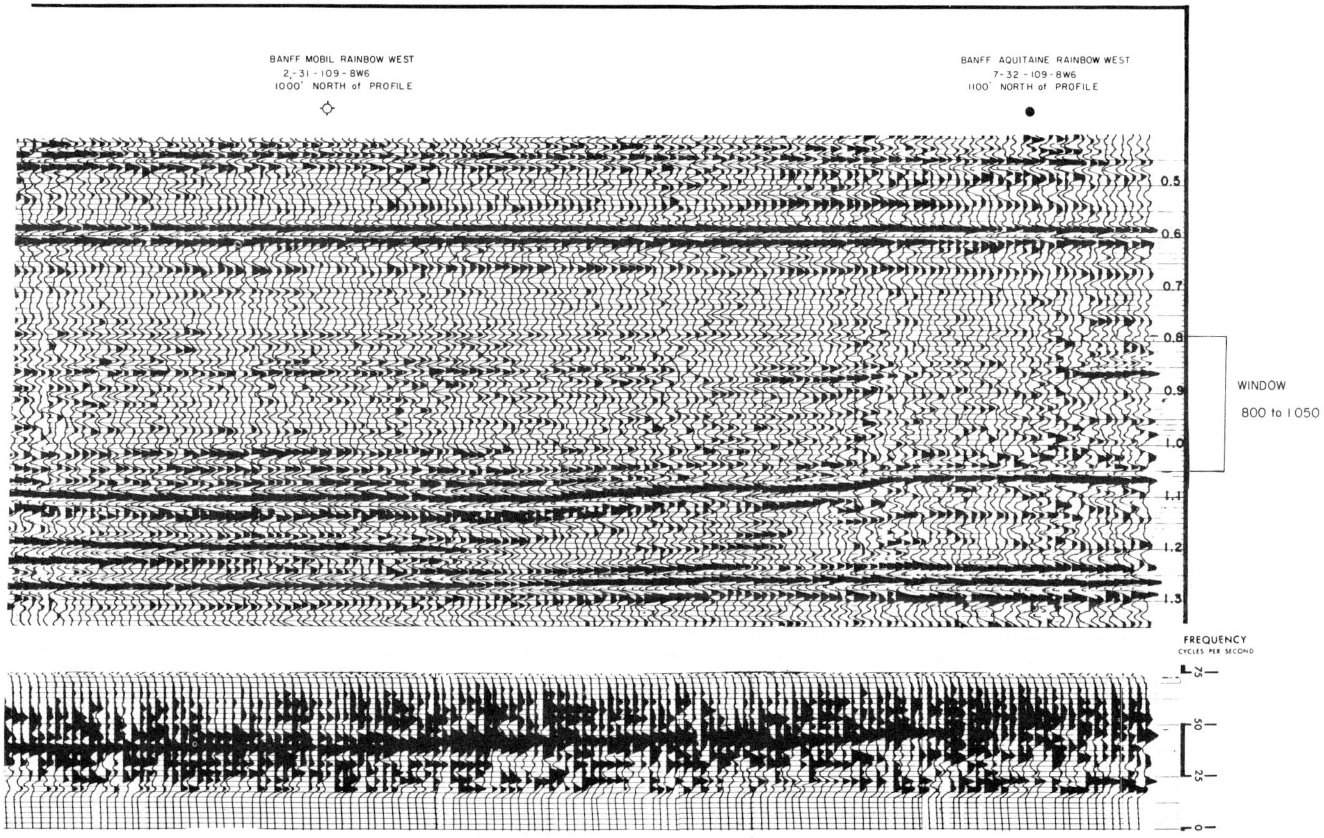

FIGURE 9.6

An amplitude spectrum of each trace within the specified window yields a continuous frequency plot, which may be seen to rise in frequency over the reef.

chosen for analysis will include other, extraneous, effects occurring above or below the zone of interest. Narrowing the window across the zone of interest will avoid this risk but then, in a Catch 22 effect, lose resolution. Recalling sample theory, short sample lengths in the time domain produce large steps between frequency sample points in the frequency domain. The frequency changes resulting from a thin reservoir bed may be very subtle, so closely spaced frequency sampling is required, but in the Fourier transform this can be achieved only by lengthening the time span of the sample window.

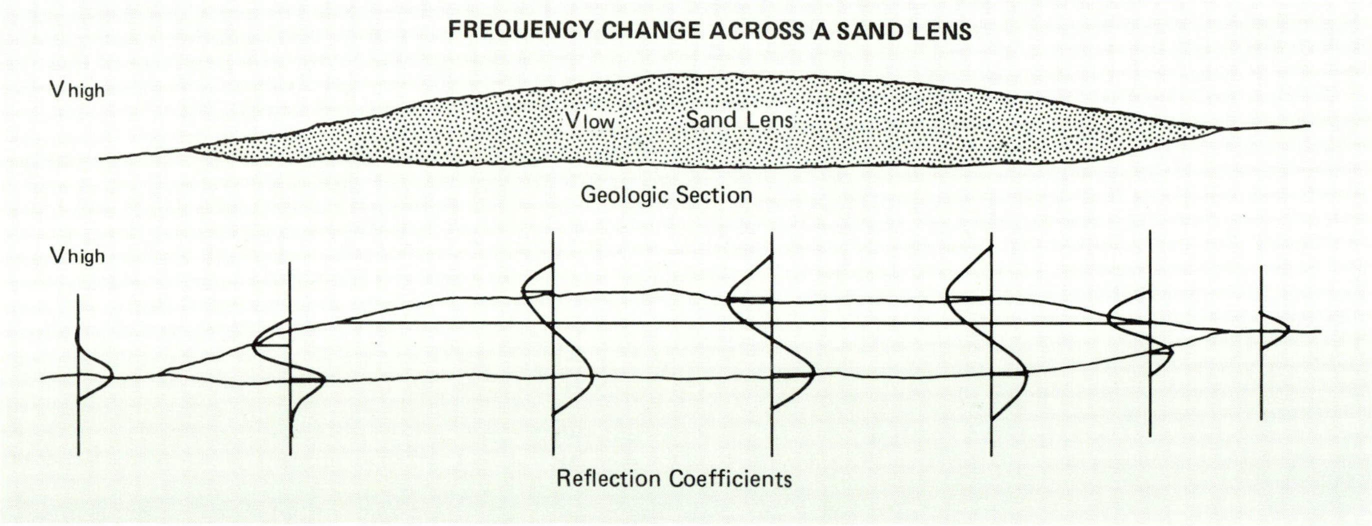

FIGURE 9.7

As the sand bar thickens, it creates a longer time period (lower frequency) between top and bottom.

Colour Displays

Once the signal processing is complete, the evaluation of any given seismic attribute may be dependent, to a large degree, on the method used to display the output. Bright Spot sections require a rather delicate overall amplitude balance level to be maintained, which subdues most of the information in order to allow amplitudes above a certain threshold level to be readily visible, (Fig 9.9). Conventional plotting methods, in which peak amplitudes merge with adjacent traces, often make it difficult for the human eye to recognize and discriminate between small amplitude changes, (Fig 9.10).

One solution is to use colour coded displays to enhance visibility and emphasize contrasts. However, colour is not the answer for everyone; a surprising number of people are colour blind, or have some deficiency in their color

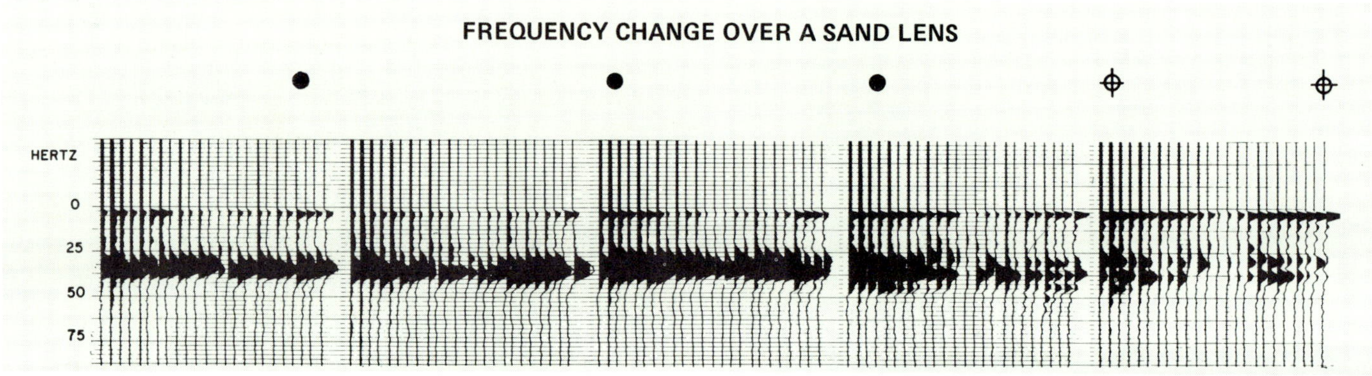

FIGURE 9.8

Amplitude spectrum of reflections across a known sand bar.

FIGURE 9.9

Plotting amplitude must be subdued in order to make the contrast of a Bright Spot clearly visible. Note the pronounced Flat Spot on the upper reflection.

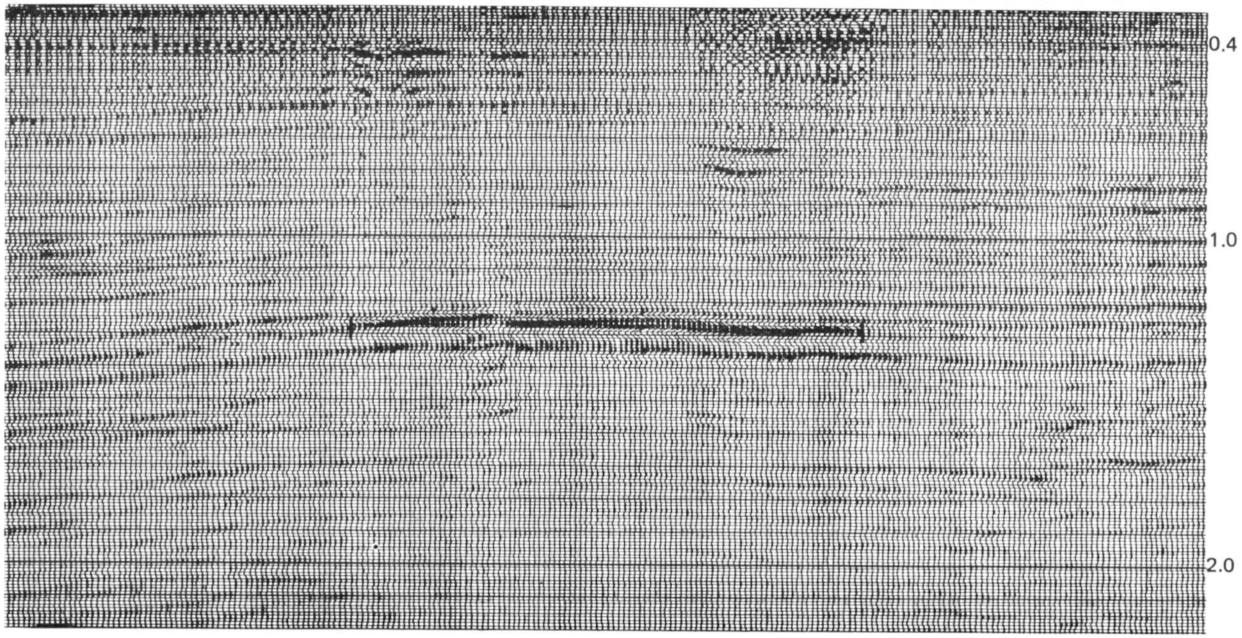

*Courtesy of Western Geophysical Company

vision. The primary use of color is to emphasize change, and effective color displays require care and skill in their preparation. A continuous spectrum of smoothly varying color changes may be more difficult to analyze than one having discrete color steps with obvious color contrasts. Intensity of color may be as important as the color itself in presenting contrasts. The average person is more cognizant of certain colors than of others, because conventional usage attaches more importance to them, (ie: the use of red for danger).

The scaling of color is, in itself, something of an art, due to the wide variation in the range and distribution of numbers resulting from different analysis techniques. For instance, reflection amplitudes have a predominance of number values approaching zero, and very few extremely high values. A color scale based on equal divisions of amplitude may result in a color display that is almost entrirely of one color. Better results may be obtained from use of a logarithmic, or db scale.

A useful tool to aid in establishing an appropriate scale is a density plot. A simple count of the number of values within each step of some selected increment is graphed. The scale increments may then be selected to contain a relatively constant number of values per color step, or use some other arbitrary division.

Reflection Amplitude

The major significance of reflection amplitude attributes has been discussed under the heading of Bright Spots. Reflection amplitude is a function of acoustic impedance in the geological section, and does have a direct, if not unique, relation to geology. Perhaps for this reason it is also the most popular of the seismic attribute displays. (Figure 9.11).

Amplitude has also been used in an effort to map very thin beds. When two reflection boundaries converge, and the interval thins to less than the limit of resolution (about ¼ wavelength) constructive interference causes a sudden

Hilbert Transforms

increase in amplitude to occur. Further thinning causes the amplitude to vary as a function of bed separation.

A normal Fourier transform computation involves operations on a set of complex numbers having two arrays, Real and the Imaginary. When a seismic signal is to be transformed, all input values are placed in the Real array, and the corresponding Imaginary array is filled with zeros.

The signal input samples are transformed into another complex number set of component frequencies. Each frequency is defined by a Real and an Imaginary number pair. The square root of the sum of the squared Real and Imaginary values defines the amplitude, while dividing the Imaginary number by the Real yields the tangent of the phase angle. It follows that the amplitude divided by the Real value will yield the cosine of the phase angle. Thus, in the frequency domain, the transform describes a series of vectors of variable length (the amplitude), which are directed about the frequency axis at variable angles, (the phase). The amplitude divided by the corresponding

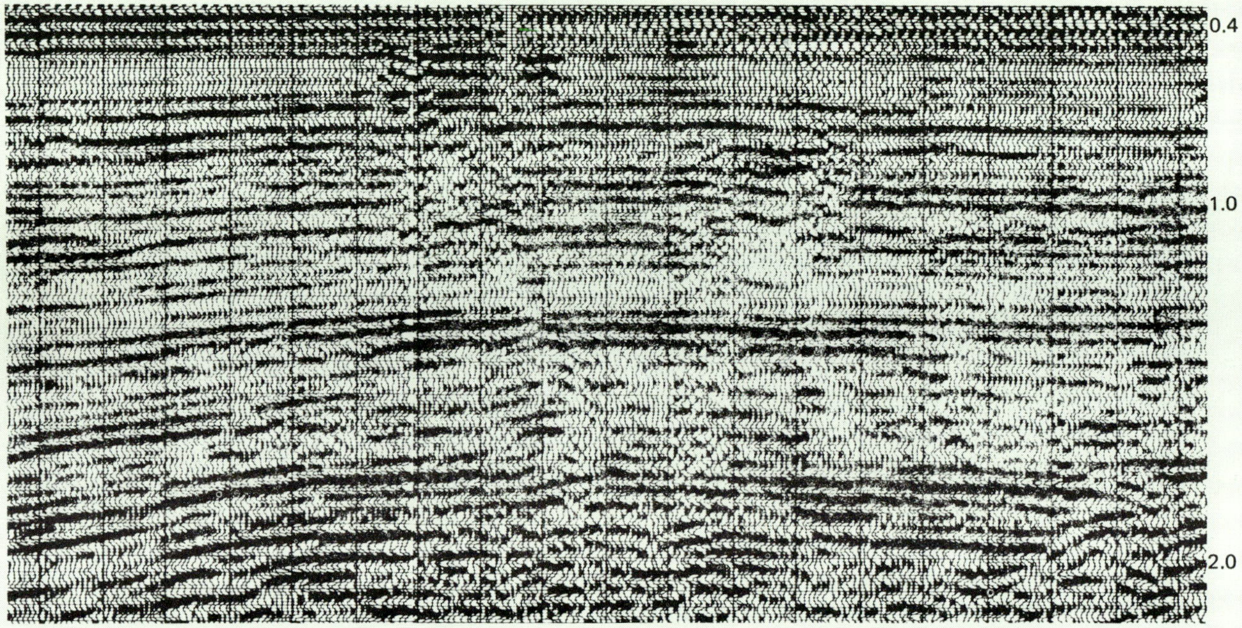

*Courtesy of Western Geophysical Company

FIGURE 9.10

Plotting at conventional amplitude tends to mask the prominent Bright Spots of Figure 9.9.

Real and Imaginary vector components yields the respective Cosine and Sine functions, **(See Figure 1.11)**

The Cosine function of any single frequency, generated in the time domain at the appropriate amplitude, yields the continuous oscillation of that single frequency component as it exists in the original input signal.

If all the single frequency components are reproduced in a similar manner and added, the result equals the original input trace **(Fig 1.6)**. Hence it sometimes is termed the Cosine transform.

A related signal results when the Imaginary values, i.e. the Sine components, are used to generate the time domain function. Any single frequency generated in this manner is merely a 90° phase shifted version of the original Cosine component, and the time domain sum of all such components is merely another version of the original trace, phase shifted 90°.

Instantaneous Amplitude

A display of reflection amplitude demonstrates the absolute amplitude of the seismic trace. Whereas this is considered to be the Cosine portion of a Fourier transform, the Hilbert transform of the trace yields a corollary Sine trace having the amplitude spectrum of the original, but with all phase components shifted 90°. A third trace is obtained by computing a new continuous series of amplitude vectors, for all Sine-Cosine sample pairs, to yield the reflection envelope, termed Instantaneous Amplitude, of the original input trace.

When plotted as a color display **(Fig 9.12)** superimposed on the original trace cross section, Instantaneous Amplitude is often the most easily assimilated display of seismic amplitude variations. It is less sensitive than reflection amplitude to local variations caused by noise or minor changes in the section. It is often used in place of, or in conjunction with, a display of reflection amplitude.

Instantaneous Amplitude is very much a function of reflection band width, since narrow band signals tend to oscillate, and decay slowly, extending the apparent time of any seismic reflection response, and hence extending the length of the reflection envelope.

Instantaneous Phase

In a manner similar to determination of the Instantaneous Amplitude vectors, the original input trace and the phase shifted Hilbert transform trace can be used to determine the phase angle of the vector computed for each sample pair. A continuous display of the resulting phase angle, as a function of reflection time, is termed the Instantaneous Phase (Fig 9.13). A phase plot amounts to a seismic trace from which all reflection amplitude information has been removed. It can be displayed on a conventional black and white section plotter, with amplitude representing phase (i.e. maximum positive amplitude representing +180°, and maximum negative amplitude representing -180°), but is most frequently plotted in color. The Instantaneous Phase plot ignores all reflection amplitude components, and hence illustrates only the continuity of reflections. For that reason, it is sometimes known as a continuity plot.

Instantaneous Phase has its greatest value in structure mapping to show the pattern of reflections, regardless of changes in lithology. Seismic reflections almost invariably represent time lines, so the patterns revealed by instantaneous plots can also be helpful in stratigraphic interpretation.

Instantaneous Frequency

The Instantaneous Phase angle calculated above will wrap around the polar axis of the sampled time series in a more or less regular manner. Between any two successive samples, the amount of phase shift can be determined. If the change in angle between any two adjacent samples is multiplied by the sample rate, (in samples per second) and divided by 360°, it will yield a number which is defined as Instantaneous Frequency. (Fig 9.14).

Instantaneous Frequency is fundamentally a measure of the local reflection frequency of the seismic trace, and hence should serve to accentuate such features as changing bed thickness in the seismic section. It should help delineate the onset of major reflection boundaries and sudden changes in depositional patterns in the seismic section.

Interpretation

Seismic attributes have been around in one form or another for some time and many examples are available of their ability to highlight changes of significance in the seismic section. However, except for amplitude, (including Bright Spot), they have never become very popular, nor are they used extensively in interpretation. The reason for this seems to lie in the fact that most of them cannot be tied directly to geology: only Reflection Amplitude, and possibly Instantaneous Frequency, can be related to a quantitative function of the geologic section.

Hence, the primary use of seismic attributes is to emphasize or draw attention to unusual or anomalous changes in the seismic section, which, if deemed significant, can be analyzed by some other method to determine their origin. With experience, some condition may be related pragmatically to known geology, such as a Bright Spot to gas filled porosity, for example, and may then become a direct indicator of that condition.

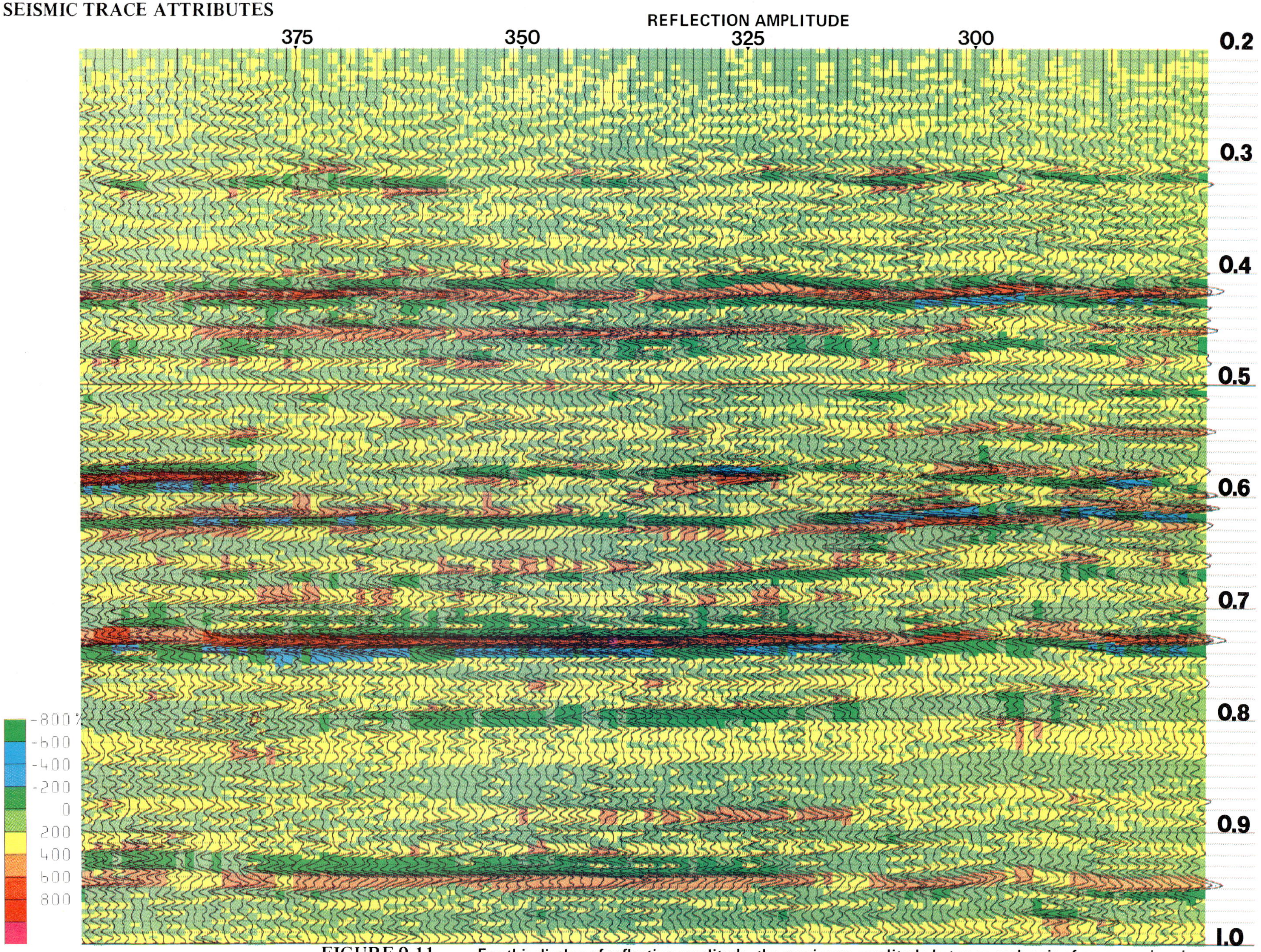

FIGURE 9.11 For this display of reflection amplitude, the maximum amplitude between each pair of zero crossings is measured, and a continuous band of color coded to the amplitude, is applied.

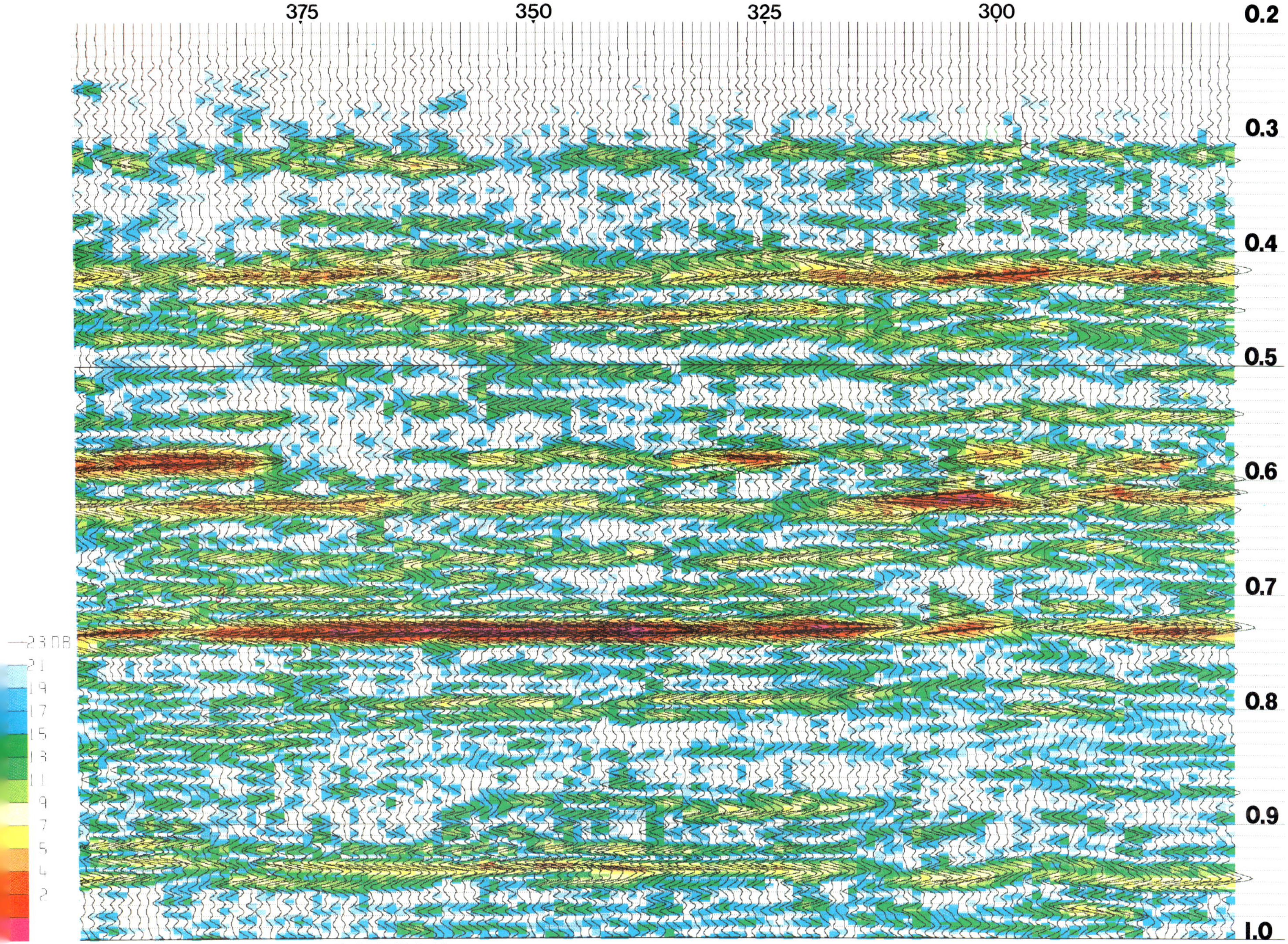

FIGURE 9.12 Amplitude of the Hilbert Transform is displayed in color. When combined with the conventional traces it yields amplitude of the reflection envelope, independent of reflection phase.

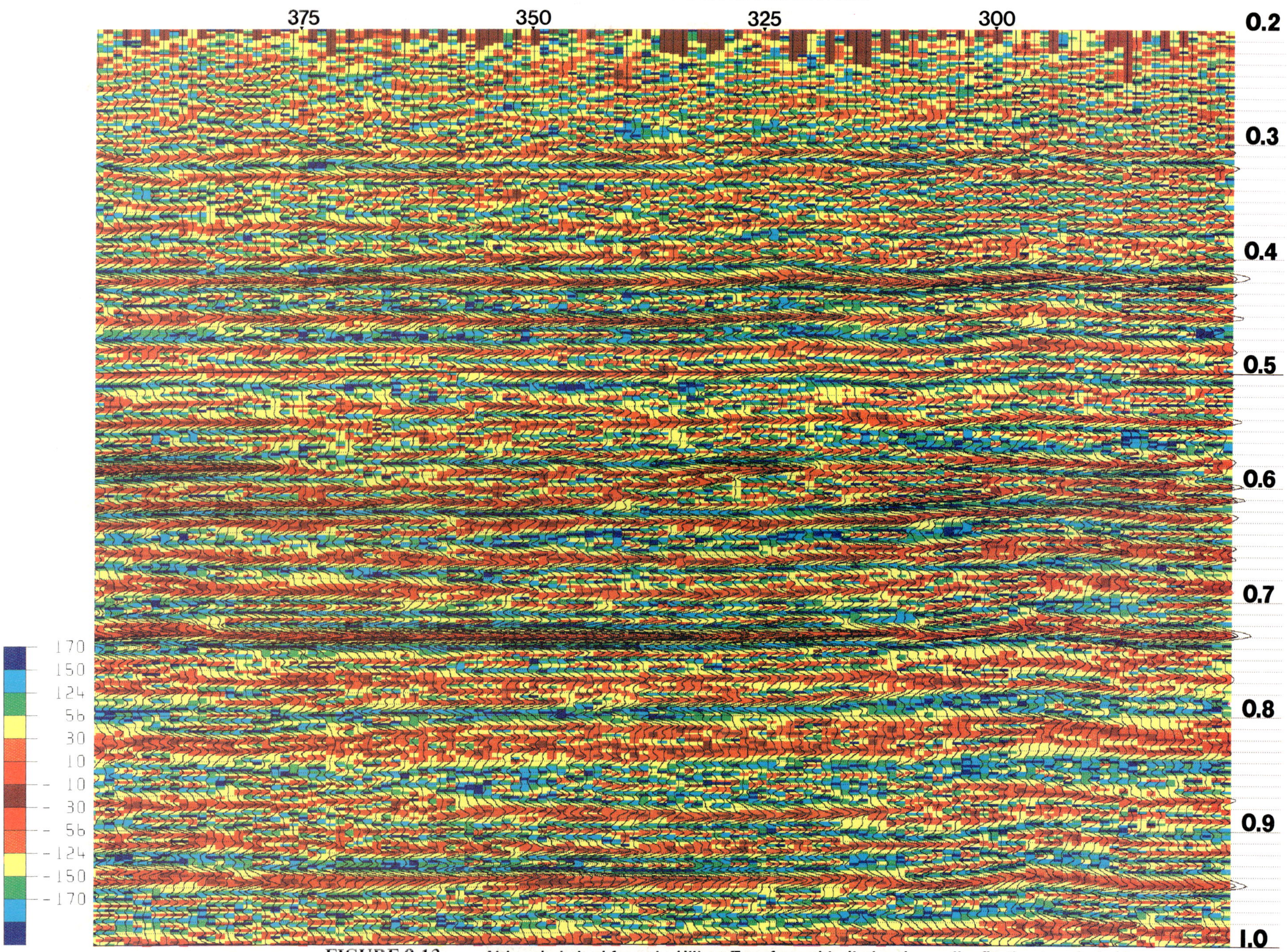

FIGURE 9.13 Although derived from the Hilbert Transform, this display shows all reflection troughs in blue, and all peaks in red, separated by zero crossings in yellow.

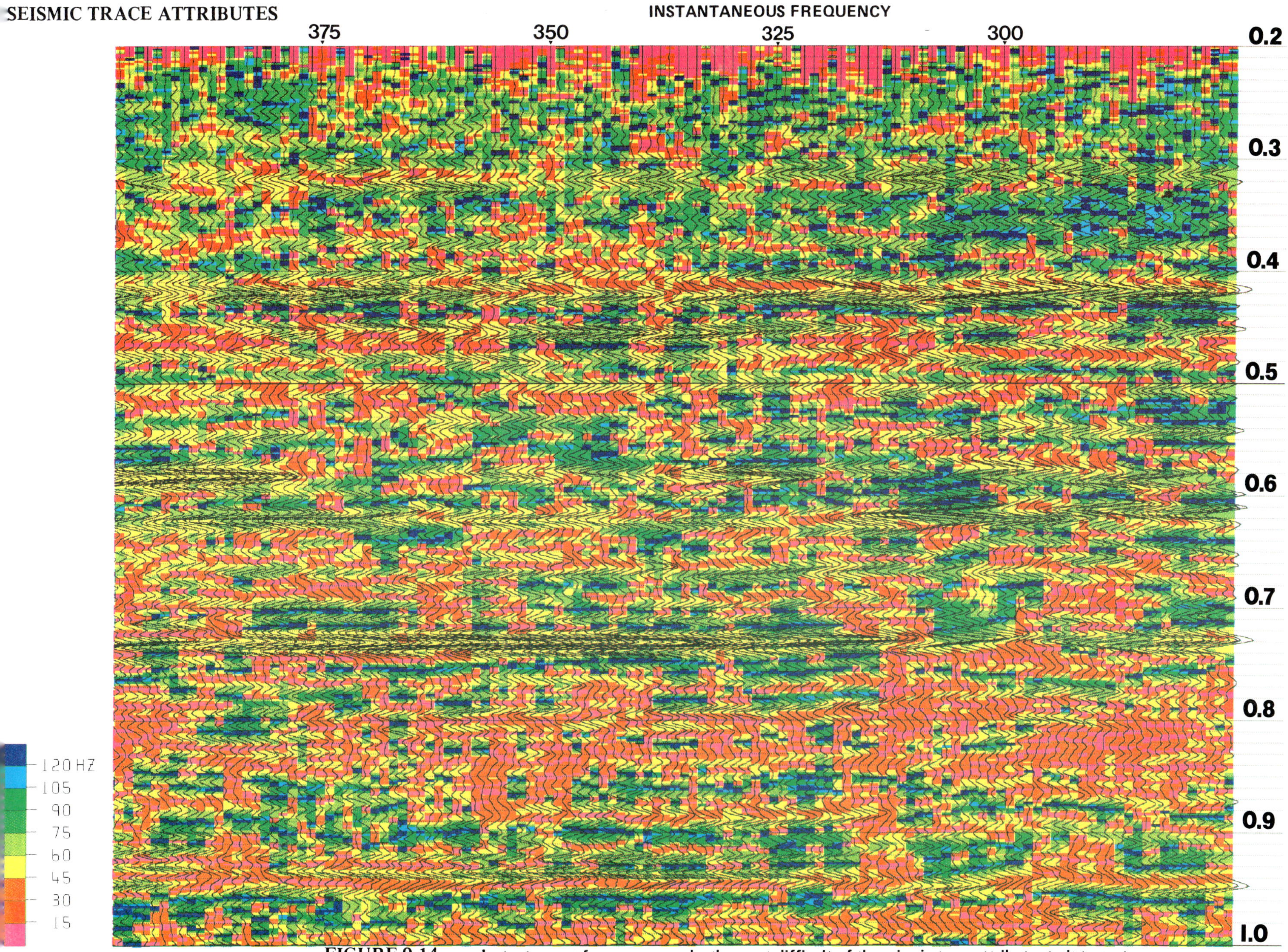

FIGURE 9.14 Instantaneous frequency may be the most difficult of the seismic trace attributes to interpret, since it is a derived function which is very sensitive to frequency changes.

Acoustic Impedance

Seismic reflections are generated at the boundaries between layers of contrasting velocity and density within the earth; as demonstrated earlier by the simple model for developing reflection coefficients from density and velocity logs. The calculation of reflection coefficients produces an effect similar to differentiation.

The reverse process is discussed in detail in Section 12, but a much simpler version is used from time to time to produce Acoustic Impedance Sections.

If the seismic field data are wavelet processed to the best possible approximation of reflection coefficients, **(Fig 9.15)** it can be shown that a simple integration of each trace, by summing the samples over a time series which is approximately equal to the logarithmics of acoustic impedance:

$$\ell_n \rho_i \nu_i$$

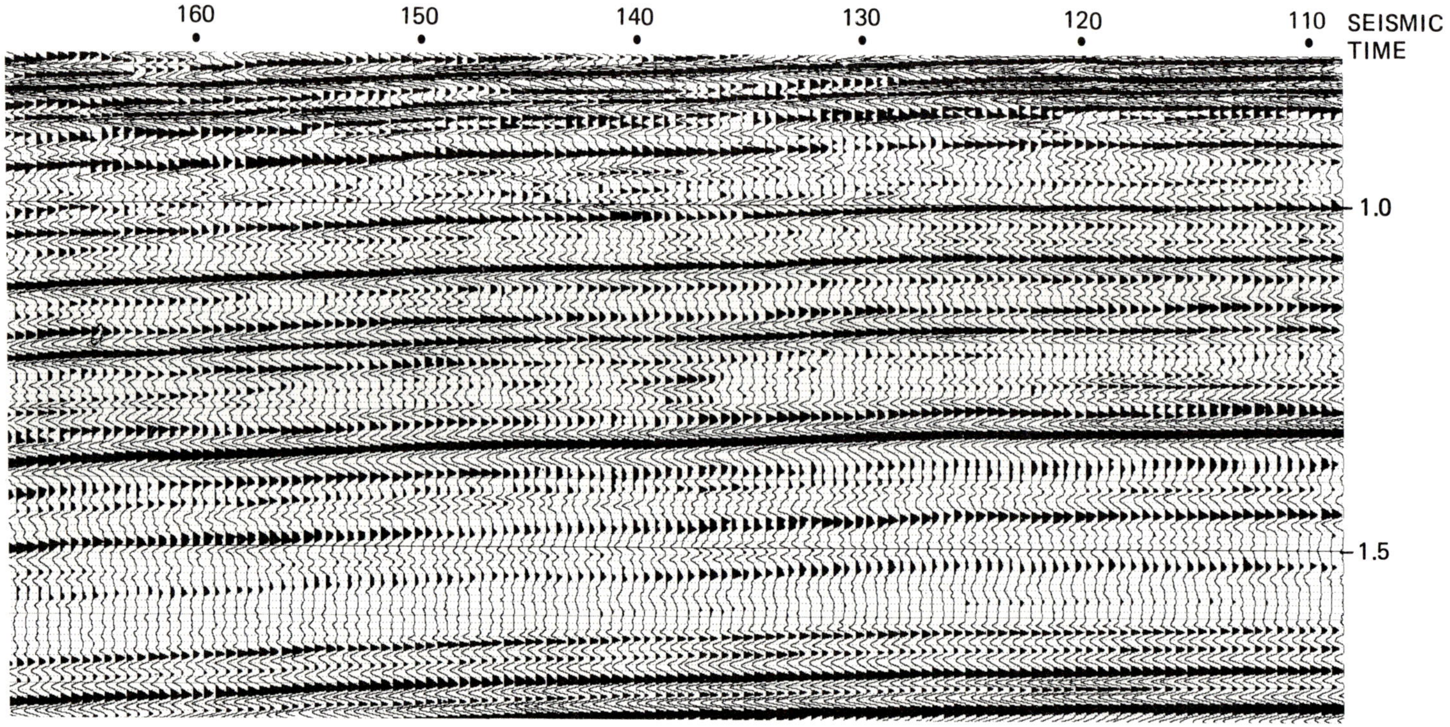

FIGURE 9.15

Wavelet processed seismic section maps boundaries between rock units.

A plot of Acoustic Impedance **(Fig 9.16)** can be a useful adjunct for seismic work. Impedance logs map geologic units, not velocity boundaries. Thus, subsurface rock units can be identified and mapped. Velocity changes within each rock unit can be identified and interpreted in terms of lithology and reservoir properties, whereas conventional seismic sections merely map the ratio of velocities between adjacent units.

Color coding is an important aid in the interpretation of impedence sections, because both positive and negative velocity shifts can be of equal importance. A drop in velocity can be be indicative of porosity, or gas, or shale in the section. A positive shift or increase in velocity, can be an important indicator of reservoir rock, and of production in sandstone units, since clean sandstones often increase in velocity.

If the input data has been properly wavelet processed, an impedance log should match a filtered integrated sonic log

FIGURE 9.16

Impedance logs map subsurface rock units, not velocity boundaries. Velocity changes within each unit can be interpreted in terms of lithology and reservoir properties.

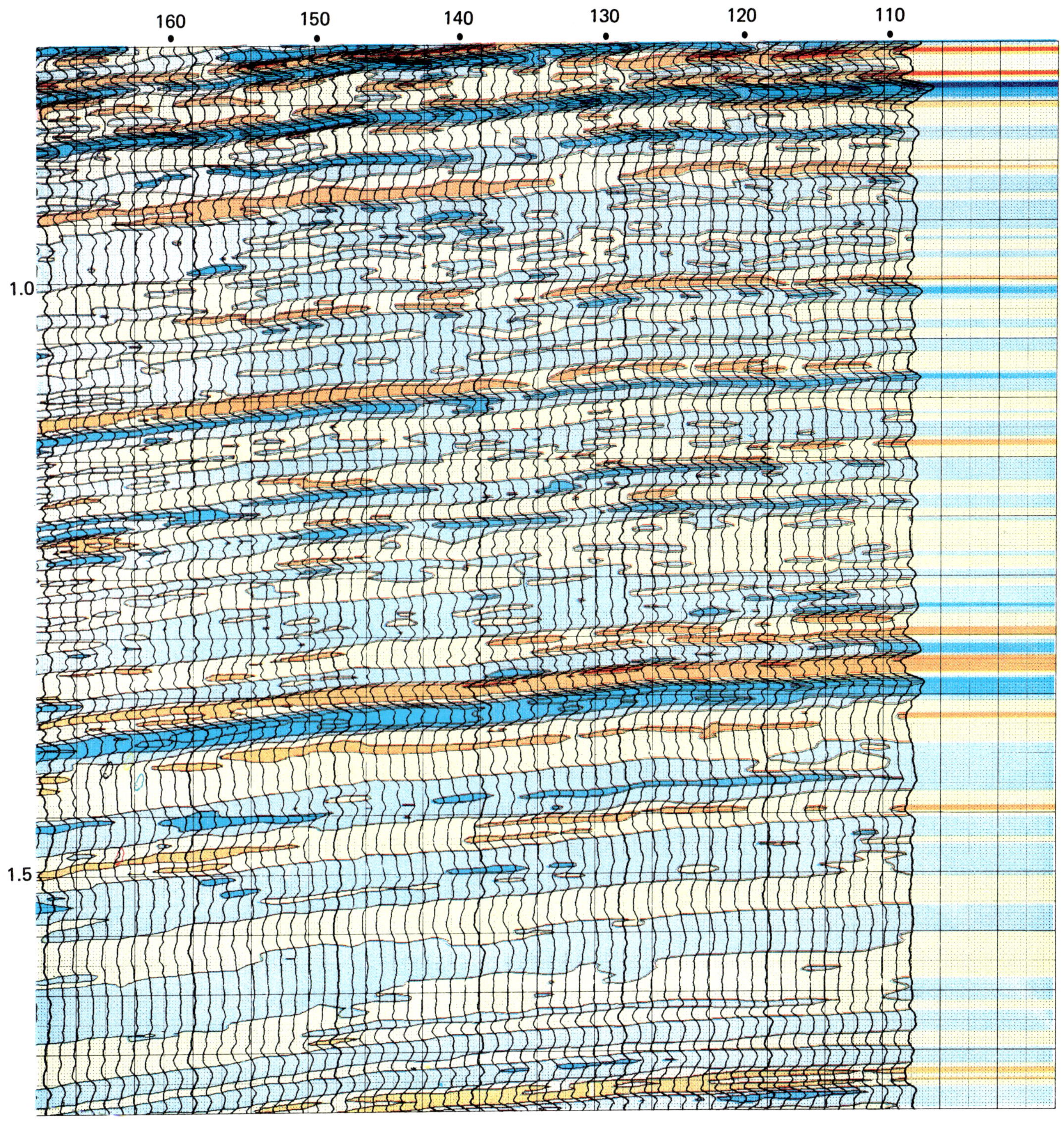

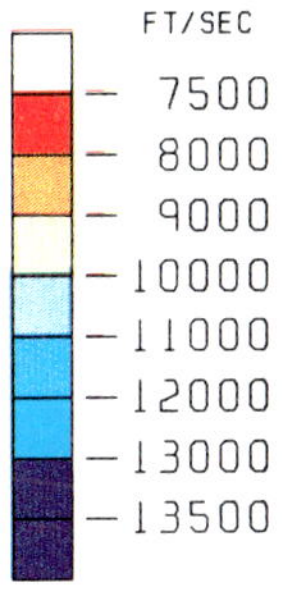

at a tie location, ensuring positive correlation of geology to the seismic data. Vertical time scales remain in time, similar to those of the conventional seismic section, and normal structure mapping can be carried out on impedance sections in the same manner as on conventional sections. The results tie to existing seismic work, but the impedance provides more precise definition of the subsurface. Results may vary as a function of data vintage.

10 MULTI CHANNEL OPERATIONS TWO DIMENSIONAL FILTERS

Introduction

Of all the tools which are available for processing seismic data the technique of band-pass filtering is the oldest. It is still useful in some cases although current good practice is to acquire and preserve the broadest possible frequency band.

A properly selected digital band-pass filter can eliminate noise very effectively when the frequency of the noise differs from that of the signal. But today most noise lies within the wider range of signal frequencies. The problem is now one of degree expressed by the relative signal-to-noise ratio. In all such cases the noise cannot be removed by band-pass filters without removing some signal also.

Deconvolution does not, as a rule, remove noise. Spiking deconvolution merely restores weak frequency components to their proper level. But if the restored components are noisy their influence on the output will increase and they will degrade rather than improve the output.

Conversely, deconvolution may also suppress highly colored (very strong) spectrum components. This tends to eliminate oscillatory noise from the signal by merely suppressing the cyclic component without restoring the correct parameters for that frequency. Most likely the phase component will be incorrect.

Predictive deconvolution may be used to eliminate multiple reflections from the seismic signal when the signal meets the assumptions of the model upon which the deconvolution method is based.

Most seismic noise elements have at least some of the same frequency content as the desired signal and, in most cases, are extremely difficult to remove by any standard single-trace filter method. This is due to the fact that most single-channel filter operations merely modify the amplitude of the frequency content of the signal.

Signal and Noise Components

Stacking is extremely popular and successful because it is highly effective at attenuating random noise on the seismic trace,even when that noise lies within the frequency range of useful seismic signals. Very few other filtering techniques can be used to separate signals from noise of the same frequency.

Another method which is not totally frequency dependent, and which can separate noise from signal of the same frequency, is the multi-channel filter, commonly identified under several names such as two-dimensional dip-velocity and/or Pie-Slice-filter.

One way to classify the signals contained on a seismic record might be based on Figure 10.1 which illustrates the manner in which the main seismic events arrive at the recording spread. For the illustration, normal moveout of the primary reflections has been ignored since it can be easily corrected. Reflections, the desired signals, arrive at discrete times. For this example they are essentially flat. The useful information includes several such events and is to be retained (Fig. 10.1a).

The remainder of the recorded information will include some coherent events which may contain useful information but which are classified as noise because they are not wanted on the final record. Other types of noise will be more or less random over time and distance and appear as uniform background "coloration" on the record.

One strong noise component may be ground roll (**Fig. 10.1b**). It radiates from the energy source across the surface of the ground to arrive on each trace at successively later times, a simple function of distance and velocity. Ground roll velocity is rather low, so the signal will appear to have a large angular attitude on the seismic record **(Fig 10.2)** Ground roll, in the past was eliminated largely by low-cut frequency filtering. All frequency filtering throws out some good signal with the bad so a more selective method of noise elimination, which could discriminate between noise and signal, would be preferable.

A similar comment might apply to other horizontally travelling waves, such as refracted energy, air blast waves and water breaks on marine records. All of these usually appear at a high angle on the seismic record, relative to reflections, and usually contain a substantial amount of the same frequencies which comprise the seismic signals. Such noise cannot be removed by frequency-dependent filtering without damage to the reflections.

The section may be plagued by diffraction noise, particularly in an area that has been tectonically disturbed. The points of diffraction generate new wave fronts which, at some point, are tangent to the reflections and from that point onward will have an increasing angular "dip". Multiple reflections will normally have greater normal moveout than the primary reflections. Hence, although they may be essentially flat over the near traces, they will have increasing angular deviation from the primary reflections on the far traces.

Multiples and diffractions will have essentially the same wavelet signature as the reflections and, therefore, are very similar in terms of frequency content. Their presence in a power spectrum or an autocorrelation will not be readily be detected. Any slight difference is averaged out with other events, leaving no substantial frequency discrimination between the noise signals and the reflection signals.

The one basic difference between the signal and noise in all these examples is the angle at which the events cross the recording. The angular difference provides the experienced interpreter with a criterion sufficient to discriminate unwanted events from the continuity of the reflection signals. But this does not mean that the section could be interpreted without difficulty. The noise may interfere to such an extent that it completely precludes reliable interpretation. An opportunity to remove the noise by data processing prior to interpretation is provided by multi-channel or F/K filtering.

FIGURE 10.1

Except in tectonically disturbed areas, most reflection signals tend to be flat after correction for normal move-out.

Several types of information are usually contained on a seismic record in addition to reflection signals.

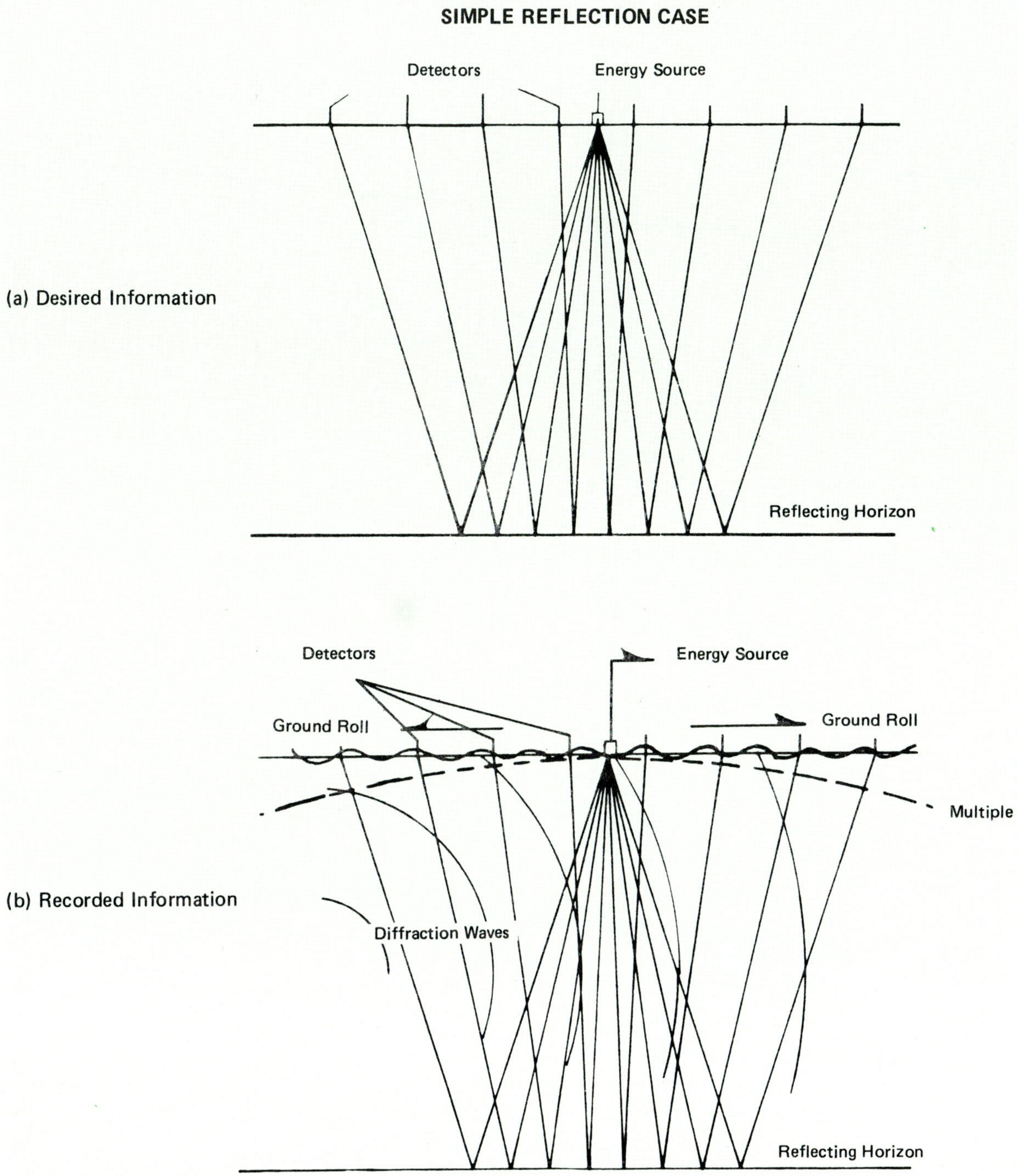

Fourier Transform

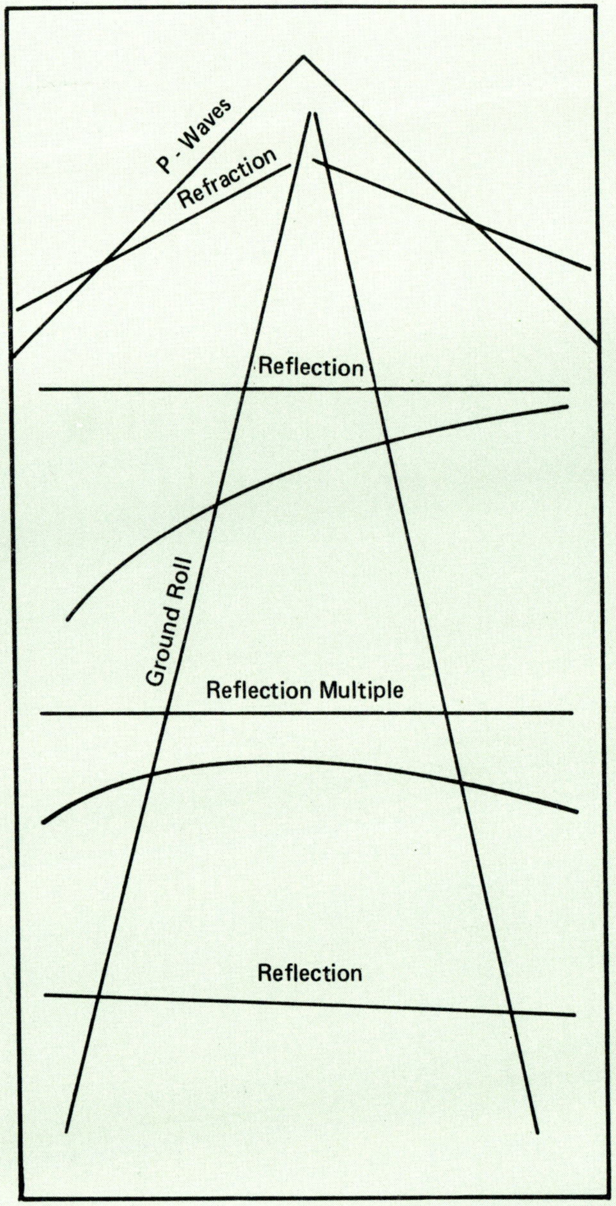

FIGURE 10.2

Many of unwanted events on a seismic record arrive with some angular difference from the reflections.

The Fourier transform will convert any seismic trace into its individual frequency components, each defined by a cosine wave in terms of its amplitude and phase. The pure cosine wave corresponding to any single given frequency component, can be examined in isolation from all others.

Every trace on a given seismic section can be separated into its component frequencies in a similar manner. Then a number of seismic sections could be constructed, each of traces having a single frequency only.

Each trace on a seismic section is a series of discrete samples, or numbers, at uniform time increments, almost invariably oriented to some horizontal zero-time datum. A seismic section is, therefore, a matrix of samples spaced at uniform vertical time increments and uniform horizontal space increments (trace spacing). The section matrix can be analysed horizontally along each row, rather than down each column. The row of samples across all seismic traces at each time increment can be treated as a continuous series in a similar manner to the vertical series representing a single trace.

Let us now consider horizontal analysis across a section of single-frequency traces. It is assumed that trace spacing is uniform, that normal moveout has been corrected and that any required surface static time corrections have been applied. **(Fig 10.3)**

If the monofrequency section is completely flat and uniform the amplitudes of all samples along each horizontal time line will be identical. The value will vary for different time lines. The resulting series of horizontal samples could be plotted as a continuous signal by joining the set of horizontal sample points with a continuous smooth line, in this case straight and flat.

More commonly the signal obtained in this

manner is a smooth, continuously varying function which lends itself to frequency analysis in the same manner as the customary vertical time series.

The periods of the horizontal waves would be measured in units of distance rather than in units of time. In a time series, if the period of oscillation is T, the frequency, F, equals 1/T. Similarly, the wavelength of a horizontal period of oscillation is the distance, X, and the horizontal frequency (or wavenumber), K, would therefore be 1/X. If the distance between samples, i.e., traces, is unknown the frequency can be expressed as a function of cycles per trace.

The frequency content of the horizontal signal could be determined by the Fourier transform. A plot of amplitude versus wavenumber will provide an amplitude spectrum. In the case of the flat section, the horizontal signal would appear as a simple constant shift from the zero line (**Fig. 10.3**). This is the equivalent of a signal of zero frequency. As long as the reflection energy is in phase (flat dip) and the amplitude of each monofrequency trace is uniform, the output of sampling along any time line will always be a "DC" component, varying only in amplitude from one time level to the next. The frequency of a DC signal is zero, so the point would be graphed with the measured amplitude at zero frequency.

Design

A more useful form of graphic display would be to plot the horizontal (spatial) frequency against the vertical (temporal) frequency. To distinguish between these two measures of frequency K has come to represent the horizontal frequency and F the vertical. A measured amplitude would be plotted at the intersection of the corresponding F and K co-ordinates. The amplitude value would be common to both the vertical and horizontal sampling, and therefore could be plotted as an elevation, above the F/K plane. The F-axis will have the familiar Nyquist frequency but the K-axis will have a + and a - Nyquist frequency to account for positive and negative dips on the time section.

If the section of single-frequency traces had a small amount of dip (**Fig. 10.4**) each trace of the series would be successively shifted by a slight time increment. Sampling along the horizontal time line would yield results quite different from the first example (flat dip). Each of the vertical traces is now slightly out of phase with the adjoining traces. As a result the series of horizontal samples will not be constant. If the rate of dip is uniform the change from sample to sample will exhibit a systematic pattern. The constant-time increment on each vertical trace due to the dip yields a constant trigonometric phase increment in this systematic pattern. These horizontal samples, when recast as a horizontal trace, will, in fact, have a low-frequency cyclic or sinusoidal character.

The zero horizontal frequency component in the first example was plotted on the F-axis (zero on the horizontal frequency axis) at the appropriate temporal or vertical frequency value. The corresponding low wave number, K, in the second example will be plotted where this K value intersects the same vertical frequency as the previous example. The amplitude value will again be an elevation above the F/K plane.

Increasing the amount of dip (**Fig. 10.5**) and holding the vertical frequency constant, as in the first and second examples, causes the horizontal frequency to increase. Evidently, for any given vertical frequency, the horizontal frequency is related directly to the amount of dip. Measurement of the horizontal frequency as a function of vertical frequency provides a reliable means of separating and measuring the various dip components in a seismic section. Note that the angle of this dip component is not the angle of geologic dip but the slope on the time section in units of time and distance.

To this point, all examples have used a constant vertical frequency while increasing the amount of dip to observe the corresponding change in horizontal frequency. For the next example (**Fig. 10.6**) the dip is held constant for

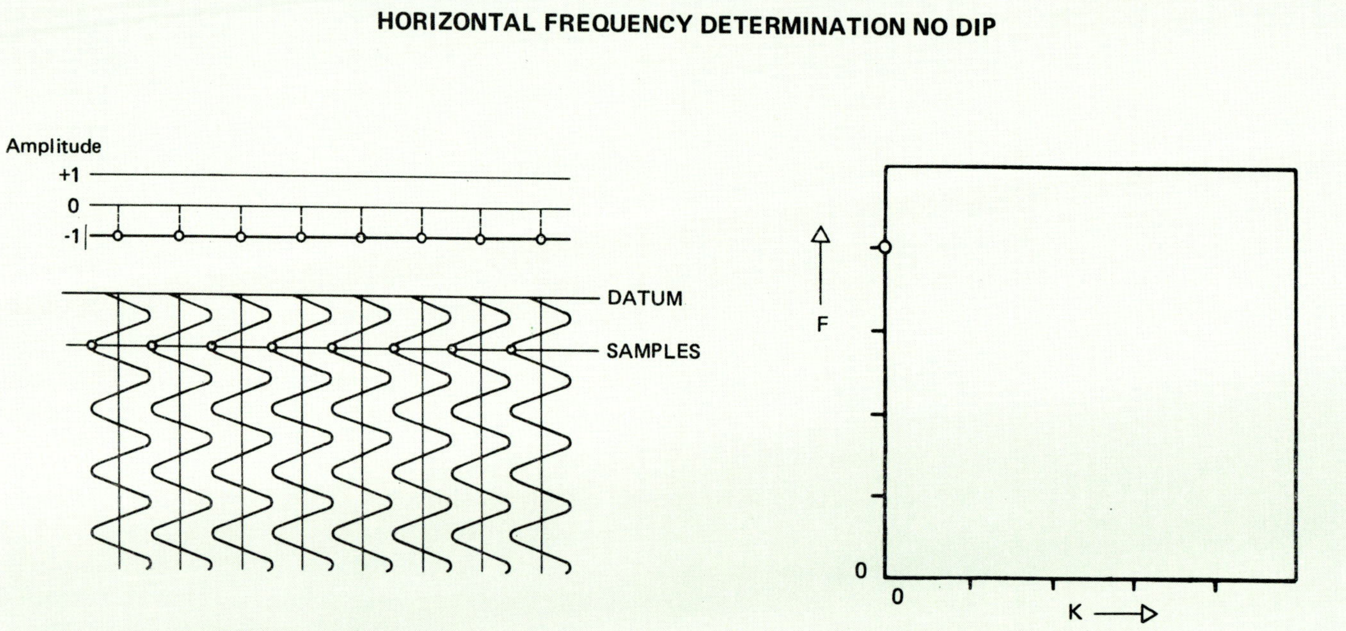

FIGURE 10.3

Multiplex sampling along a time line of mono-frequency components from a flat seismic section yields a horizontal data set having zero frequency.

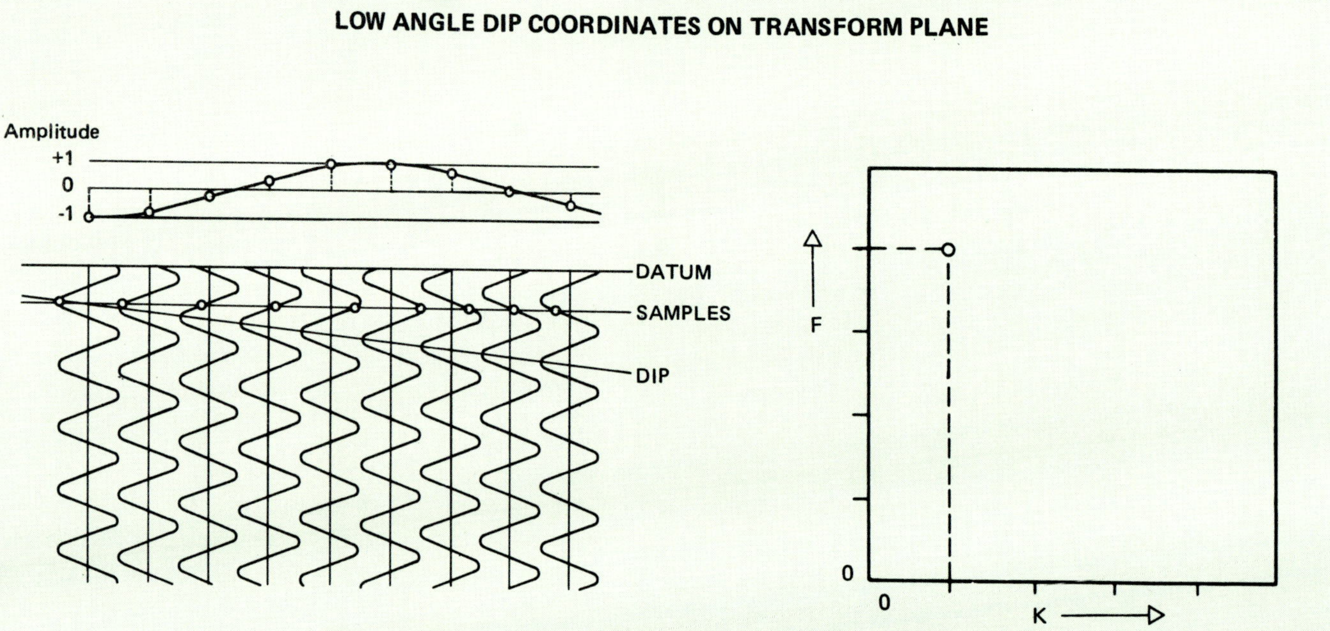

FIGURE 10.4

A small amount of dip yields a low frequency horizontal signal which can be plotted as amplitude on horizontal and vertical frequency coordinates.

HIGH ANGLE DIP COORDINATES ON TRANSFORM PLANE

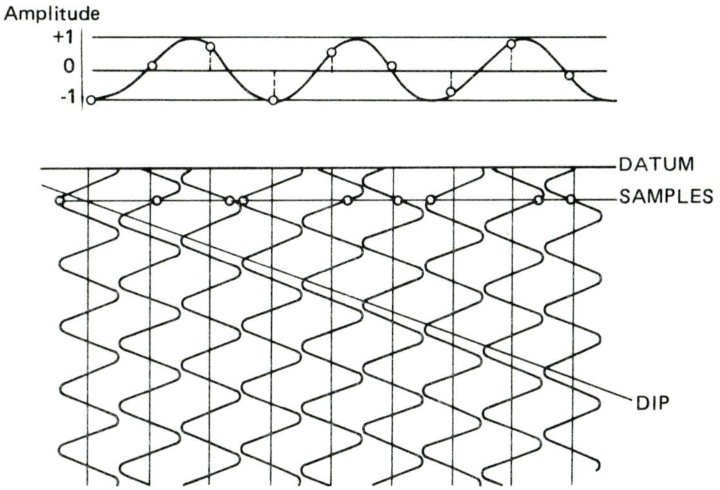

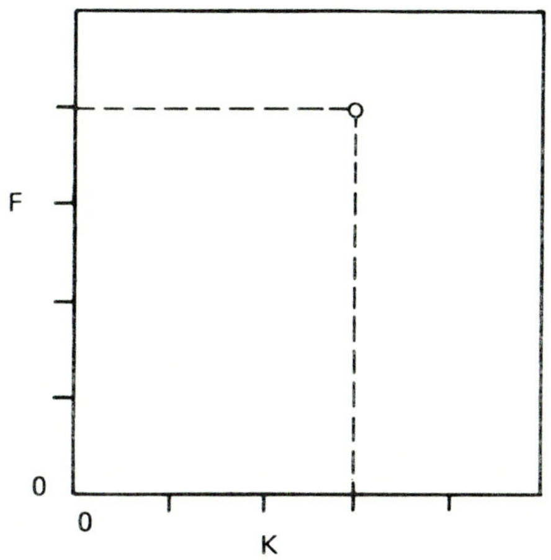

FIGURE 10.5

Increasing the dip on the same mono-frequency section of Figure 10.3 causes a corresponding increase in horizontal frequency.

HORIZONTAL FREQUENCY IS A FUNCTION OF VERTICAL FREQUENCY AND DIP

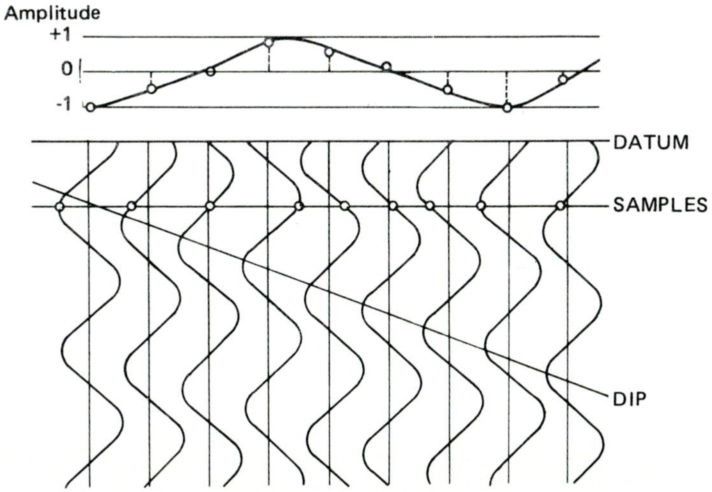

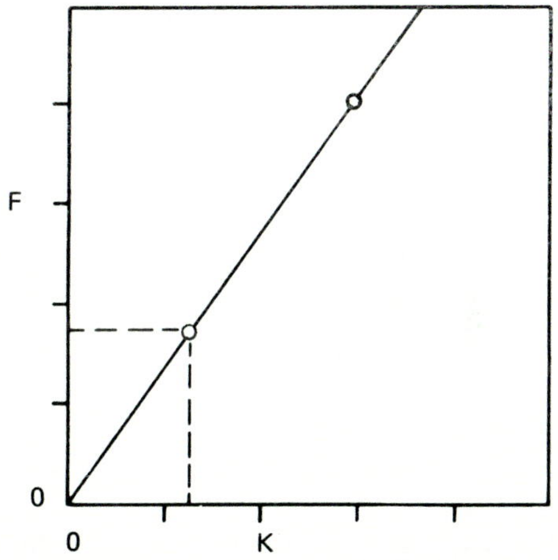

FIGURE 10.6

Horizontal sampling of a lower frequency data set having the same dip of Figure 10.5 yields a lower horizontal frequency. The F,K, frequency coordinates for a constant dip fall on a straight line.

a chosen, lower vertical frequency. As a result the horizontal frequency is also lowered, and the coordinate point moves to a new position of the plot. In this case, although the vertical time shift of the event between traces has not changed, the lower vertical frequency results in a smaller phase change between horizontal samples and hence yields a lower horizontal frequency. Thus, for a given constant amount of dip, the horizontal frequency will be dependent upon the vertical frequency. Doubling the vertical frequency will double the horizontal frequency, resulting in a direct linear relationship between the vertical and horizontal frequencies.

It is therefore possible to graph (**Fig. 10.7**) the relation between horizontal and vertical frequency for given amounts of dip. The best way to express the amount of dip is in terms of Δt, the incremental time shift from trace to trace. This is related through a velocity function to the amount of actual physical dip in the ground but, since the velocity function can be very complex, the Δt method is a convenient way of defining the dip on a seismic time section. Note that horizontal (flat) dip on the time section is transformed to points on the vertical F - axis on the F/K plane, and dips measured from the horizontal in the time section are transformed into vertical angles measured from the vertical or F - axis.

A chart may then be constructed (**Fig. 10.8**) to display the correspondence between Δt stepout per trace, the angular expression of the F/K plane and frequency in each direction. In this case, the angular relationships expressed in degrees on the chart refer only to the F/K plane of the chart and do not represent dip in the earth. Note also that the chart is valid for only one time-sample rate.

Charts similar to **Figure 10.8** may be constructed for other temporal sample rates. Regardless of the sample rate or number of traces, for a square sample matrix, the 45° line on the F/K chart will always represent an increment of one time sample per trace. For example, if processing is to be done at 2 ms per sample, the 45° line would represent 2 ms of dip per trace.

What we have developed is a relatively simple basic concept of the amplitude spectrum of the 2-dimensional Fourier transform. In real life the determination is somewhat more complicated. As indicated, dip components are determined by measuring the horizontal frequencies for each of the vertical frequencies in order to provide redundancy and statistical reliability in the measurement. The seismic dip is usually not constant, nor it is uniform over the length of the section, so all combinations of horizontal and vertical frequencies must be measured. The number of combinations becomes very large. But, fortunately, the analysis can be treated as a double integral and handled by approximation with a two-dimensional Fourier series. In practice, vertical transforms are computed for each trace and the resulting frequency elements sampled horizontally. A transform of each horizontal row, corresponding to each frequency element, accomplishes the operation. A two-dimensional Fourier transform will provide a set of coefficients for every combination of frequencies from zero to the aliasing frequency in each direction. From each of these, amplitude and phase can be computed to provide the required description in the frequency domain.

Whereas the amplitude spectrum of the single channel transform can be described by a 2-dimensional coordinate system and plotted on a histogram of amplitude versus frequency, the amplitude spectrum of a 2-dimensional transform requires a three-dimensional coordinate system, F, K and A, to represent a point in the system. Amplitude is mapped as elevation at frequency coordinate locations and the transform may be contoured to illustrate the amplitude spectrum surface (**Fig. 10.9**). Contouring has been suppressed in the central portion of the illustration where the elevation is extreme.

Two transform quadrants are mapped because the normal sine/cosine relationship used in development of the

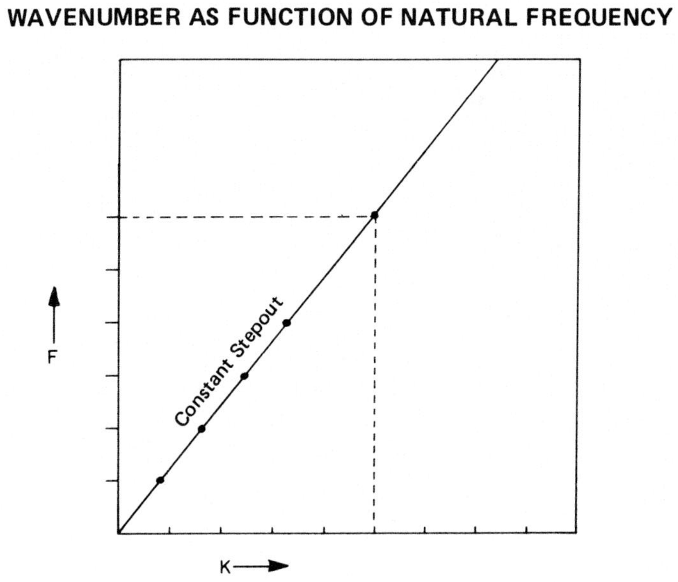

FIGURE 10.7

Seismic time sections have dimensions of time and distance, but dip can be expressed in terms of time shift or stepout per trace. A constant stepout across several traces yields a straight line on the F,K, plane.

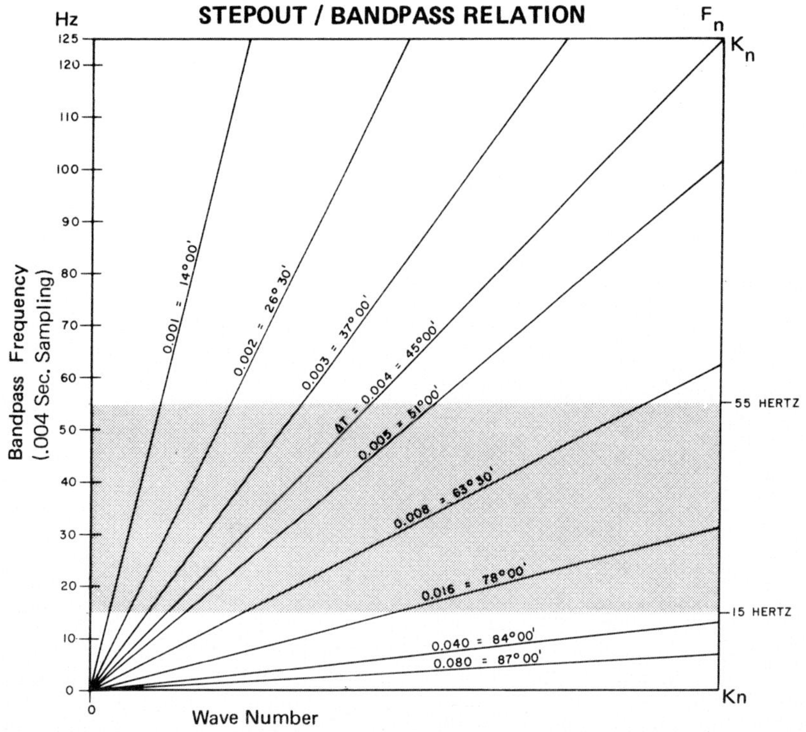

FIGURE 10.8

For any given time sampling rate, a time section dip of one sample per trace on the diagonal line extending from 0,0 to K_n, F_n on the transform plane. If the time-domain input had the same number of horizontal and vertical samples, the angles on the transform plane will be 45 . This angle is only indirectly related to geologic dip.

Frequency Relationships

Fourier transform provides the ability to distinguish between right and left dip. In fact, the full transform is bi-symmetric in four quadrants. But only the upper half need be referred to.

The contour values are not labelled in the example but except for the central portion, all have about the same value and appear to be randomly distributed. This is very likely due to random noise. Within the central portion of the chart, it is possible to draw two lines bounding a not-quite symmetrical wedge enclosing most of the contours rising above the background level. If the chart of **Figure 10.8** were laid over **Figure 10.9** the two bounding diagonals would correspond to seismic-section dip slightly greater than 1 ms per trace. Evidently a rather flat seismic section was used as input.

The lack of contours in the area below 10 Hz and above 50 Hz shows that the data were band-pass filtered at some prior stage. All information, useful signal or otherwise, outside the passband has been eliminated from the F/K plane. Perhaps a better illustration of the nature of the amplitude spectrum can be obtained from a perspective view of the transform surface. **Figure 10.10** is a view of the left dip quadrant of **Figure 10.9**, as seen from the upper left hand corner. It clearly illustrates the random nature of the background noise within the frequency passband and the concentration of dip information near the vertical or temporal-frequency axis.

The relatively simple time-domain/frequency-domain reciprocal relationships which exist for the simple single-channel transform become slightly more complicated in the two-dimensional case, but should be studied if the effects of process operations

TWO DIMENSIONAL AMPLITUDE SPECTRUM OF A SEISMIC SECTION

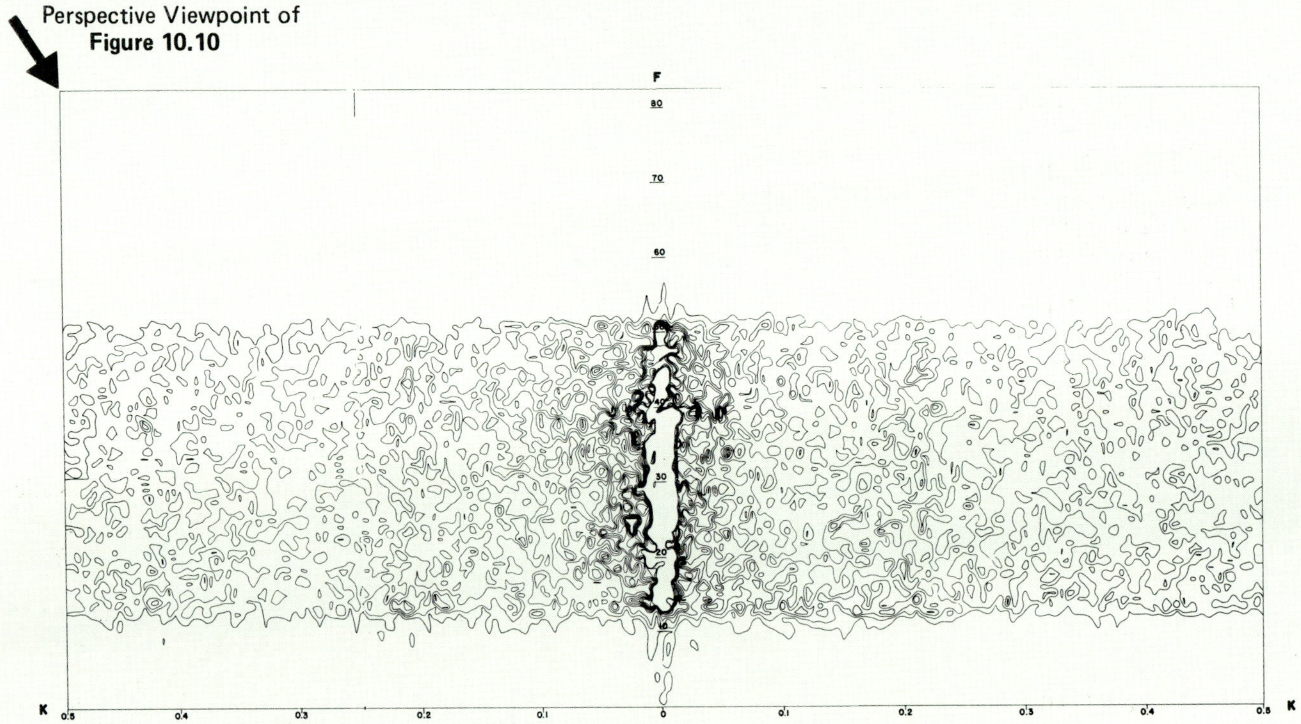

FIGURE 10.9

Contours represent frequency/ wave number amplitude values on the 2-dimensional amplitude spectrum of a seismic section. Contour values are not labelled, but are low over all regions except the central portion. Contours in the core have been suppressed due to the extremely high values. The data has been frequency filtered to a band-pass of 10-55 Hz.

are to be well understood. For instance, in the single-channel case a spike in the time domain is transformed to a continuous line across the amplitude spectrum in the frequency domain. In the two-dimensional case a spike transforms to a platform spread over the entire F/K amplitude plane (Fig. 10.11). The location of the spike in the time domain will be determined by the phase spectrum (not shown). These relationships deserve some study. A small amount of effort will go far toward understanding the fascinating reciprocity that governs transform pairs (Figs. 10.11 and 10.12).

Filter Operations

The procedures for application of filters in the 2-dimensional frequency-domain transform are a direct extension of operations performed on the one-dimensional transform. Primarily, they involve modification of the amplitude spectrum. A band-pass filter on the single-channel transform can be applied by simply setting the value of the unwanted components to zero. The identical operation is performed on the two-dimensional plane (Fig. 10.8) by setting all unwanted amplitude values, outside the passband, to zero. This may be done in either frequency (as illustrated by the shaded area) or wavenumber. Horizontal or frequency filters can be used to eliminate ground roll and other horizontally travelling low-frequency phenomena. But the most powerful filter is one which takes advantage of the unique ability of the two-dimensional transform to classify dip components.

The two-dimensional transform converts time-domain dip, measured in terms of Δt stepout per trace, into frequency-domain dip, measured in terms of vertical angles between the vertical or temporal frequency axis and lines

FIGURE 10.10

A perspective view of the amplitude spectrum of Figure 10.9. The corresponding time section is evidently very flat, since most of the amplitude is confined to very low horizontal frequencies.

PERSPECTIVE VIEW OF LEFT DIP QUADRANT,
TWO–DIMENSIONAL AMPLITUDE SPECTRUM

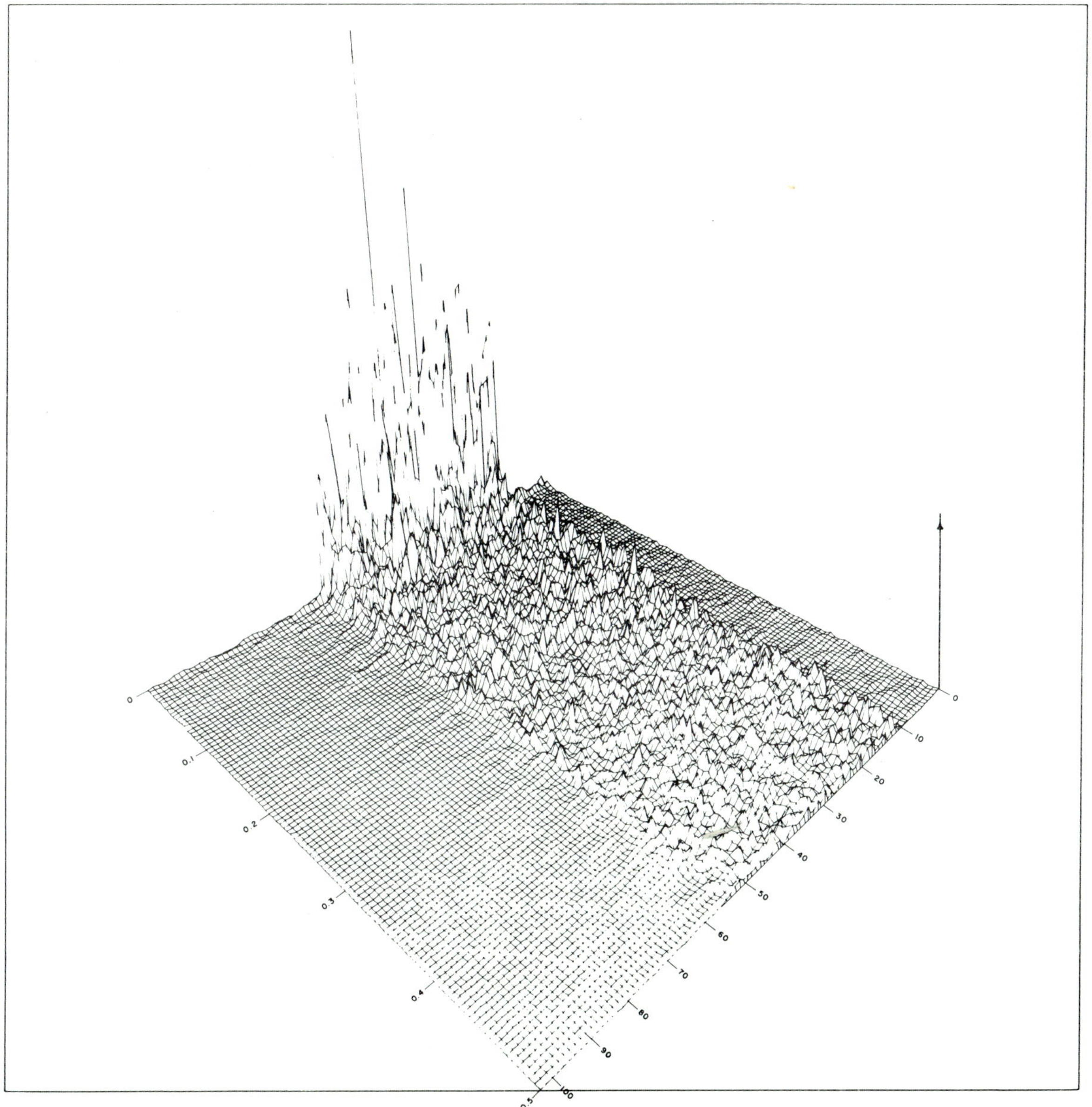

FIGURE 10.11

Fixed relationships exist between time domain and frequency domain expressions of typical signal phenomena.

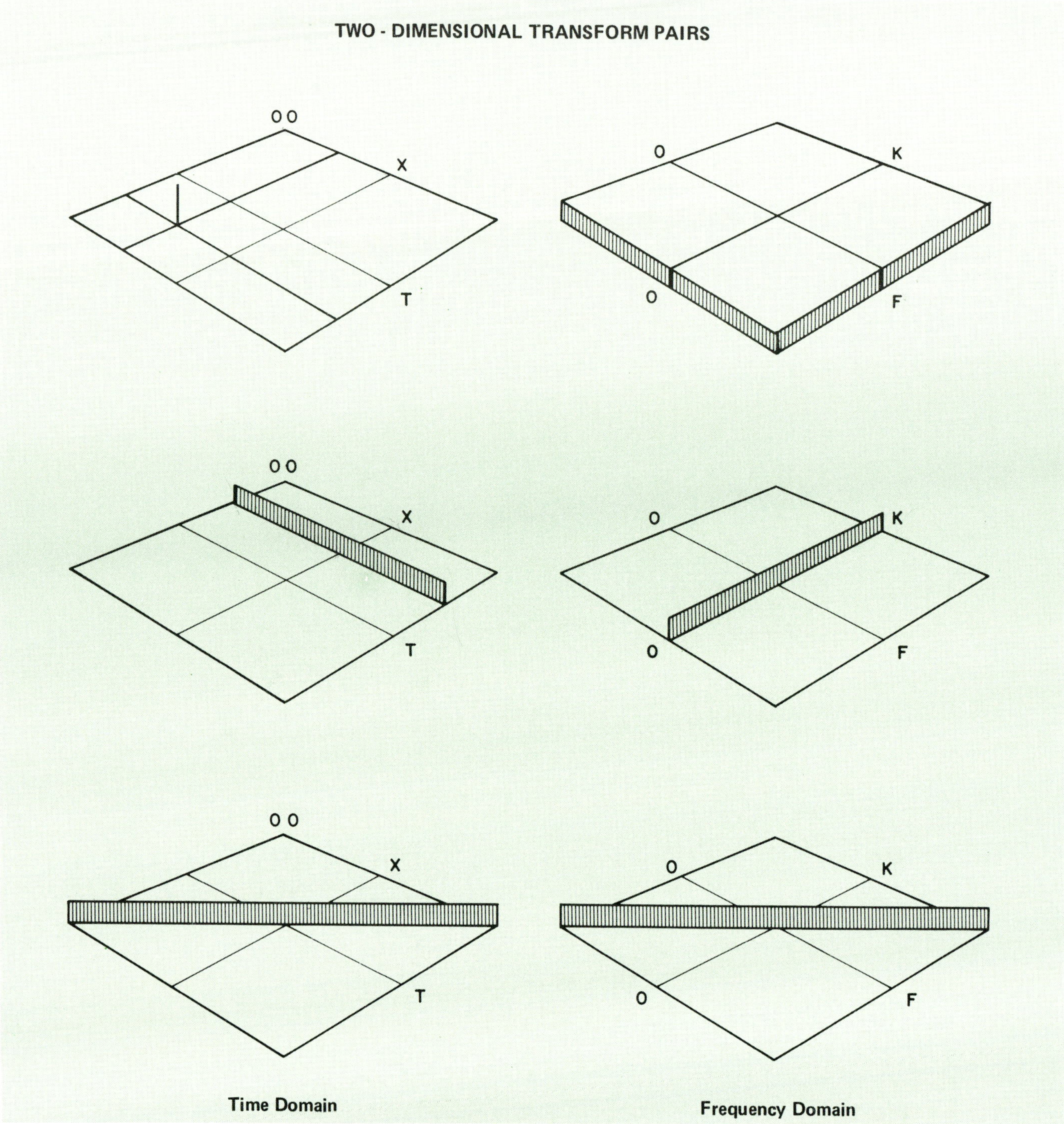

FIGURE 10.12

TWO - DIMENSIONAL TRANSFORM PAIRS

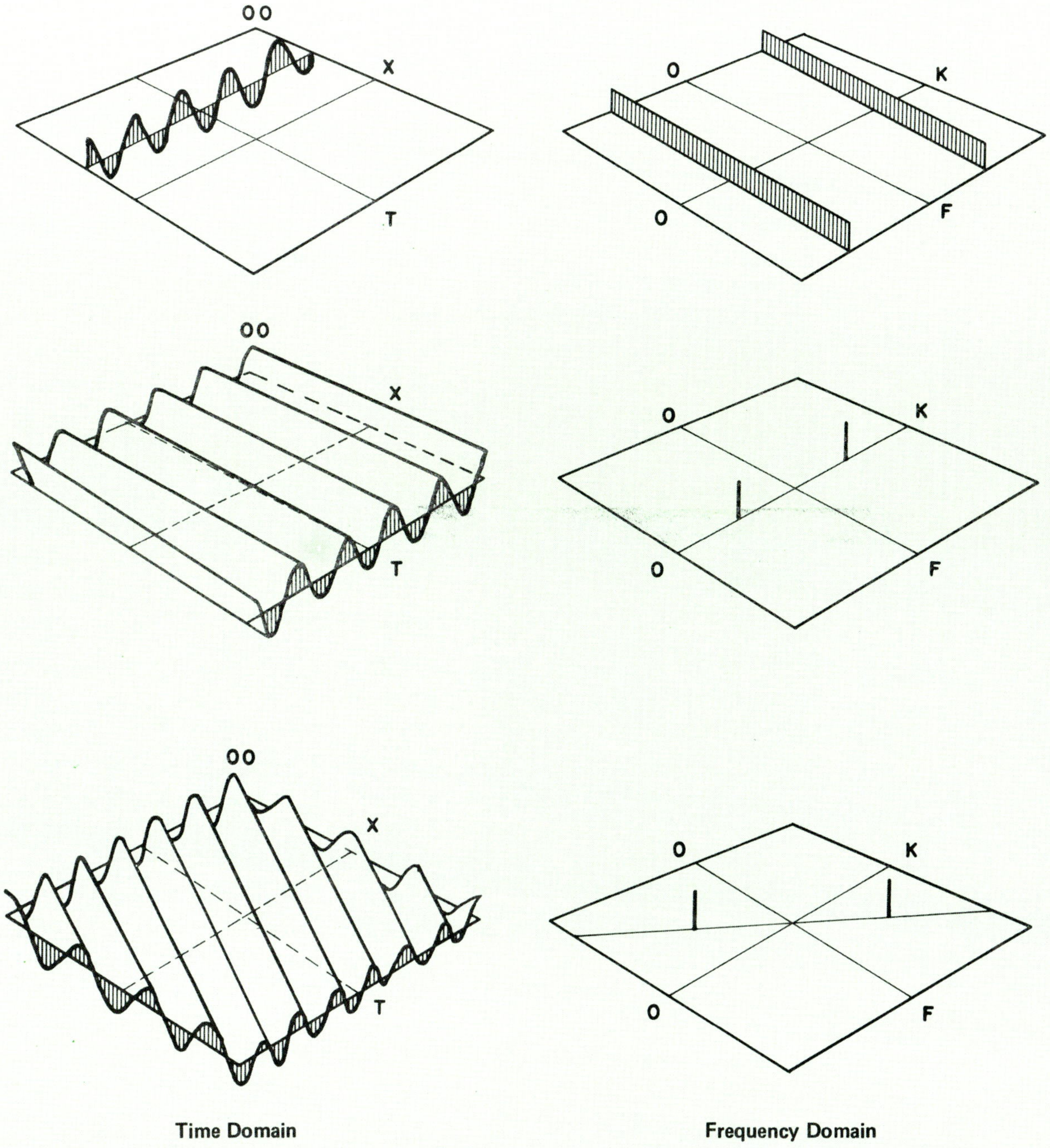

radiating from the origin of the F/K plane. There is no direct relationship between coordinate locations on the time section and coordinate locations on the transform plane. All fragments of time-domain dip at a given Δt stepout value are consolidated into a single vector or radial line on the transform plane. Thus, regardless of their location in time and space, all dip components having a given stepout per trace can be eliminated by removing the corresponding dip vector from the transform plane. If, as is more likely, it is desired to eliminate from the time section all events which dip more than some specified rate, this may be done quite simply by setting to zero all values of the transform lying outside the appropriate transform angle or vector. A simple chart can be constructed to relate the dip-filter parameters to the stepout of events on the seismic section (**Fig. 10.13**). Roll-off at the filter boundaries may be included to avoid side lobes which have an angular or dipping expression on the time section.

In this procedure, simple temporal frequency is no longer the limiting parameter of filtering. Now that portion of vertical and horizontal frequencies which combine to produce unwanted dip angles on the seismic section can be eliminated. Therefore, it becomes possible to eliminate noise having frequencies identical to the deisired signal when that noise exhibits a different dip angle than the signal, eg., ground roll and diffractions.

The idealized response of an single-channel frequency band-pass filter is a rectangle having an amplitude of 1.0 over the passband and zero elsewhere. The response of the dip filter has an added dimension in the transform space, and usually has the shape of a wedge. From this shape was derived the well known term "Pie-Slice" filter for multichannel operations.

Synthetic Examples

To illustrate the effects of two-dimensional dip-filter operations a test section has been developed, (Fig. 10.14a) containing a fan of information dipping over a fairly broad range. A number of dip filters of different parameters were then applied to the data. In (Fig. 10.14b) a single-component notch filter designed to eliminate flat dip, with a fairly sharp rolloff from 0° to 5° on both sides, has been applied to the section. Figure 10.14a illustrates a dip-filter passing 10° and 10° left dip, with a gradual rolloff.

(**Fig. 10.14d**) has been filtered with a sharp reject band extending from 0° to 15° left dip only. The results illustrate the very sharp cutoff characteristics and the high degree of discrimination which may be obtained. Note the excellent attenuation in the reject portion of the section, although some weak side-lobe effects are also visible.

It is not difficult to see that this operation has a valuable practical application as a dip filter, to pass or reject a given band of dips on a seismic time section. The flexibility of the method is extremely broad provided the geologic structure is not itself complicated by a wide range of dip components. The technique is not limited to flat dipping events but, at least in theory, can be applied to any dip orientation in which a separation exists between the dip exhibited by signal and noise.

Regular frequency-bandpass filters may be combined with the dip filter and applied simultaneously to filter an entire seismic section in both dip and frequency. The need for bandpass filtering may be diminished by the dip filter because major coherent noise signals usually have a definite directional component, which can be eliminated by the dip filter.

Laserscan

Mechanical implementation of this process was cleverly conceived and applied long before digital processing. An optical process, termed the Laserscan, was developed by the United Geophysical Company and the Conductron Corporation, and several examples of its operation, are outlined in an outstanding paper by Dobrin, Ingalls, and Long, in Geophysics, December 1965. The many examples contained in the article, together with the detailed explanation,

provide an excellent exposition of the process. The basics of the optical processing described in the paper differ from digital processing only in that analog optical methods are used to obtain the transform and to apply the filters.

Unfortunately, the resolution obtainable with the optical process was at that time somewhat limited due to the quality of lens optics and diffraction gratings. Limitations of resolution of the optical process precluded it from reproducing the individual filtered seismic traces on the section after the filter application.

This becomes a trivial matter when the operation is performed by digital methods. The sample values used for digital computations can be carried to almost any desired degree of resolution, eliminating any quality restriction. The digital process also permits other modifications to be made to the 2-dimensional amplitude spectrum, making it potentially feasible to perform other signal process operations in the F/K domain.

Equivalent Operations

Although the 2-dimensional filters described here have been described as operations in the frequency domain, it should be remembered that similar operations can be carried out, at least in theory, in the time domain. The basic principles outlined here apply to all dip filters, variously described as 2D filters, velocity filters, dip filters, time/space filters, moveout filters and the Pie-slice process. Practical limitations interfere when trying to perform truly identical operations in both domains. For example, a single-channel bandpass filter in the time domain truly equals the simple frequency-domain application only when the time-domain filter operator is as long as the total length of the signal to be filtered. Shorter operators can only approximate the frequency-domain operation because they sample at large increments of ΔF. In the same manner a 2-dimensional operator in the time domain is a matrix, which must be the size of the input sample matrix if it is to equal the frequency domain operation.

An extension of the theory can be used to develop a three-dimensional transform. This would provide the description of seismic information obtained from a 3-dimensional seismic survey. Continuous sampling in X, Y, coordinates over the ground surface allows dip components to be migrated properly in 3-dimensional space. True three-dimensional coverage and application of the transform theory would require mapping of the amplitude spectrum as a hypersurface in 4-dimensional space, a difficult surface to display. Nevertheless the application of dip filters specifically oriented to spatial structure is quite possible. Holograms utilize a type of 3-dimensional transform mapped on a 2-dimensional plane.

Aliasing

All 2-dimensional filters can be seriously affected by a problem of aliasing in the horizontal direction. Aliasing is a well known effect in single-channel operations caused by an insufficient sampling rate. (See Fig. 11.13). FIG. 10.15

In 2-dimensional analysis, similar frequency aliasing may occur in both vertical and horizontal directions. But the most common problem is the result of dip aliasing, when the rate of dip is so great that the time stepout between adjacent traces produces a phase shift greater than one-half cycle of any vertical frequency in the frequency pass band (**Fig. 10.16**).

If the Δt shift per trace exceeds the limit described, the dip measurement is reproduced as a lesser dip in the opposite direction, ie, the horizontal frequency is aliased to a false lower wavenumber.

For any given amount of dip, the horizontal frequency will be correctly defined until the dip shifts some particular vertical frequency one-half cycle out of phase. This point will be represented by a horizontal frequency of 50 cycles per 100 traces or ½ cycle per trace (**Fig. 10.16**). On the next higher vertical frequency the same amount of dip causes the horizontal frequency to "fold back" to an alias frequency slightly less than ½ cycle per trace (**Fig. 10.17**). The sign of the horizontal frequency changes so the dip appears at a reduced angle on the time section, in the opposite direction. As the vertical frequency increases the horizontal frequency and the spurious

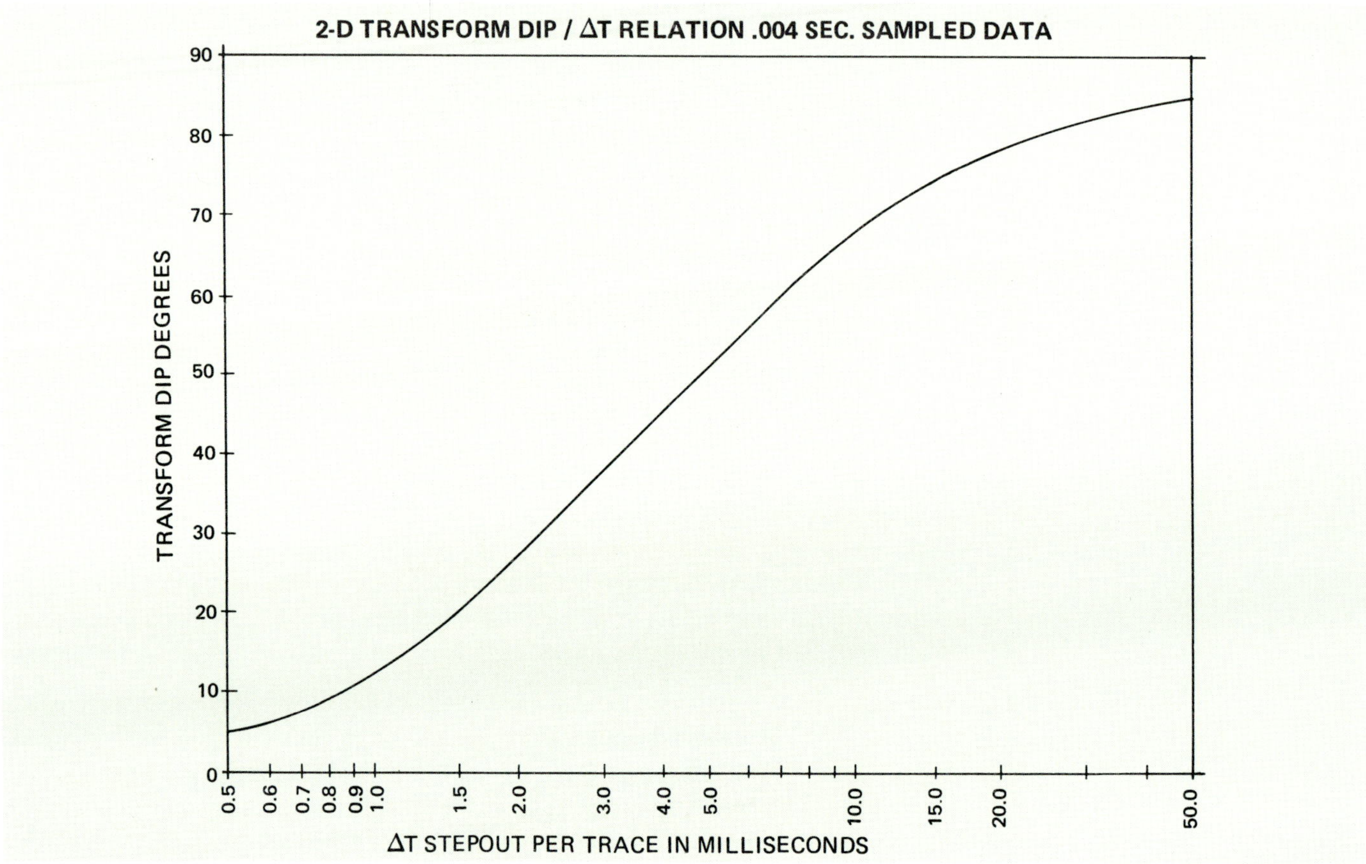

FIGURE 10.13

The chart illustrates the relation between time step-out per trace on a seismic section and a dip angle measured against the vertical axis (zero horizontal frequency) of the 2-D Fourier transform plane. This chart is valid only for a square input matrix of data sampled at 4 ms.

dip angle will appear to decrease. As the vertical frequency continues to increase, the horizontal frequency will reach zero (1 cycle per trace = 0 cycles per trace) and then change sign again and continue toward the aliasing frequency as in the first cycle. The apparent dip will also reach zero and then increase as in the primary cycle. Thus, since the horizontal sampling rate (trace spacing) determines the horizontal Nyquist frequency the range of time-section dips that can be represented correctly on the transform plane has a fixed maximum, in either direction, for a given temporal Nyquist frequency or vertical sampling rate.

If the vertical frequency is held constant and the dip in increased the horizontal frequency will range between the + and - Nyquist frequencies (**Fig. 10.17**). The aliasing or folding will always occur when the vertical displacement between traces equals an odd number of half-wavelengths of the vertical frequency. Obviously, the higher the vertical frequency the greater the danger of dip aliasing.

The effect, in the frequency domain, can best be illustrated by swapping the sides of the transform (**Fig. 10.18**). In the past, much of this aliasing fell above the limits of the normal frequency band-pass and did not interfere with the results. The higher frequencies recorded in modern seismic operations have elevated horizontal aliasing to a serious, but sometimes not recognized, problem.

The same sampling rules that govern time-series aliasing apply with equal rigor here. If higher frequencies are to be

FIGURE 10.14

The effect of applying 2-dimensional dip filters to the model input data of Figure 10.14a.

TWO - DIMENSIONAL DIP FILTERS

INPUT DATA

Figure 10.14a

BANDPASS 0° - 10° ROLL-OFF 10°

Figure 10.14c

REJECT 0° ROLL-OFF 5° R/L

Figure 10.14b

REJECT 0° - 15° LEFT. ROLL-OFF 0°

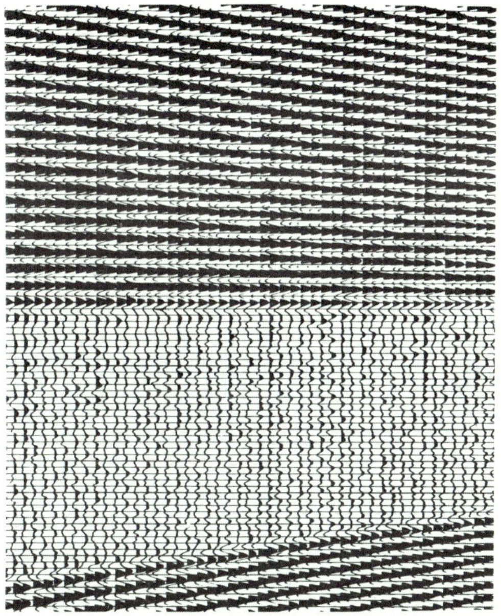

Figure 10.14d

FIGURE 10.15

Insufficient sampling yields an aliased data set having spurious frequencies.

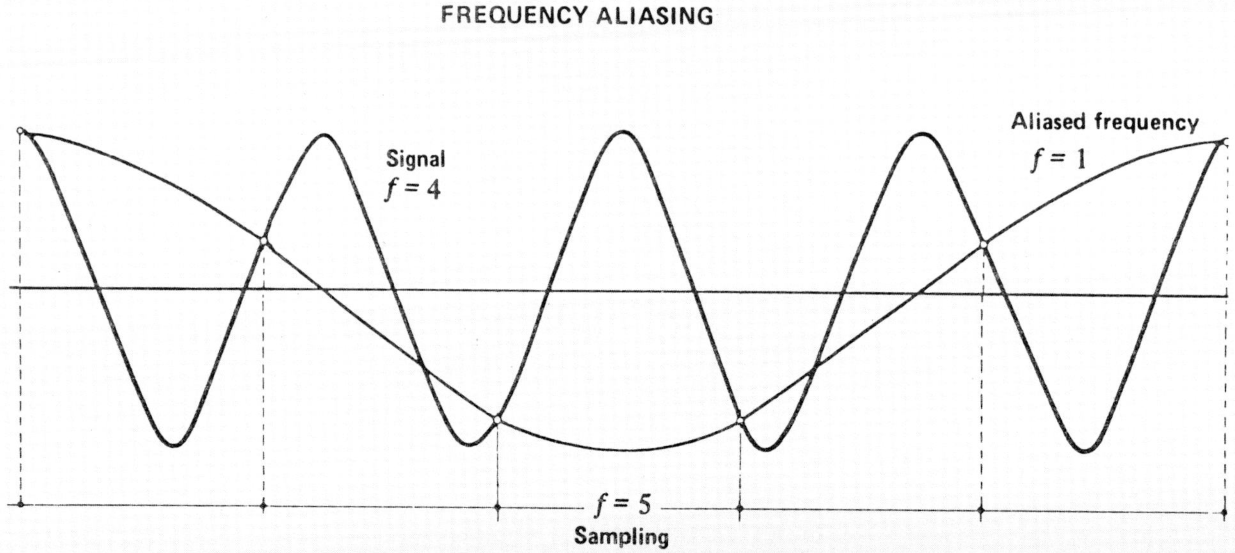

FIGURE 10.16

A stepout of one-half cycle per trace yields the horizontal Nyquist frequency.

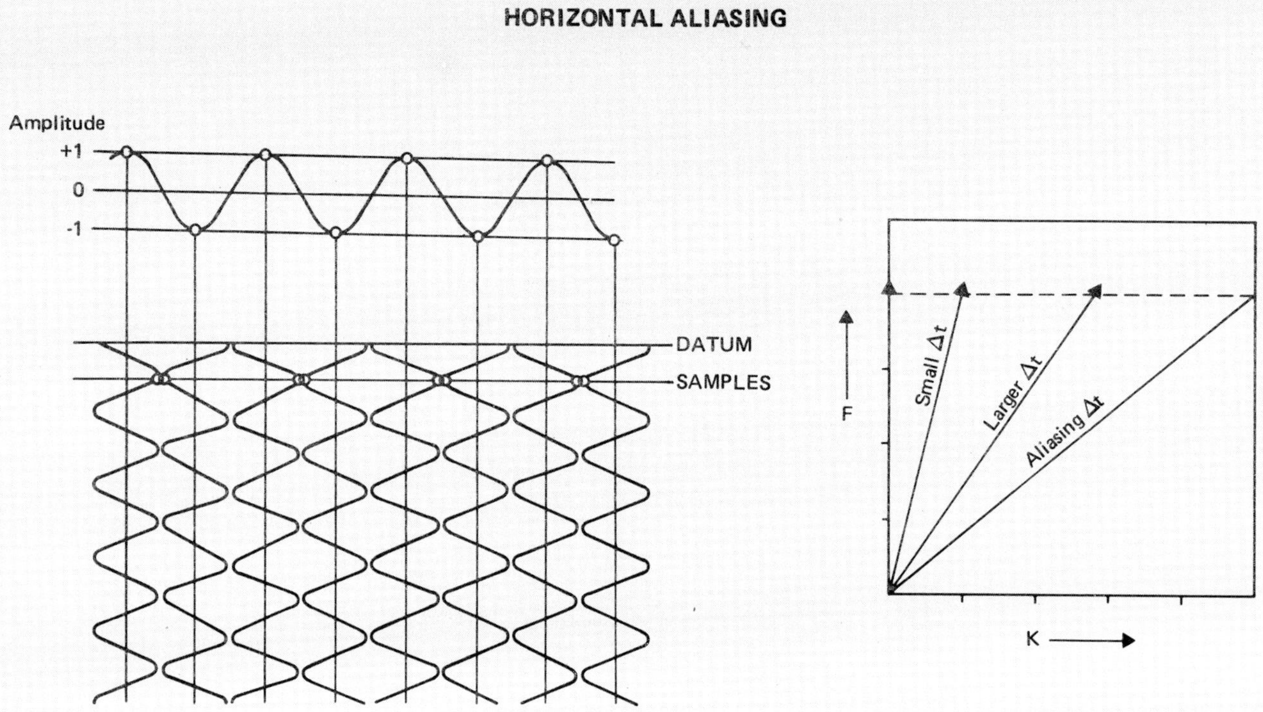

used, the only means to avoid aliasing is more frequent sampling in the horizontal direction, i.e., closer geophone group spacing in the field, a requirement which increases the cost of data aquisition but which is absolutely necessary if the horizontal integrity of high-frequency information is to be maintained. Obviously, both vertical and horizontal sampling rates are of equal importance for 2-dimensional operations. Even if a 2-dimensional filter operation is not contemplated a form of visual aliasing may be seen when a conventional seismic section is examined, creating the impression of high-frequency noise.

Vertical frequency aliasing can usually be avoided with relative ease by increased sampling of the time series and, in any case, the alias filter used in the recording instruments ensures a margin of safety. But not often is adequate attention given to horizontal aliasing, since most signal processing is applied only to single-channel data and the cost or difficulty of reduced geophone group spacing can be considerable. Closer group spacing requires either shorter spreads or an increase in the number of channels recorded over the same spread. In the first case, recording spreads may become too short to obtain the proper geometry for adequate separation of primary and multiple reflections. The solution is to increase the number of instrument channels, keeping an effective spread length but reducing the individual trace spacing.

With a knowledge of sampling theory the sampling rate necessary to avoid aliasing to be determined. The trace spacing required to avoid aliasing is a function of both dip and frequency. If the highest frequency to be recorded is 100 Hz, the maximum Δt stepout per trace that can be accommodated is the length of one-half period or 5 ms per trace. In areas of steep dip, the dip on conventional spreads often amounts to 25 ms per trace or more at 200 ft separation. In this case, about 40 ft trace separation would be required to avoid aliasing.

Surface waves, which travel along the ground through the spread are also recorded with the seismic reflection data. To record them properly without aliasing requires knowledge of their wavelength, which is dependent upon frequency and velocity. One cycle of a horizontal frequency will have some wavelength in distance equal to the time period multiplied by the local velocity. For instance, a 50 Hz wave moving laterally along the surface of the ground at 10,000 ft/sec will have the 50 Hz spread out over 10,000 ft. The wavelength will then be 10,000 ft divided by 50 or 200 ft. This is equivalent to a horizontal frequency or wavenumber of 1/200 or .005 cycles per foot. (A lower velocity, say 5000 ft/sec, would develop a wavelength of only 100 ft for the same frequency). Trace spacing necessary to prevent aliasing can be no greater than one-half the wavelength.

Wavenumber

The horizontal frequency (wavenumber) can be defined by the simple formula of frequency in Hz divided by velocity. The maximum spacing for proper recording in a given area (2 samples per cycle) must be one-half the reciprocal of the largest wavenumber, (shortest wavelength) in other words, half the slowest horizontal velocity divided by the highest temporal frequency (Fig. 10.19).

Similar relationships link the horizontal and vertical direction components of all coherent signals recorded on the field spread. The horizontal frequency and the vertical frequency both contribute to the definition of those events which arrive on the section with some angle of dip. Some convention is required to tie the different units of frequency together. Take for example a spread having a spacing between geophones of 200 ft. Consider an event, with a frequency of 50 Hz, which arrives in such a way that it reaches the second geophone .010 sec later than the first. Regardless of the true velocity of the wavefront, the apparent velocity between two geophones will be represented by 200 ft in .010 sec or 20,000 ft/sec. The frequency of the signal divided by the apparent velocity will give the wavenumber, in this case .0025 cycles/ft. Apparent velocity is the parameter which relates the time- and distance-frequencies.

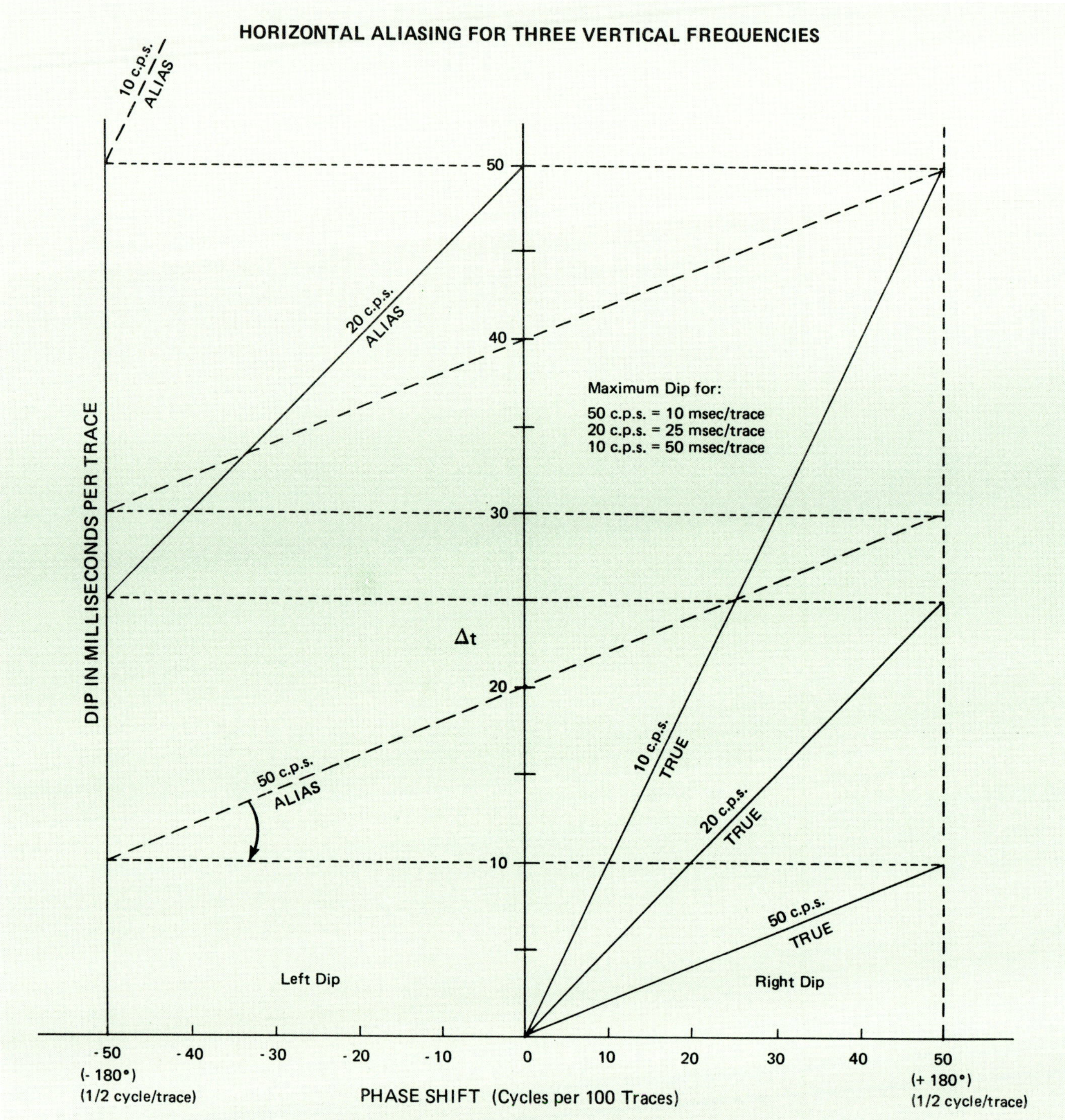

FIGURE 10.17

Dip aliasing, the result of inadequate horizontal sampling varies as a function of vertical frequency and stepout per trace.

FIGURE 10.18

Transposing right and left halves of the amplitude spectrum illustrates the "wraparound" effect of dip aliasing.

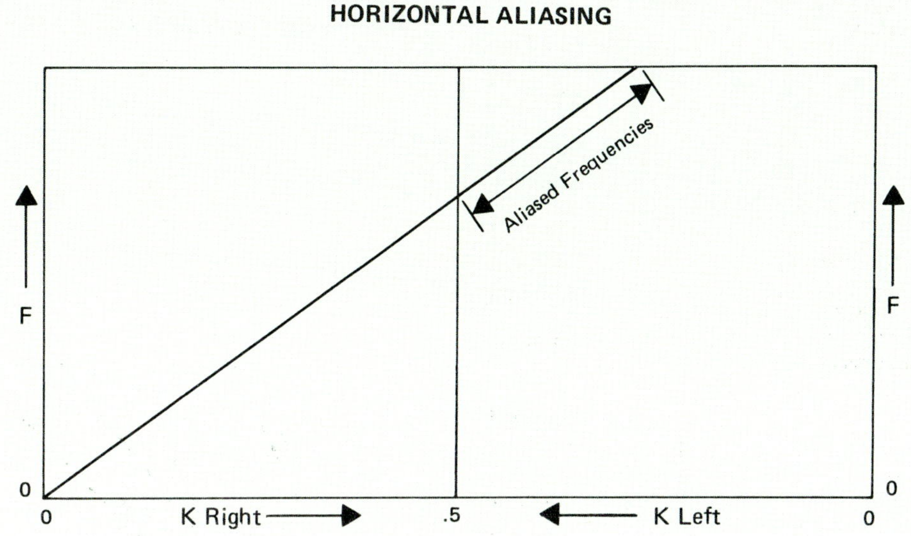

For this reason, 2-dimensional filters are sometimes called velocity filters. All have the same purpose, of discriminating against dips which cross the seismic data with stepout greater than a certain range.

This brings to attention another important point, the influence of bandwidth. In the single-channel case a narrow pass-band of frequencies will produce a signal that tends to oscillate at the dominate frequency. For a similar reason, broad-band dip filters, expressed in transform degrees, are essential if disturbing side effects are to be avoided. Narrow-band dip filters, (eg: left and right 5°) lose definition and may cause oscillations at the dominant dip angle just as the output from a narrow-band frequency filter oscillates in time. Some of this effect is always evident in the visual appearance of a dip-filtered section. The removal of many distracting unwanted dip components allows the remaining elements to stand out much more obviously.

Practical Applications

The seismic section in Figure 10.20 is six-fold multiple coverage, which has been deconvolved and band-pass filtered before stacking. Continuity is fair but not certain. A static error is apparent over four traces on the left side. An anomaly is believed to exist on the right side in the deep part of the section.

The data were subsequently processed through a 2-dimensional filter passing right and left dip to 2 ms per trace, producing the result illustrated in **Figure 10.21.** Continuity is substantially improved and apparently most of the noise has been eliminated from the section.

Figures 10.22a and **10.22b** represent separation of the **Figure 10.20** section into all the right-dip components and all the left-dip components, respectively. The possible static error noted on the input **(Fig 10.20)** now is seen to extend over a greater number of traces than indicated on the original section.

Rejection of dip components will almost certainly remove some legitimate dip components along with the noise in all but the flattest sections. The filter will produce smooth, continuous horizons, but may remove fine detail from the section. The two output sections clearly appear to have different dip slopes, yet if the two sections were summed, the result would exactly equal the input, except for a scaling factor. There are two points to be made here: No.1; Just as single-channel signals are made up of many frequency components, multi-channel signals, i.e., seismic (geologic) sections are made up of many dip components. No.2; Use of filters without a thorough understanding

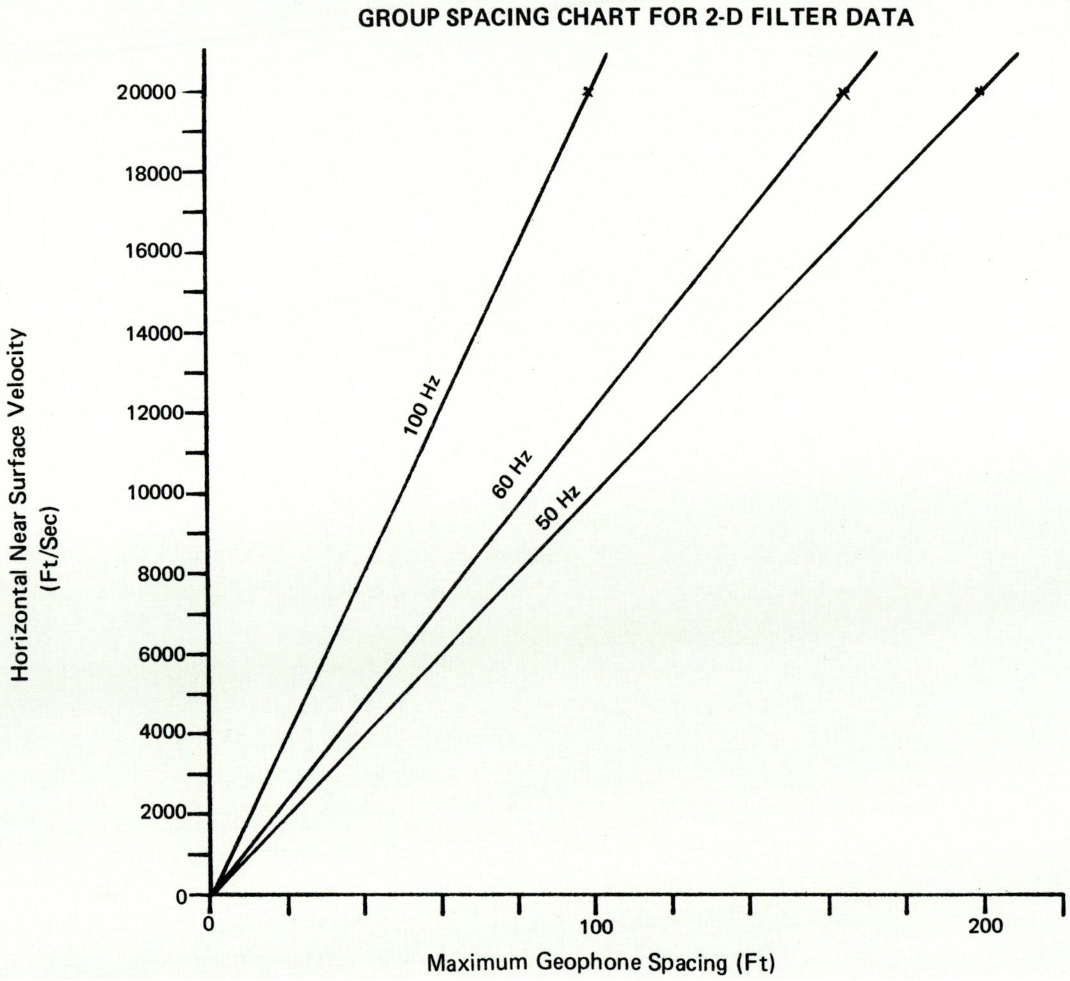

FIGURE 10.19

Horizontal aliasing can be prevented by reducing horizontal trace spacing, as a function of frequency and velocity.

of their action can yield misunderstood results. The source of the dip components on a seismic section can be either geology or wave phenomena associated with the data acquisition process. If the first is signal and the second is noise, the object of the filter, as in a frequency bandpass filter, is to remove the maximum amount of noise but a minimum amount of signal. Invariably these objectives overlap, and it will be almost impossible to remove the noise without removing some signal. The question then becomes one of what is an acceptable compromise.

Mixing

At first glance, the output of the 2-dimensional filter appears to be mixed. Mixing is a running average of seismic traces, which itself is horizontal convolution with a boxcar bandpass operator if the mix is not weighted, whereas the 2-dimensional filter is a rejection process. Evidence of this is provided by the examples of Figure 10.14. The results obtained in the examples of Figure 10.14d, are impossible to obtain by any mixing procedure. The 2-dimensional filter parameters of Figure 10.14b are similar to those used to obtain Figure 10.21.

INPUT DATA STACK

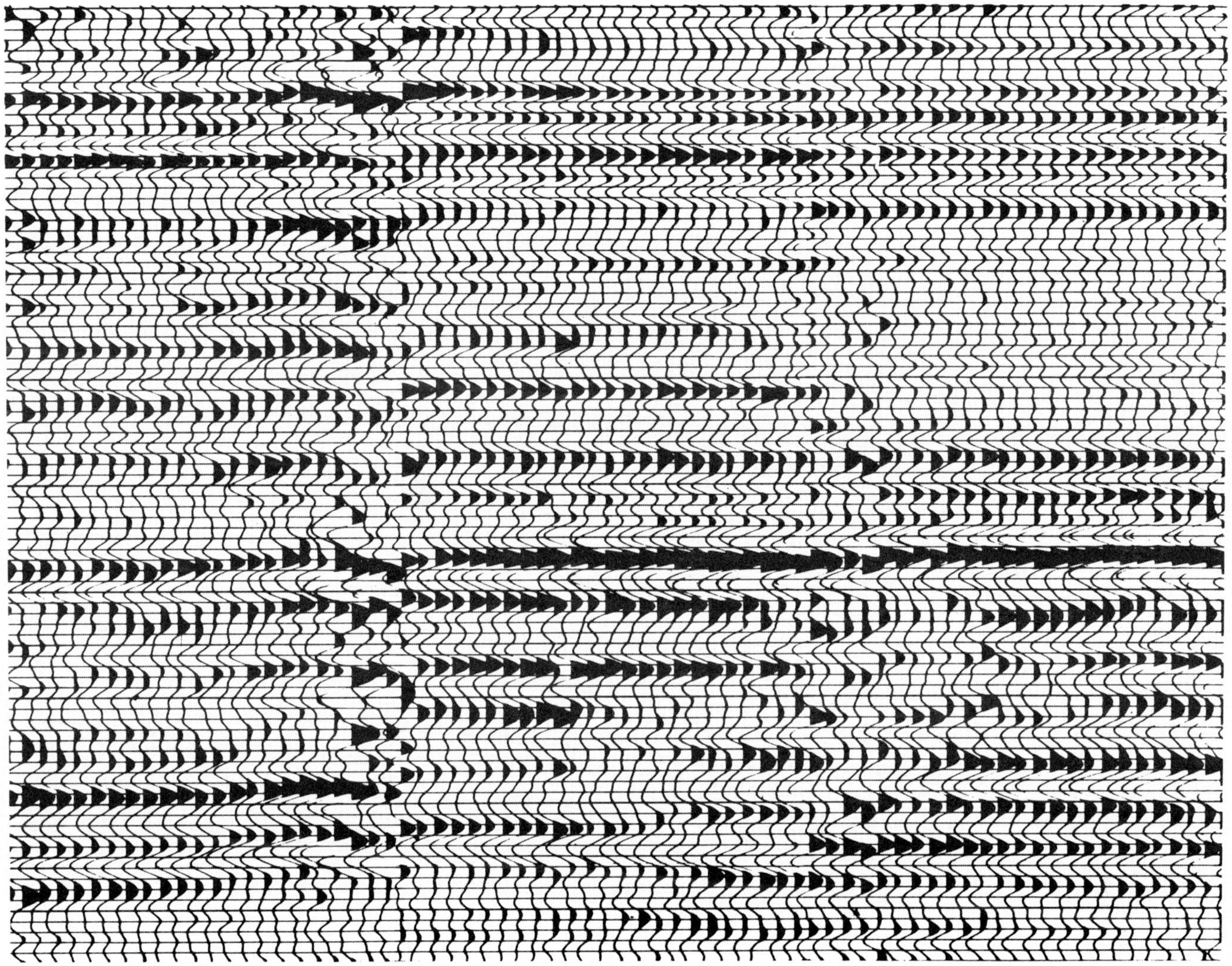

FIGURE 10.20

A stacked seismic section prior to dip filtering. Note the static errors across several traces, and the uncertain correlation in the lower right hand corner.

FIGURE 10.21

The section of Figure 10.21 after applying a 2-dimensional dip filter passing 0°- 14° on the transform plane. Note the elimination of the static errors, and the increased clarity of reflection continuity.

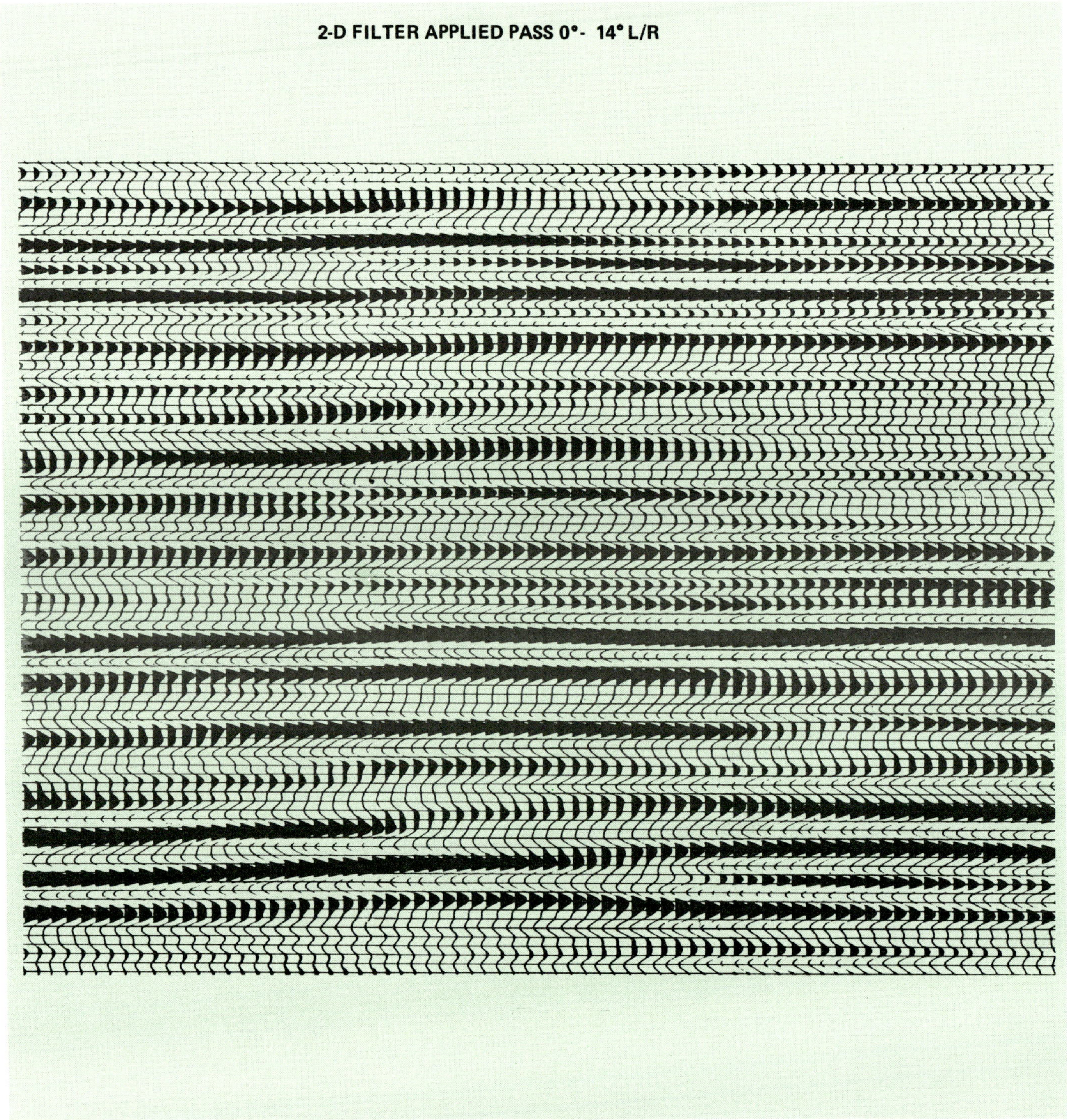

FIGURE 10.22

The right and left dip components of the input data in Figure 10.20.

The application of a two-dimensional filter is a rejection process, which is quite easy to demonstrate by applying the filter twice, reversing the pass and reject parameters. The reject portion from the original operation can be displayed **(Fig 10.28)** and, assuming no differential scaling between the pass and the reject output, the pass portion **(Fig 10.27)** and the reject portion can be vertically stacked, i.e., simply summed at the corresponding sample positions, to reproduce exactly the input.

Mixing on the other hand, is a simple average of several traces and, at first glance, would appear to have no rejection. In fact, it is for this reason that most geophysicists abhor mixing. They believe it introduces adjacent elements into the information from a subsurface point and, correctly, that it "smears" subsurface information over several traces. In fact, the Fresnel effect already assures substantial "smearing" of the subsurface data, but this is beside the point.

The operation of mixing can be readily described, and its effects understood, by resorting to filter theory. A simple mix, say a running mix of five traces, is obtained when two adjacent traces on each side plus the center trace are summed to produce a new center trace output, and the operation repeated every trace on the seismic section. This is nothing more than application of a horizontal filter, with coefficients of 1,1,1,1,1. This boxcar operator, convolved along horizontal time lines on a seismic section, will produce exactly the same response as the mixing operation.

The response of this type of operator is well known from work with geophone arrays, which, for an untapered array, have the same type of response. Basically it is a horizontal frequency filter having rather uneven response characteristics, notching out some frequencies while passing others with increasing attenuation toward

FIGURE 10.23

The horizontal frequency response of an 8-trace mix.

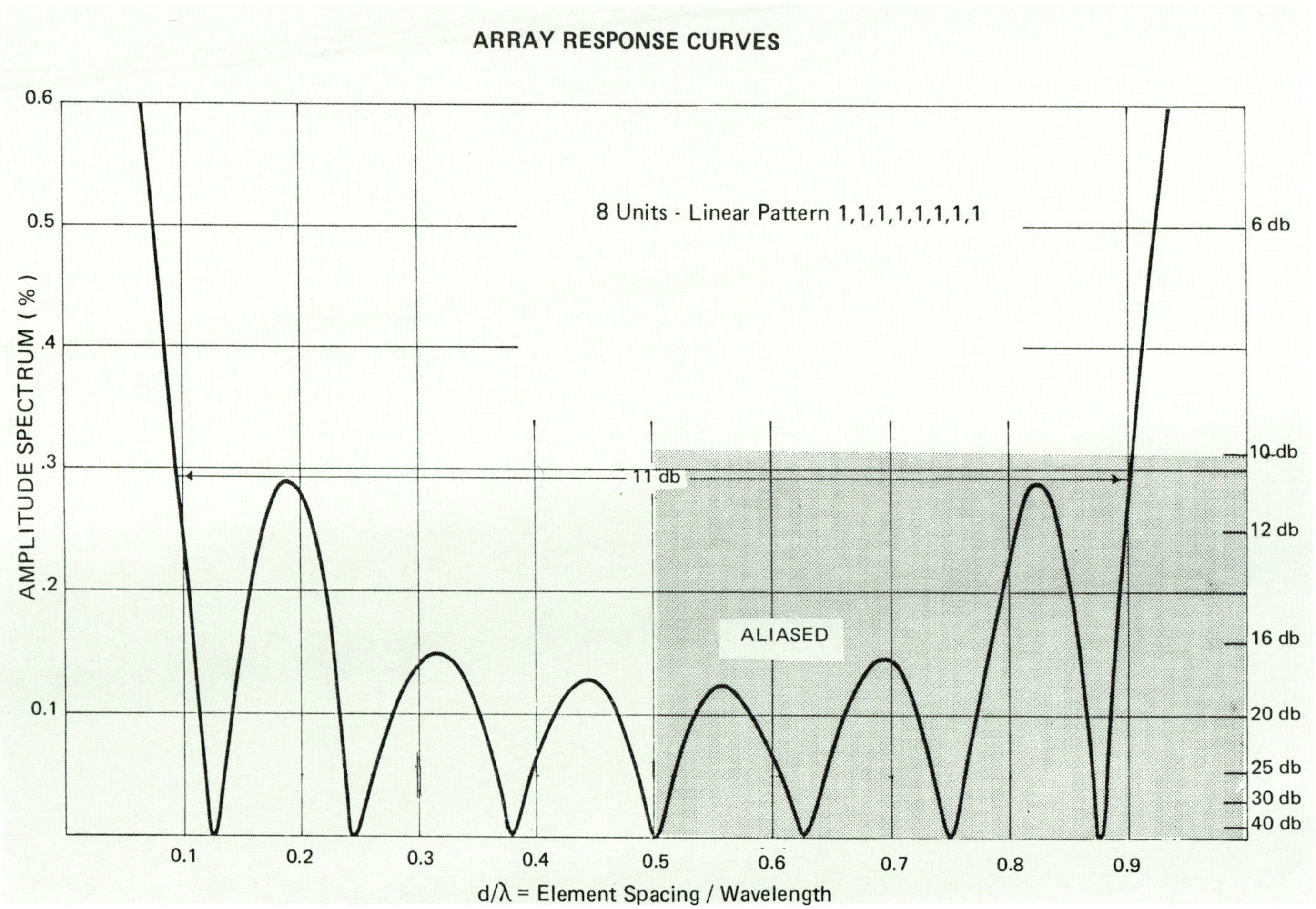

frequencies **(Fig 10.20)**.

It should be remembered that the removal of any frequencies from the bandwidth, particularly at the higher end, causes the time response of the wavelets to be broadened and, usually, side lobes to be developed. This filtering applied to an impulse, would broaden it over several samples. Filtering a step signal has a similar effect. A sudden termination of a reflection, such as that produced by a fault, appears as a step function on horizontal sampling. Application of any band-limiting horizontal filter will spread the response over some number of traces, to yield the "smearing" effect so undesirable to interpreters. Yet, from a philosophical view, this is inconsistent. A vertical frequency bandpass filter has exactly the same response in the time dimension, "smearing" or mixing the response of one horizon with those within several tens of milliseconds above and below.

Mixing is no more than horizontal convolution with a short operator. Provided the response of that operation is understood, **(Fig 10.23)** the operation should be no more unacceptable than vertical convolution for filter operations. In this case, information representing several normally unrelated ages of deposition above and/or below each sample is combined in a weighted average, by the operator, resulting in reduced definition at each level in the sedimentary section. The unsatisfactory filter response of the unweighted mix, though, should certainly indicate the need to use a weighted mix, which would have more of the sharp cut-off characteristics of a good bandpass operator. Even a tapered mix, with the weights decreasing as a function of distance, is far superior to a straight mix.

Some of the response characteristics of two-dimensional filtering, which removes selected combinations of vertical and horizontal frequencies, can be considered analogous to the effects of frequency filters, but with the filter rejecting dip components instead of frequencies. A single impulse spike in the time domain spreads over the entire transform plane in the 2-dimensional frequency domain **(Fig 10.11)**. If any portion of the frequency domain response is removed, and the reverse transform taken, the time domain impulse will no longer be a spike but will spread out over several adjacent samples. Truncation of the horizontal "fence" illustrated in **Figure 10.11**, in the time domain, will turn the vertical plane in the frequency domain into a vertical wedge with side lobes, while any truncation in the frequency domain will also have a spreading effect in the time domain.

Part of the object of a filtering process is to attenuate or remove noise and thereby enhance the signal. The ideal result would be a complete noise-free section. A noise-free section may be simulated by synthetic seismograms constructed from acoustic logs. A number of synthetic seismograms at widely spaced locations are reproduced in **Figure 10.24**.

Note that in spite of the considerable distance between these, there is very little change of character. At the normal bandwidth of conventional seismic sections, only the gross geologic features remain in this rather uniform section. The local detail has been removed with the higher frequencies which were filtered out. Evidently the gross characteristics of the section change slowly in a lateral sense, and maintain a remarkable degree of smooth continuity in the corresponding seismic reflections. The synthetics are noise-free and represent the output of an ideal filter. Under fortunate conditions the two-dimensional filter will eliminate sufficient noise to produce output approaching the characteristics of the ideal noise-free section.

FIGURE 10.24

Synthetic seismogram model of a noise-free seismic section.

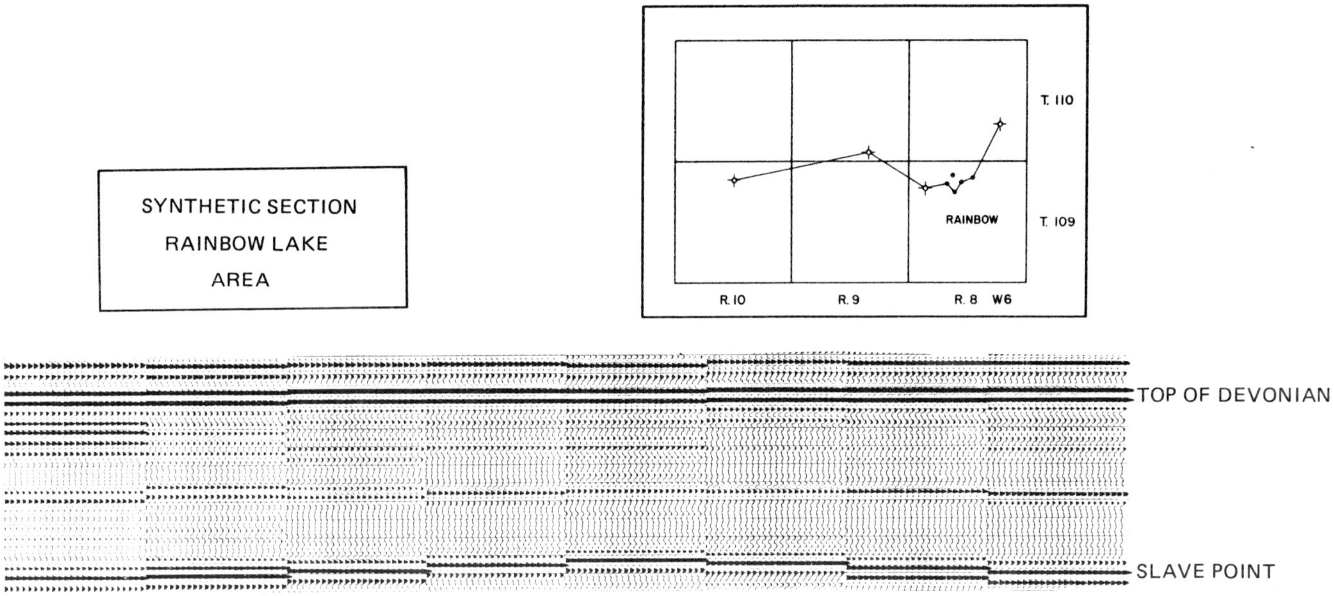

However, the synthetics have a relatively narrow pass band, due to the filtering effect of the convolution operator. This can be verified by filtering the well logs to the same frequency. But over a broader band, there is considerable change even between the closely spaced wells. Any filtering technique requires some knowledge of the data in order to select the best parameters. Indiscriminate selection may eliminate legitimate information in the filtering process. **Figure 10.25** illustrates a well defined reef anomaly having characteristic sharp dip at the reef front. The line closely approximates the west half of the synthetic section of **Figure 10.24**. Application of the 2-dimensional filter has eliminated some noise to provide the improved output of **Figure 10.26**. The parameters of the filter have been further restricted to develop the output of **Figure 10.27**, in which the continuity over the reef edge has been destroyed. In this case, while a definite anomaly remains, the interpretation might be different.

A valuable advantage of the procedure is the ability to examine the noise removed by the dip filter. Coherent noise patterns are sometimes revealed more clearly in this manner. The noise which was removed from **Figure 10.25** to obtain **Figure 10.27** is reproduced in **Figure 10.28**. The input data was relatively free of noise, so the noise level in the data is low. A rather astonishing coherent pattern is revealed which would not be detected on the original data under normal conditions. The angular pattern is apparently the simple surface multiple of the strong flat shallower reflection from a steeply sloping ground surface.

CDP Trace Gather Filtering

A velocity type filter may appear to be useful for improving the quality of data in the common depth point trace gather group prior to stacking. The desired continuity of a common depth point gather group, assuming correct normal moveout and static corrections have been applied, is always flat. Thus, no other knowledge is required, and a velocity filter could be applied to eliminate all but the flat components from the gather group. In fact, this would be a wasted effort. The output of a velocity filter which eliminates all but flat components, when stacked is exactly the same as the simple stack of traces, assuming the same number of traces in both cases.

The reason for this rather suprising conclusion is that the ability of the 2-dimensional filter to discriminate against unwanted angular components is limited when the number of traces (samples) is small. In this case, the same level of rejection is obtained as the cancellation obtained by the simple stack. The two are the same because, unfortunately, 2-dimensional filters can remove only the noise components which are periodic across the trace gather group. By nature the mean value of these components is zero and they will normally cancel out in the stack process. Hence, a simple stack provides signal enhancement equal to that obtainable with the 2-dimensional or velocity filter applied to the same trace-gather group.

Time Domain Operations

Prior to the introduction of high speed Fourier transform algorithms, it was unfeasible to attempt the full 2-dimensional transform computation and, therefore, it was not possible to apply the filtering process in the frequency domain. An approximately equivalent but very laborious process can be done in the time domain.

If the operation is to be done in the time/distance domain rather than the frequency domain some method must be found which will operate upon the vertical frequency in a manner dependent upon the horizontal frequency.

Single channel convolution filtering requires the total number of points of the operator to be cross multiplied with an equivalent number of points of the signal in order to obtain one point of output. Two-dimensional operations require these two operations to be carried out simultaneously, so in the time/space domain the reqired operator is a matrix. The operator matrix is multiplied with the appropriate corresponding points on a group of sampled traces. The products are summed to obtain one point of output. The next output point can

IM

be obtained by shifting the matrix one position either vertically or horizontally. A regular procedure might be to do all the vertical shifts first, to obtain one output trace, then shift horizontally one position and repeat all vertical operations to obtain the output trace.

The process becomes very lengthy; in fact, compared to single-channel operations with an operator "n" points long, the number of multiple-adds for multi-channel operation goes up by a factor of approximately n^2 if a square matrix operator is used.

Velocity filtering is a simultaneous vertical and horizontal bandpass filtering operation, so it is equally important to use an operator of adequate length in the horizontal dierction as well as the vertical.

One might consider that the two-dimensional filtering could be handled in the time domain by first filtering vertically in the conventional manner, then by sampling the section horizontally and applying a horizontal bandpass filter one row at a time. An effective operation cannot be done in this manner. It has been demonstrated that for a given dip, horizontal frequency is a direct function of the vertical frequency, hence a fixed bandpass in the horizontal direction will reject dip of varying amount, depending upon the vertical frequency content. It is not possible to do an effective job in this manner, except as an approximation to very flat-dip filters.

Pie-Slice

One of the first practical time-domain approaches to velocity filters was the Pie-Slice process which uses matrix convolution to handle the problem of varying the horizontal frequency as a function of the vertical frequency. The Pie-Slice process as described by Embree et al (1963) uses one set of fixed convolution operators having parameters which are a function of the distances between traces and the dip which is to be removed. The set of operators forms a symmetrical matrix in both directions. In the interest of economy, the matrix is not square, although it can be.

Say, for example, it is desired to pass a fairly broad range of dip, equal to one vertical time sample unit per trace interval, and reject all other dips. The slope of the dip on the F/K plot will be 45°, the diagonal intersecting the temporal and spatial frequencies at the point where the two aliasing frequencies intersect.

The response is symmetrical, right and left, in both time and frequency domain. The appropriate operators are designed, one for each column of the filter operator matrix, with the operator for each column varying as a function of the distance from the central trace. The operators are not normal bandpass operators but rather are rejection operators. Each appropriate matrix column operator is convolved with the corresponding trace of the seismic section as a single channel operator. All of the resulting filtered traces from one set of operators are summed, to produce the single two-dimensional filtered output trace. The matrix (the set of operators) is shifted horizontally and the process repeated.

The result achieves an effective approximation to the desired response and the final output will be quite effectively filtered. The response is generally good, but the response of the output only approximates the true diagonal line of rejection (**Fig 10.29**). Although Pie-Slice is based on the division of a spectrum, its precision is limited by the bandwidth of the horizontal and the vertical frequency sampling, which is controlled by the operator length.

The production of each single output trace filtered in this manner requires the application of individual operators to (normally) 12 input traces, and involves a monumental number of filtering operations. A later paper by Treitel, Shanks and Frasier (1967) pointed out that since the response was completely symmetrical, pairs of traces could be first summed and then filtered, cutting the number of operations in half.

Subsequently, it was shown in the same paper that a similar effect could be obtained by applying a single operator to each trace, followed by subsequent time shifting of the traces before compositing. This reduces the amount of

FIGURE 10.25

Stacked seismic section across a buried reef.

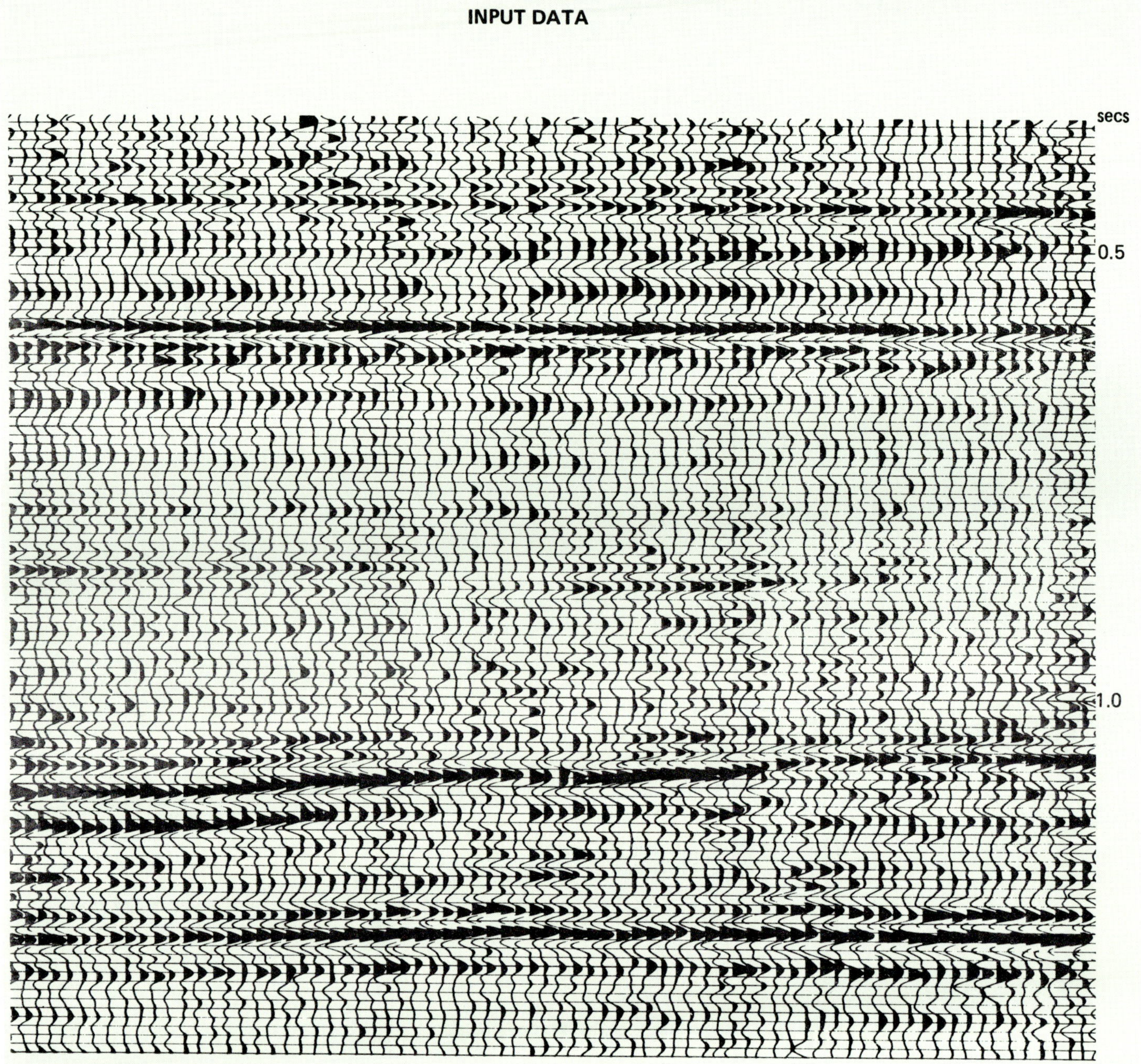

FIGURE 10.26

A 2-dimensional dip filter applied to the data of Figure 10.25 improves the continuity and makes amplitude relationships easier to see.

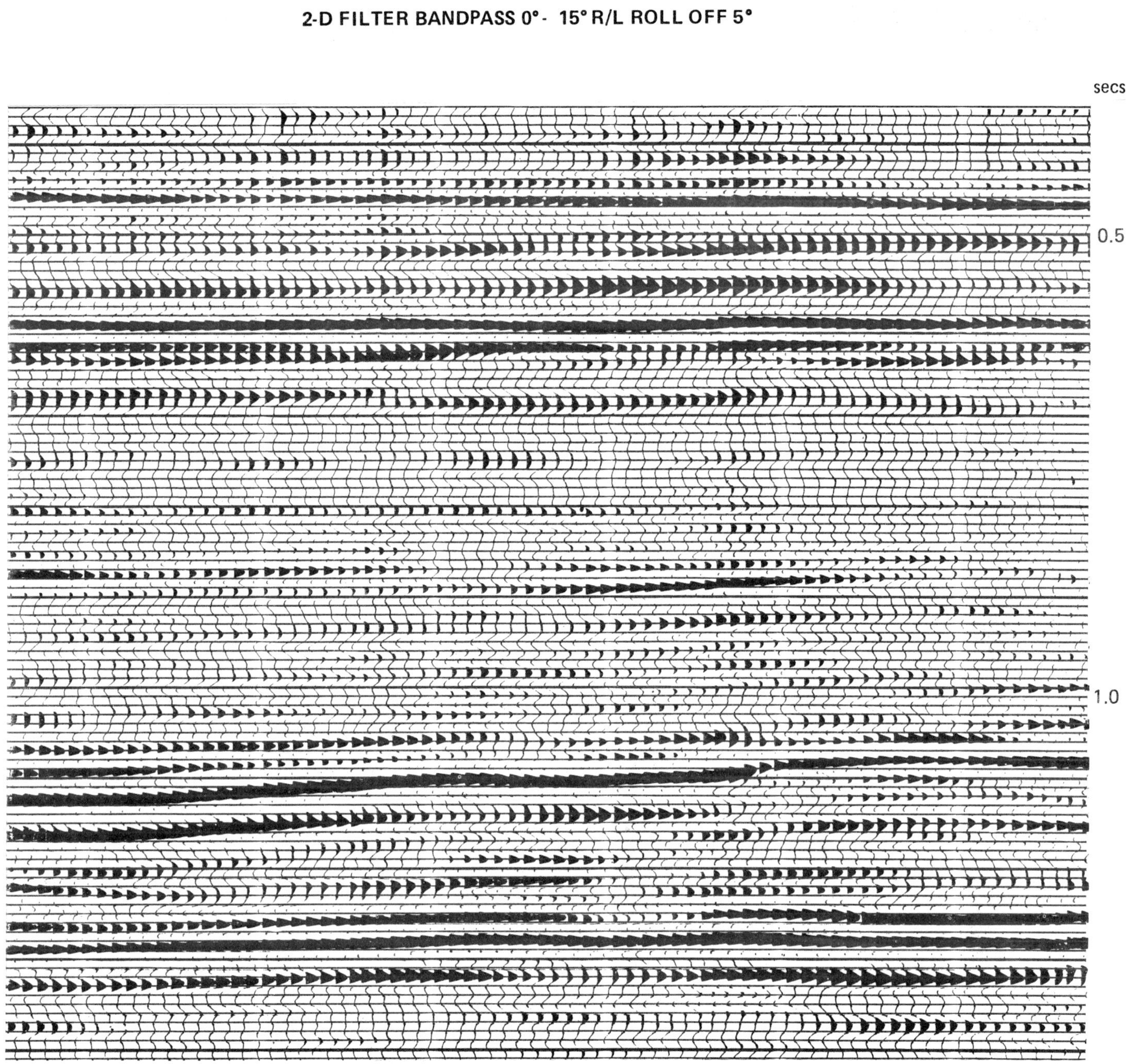

FIGURE 10.27

An excessively narrow band dip filter has removed some legitimate dip information outside the dip pass band of the filter and destroyed the continuity of the reflection across the top of the reef.

2-D FILTER BANDPASS 0° - 7° R/L ROLL OFF 3°

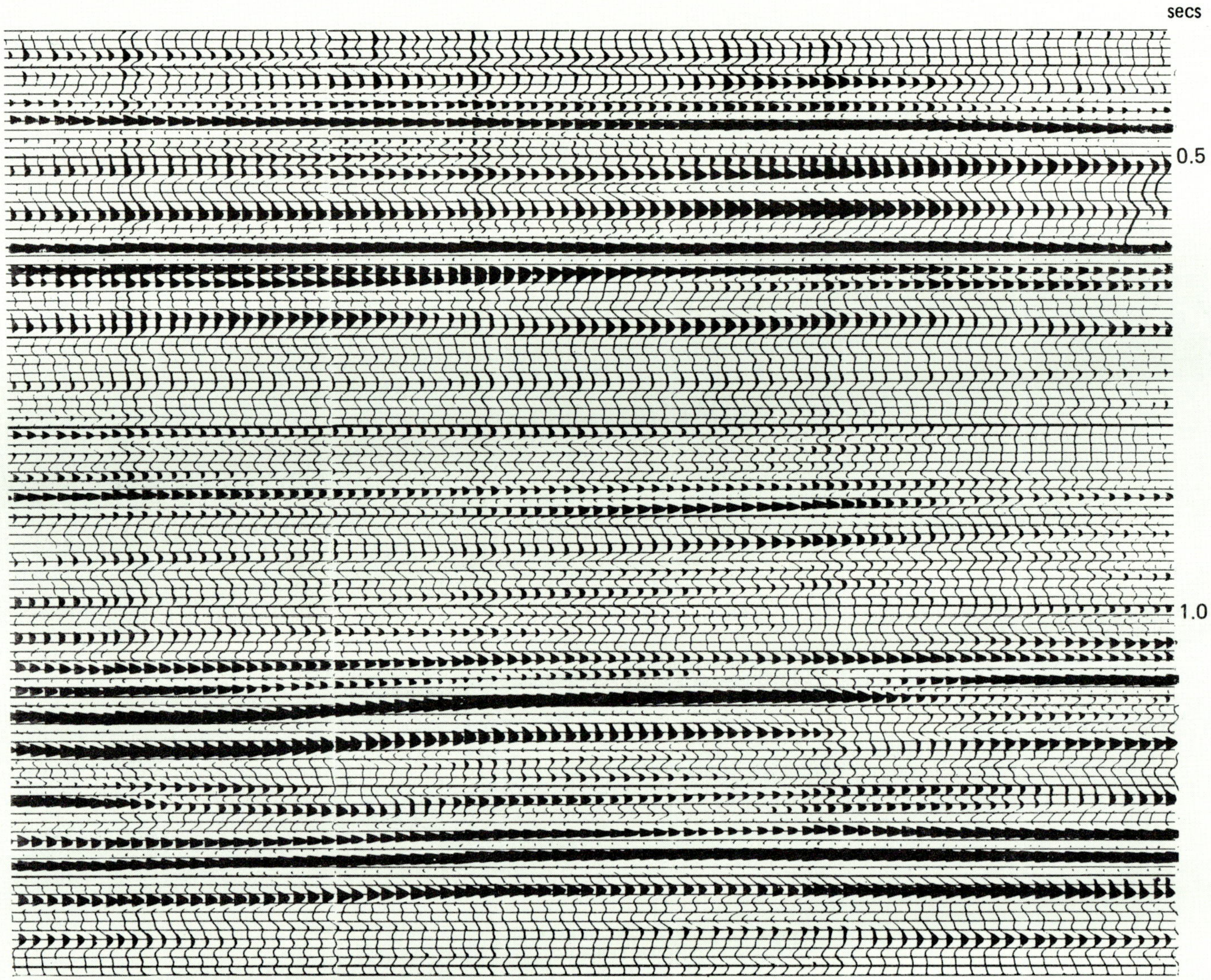

NOISE SECTION

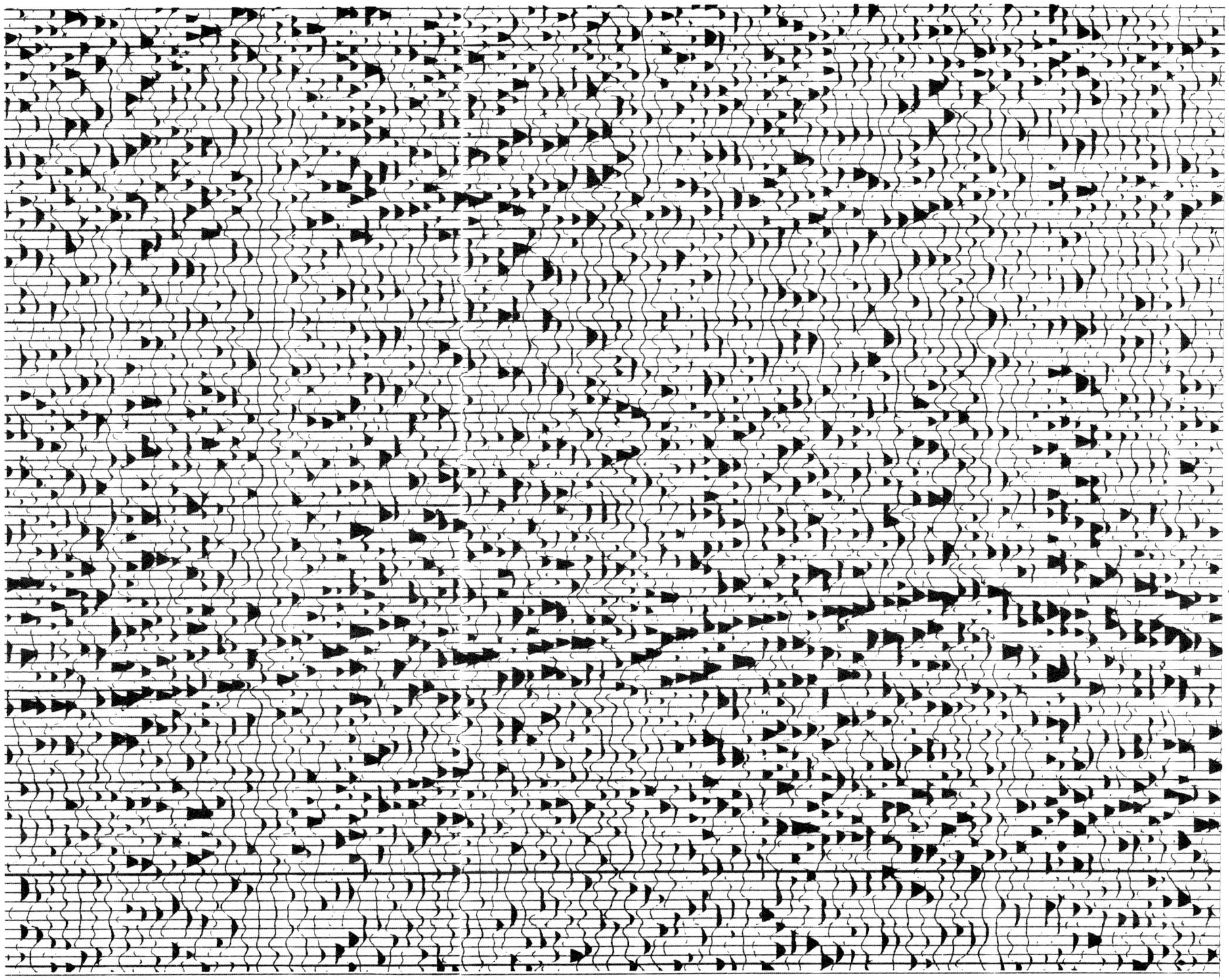

FIGURE 10.28

Interchanging the pass-rejection parameters of the dip filter applied to Figure 10.27 yields output of the noise removed from Figure 10.25 to obtain 10.27. The pronounced steep dip across the lower third of the record is a surface multiple between the strong reflection near the top of the input and the steeply dipping ground surface.

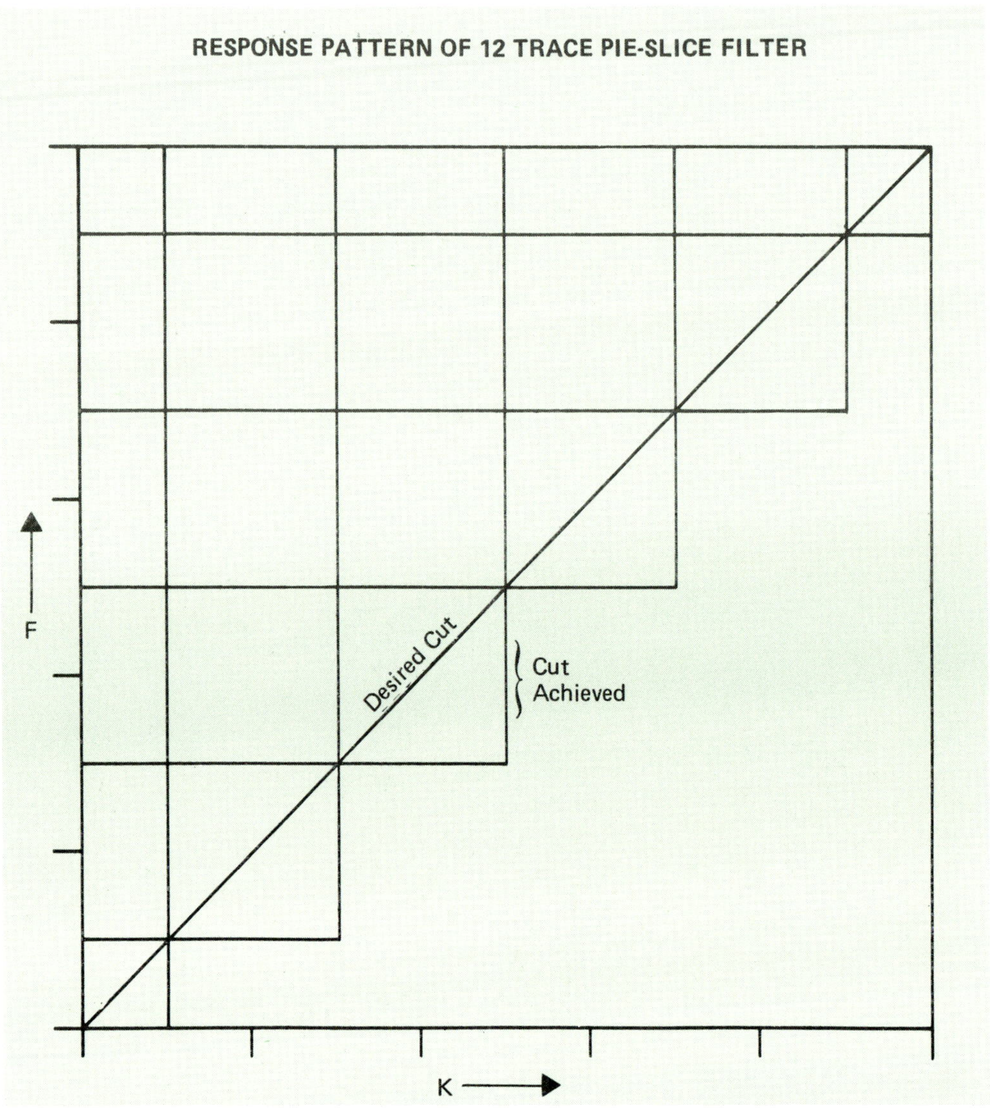

FIGURE 10.29

Time domain 2-dimensional filters with limited operator length achieve only an approximation of the desired rejection angle, due to the large frequency sample increment corresponding to short operators.

computational time substantially. Reduced computational time is realized but the sharpness of response is further reduced.

Simple velocity filtering, as has been pointed out, offers no advantage as a stacking filter when applied to a single gather group. However, other forms of multichannel filtering may be used to some advantage. Least-square statistical techniques can be used to design optimum filters to meet some response criteria for each set of traces. As might be imagined, it is also a very lengthy procedure, both to design and apply. For this reason multichannel operations with optimum matrix operators have not been widely used. The cost becomes rather high.

At one time, a great deal of effort went into the design of recursive techniques, which offer great savings in the time of application, but the problem of stability, which is serious in the single-channel case, becomes more severe in the multichannel case.

The time domain methods can be effective, but a great number of operations are required to obtain acceptable results. Velocity filters may be applied with relative ease in the frequency domain and an entire section may be filtered in one operation. The cost is relatively low and the results are very precise because, in any frequency-domain operation, the effect of a filter applied to the transform plane is equal to that of a time-domain operator of the same length as the signal.

The transform filter may be oriented or "steered" to fit particular dips. If a section has a number of differing dips, it may be treated in a number of time-varying and/or space-varying segments with a filter to fit each one. This is useful in the case where the rate of dip changes frequently, accompanying structural changes along the line. In the vertical sense, in a situation where steep-dipping beds are overlain by an unconformity, a slice may be made to change the dip angle at some particular point. This points out the serious major limitation of velocity filters. Their use is severely limited in areas of considerable structural activity, since the power of rejection is purely based on dip.

Conclusions

The results of two-dimensional-transform filtering illustrate dramatic improvements in continuity. Multichannel filters offer one of the greatest opportunities for improved data enhancement, primarily because they offer the ability to separate noise and signal of the same frequency. They also require thorough understanding of their development, application and effects.

11 AUTOMATIC MIGRATION

Introduction

It is a well known, but often ignored, fact that the events which appear on a seismic section rarely correspond to their true position in space. Correspondence is possible only if the reflecting horizon is truly a horizontal plane surface. Under all other conditions the reflection must be migrated in order to position it properly in space. Reflected energy obeys physical laws and, in the simplest dip case, the travel path back to the origin will be perpendicular to the dip of a reflecting bed, rather than vertically as it is plotted on the seismic cross section.

In areas of moderate dip, say up to 10°, the seismic section approximates the structure with sufficient accuracy to be acceptable. In areas of steep dip a reflection point plotted below the shot point may actually be removed laterally several hundreds of feet from the vertical line below the shotpoint. Even this problem may not be too serious in areas of fairly simple structure, because the error diminishes with decreasing dip and the geologically important crest of an anticline will be correctly located.

In more complex areas, where faulting, severe non-symmetrical folding, and sharp synclines are present, diffractions and double image reflections may be distorted to the point where the resulting seismic section bears little or no resemblance to the actual structure. This situation is not uncommon and it can be very difficult to interpret true structure completely from a section that has not been migrated properly.

Complicating the picture is the fact that diffraction patterns are recorded with the reflections. This phenomenon is most prevalent in faulted areas but is common to a certain extent to all. The termination of a reflection can grade smoothly from a reflection into a diffraction. When this occurs all or part of the diffraction might be interpreted as dipping reflected energy. This is erroneous and may result in failure to recognize faulting, or in improper location of the fault plane, with the result that a well may be drilled on the wrong side of the fault. Synclinal folds with a radius smaller than the distance of the reflector to the surface produce an apparent structural inversion which can be mistaken for an anticline or fault diffraction.

Prior to automatic migration by digital computer, seismic section dips could be transformed to an approximation of the true section only by manual migration. Two reflecting points on an apparent dip segment can be timed, and, if the velocity function is reasonably well defined, a series of rather lengthy calculations will yield the co-ordinates in space along the line of section of the points from which the reflection originated.

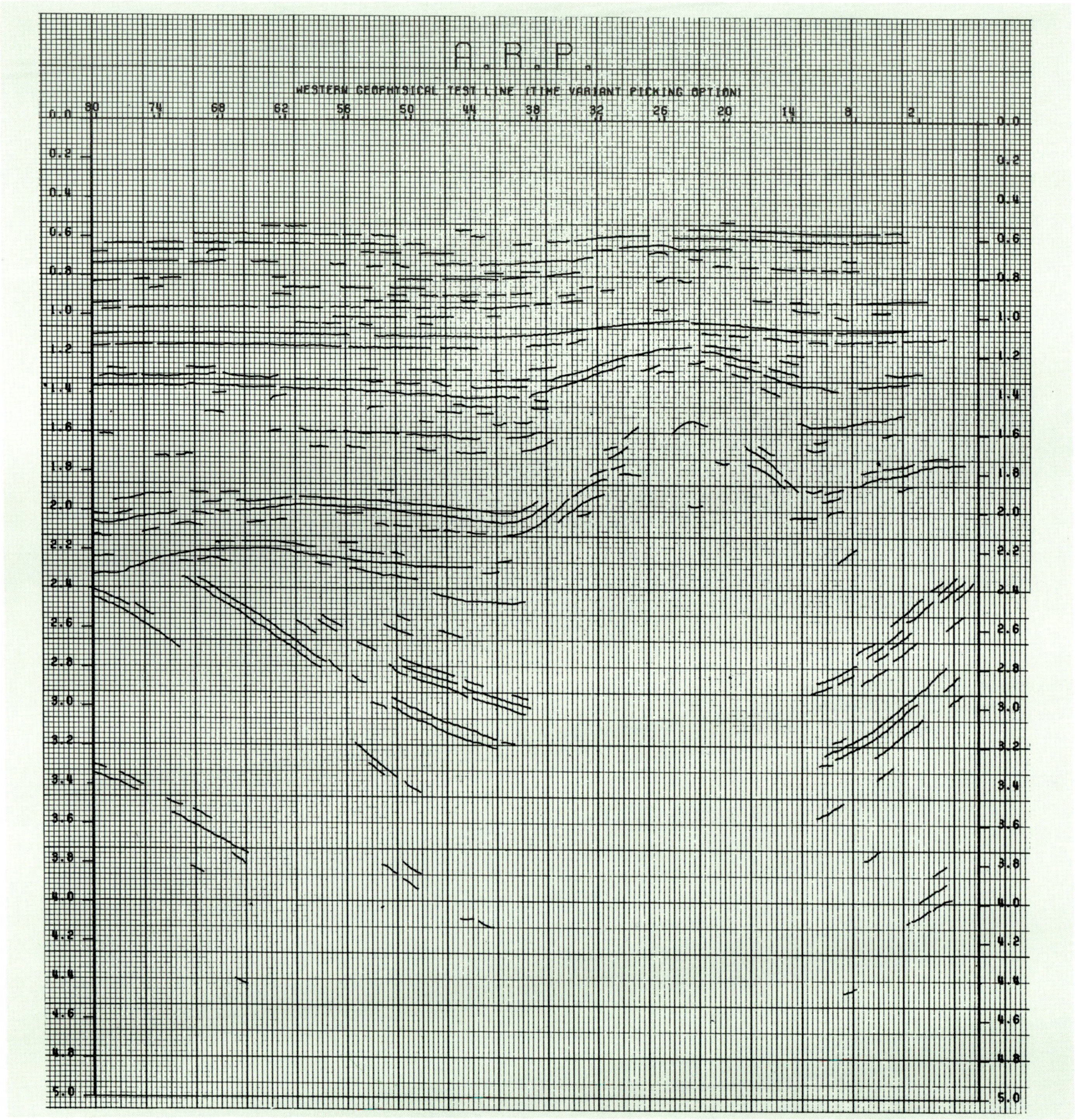

Courtesy of Western Geophysical Company

FIGURE 11.1

Early use of a computer to migrate manually picked reflection horizons.

One common practice was to draw up a raypath chart using some relatively simple velocity function for the area. Reflections could be migrated by locating the appropriate reflection time and the amount of dip, expressed in two-way time, on the chart, and drawing in a dip segment tangent to the computed wavefront. Some of these steps were automated when digital computers first came available, and routines to compute and draw a migrated dip section (as illustrated in **Figure 11.1**) were a welcome relief to interpreters.

A factor which caused errors in early migration operations was the lack of adequate velocity information, a problem that has not yet been completely resolved. Velocity surveys were obtained at infrequent intervals. Quite often the only available source of velocity data was a single distant borehole well velocity survey, or the occasional velocity profile.

Other errors which may affect the migration process arise from the use of a straight line, or at best, a simple curved raypath assumption, which, in areas of layered velocities of high contrast, can be greatly in error.

The basic problems of migrating reflection and diffraction information were not due to lack of knowledge. On the contrary, as is the case so often in seismic development history, the theory and the physical reasons behind most of these phenomena were well known, having been expounded many years before, principally by Huygens about the year 1650. In fact, the same basic principles constitute the foundation for modern holography. The difficulty in utilizing the knowledge lay in the inability to extract all of the information contained in the seismic signal, fractionate it into the appropriate components, and reassemble it into migrated sections. With the advent of the digital computer and, aided by the additional information provided by the common-depth-point method, the operation has now become a relatively simple, if not completely reliable task.

Unfortunately even on high speed digital equipment considerable computing time is required because of the immense number of operations involved. Also, if migration is applied as a two-dimensional process, line intersections very likely will not tie on migrated data. For these reasons migration of all reflected seismic data is not yet general practice. Not only is the migration process expensive, but the calculation of the velocity data required for effective migration is another considerable cost item. Nevertheless, the ability to perform automatic migration is a major advance in the seismic survey process. Improved and reliability of special hardware, and improved algorithms, continues to perform this highly desirable process at less cost and a faster rate than could be done only a short time ago. Certainly, migration must be considered an advance in seismic digital processing fully as important, as say, filtering or deconvolution.

General Principles

Assume a geological section homogeneous above the first reflector, and a velocity function such that, for the purpose at hand, time and distance may be interchanged.

If a source and receiver are placed at the same location A on the surface of the section, **(Fig 11.2)** a seismic reflection received from the first reflection boundary would be plotted vertically below the source at point A′. In addition to a point directly below the receiver, the reflection could originate from any other point on the half circle having a radius AA′. Any reflector tangent to the circle could reflect energy back to the origin.

In fact, the radius of reflection not only describes a half circle but actually sweeps out a spheroid perpendicular to the line direction. Unless additional information is recorded, e.g. a cross spread perpendicular to the surface line, the lateral component will be unknown so only the component along the line of section can be considered.

A similar recording from a second point B', **(Fig 11.3)** will produce the same results but, if the bed is dipping, the travel time will be shorter and will appear at B . The horizon will then appear to have dip A′ B′.

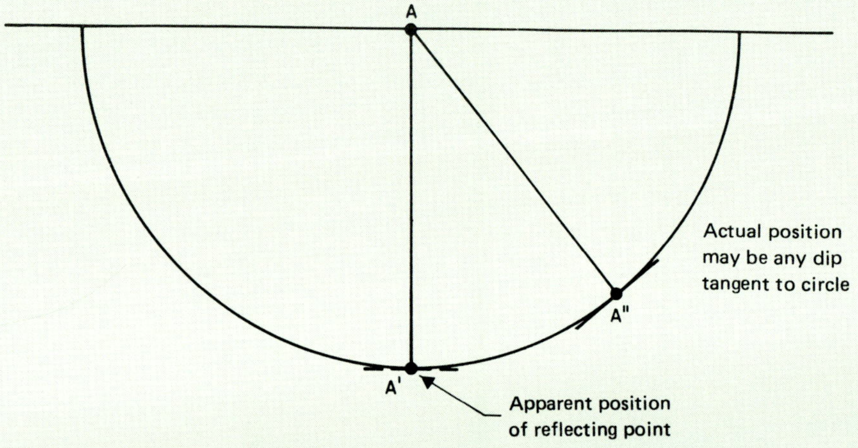

FIGURE 11.2

A reflection point plotted directly below the shot point on a seismic section, may, in fact originate anywhere on the circumference of a half-circle swept out by the radius AA′.

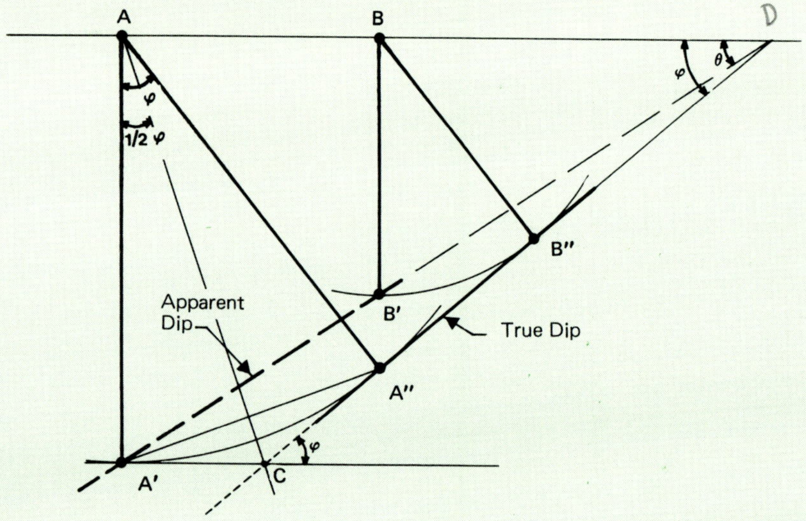

FIGURE 11.3

True dip is the line tangent to the two reflection source circles. Note that the length of the dip segment shortens from AB/cos θ to AB cos φ.

Under the special conditions defined for our model, half-circles which represent the points from which A′ and B′ may originate can be struck off. A line drawn tangent to both will provide the plane A″B″ representing the true position of the reflector in the line of section. Note that the projection of the apparent dip and the true dip intersect at the surface datum. If the apparent angle dip is θ, and the true dip angle φ then: $\tan \theta = \sin \varphi$.

Under the conditions outlined, it is not possible for any unmigrated seismic section to have a dip greater than 45°. A vertical reflector at B would have a reflection time equal to AB plotted vertically below A at A′ **(Fig 11.4)**. Thus AB and AA′ would be two equal sides of a right triangle, making the apparent dip 45° for a true dip of 90°. The angular relationship between apparent dip and true dip, $\tan \theta = \sin \varphi$ confirms this, and shows the angular difference between apparent and true dip to increase with increasing dip.

The continuous reflector will have a vertical trace plotted for each geophone position. To locate it properly in space, the reflection times recorded for all intermediate locations could be swung on arcs to develop the tangents marking the position of origin. If the intermediate locations are sufficiently close, the result of swinging the several arcs will be a continuous wavefront. **(Fig 11.5)**.

This is exactly the principle expound-

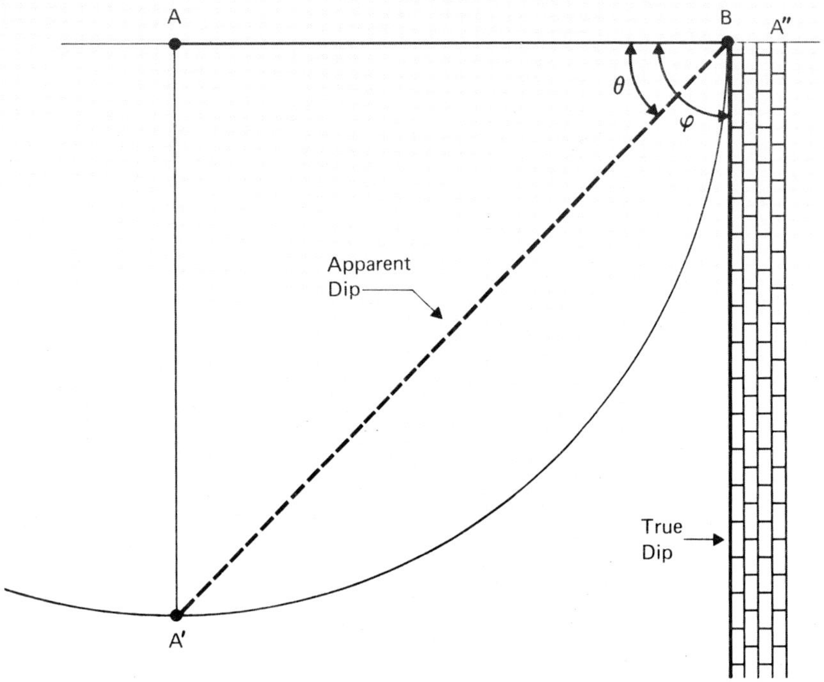

FIGURE 11.4

In homogeneous space a true vertical dip will have an apparent dip at the maximum dip value of 45°.

Diffractions

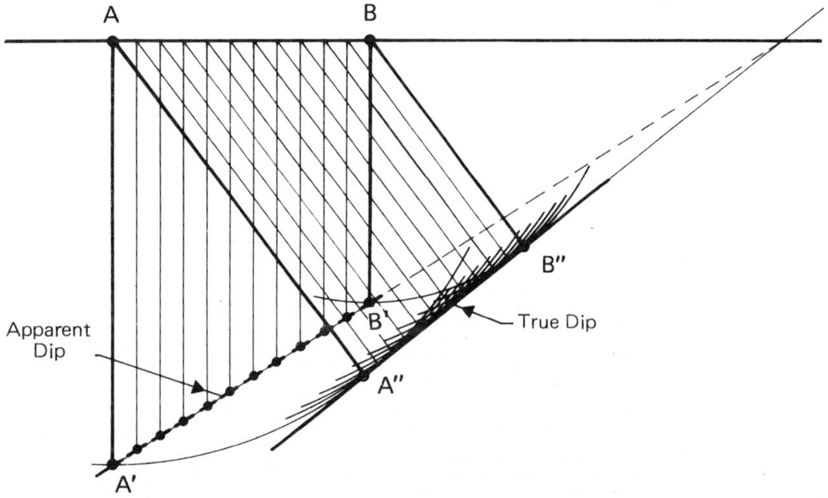

FIGURE 11.5

If several closely spaced points of apparent dip are all swung along radii, the resulting tangent wave front will follow the true dip.

ed by Huygens. It is the basis of the original automated migration technique. The simple basic process of moving each reflected point to every other possible reflecting point, modified to consider actual physical conditions, will provide an effective migration of the data, if the velocity has been adequately determined.

The procedure described assumes a source and receiver at the same position. In practice most receivers are located some distance from the source. Again, assuming a constant velocity to the reflector, a similar development can be applied. The only difference is that the constant travel time to any point of reflection no longer describes a circle but an ellipse. **(Fig 11.6)**. The returning reflection may come from any point on the ellipse where a dipping bed is tangent to it.

A complicating factor is the presence of diffracted energy. Whereas reflected energy requires that the angle of incidence equal the angle of reflection, a diffracting point accepts energy from any angle and radiates it along a spherical wavefront as though the diffracting point were a new energy source.

This means that a receiver may theoretically record diffracted energy from any point in space. However, it should be evident that for any given constant travel time between points A and B the diffractor must also lie at some point on the same ellipse described for the reflection points. **(Fig 11.7)**.

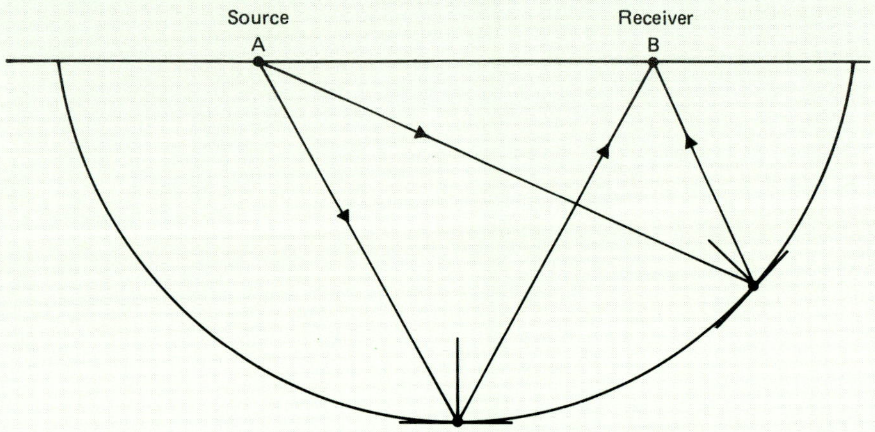

FIGURE 11.6

When source and receiver are separated the reflection tangent traces out an ellipse.

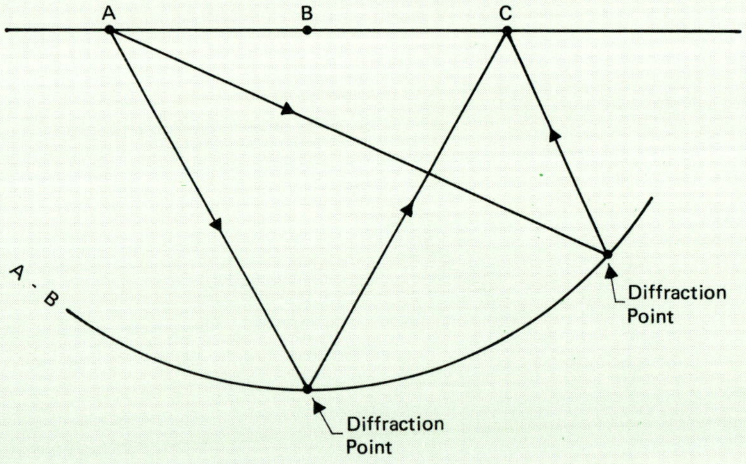

FIGURE 11.7

Diffraction source points are located on the same ellipse as reflectors.

This again confirms Huygens' basic theory, which states that any continuous reflector may be treated as a series of closely spaced diffraction points. Hence, in the development of a practical application there is no need to discriminate between the type of energy recorded; it may be either a reflection or a diffraction.

The difference will be in the results of the process. If the reflected signals recorded at all stations on the section are swung along the appropriate ellipse as indicated, the closely spaced tangents will sweep out one or more continuous wavefronts. But energy recorded as a diffraction pattern over several stations will swing to intersect at the common point source of the diffraction in space. **(Fig 11.8)**. Thus the identical migration procedure solves for the origin of both reflected and diffracted energy appearing on the seismic section.

The simplicity of this basic process is deceiving. Practical application is, in fact, one of considerable complexity. For ideal results, it requires much more precise control of the data and process parameters than is generally available through the seismic process. The paradox arises from the fact that the lithologic section is far from homogeneous and the velocity of the section almost always varies continuously, both vertically and horizontally, in space. The areas in which the process is of greatest value, e.g. the most complex ones, are also those in which

Velocity Compensation

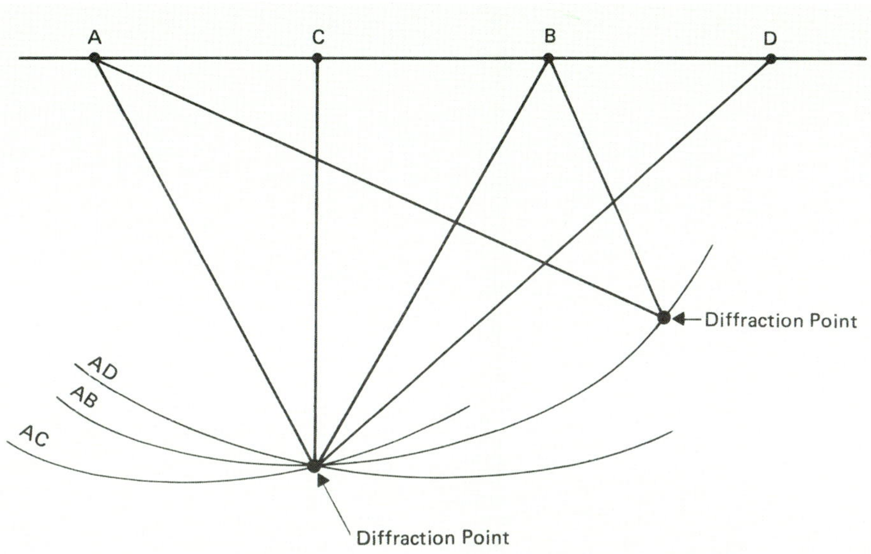

FIGURE 11.8

Migration of diffractions converges to the point source origin.

the velocities are the most variable and are the most difficult to determine.

Both input and output data for the migration process are digital seismic sections, composed of an array of time samples which make up a two-dimensional space-time grid. The dimensions of each are equal to the trace spacing in the horizontal direction. Conversion to the uniform space is obtained by proper application of the appropriate velocity function.

First assume a velocity function which varies only in the vertical direction. If the velocity can be determined for all points on the section, the depth of each time sample on the vertical column may be calculated. Then, for the simplest case, by a sample application of geometry, the distance of a straight line travel path from the source at the surface to each point on the grid and back to a receiver can be calculated. Next, each path is divided by the appropriate velocity, to obtain the corresponding travel times from origin to the receiver through each grid node.

It is thus relatively simple to trace, for each vertical time, the constant time front through the grid. The shape will no longer be spherical nor an ellipse, but will be correct for the velocity function.

The migration process is then done by moving each vertical time sample along its constant time line to the intersection with each column within the radius of the time line, adding the migrated value to the sum of any values found there.

This process is repeated for every combination of origin and distance which has been recorded in the field. The result should be a new output trace for each input trace of the section. The output traces will represent the sum of the input data plus all other data points which migrated to that location.

As before, reflecting horizons will appear as wavefronts. The diffraction responses will sum and indicate the diffraction points. **(Fig 11.8).**

The majority of the positions to which the input information will have been migrated will not be valid, i.e. not the proper position in space. Therefore, the sum total at any given output position will contain a great number of input points which have been incorrectly relocated. Fortunately these will tend to cancel out, leaving only that information which is correctly positioned on the output section. Evidently, in this operation, the total number of computations and data sample movements is enormous.

In practice the process is further complicated by the fact that velocities almost always vary horizontally as well as vertically. Reliable, continuous measurement of the velocity within the lithologic section is not possible by direct means so automated indirect determinations are used. The section on Velocity Analysis treats this matter and il-

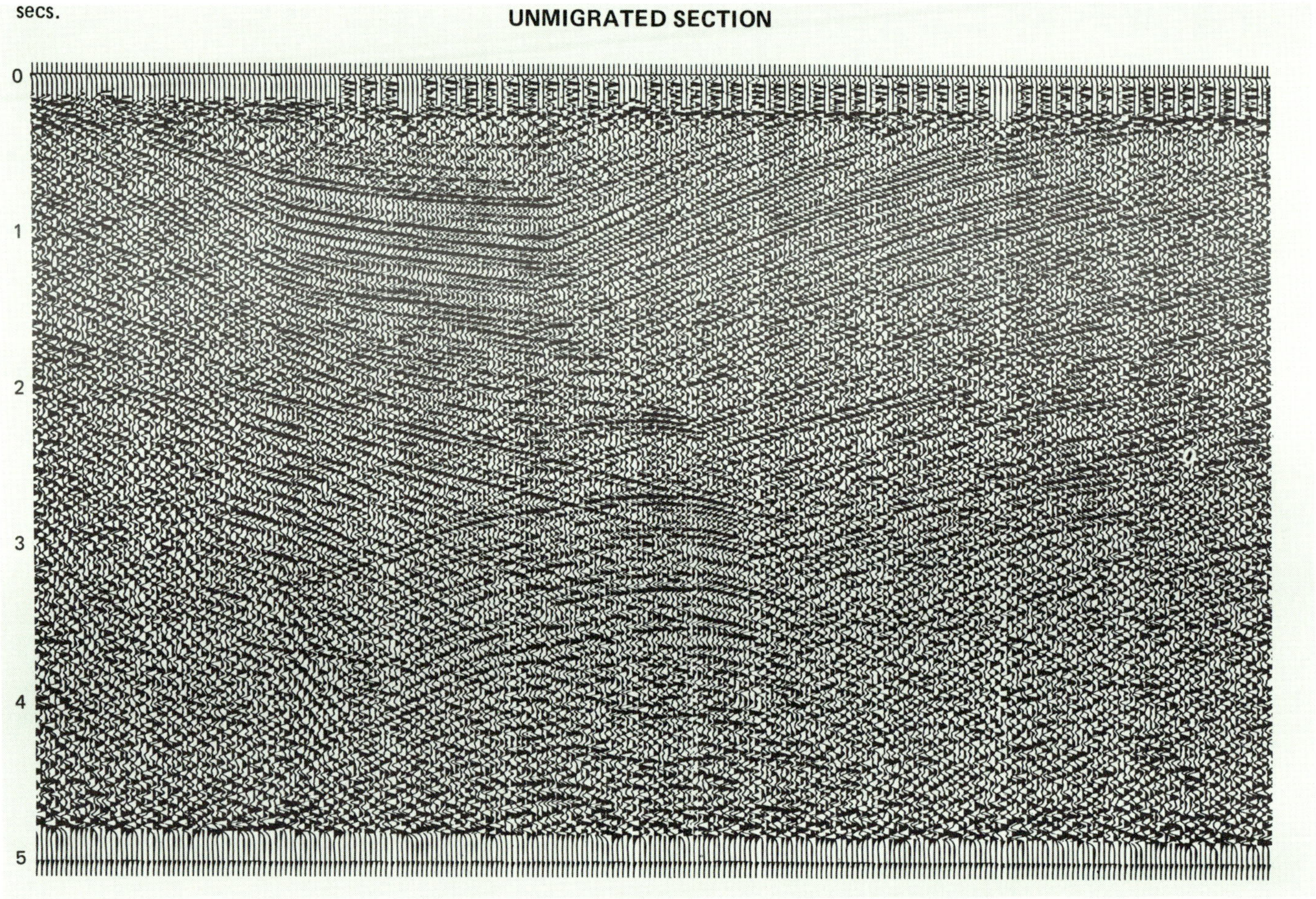

FIGURE 11.9

The unmigrated section has an apparent anticline at depth.

lustrates some of the many problems which are associated with the automated determination.

In particular, due to the difference between the straight line travel path assumed through the section and the actual path, which follows Snell's Law, the velocity function determined is only an apparent velocity. The discrepancy between the straight line path and the actual travel path is also a major consideration in the migration process. Fortunately, the stacking velocities which are determined from automatic velocity determinations compensate to a certain degree for the difference between the true ray path and the straight ray path assumption, but may be distorted by the effects of anisotropy, if present.

Another error arises from the fact that the velocities obtained are applied to reflection times considered to be vertical. In fact, in the case of steeply dipping beds, the calculation of such velocities thus may include a large horizontal component and actually correspond to a much shallower time than the reflection time would indicate. This will distort the output and introduce errors which tend to degrade the quality of the final product.

Most geophysicists are thoroughly familiar with diffraction phenomena, and usually associate them with faulting, but similar diffraction patterns may be generated from simple pinchouts and the banks of buried channels. A somewhat

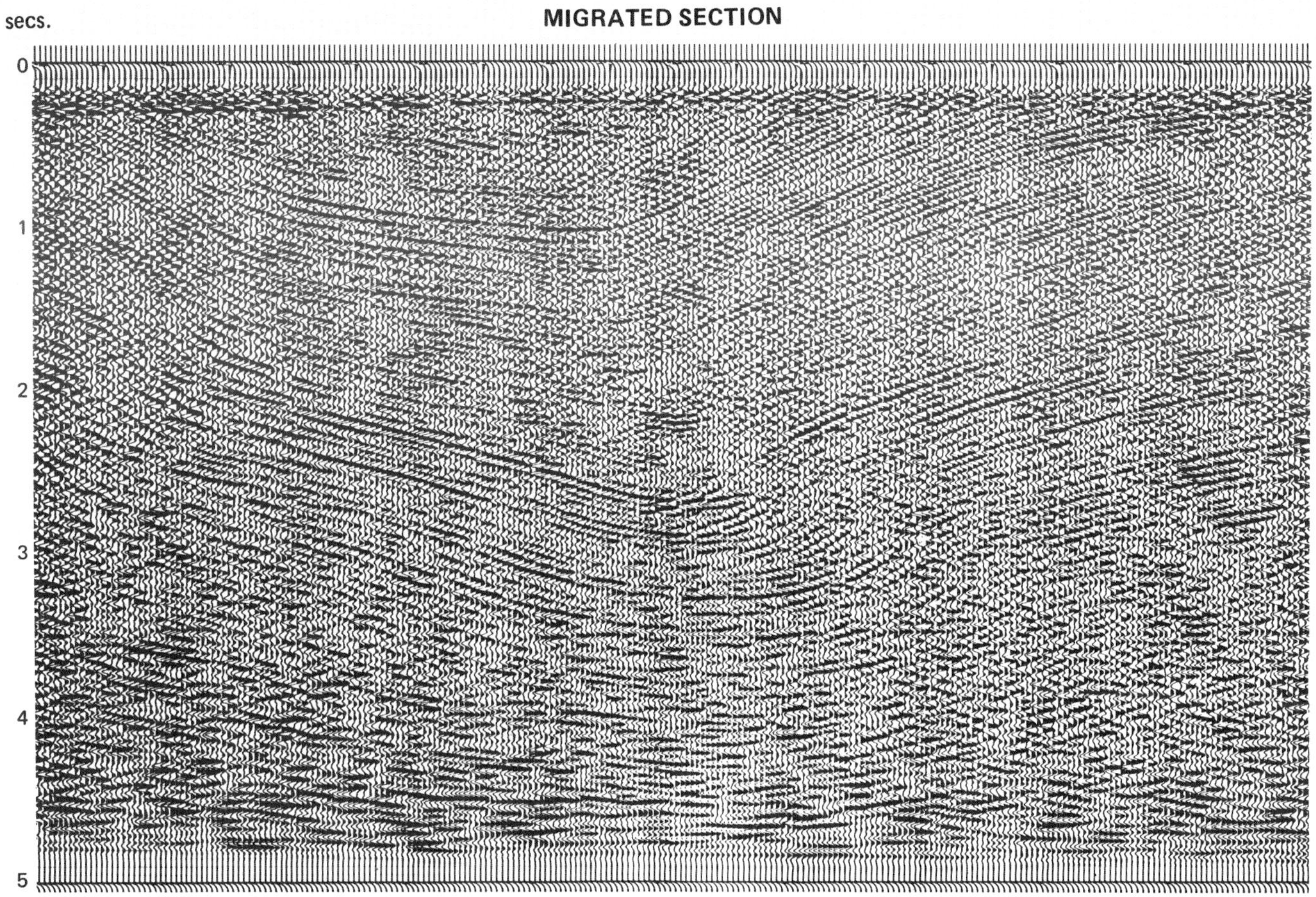

FIGURE 11.10

Migration of the section of Fig 11.9 shows the apparent anticline to be a syncline having a short radius.

different cause produces a similar appearing phenomenon, generated within synclines in which the turning radius is below the surface of the section. **Figure 11.9** is an example of such a section, which also appears to be faulted. At depth anticlinal reflections appear, which most geophysicists would recognize as spurious. Yet, it is always difficult to ignore any anticlinal feature so the section would elicit considerable attention, possibly even a borehole test to be sure.

The same section migrated **(Fig 11.10)** loses all traces of an anticline at depth. Rather, a very sharp syncline is illustrated, faulted on the right flank.

The next example, **Figure 11.11**, appears to map two overlapping anticlines, an unlikely, if not impossible, situation. Some geologists would opt to drill on the crest of the major feature, in which case migration would not be needed since the crestal position of an anticline does not migrate. A flank test would be another possiblility, due to the presence of onlapping sediments presumably trapped against the structure. In this case, migration would be necessary.

As predicted, the migrated section **(Fig 11.12)** shows no appreciable change at the crest of the anticlines. However, an interesting graben, separating the two, is now revealed. Both anticlines appear to be keystone features, but else-

FIGURE 11.11

The unmigrated section has two apparently intersecting anticlines.

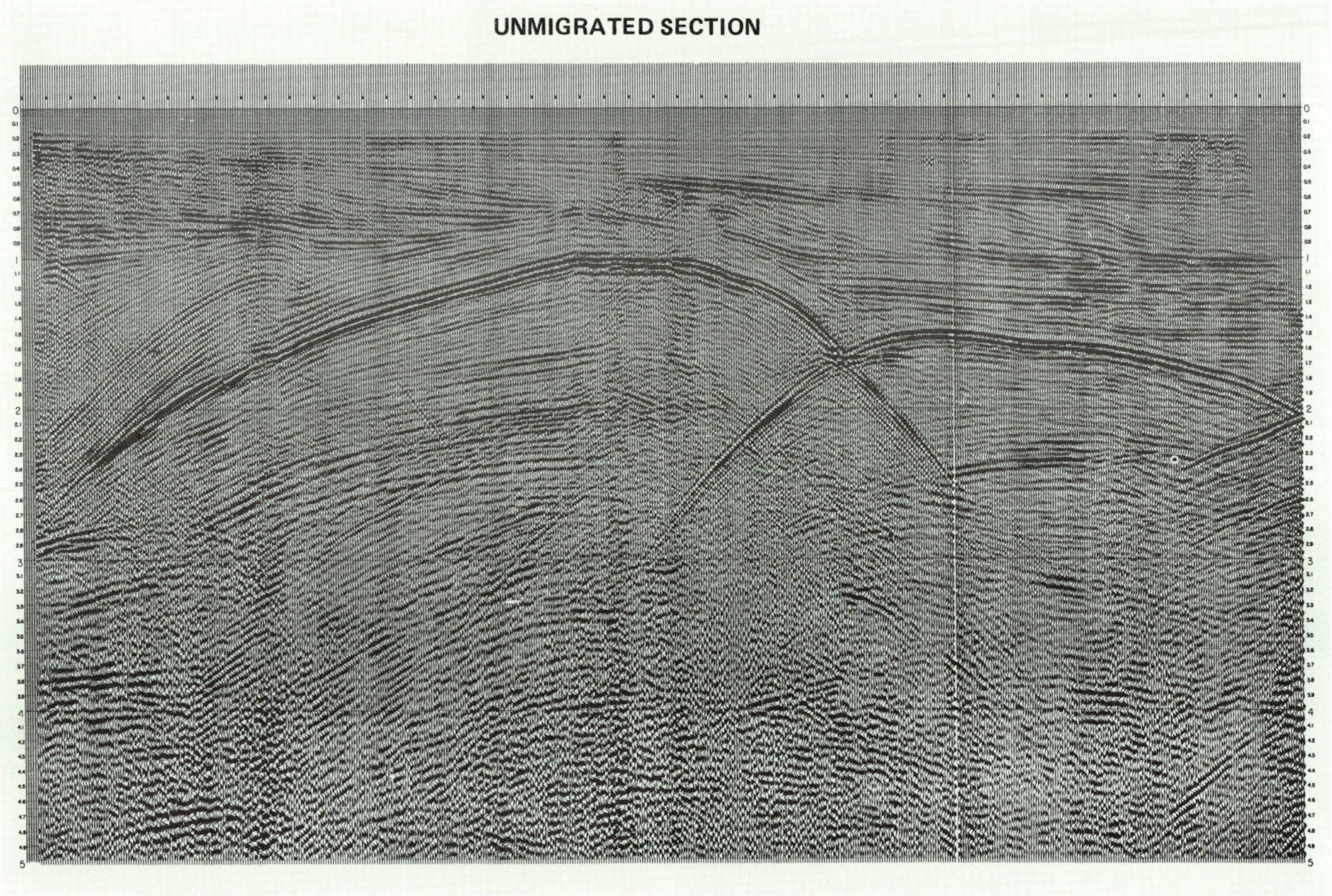

where the deeper section is poorly defined. Velocity in this zone may have been incorrect, or possibly there are no continuous reflectors, the apparent reflectors on **Figure 11.11** being diffractions.

The barren zone on the left side of the migrated section near 2.0 sec may simply have dips so steep they cannot be recorded within the limited extension of the line to the left.

Figure 11.13 includes some of the features of the previous two examples; including two apparent deep anticlines, but no clear indication of the location of faulting. The migrated version, **Figure 11.14**, is very interesting, since both anticlines have been converted to small synclines.

Wave Equation Migration

One problem with the Huygens method of wavefront migration, as can be seen on the examples, is that the process acts as a heavy filter, eliminating much of the high frequency information and often distorting the waveform. To overcome this problem a major research project was initiated under the direction of Dr. J.F. Claerbout at Stanford University, resulting in very successful methods of direct wave equation migration.

This analytical approach applies partial differential equations to the wave field to derive a set of operators which will consolidate diffractions and provide correct dip relationships on the output. The equations are lengthy, and

MIGRATED SECTION

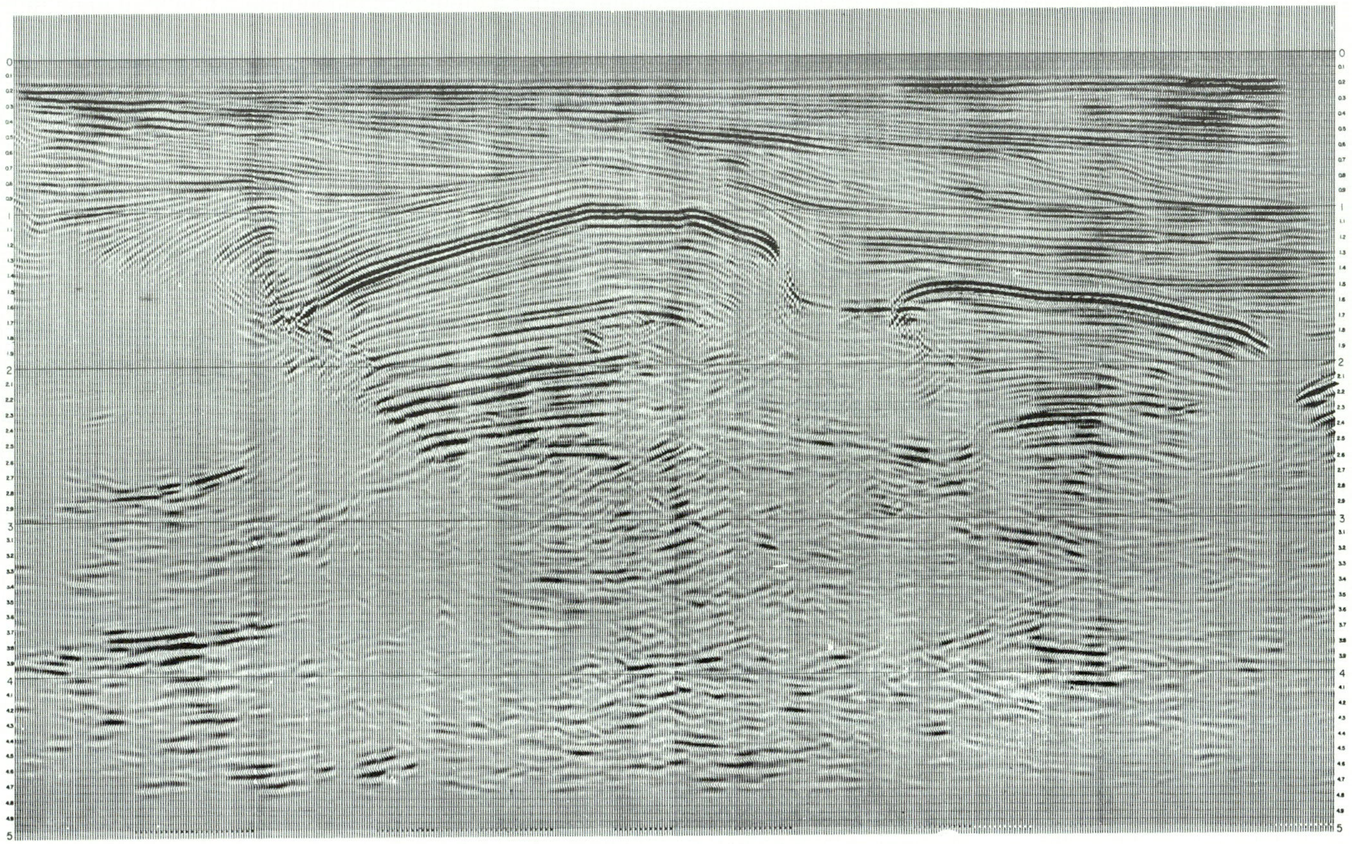

FIGURE 11.12

The migrated section of Fig 11.11 shows a much more probable structural solution.

the operation difficult to understand, but the process has the advantage of causing very little distortion of the wave shape.

Perhaps the process can be illustrated best through use of a model. The usual reflection model is of a surface source transmitting signals which bounce back to the surface along paths perpendicular to the reflection plane. **(Fig 11.3).**

The conventional migration model assumes all subsurface points to be diffraction points. Each diffraction point acts as a new signal source, radiating signal in all directions. Migration is achieved by swinging each point on a seismic section to every possible diffraction point in the subsurface.

In common with many other advances in seismic signal processing, the development of wave equation migration depended on a new concept, including a new and ingenious model of the physical process involved.

Imagine the subsurface section to be an orthogonal grid of small compartments, each measuring one time sample per side. **(Fig 11.15a).** Each contains a tiny explosive charge equal in amplitude to the reflection amplitude at that point. At some time zero, all charges are detonated simultaneously. If a video camera were to record the wave front action, at increments of one time sample, circles, like expanding ripples on a pond, would be seen to radiate outward from each signal source to form a complex wave field. **(Fig 11.15).**

FIGURE 11.13

The unmigrated section maps a complex combination of reflections and diffractions.

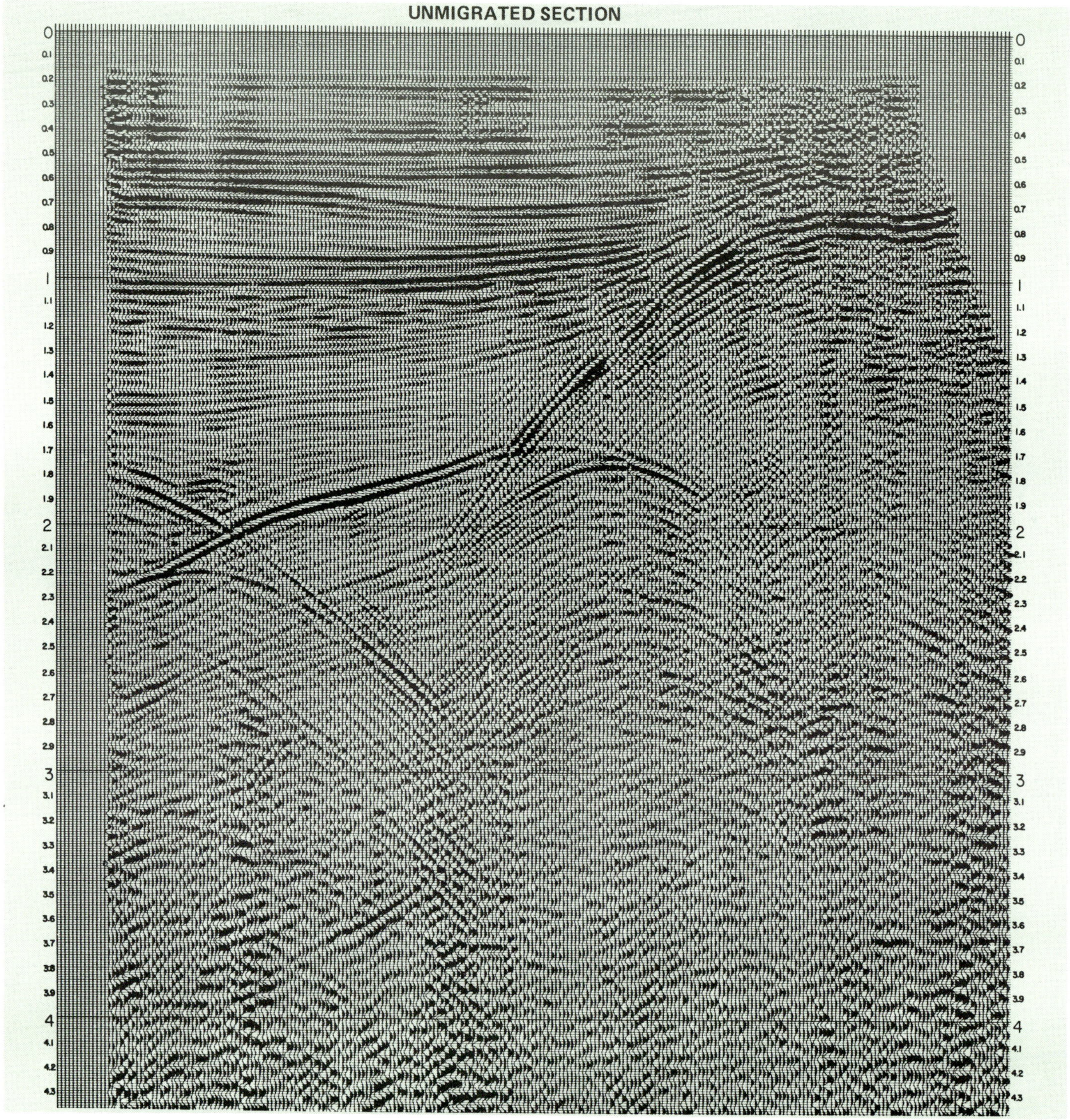

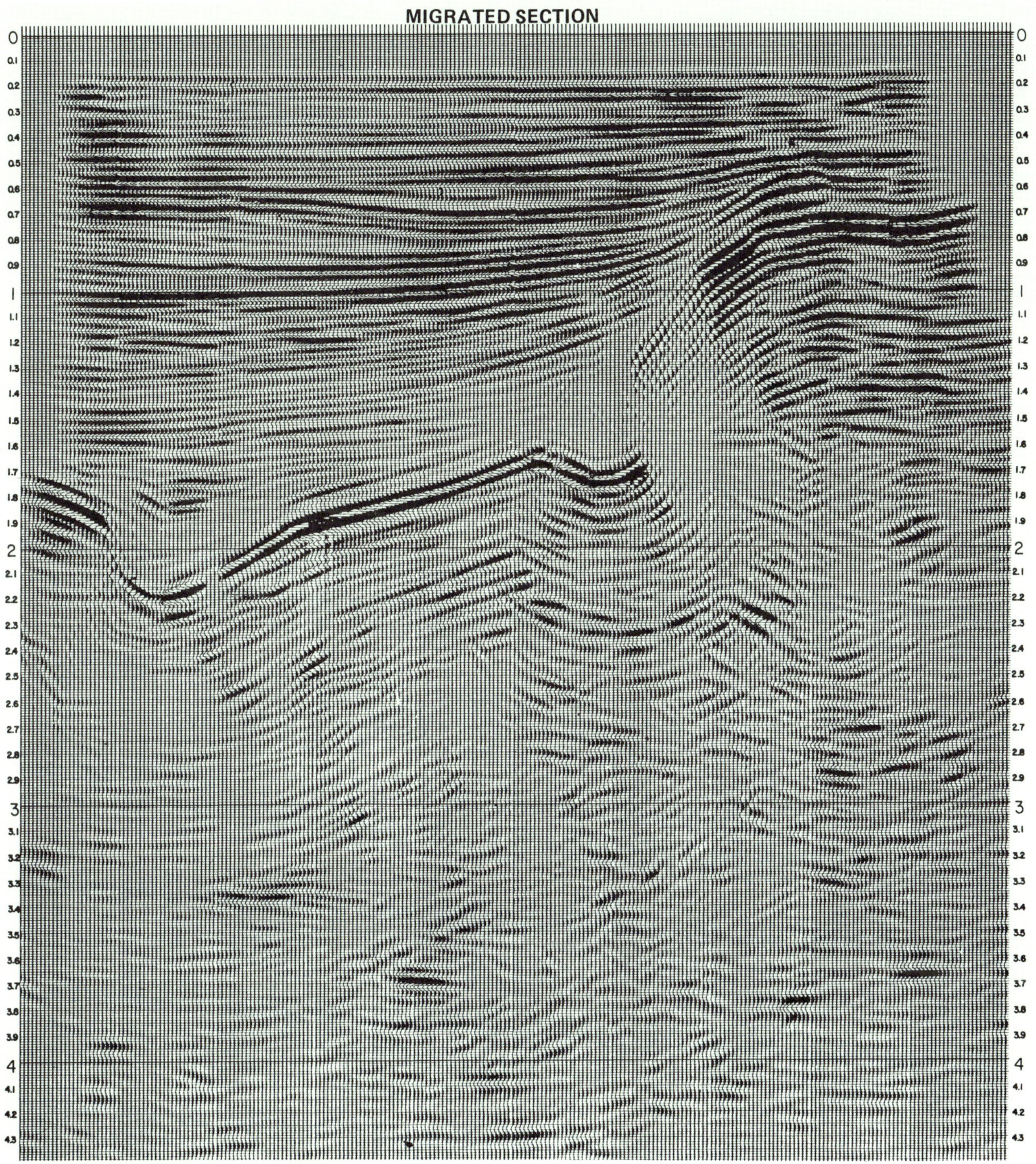

FIGURE 11.14

Two apparent anticlines on the unmigrated section of Fig 11.13 now appear in their true form of narrow synclinal troughs.

FIGURE 11.15

The sequence illustrates the expanding wave field generated by point sources of radiation, as would be seen at uniform time increments.

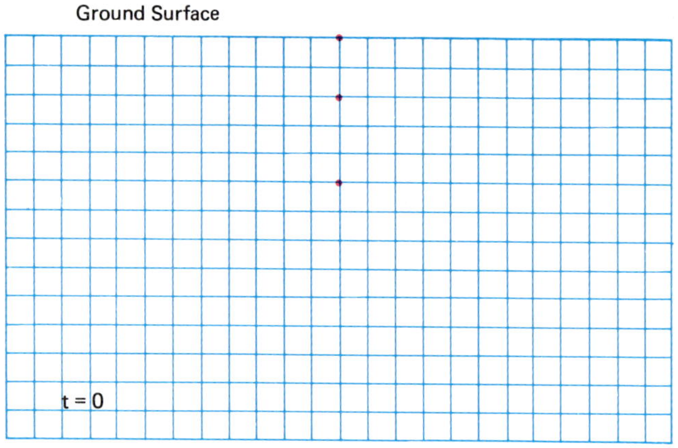

Fig. 11.15a

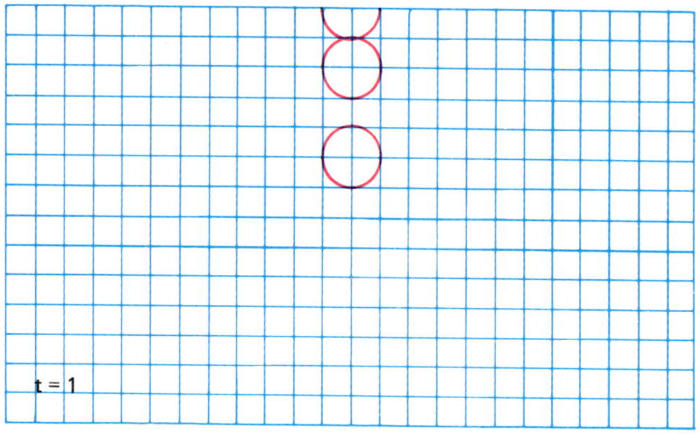

Fig. 11.15b

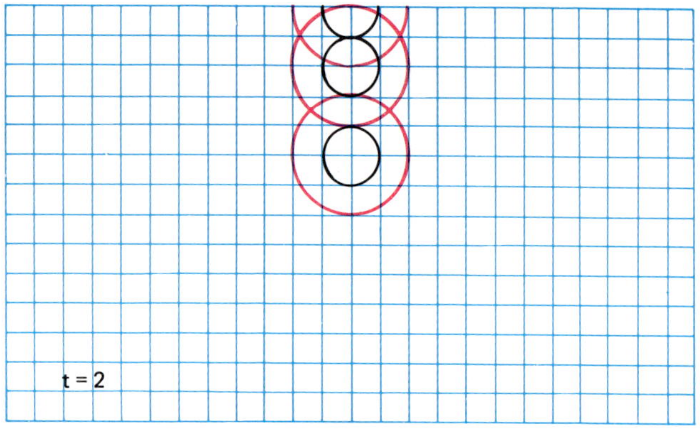

Fig. 11.15c

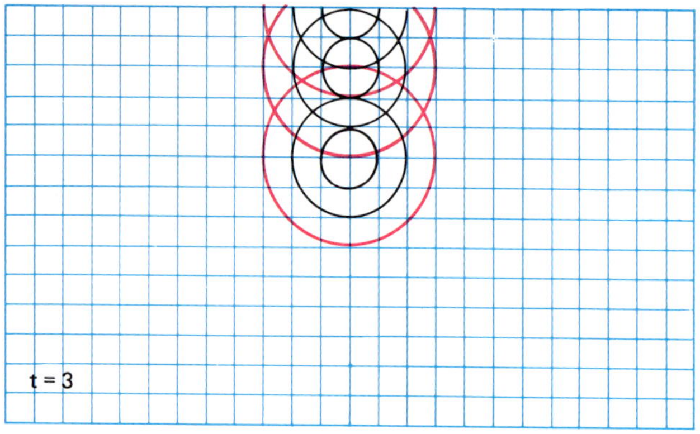

Fig. 11.15d

The camera could be reversed, and we would then see the wave field collapse one step at a time, back to time zero.

Now, to more closely model the real situation, the entire section cannot be viewed at one time. Observations are to be made from a series of geophones extending over the length of the section at the surface of the ground, **(Fig 11.16a).** Since the line of observation can record only impulses received at the instant of each time slice, the expanding wave front from any point source would appear at the surface to move laterally away from the point of intercept with each successive time increment. All geophones are scanned once at each time increment and plotted, developing an image produced in raster form, one line at a time, like space pictures from a satellite, or the output of an electrostatic plotter attached to a computer, **(Fig 11.16b-h).**

FIGURE 11.15

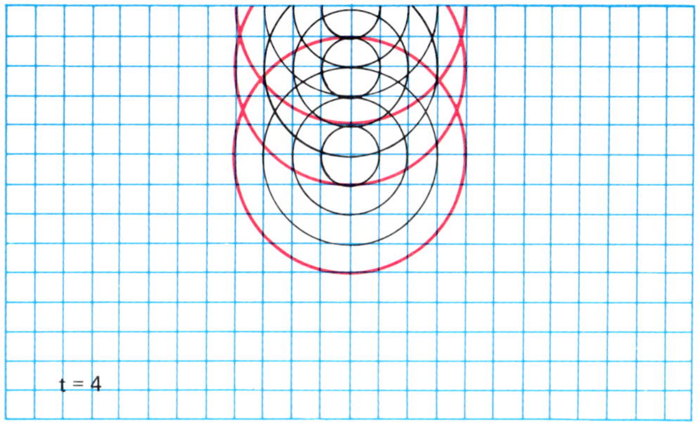

Fig. 11.15e

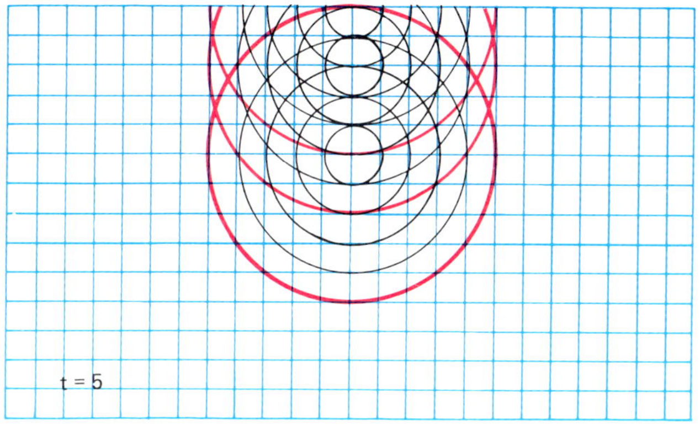

Fig. 11.15f

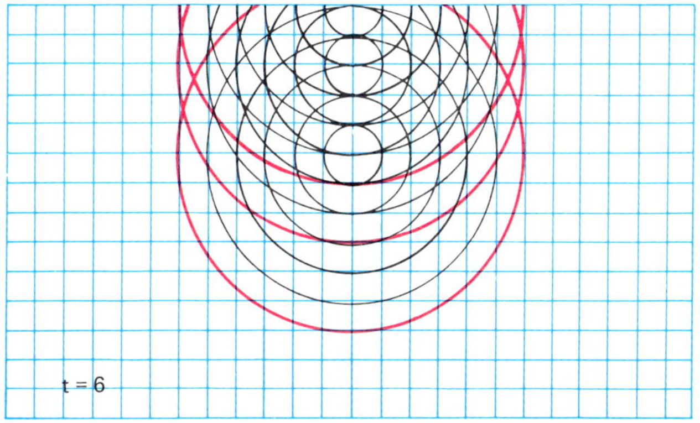

Fig. 11.15g

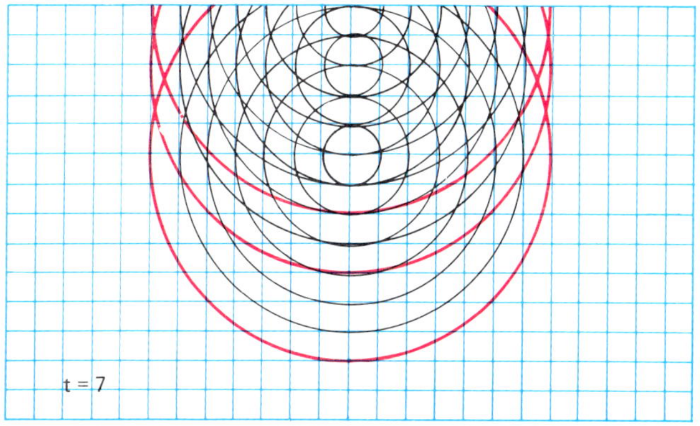

Fig. 11.15h

Plotting the image would start at the surface datum, and line by line, record events from deeper and deeper levels to eventually yield a conventional cross section.

Wave equation migration reverses this process, to produce a revised image by simulating the following:
If a layer of surface material equal to one sample time increment was removed, the geophones lowered to the new level, and the shot repeated, the recorded image would be similar to the first, but all events would arrive one sample time earlier and have a slight change in the geometry, **(Fig 11.17b)**.

Next imagine the above process repeated several times. Starting with the section recorded at the surface, a new section, as it would have appeared had the observation datum been one time increment deeper, is calculated. Repeating the process, a series of section images is computed, each being the result of moving the recording plane one time sample deeper in the section.

FIGURE 11.16

Sampling the expanding wave field of Fig 11.16 at the surface at the same time increment will generate the typical diffraction patterns observed from point sources.

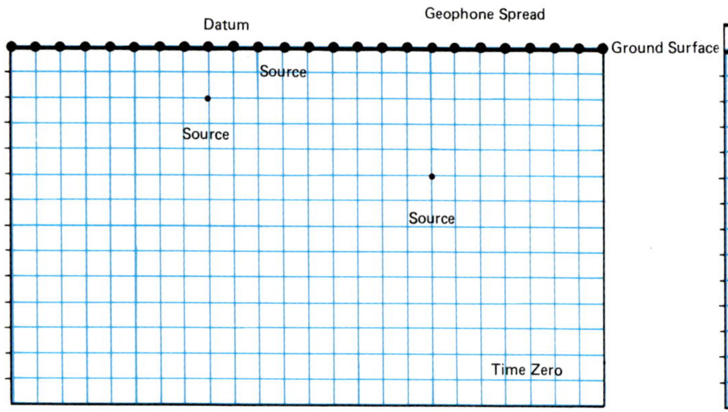

Fig. 11.16a

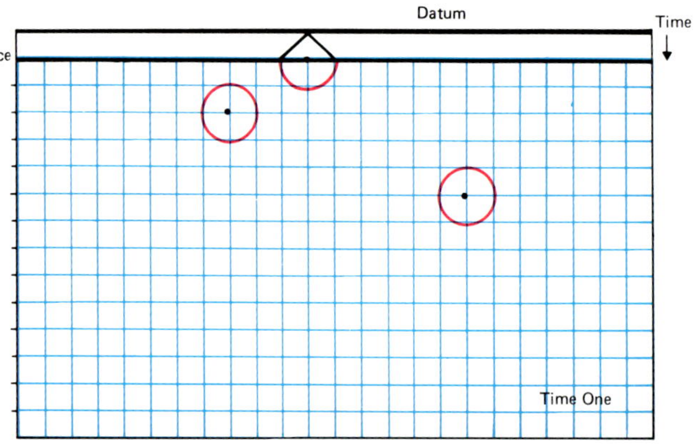

Fig. 11.16b

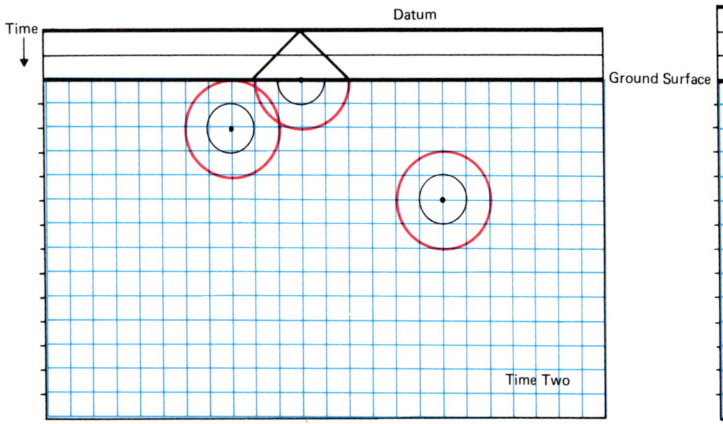

Fig. 11.16c

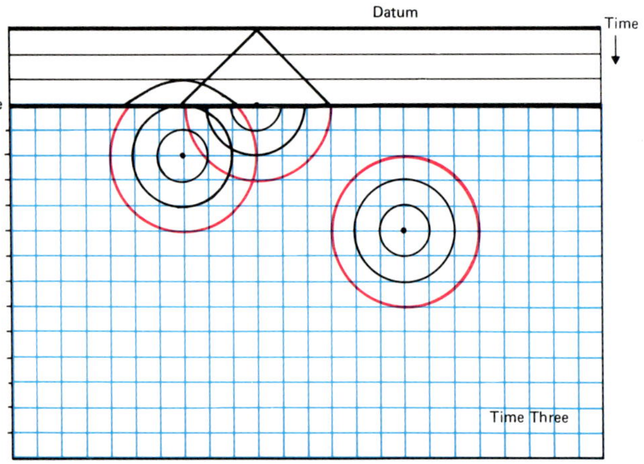

Fig. 11.16d

The diffraction parabolas would collapse into successively smaller features, until, as the datum passed through each source point, the single point response at time zero from that level would be shown, **(Fig 11.17a,c &f).** Only the top, (zero) line of each new display, is retained. Through successive iterations, the datum is gradually floated down to the bottom of the section.

FIGURE 11.16

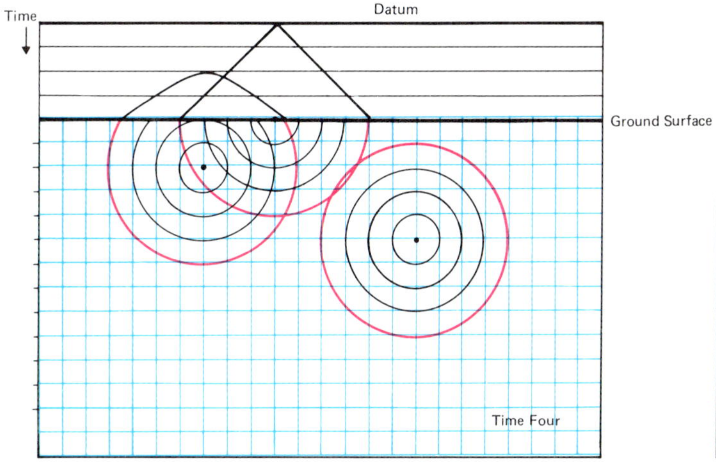

Fig. 11.16e

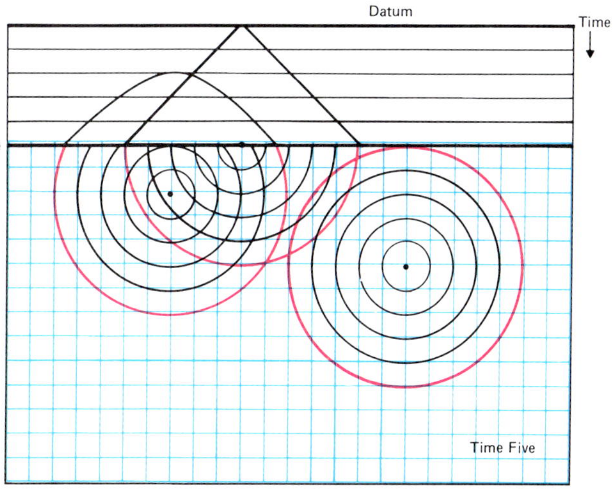

Fig. 11.16f

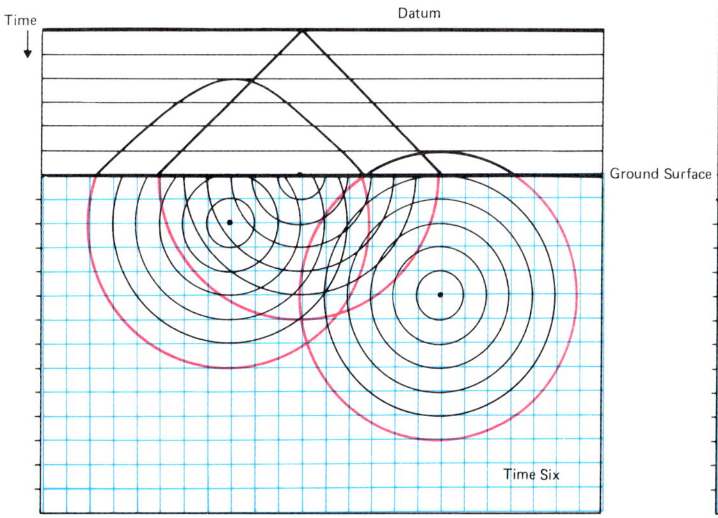

Fig. 11.16g

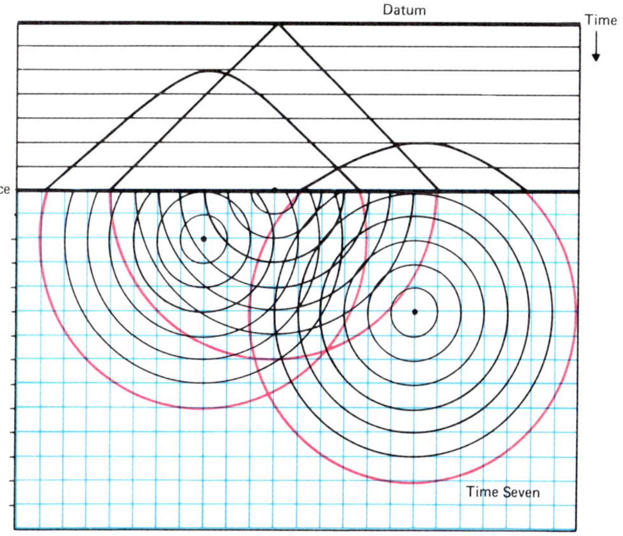

Fig. 11.16h

At the end of the process, with all wave fronts collapsed to their point sources at the appropriate levels, a single section, the equivalent of a migrated section, would remain with the amplitude values at each point equal to the shot source amplitude, (reflection coefficient).

The computing algorithm is a form of the downward continuation process, an elegant extension of the type of

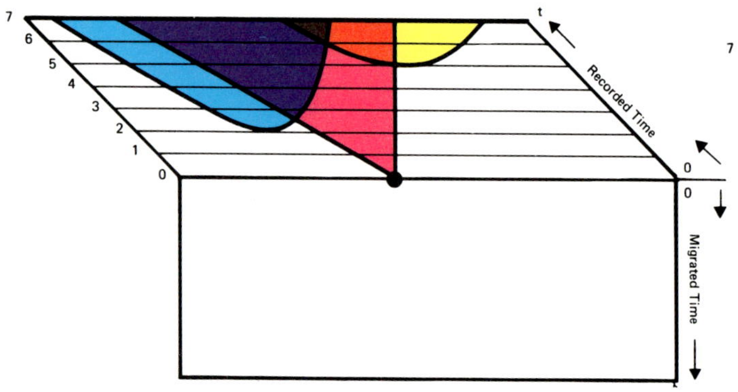

Fig. 11.17a

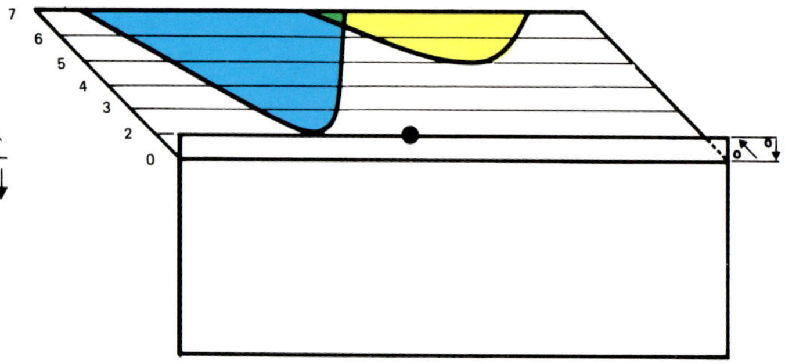

Fig. 11.17b

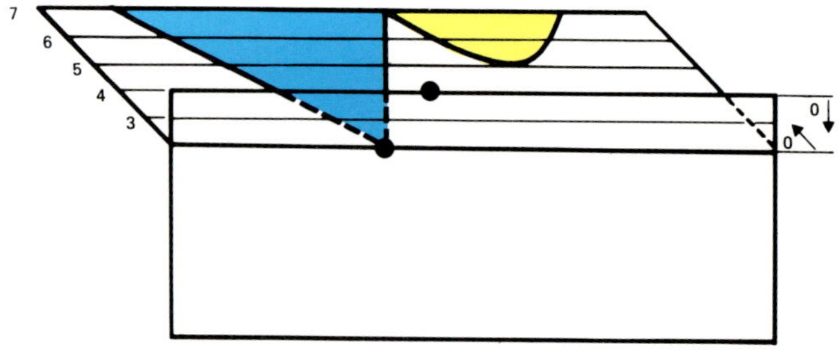

Fig. 11.17c

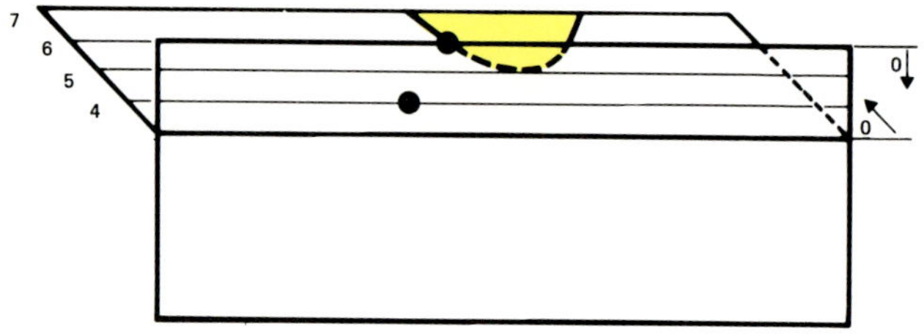

Fig. 11.17d

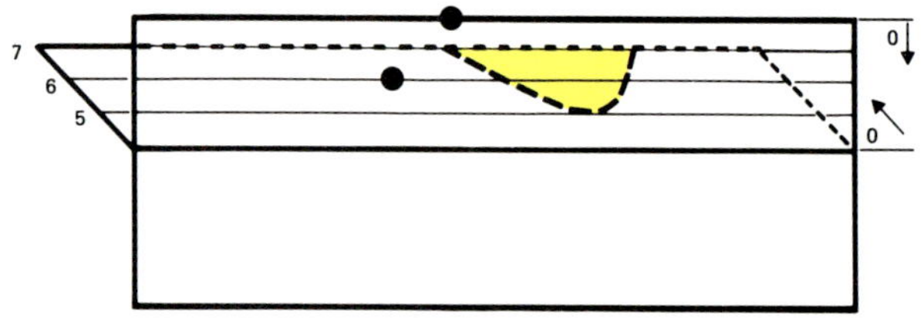

Fig. 11.17e

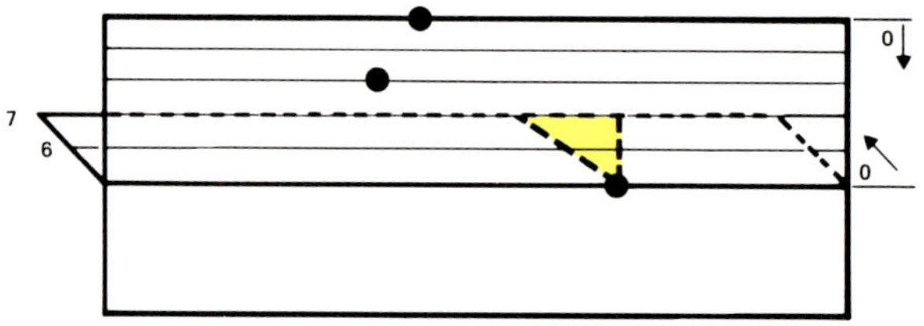

Fig. 11.17f

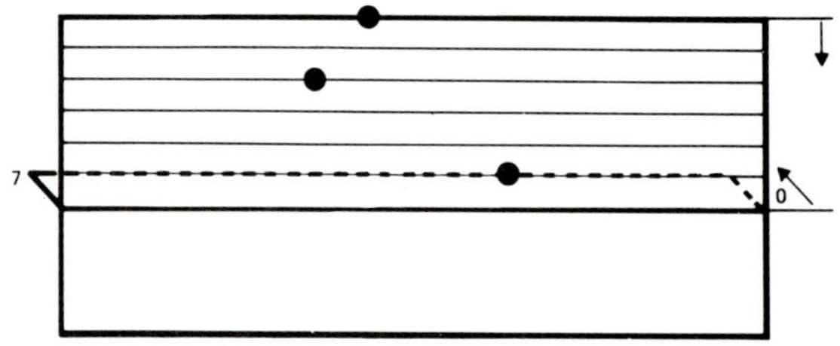

Fig. 11.17g

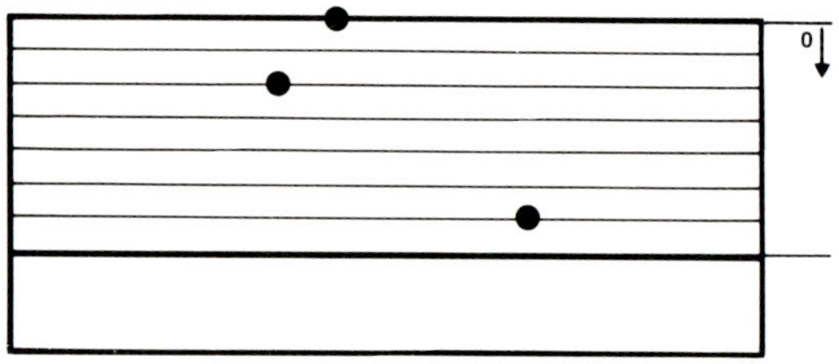

Fig. 11.17h

FIGURE 11.17

Wave equation migration collapses the wave field response, recorded in Fig 11.16, by successively moving the recording plane datum down one time increment and calculating the response that would have been recorded at that level. The datum response at each increment is preserved as one time line of the migrated section.

DIP MIGRATION

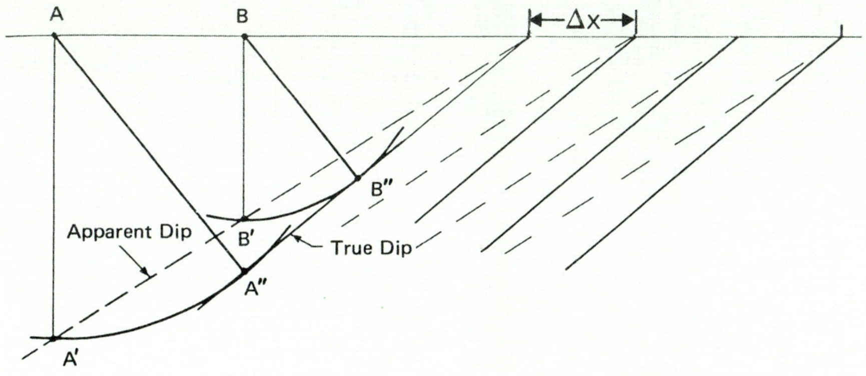

FIGURE 11.18

Migration of apparent dip to true dip rotates each dip plane about a point on the surface. The horizontal spacing between beds, remains constant, so the horizontal frequency does not change, but the vertical spacing is increased with a corresponding decrease in vertical frequency.

mathematical process used to sharpen potential field observations.

A set of operators, one for every time level increment, is needed to convert each level to the next deeper time level response. In a true time space, a single set of operators, one for each level, would be adequate, cycling iteratively down through the section, discarding the bottom operator after each iteration. In practice, because the hoizontal sampling is distance, not time, velocity changes require appropriate adjustments. When there are lateral changes in velocity, as there often are, distortion may result which cannot always be readily accommodated.

FREQUENCY DOMAIN MIGRATION

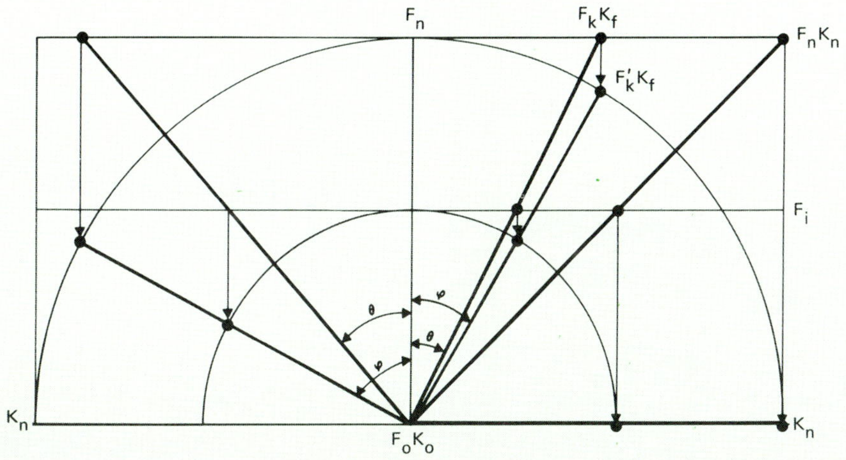

F/K Migration

FIGURE 11.19

Migration is accomplished by shifting existing points on the F/K Fourier transform plane down frequency to rotate the dip angle an amount corresponding to the relationship tan θ = sin φ.

The design of the operators includes fairly lengthy calculations, and second order effects are often ignored or approximated for reasons of economy and speed. This may result in a limitation of the amount of dip that can be migrated without distortion.

The most recent development is direct migration on the F/K plane of the two-dimensional Fourier transform. This procedure, first described by Stolt, involves a simple repositioning of frequency domain coordinates. Following is a brief description of the basic model and procedures. The reader is referred to Chun and Jacewitz for a clear and more comprehensive discussion.

The method exploits the property of

FIGURE 11.20

This unmigrated section has numerous diffractions generated by faulting and other discontinuities.

BEFORE MIGRATION

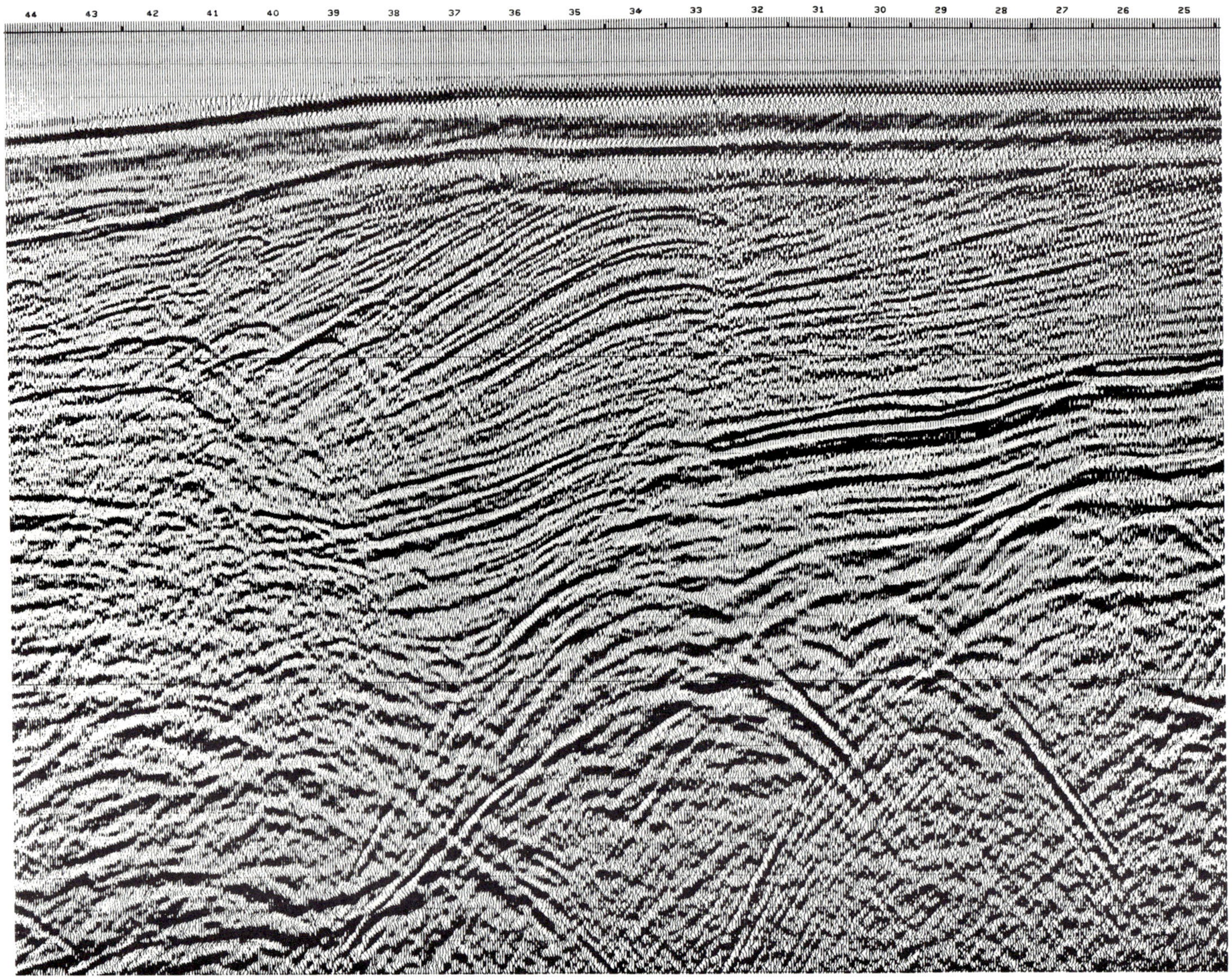

the two-dimensional transform to sort and consolidate all dips on a seismic section into corresponding common angular vectors on the F/K plane (Section 10). For development of the model, assume, as in previous discussions, a seismic time/space having velocity such that horizontal and vertical scaling is equal. Then, all dip angles which exist on the unmigrated seismic section have the common relationship:

$$\sin \varphi \text{ (true)} \quad = \quad \tan \theta \text{ (apparent)}$$

AFTER MIGRATION

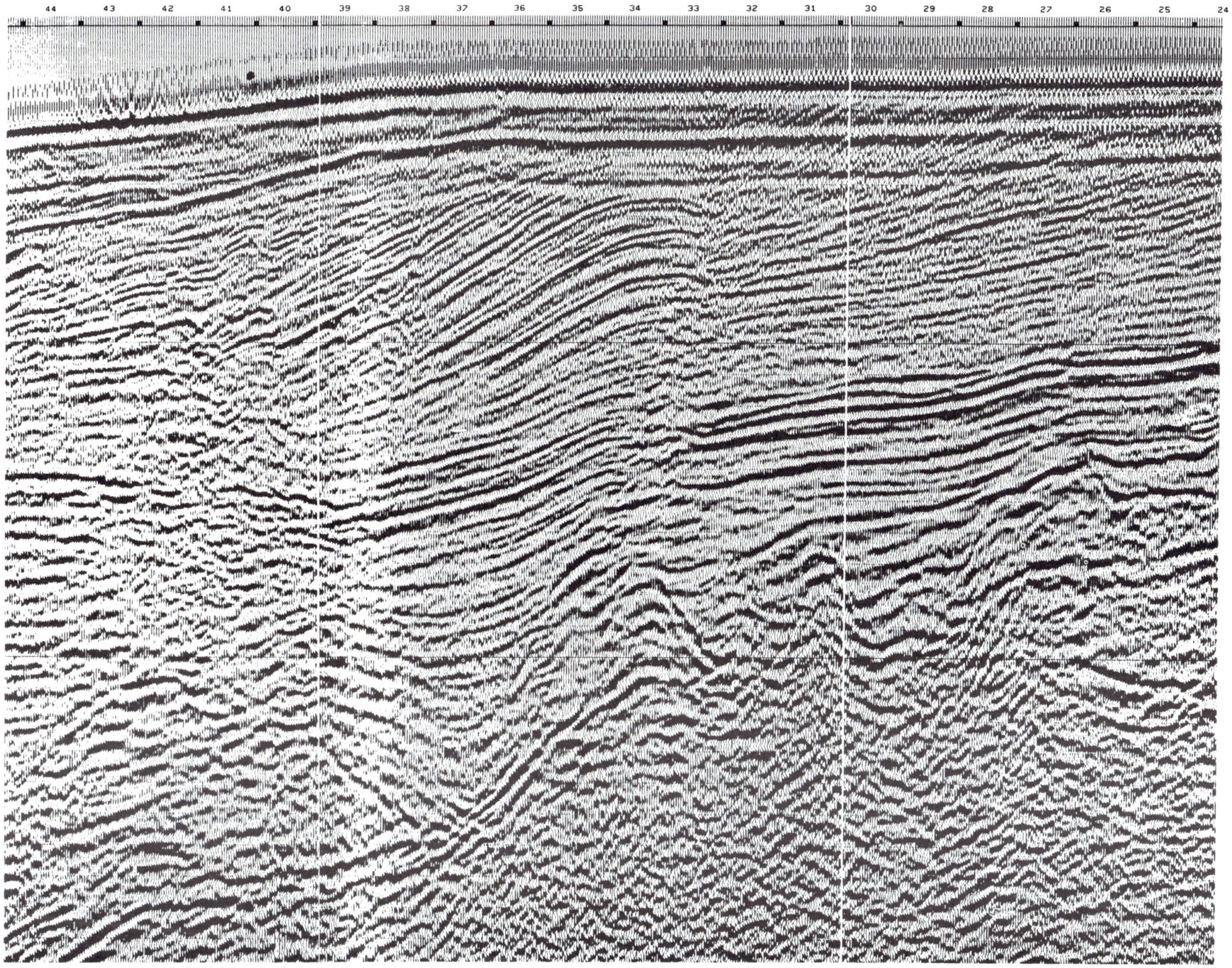

FIGURE 11.21

The reflection data of Figure 11.20, after frequency domain migration are properly positioned, yet maintain the fidelity of the reflection wavelet.

A time domain section can be transformed to the frequency domain to yield the corresponding F/K plane. In the process all dip components having a common angle on the time domain seismic section are consolidated into a single dip vector in the frequency domain. Migration on the time domain seismic section involves a simple dip plane rotation from the apparent dip angle to the steeper true dip angle, and a similar feature can be exploited into the frequency domain. In the time/space domain, as shown in **Figure 11.18**, apparent and true dips, when extended, meet at a common point on the surface. On th F/K plane apparent and true dips all converge to a single point at the $F_0 K_0$ coordinate position, **(Fig 11.19).** Rotating each vector on the F/K plane from the apparent dip angle to a

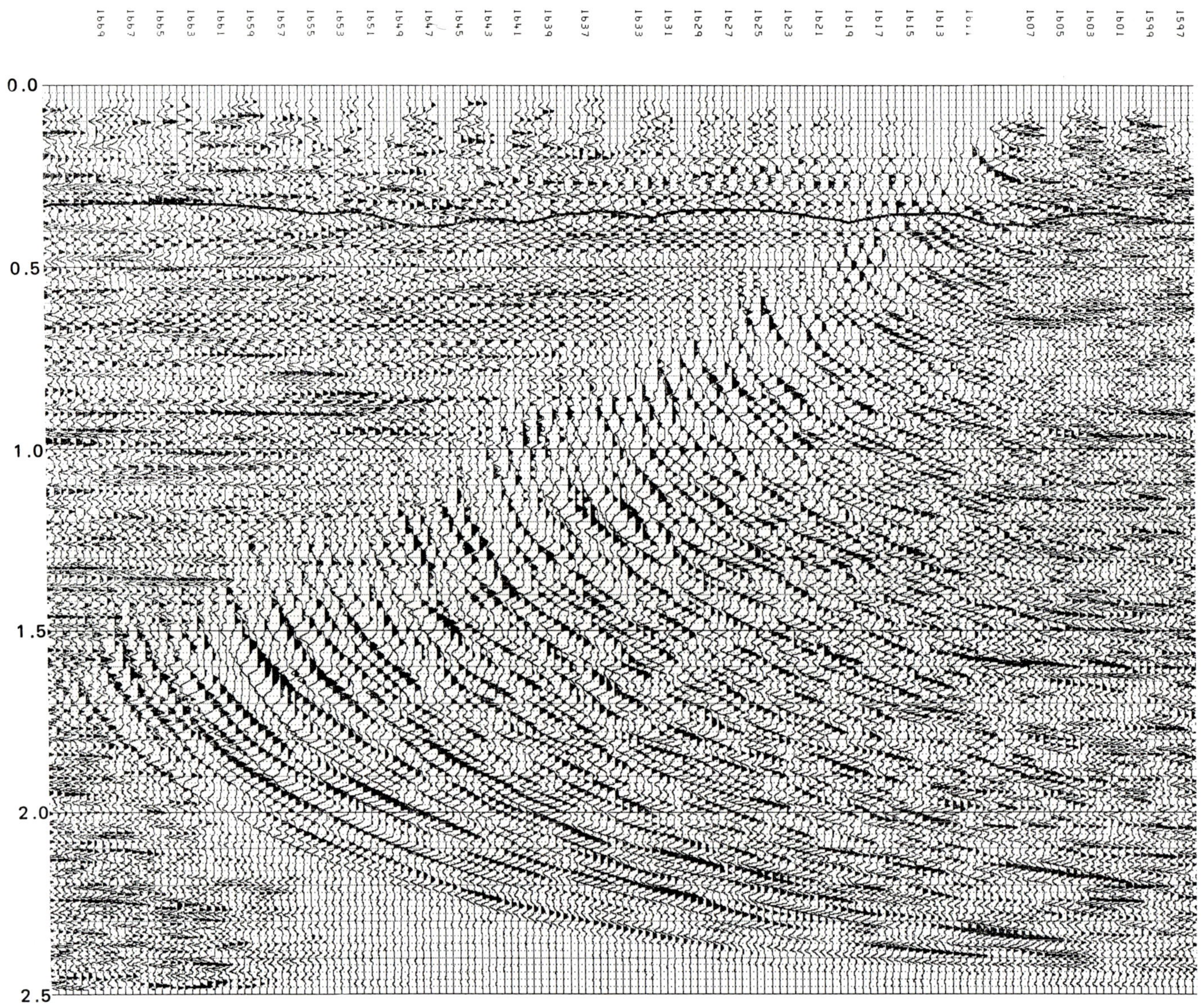

FIGURE 11.22

An extreme example of 'smiles'; the result of an artificial horizontal discontinuity in the seismic reflection data.

new angle, corresponding to true dip, will, when the reverse transform is taken, have migrated all apparent dips to their true dip angle.

All apparent dips on a time domain section, projected to the surface datum intersect the corresponding projection of the true plane, since such points remain fixed at the surface. In the process of rotating dip planes from the apparent to the true dip slope, the horizontal space between any two equal dip planes **(Fig 11.18)** remains constant. Therefore, the horizontal frequency will remain unchanged, whereas the vertical (temporal) frequency will decrease, Hence, in the frequency domain, migration can be accomplished simply by applying a vertical downward shift of all temporal frequencies **(Fig 11.19)**, and scaling of the amplitudes by a small factor.

If F is the vertical frequency to be migrated, and K the corresponding horizontal frequency, the frequency coordinate position is migrated from F_k, K_f to F'_k, K_f where:

$$F'_k = (F_k^2 - K_f^2)^{1/2}$$

The solution involves a single specific shift for each F/K coordinate point. Hence no distortion of the waveform occurs except the appropriate slight decrease of vertical frequency. The method is simple, yet precise, and may obsolete all others. It is an advance made possible by the invention of Fast Fourier Transform. A relatively simple extension of the theory provides three-dimensional migration as well.

Figs. 11.20 and 11.21 compare the input and the output of a frequency domain migration program. In spite of considerable migration of both diffractions and dips, the waveform and character of the output section remain essentially unchanged, and there is none of the filtering effect of the earlier spatial method.

All seismic migration techniques suffer from limitations imposed by the ability to determine the correct velocity. The basic theory of migration has in each case, been developed for a homogeneous space, having uniform dimensions in time and distance, rather than the variable dimensions of the actual case. Velocity is required to correct the distortion.

Unfortunately, as is the case with many data dependent seismic process functions, velocity information is least reliable in areas where the need is greatest. In areas of steep dip, or extensive faulting, velocity determination is difficult and stacking may suffer also, all of which affects the quality of migration operations.

A particular concern of F/K migration is the introduction of pronounced upward arcs, not unlike inverted diffraction patterns, **(Fig 11.22)** sometimes termed smiles. These are caused by artificial horizontal discontinuities in the data, such as those which occur at line endings, or as the result of noise bursts, static errors, or incorrect amplitude recovery. They are artifacts of processing, and except for line endings, usually are the result of improper or inadequate processing prior to migration.

12 SEISMIC DATA INVERSION

Introduction

The accumulation of petroleum in the subsurface is influenced by many subsurface geologic factors, including rock type, lithology, age, depth of burial, porosity, permeability, fluid content, and structure. Information about any of these can be of great value to oil finders and producers.

The most reliable source for this information is by direct examination and testing of rock samples obtained from well cores or cuttings. When rock samples are not available, borehole logs may be used to obtain subsurface information by indirect means. Some of the physical properties of rocks can be measured and then interpreted in terms of geologic information. Such log measurements can also be used to map structure and facies changes by correlation of formation tops across a suite of logs from several boreholes.

Mr R.E. Wyman, in his excellent AAPG Continuing Education Lecture, demonstrates the links between geology, geophysics, and several petrophysical measurements, by use of a "Rosetta Stone" which compares several types of measurements of the same geologic column.

One type of log, the sonic, yields a continuous profile of the transmission velocity of sound through the rocks which make up the geologic column. It is one of the more useful log types because, as has been shown by Faust (1951), rock velocity varies directly as a function of three important geologic components; lithology (including porosity and fluid content), age, and depth.

Unfortunately, an expensive borehole must be drilled to obtain either rock samples or log information. The only presently known form of detailed indirect subsurface measurement, or remote sensing instrument, is the seismic survey. Seismic recordings made at ground surface measure reflection amplitude, another geologic property related to sonic velocity.

SONIGRAM

VELOCITY ANALYSIS

LOCATION LSD 6 SEC 12 TWP 27 RGE 1 W 5 M

WELL NAME TIPPERARY ET AL CROSSFIELD

COMPL. DATE 5 JAN 71 **DATA SOURCE** BHC SONIC LOG

K.B. 3566 **G.L.** 3553 **TD.** 9000 **TD. corr.**

AREA **AREA CODE**

INTEGRATED FROM 1220 **TO** 9010 **TOTAL** 7790

INTEGRATION ADJUSTED TO

GEOLOGIC FORMATION	DEPTH KB	ELEVATION	INTERVAL THICKNESS	INTERVAL TRANSIT SECONDS (TWO-WAY)	INTERVAL TIME R.M.S. VELOCITY	INTERVAL HALF VEL.	REFLECT. TIME SECS (TWO-WAY)
	1220	2346					•000
			510	•090	11•28	5•641	
KNHL TF	1730	1836					•090
			1803	•331	10•97	5•453	
BELLY R	3533	33					•421
			957	•151	11•92	6•353	
LEA PRK	4490	−924					•572
			95	•017	11•69	5•622	
MILK R	4585	−1019					•589
			285	•045	11•97	6•352	
COLO	4870	−1304					•633
			780	•127	12•06	6•163	
CARD ZO	5650	−2084					•760
			86	•014	12•10	6•119	
CARD SS	5736	−2170					•774
			304	•049	12•16	6•210	
2 W S	6040	−2474					•823
			467	•078	12•12	5•970	
B F S	6507	−2941					•901
			67	•013	11•88	5•148	
VIKING	6574	−3008					•914
			270	•038	12•20	7•108	
BLAIR	6844	−3278					•952
			483	•070	12•41	6•893	
L BLAIR	7327	−3761					1•022
			256	•033	12•81	7•849	
SHUNDA	7583	−4017					1•055
			202	•022	13•46	9•087	
PEKISKO	7785	−4219					1•077
			288	•030	14•12	9•681	
BANFF	8073	−4507					1•107
			482	•055	14•45	8•707	
EXSHAW	8555	−4989					1•162
			8	•001	14•43	7•095	
WABAMUN	8563	−4997					1•163
			295	•030	14•88	9•693	
CROSSFD	8858	−5292					1•194
			152	•015	15•32	10•011	
TD INT	9010	−5444					1•209

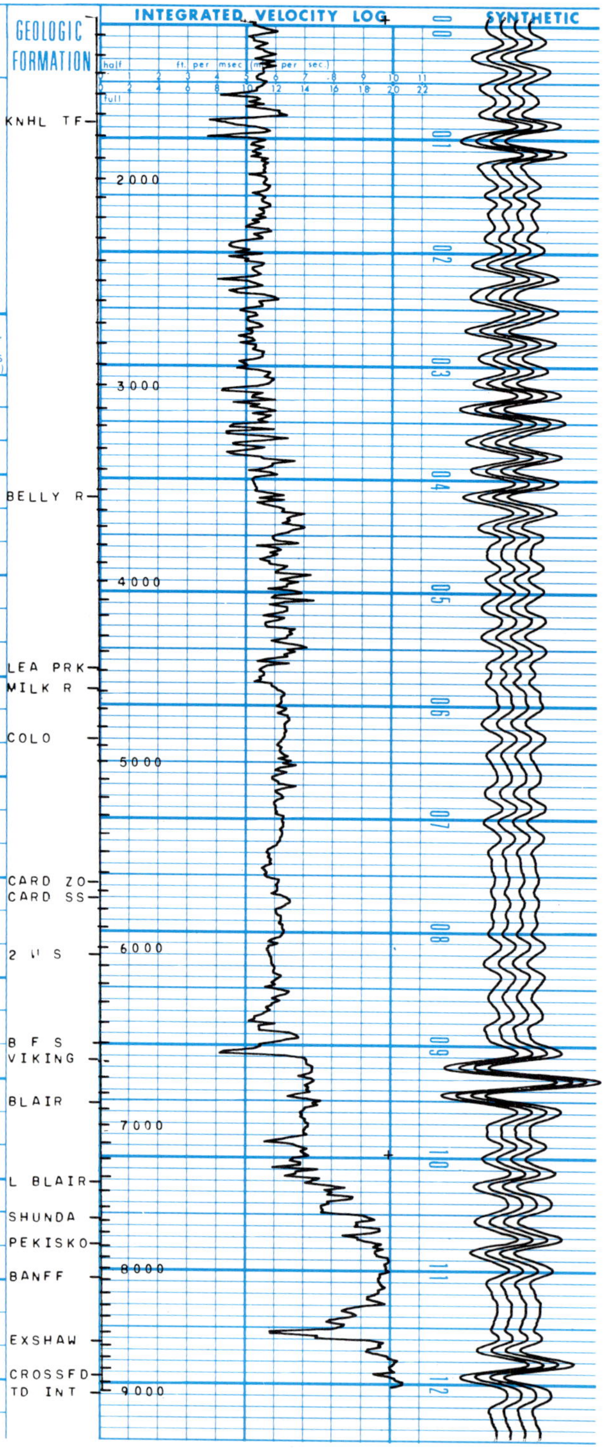

SONIGRAM courtesy of DIGITECH LTD.

FIGURE 12.1

A synthetic seismic trace calculated from an integrated sonic log is used to tie geologic information to a seismic survey.

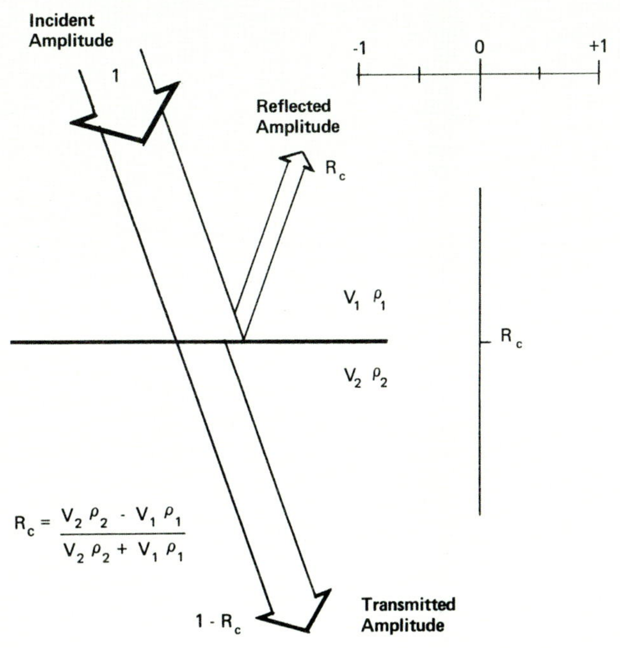

FIGURE 12.2

An arriving impulse is split into a reflected and a transmitted component, a function of the reflection coefficient.

The contrasts in velocity and/or density which occur at the boundaries between distinct types or ages of rock usually coincide with major geologic markers. (**Fig 12.1**). These contrasts reflect seismic signals and the boundaries map subsurface structure. A seismic survey is the most economical source of detailed subsurface information. Its major use is mapping structure, but the seismic signal carries other components pertaining to rock type, porosity, and, under certain conditions, fluid content. Both seismic and sonic/density log data contain similar information expressed in a different form; a seismic trace measures reflection amplitude, which varies as the ration of velocity contrast across rock boundaries, whereas a sonic log measures rock velocity directly.

At least some of the log information can be extracted from seismic traces by reversing the procedure used to model synthetic seismic traces from sonic logs. Synthetic sonic logs, termed Seislogs, can be computed from specially processed seismic traces to unlock some of the additional geologic information buried in the signal. The sonic log and the seismic trace (**Fig 12.2**) are related by reflection coefficients (**Fig 12.3a**) which generate seismic reflections at boundaries in the subsurface where contrasts in velocity and/or density exist. If density, ρ, and velocity, V, are known, the reflection coefficient, R, at any boundary is:

$$R_i = (\rho_{i+1}V_{i+1} - \rho_i V_i) \;/\; (\rho_{i+1}V_{i+1} + \rho_i V_i) \qquad (1)$$

The contribution of density is often small and very nearly a linear function of velocity so it is not uncommon to compute the reflection coefficients from velocity only when density information is not available. For most cases this will be adequate. Therefore, a reasonable approximation of the reflection coefficient formula is simply:

$$R_i = (V_{i+1} - V_i) \,/\, (V_{i+1} + V_i) \qquad (2)$$

The value is always between +1.0 and -1.0, and will normally be small. This simple model ignores multiples. If a sonic log is integrated to a vertical time scale, sampled at some small interval, say 1 ms, the reflection coefficient may be calculated at the boundary between each pair of adjacent samples. **(Fig 12.3a)**. Therefore, reflection coefficients are merely a direct transformation of a sonic log integrated over time. They occur at the same interval as log samples, and the series yields the continuous earth signal of **Figure 12.4**. Whereas the sonic log measures velocity directly, the seismic reflection coefficients are ratios of velocity change.

As defined here they represent the perfect seismic reflection record that would be obtained from an ideal seismic

FIGURE 12.3

FIGURE a

A reflection coefficient exists at each boundary between two different velocities. For a uniform series of samples, the equivalent series of reflection coefficients can be calculated, with equal resolution.

FIGURE b

Joining all of the reflection coefficient amplitude values results in the earth signal, a continuous, ideal seismic reflection signal, free from distortion, transmission losses, and noise.

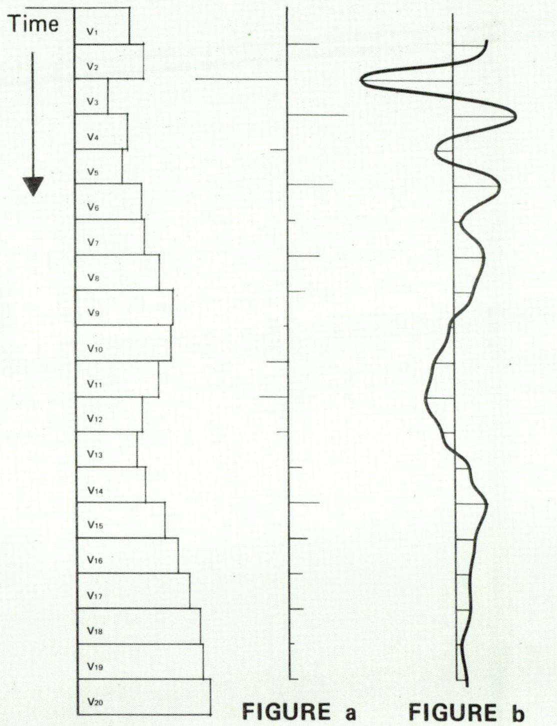

FIGURE a FIGURE b

LAYERED MODEL REFLECTION COEFFICIENTS

$$Rc_1 = \frac{V_2 - V_1}{V_2 + V_1}$$ Basic approximation

$$Rc_1 = \frac{\rho_2 V_2 - \rho_1 V_1}{\rho_2 V_2 + \rho_1 V_1}$$ Better approximation

ρ = Density

system, one free from transmission losses, noise, multiplies, and the other myriad phenomena associated with the recorded seismic signal.

A conventional seismic record is obtained when a wavelet, produced from a surface energy source, is reflected from each of the reflection coefficients below a point on the earth's surface, and is recorded with additional noise and distortion. The goal of seismic data processing is to eliminate all noise from the recorded seismic signal, and to compensate it for all transmission losses and for the distortion associated with the field process, to yield the original earth signal; i.e. the reflected coefficients, a goal which is not often reached, for a variety of reasons.

Nevertheless, if reflection coefficients are available, it is possible to carry out the reverse procedure, computing a synthetic sonic log from seismic reflection data.

Rearranging the terms of equation (2) gives ;

$$V_{i+1} = V_i \cdot (1+R_i) / (1-R_i) \qquad (3)$$

The first velocity value must be known, after which all subsequent values can be calculated iteratively. A simple additional computing step will convert the initial results to the more familiar dimensions of depth and transit time common to most borehole sonic logs.

A number of benefits are obtained from the conversion **(Fig 12.5)**. The sonic log form of information maps individual lithologic units as they appear on the sonic log, rather than the velocity boundaries of the seismic trace. The log parameters of depth and transit-time are in the geological, rather than the geophysical domain, making the data more readily comparable to other geologic information, including regional studies, well logs, and geologic data base files.

There are other advantages. Mapping geologic boundaries not only yields structural information, as does a seismic section, but has the advantage of supplying depth directly, without the velocity distortion common to time sections. More importantly it has the additional capability of simultaneously mapping changes which occur within the stratigraphic units.

Another more technical advantage with practical results is that the frequency distribution of the synthetic log spectrum approximates that of the sonic log more closely than does that of the seismic trace, so the high frequencies contribute to, but do not dominate, the subsurface section.

A complete inversion process would require reduction of the entire wave field, in a manner similar to migration, an extremely difficult task. Even the simple plane wave model process described here is fraught with a number of

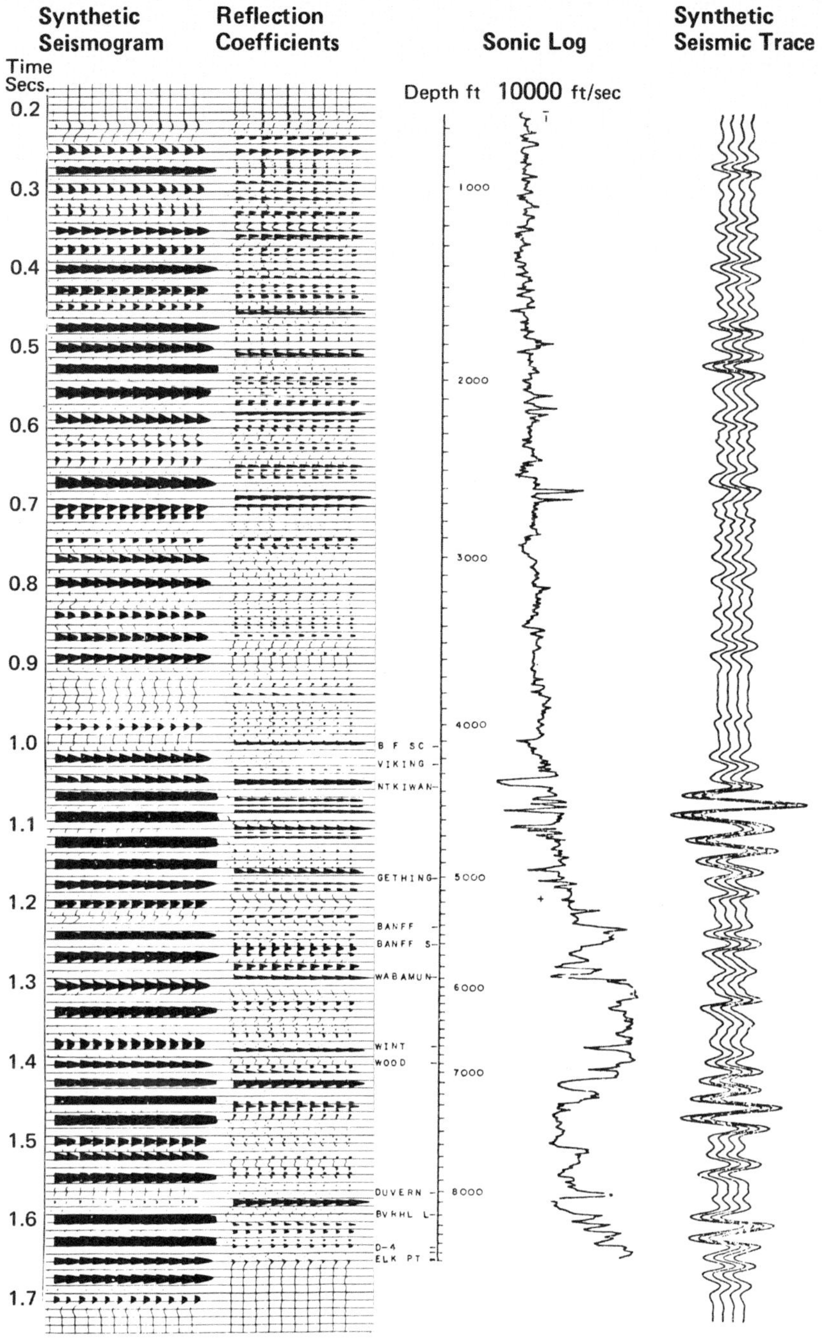

FIGURE 12.4

Stages in transformation of a sonic log through reflection coefficients to synthetic seismogram.

difficulties, most often related to the properties of seismic data.

For example the resolution of the seismic trace is far inferior to that of a sonic log, because of the earth and the recording instruments filter seismic signal frequencies to a relatively narrow bandwidth.

Figure 12.4 compares a sonic log, the corresponding series of reflection coefficients, and a model of the seismic trace. The original sonic log was sampled in vertical increments of depth but these have been integrated to yield time samples, spaced to match the uniform series of discreet closely spaced digital samples of seismic recording. The seismic trace has been plotted twice, first to the right of the log by a line plotter, to best demonstrate the amplitude relationship, and on the left, together with the reflection coefficients, by a conventional plotter. (Polarity of the seismic plotter display is reversed, to match the usual manner of display).

The reflection coefficients correlate closely to the log as they should, since they are derived directly from it. For instance, two small spikes on the log result in two small spikes on the reflection coefficients. The relatively uniform lithology logged from 3000 ft to 4000 ft yields weak reflection coefficients, while deeper, the coefficients faithfully follow varying changes in thickness and velocity on the log. Thus, the log. and the reflection coefficient series, which is merely a transform of it, have equal resolution. Resolution is lost only from the seismic trace. If the seismic trace is to be used as the source data to model the sonic log, resolution must be restored.

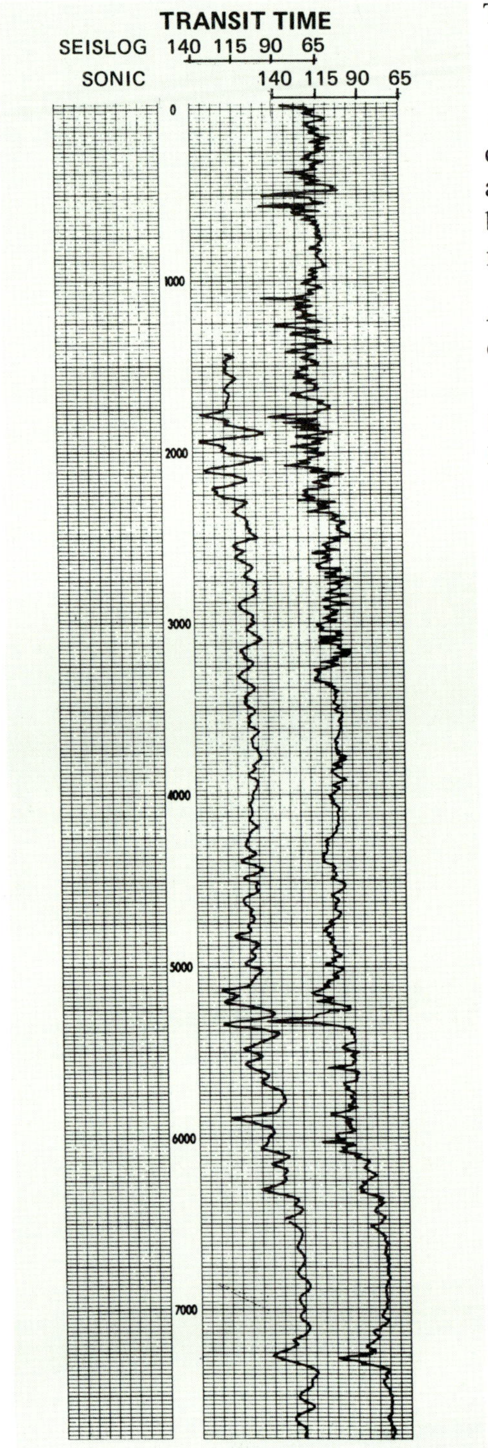

FIGURE 12.5
Comparison at a tiepoint: synthetic sonic log (left) borehole sonic log (right)

The difference in frequency content between the reflection coefficients and the seismic field signal may be confirmed by comparing amplitude spectra (**Fig 12.6**) of the two signals. Reflection coefficients are rich in all frequencies, particularly in the highs, but the seismic trace spectrum is dominated by frequencies confined to a relatively narrow band between about 20 and 50 Hz. Higher frequencies have been severely attenuated. At both ends of the scale, zero amplitude indicates frequencies eliminated from the spectrum.

An important procedure to help repair the damage to the spectrum is deconvolution. Each frequency component may be considered as a vector of length equal to the amplitude, directed to the angle indicated by the phase. Deconvolution can restore impoverished amplitude spectrum components **(Fig 12.6)** to their proper amplitude **(Fig 12.9)**, but not those which are completely missing. Some remnant of amplitude must remain in order to define the phase. Should the amplitude vector drop to zero, then the phase component is lost.

Like the amplitude, the phase angle of each frequency component can be distorted by noise or as a result of transmission through the earth. Phase variations can modify the appearance of a signal much more than amplitude. Hence, compensation for phase distortion is of considerable importance also. Conventional deconvolution methods can selectively restore amplitude components but make no comparable correction to the phase.

Newer deconvolution techniques, generally classified as wavelet processing, seek to determine and correct the phase distortion suffered by each frequency component in transmission through the earth and the recording instruments. These processes, when coupled with improved field techniques enable a reasonable, although band limited, approximation of reflection coefficients to be obtained from a field recorded seismic signal. This is important, for although band limited seismic signals serve very well to map geologic structure, only when amplitude and phase distortion have been removed and the frequency bandwith approaches that of the reflection coefficients does the displayed data truly represent the stratigraphy and lithology of the geologic column in detail. The capacity of any signal to carry information is a function of these components : signal length, frequency bandwidth, and amplitude (**Fig 12.8**). When most of the high frequency components are attenuated, or are missing from the seismic signal, only the major reflectors are visible on the trace. Detailed seismic reflections are often weak, or infrequent.

As a result, many geophysicists and geologists model the geologic section by a few widely spaced discreet reflection coefficients, corresponding to

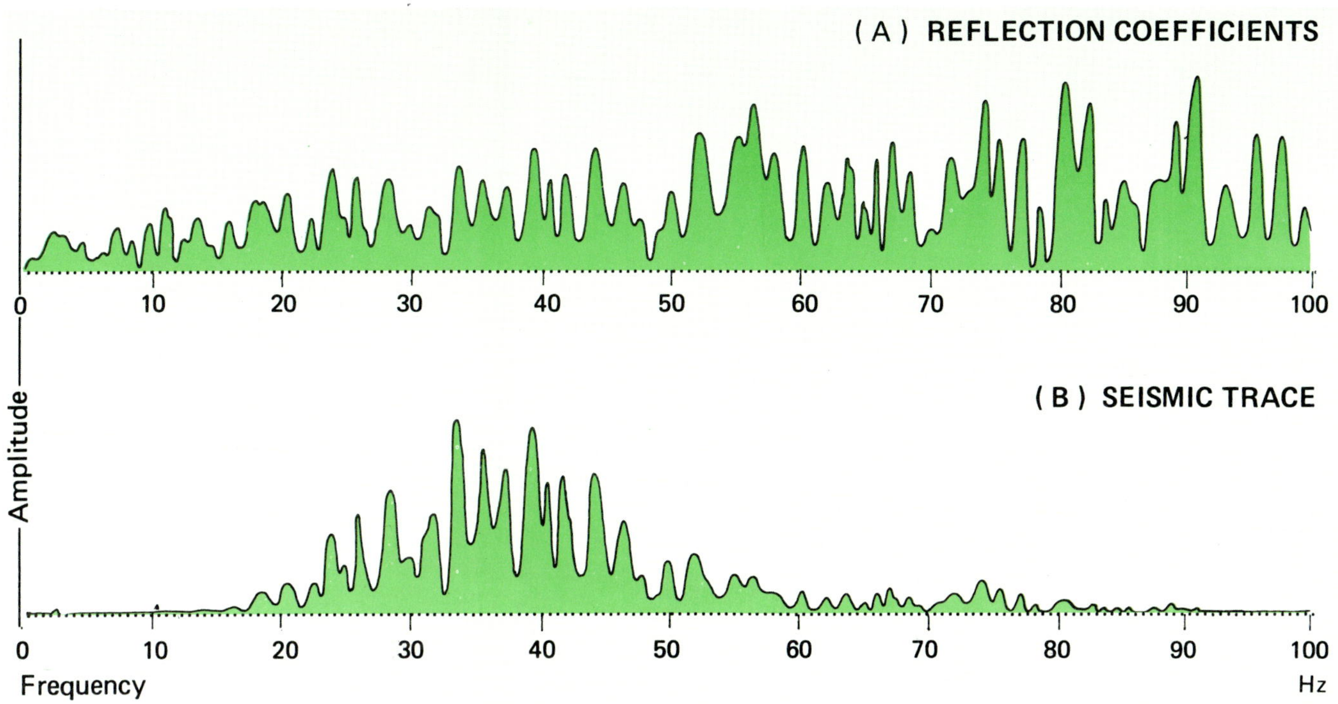

FIGURE 12.6

Amplitude spectra measure the relative amplitude of each frequency component in a signal. Reflection coefficients are rich in all frequencies, but many frequencies on a seismic trace are missing or attenuated.

major reflectors or principle geologic boundaries and ignore the information between. This patently does not represent the real case, for as we have seen the series of reflection coefficients, which generates the seismic trace in the earth, is a continuous signal which has the same resolution as the sonic log, and indeed has a one-to-one relationship to it.

Unfortunately, in many geologic sections, when seismic reflection signal bandwidth is increased, and the spectrum weighted to approximate that of reflection coefficients, so much high frequency detail becomes visible (**Fig 12.9**) that the reflections become dominated by it, and the seismic sections are difficult to follow or to interpret. At first glance these two sections do not appear to tie. But refer to the model of (**Fig 12.4**), and compare the differences between the reflection coefficients and the synthetic seismogram to the differences between the two types of process output.

Between 0.7 sec and 1.0 sec on the model, the reflection coefficients are very small, faithfully tracking the relatively low velocity contrast within the shaly section traversed by the sonic log. The synthetic seismogram, because many frequency components have been lost or attenuated, cannot correctly reproduce the response, and develops several high amplitude reflections. A similar comparison exists near 1.0 sec on **Fig 12.9**.

Now, on **Figure 12.4**, refer to the interval between 1.0 and 1.2 sec. Here the reflection coefficients vary appreciably in spacing and in amplitude, as they follow the corresponding changes of the synthetic log. The synthetic seismogram response is quite different. It almost oscillates at a uniform frequency through the series, changing amplitude smoothly through a rounded envelope.

We intuitively relate reflection amplitude to velocity contrast, but evidently this is not necessarily the case.

The reflection amplitudes modelled in the seismogram of **Fig 12.4** are not a direct function of velocity contrast.

Once again, compare the differences of the model to the real data of **Fig 12.9**, this time to the interval below about 1.3 sec. Similar differences (obtained from a similar geologic section to that logged in **Fig 12.4**) confirm that the apparently noisy broadband data may provide a more faithful mapping of reflection amplitudes than does the conventional processing. The conventional processing provides greater continuity and is far preferred by the interpreter for structural mapping, but it does not yield a good representation of the reflection coefficients.

A similar situation usually exists on data which has been processed for purposes of conventional interpretation since the spectrum, although broadband, is likely tapered toward the high frequency end (**Fig 12.7**). In older data, the problem rarely occurred because high frequencies were severely attentuated in the acquisition or filtering processes.

Inversion is able to use such data because the response of the integration process is to boost the low frequencies and progressively attenuate the high frequencies to yield a spectrum similar to that of a sonic log. The important difference between the log spectrum and that of reflection coefficients is the relative amplitude distribution. A sonic log has excellent resolution, and its high frequency content is entirely adequate for any petroleum application, yet its spectrum, relative to that of reflection coefficients, has a pronounced decay in amplitude toward the high frequency end.

Inversion of the approximate series of reflection coefficients in fact yields acoustic impedance coefficients, since density influences the data recorded in the field:

$$\rho_{i+1}V_{i+1} = \rho_i V_i (1 + R_i) / (1 - R_i) \qquad (4)$$

If the first acoustic impedance coefficient is known, the result is an acoustic impedance log, which is a relatively uncommon log form. But if the acoustic impedance can be compensated for density, the results model the sonic log, a widely used, commonly accepted data form.

Precise density corrections require an equally precise definition of rock properties. But, since this information is never known, (except possibly at borehole ties) some data dependent compensation must be applied.

Gardner et al (1974) developed the commonly accepted equation relating density and velocity:

$$\rho = .23\,V^{.25} \qquad (5)$$

from which follows:

$$V = (\rho V / .23)^{0.8} \qquad (6)$$

the relation of acoustic impedance to velocity. While this relationship fits older consolidated rocks very well, it is less suited to very young sediments of the type common to exploration of continental shelf areas offshore.

Logging handbooks published by well logging companies contain tables of density and velocity for several rock types. If the appropriate ρV values are computed and plotted against velocity (**Fig 12.10**) the relationship $V = .308\,\rho V + 3406$ (7) can be obtained, which suggest a simple scaling and shifting operation will reasonably compensate for density. In practice the factors may vary, since the seismic data is band limited and the amplitudes will not be true.

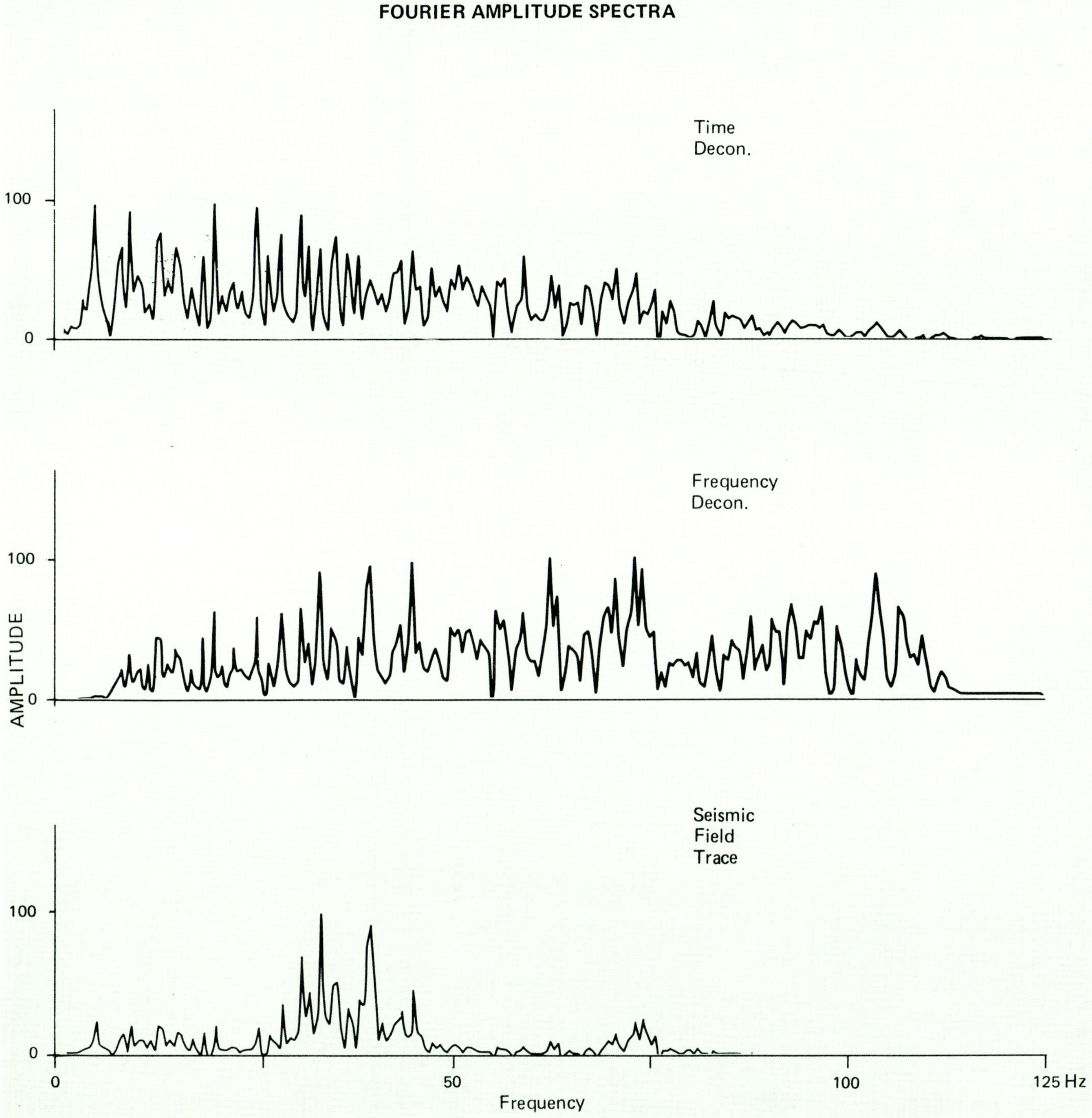

FIGURE 12.7

The information capacity of a signal is the product of its amplitude, frequently bandwidth, and length (see Fig 12.8). Deconvolution can restore diminished capacity of a seismic field trace, but different deconvolution processes may yield different results. Both the time domain and the frequency domain outputs shown have the same bandwidth but the corresponding seismic traces will be quite different, due to the different weighting of the frequencies.

Also, the linear form fits observed values for shallow unconsolidated sediments slightly better than the exponential form. A smaller shift constant may fit the Gardner curve and the values for older rocks slightly better.

Another serious obstacle to routine reverse modelling process is to compensate for the loss of low frequencies from the field data. Reflection coefficients and the sonic log are merely transforms of each other, so frequency filters will approximate the filtering action of the earth, and recording system can, with proper procedures, (Branisa, 1974), be applied directly to sonic logs to show their effects on the basic data.

The seismic signal has a relatively narrow frequency bandwidth. The lack of high frequencies may be seen to limit resolution (**Fig 12.11**). **Figure 12.12** shows the information normally lost when the very low frequency components are absent. The normal low frequency cutoff for many seismic process operations is near 10 Hz. The sonic log on the left in **Figure 12.12** has been filtered to remove frequencies from 0 - 5 Hz. The remaining signal, from 6 - 250 Hz, has good thin-bed resolution, but no basic velocity structure. Although the amplitude values are scaled in dimensions of velocity, the signal oscillates about the zero axis, with positive and negative velocities. Summing the two bandpass filtered outputs will exactly equal the original log.

Hence, filtering high frequencies degrades resolution, but filtering low frequencies removes the basic velocity structure of the log. Therefore, no matter how well band limited seismic field data is processed, inversion will yield only a series of acoustic impedance coefficients (or velocity coefficients if corrected for density), without yielding absolute velocity values, unless some way is found to replace the missing low frequency velocity components.

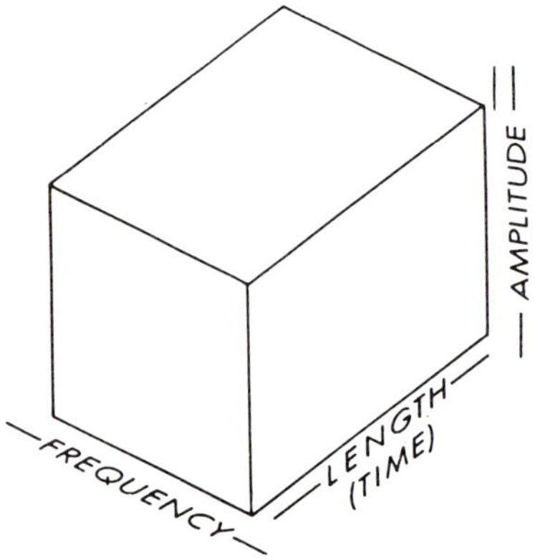

FIGURE 12.8

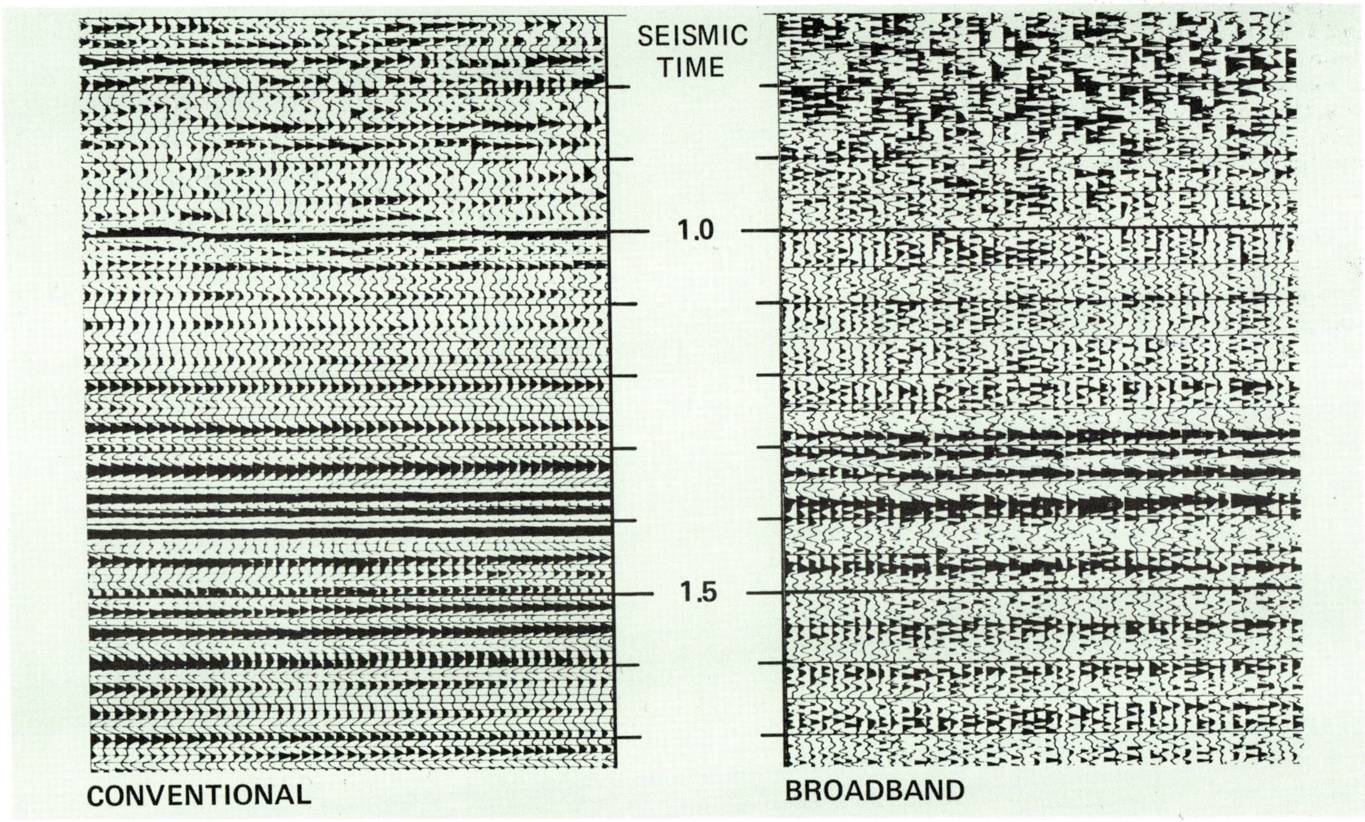

FIGURE 12.9

Broadband seismic recording and processing may not appear to tie well to conventional seismic data but should provide a more faithful representation of the subsurface. Compare this example to the interval between 0.5 and 1.4 secs of the model data of Figure 12.4.

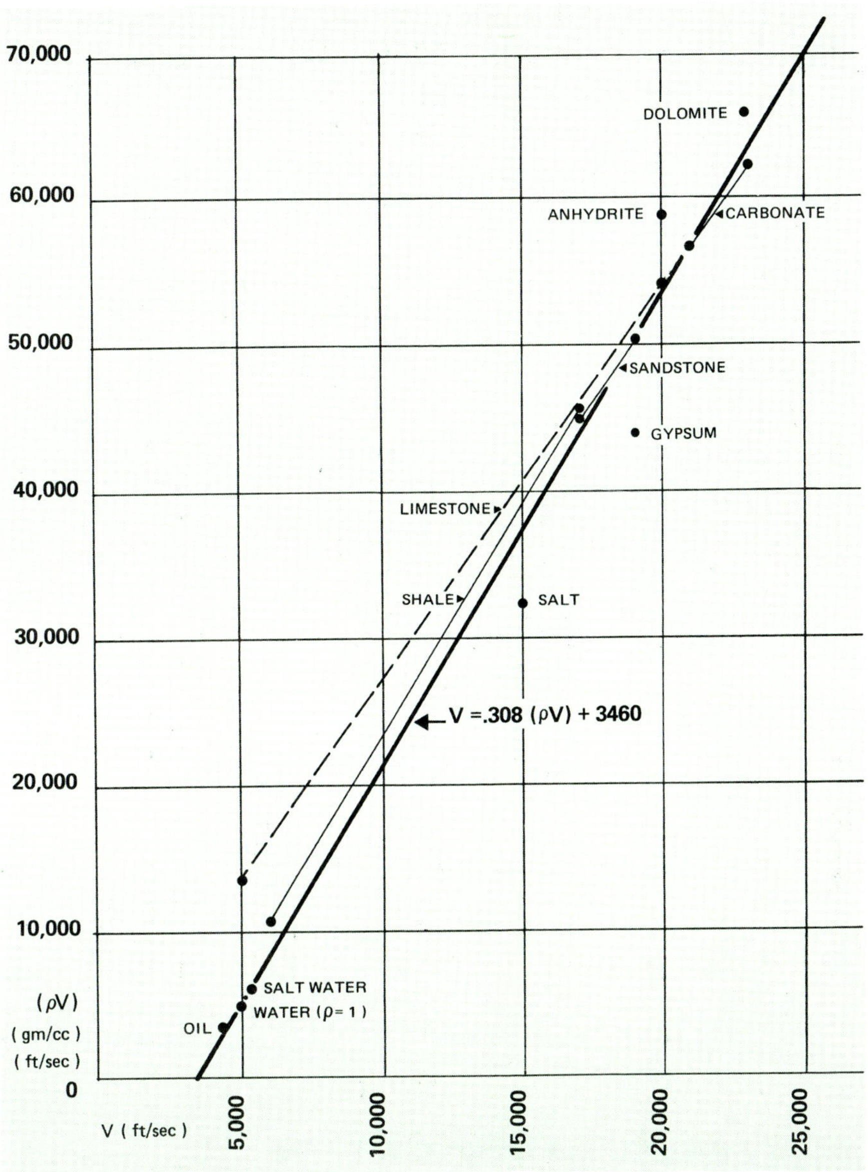

FIGURE 12.10

Velocity - Density relation of common rock types.

Velocity determination is also a prerequisite for seismic data processing, and is often a useful aid to interpretation. It can be obtained from a number of sources, including sonic logs, borehole velocity surveys, and it is carried on seismic recordings in the normal moveout distortion arising from the geometric properties of the spread. Several standard digital computer techniques, (see section 8), obtain stacking velocity information from seismic traces.

Velocity measurements, whether obtained from stacking velocity analyses, or from check shots in boreholes, are, on the average, usually spaced at relatively large time sample intervals, say 0.2 to 0.4 sec, corresponding to the spacing of major reflection events on the seismic section. By convention they are displayed as a series of interval velocity steps similar to that shown in **Figure 12.13**. Note that this concept, of widely spaced boundaries separated by uniform interval velocities, reinforces the erroneous concept of a few widely spaced reflection coefficients located only at the points where velocities change. Clearly this model is incorrect. Under this form of sampling, the model of a constant increase in velocity with depth would be replaced by one having few large reflections occurring only at widely spaced sample points.

Such long established intuitive concepts of velocity structure violate well documented sample theory. A better way to represent widely spaced samplings of continuous functions is to filter the source data to the appropriate low frequency limits. It is precisely these low frequencies we wish to obtain.

The highest frequency, F_n, that can be defined correctly in this manner for any given time series sample spacing, Δt, is $F_n = 1 / (2\Delta t)$. For 0.1 sec sample intervals the upper frequency limit that can be obtained is 5 Hz. But even at the lower frequency of 2.5 Hz, filtered velocity analysis results provide a closer fit to the log (**Fig 12.14**) than do the stepped interval velocites of **Figure 12.13.**

Velocities derived from normal moveout relationships are stacking velocities, which are nearly always higher than

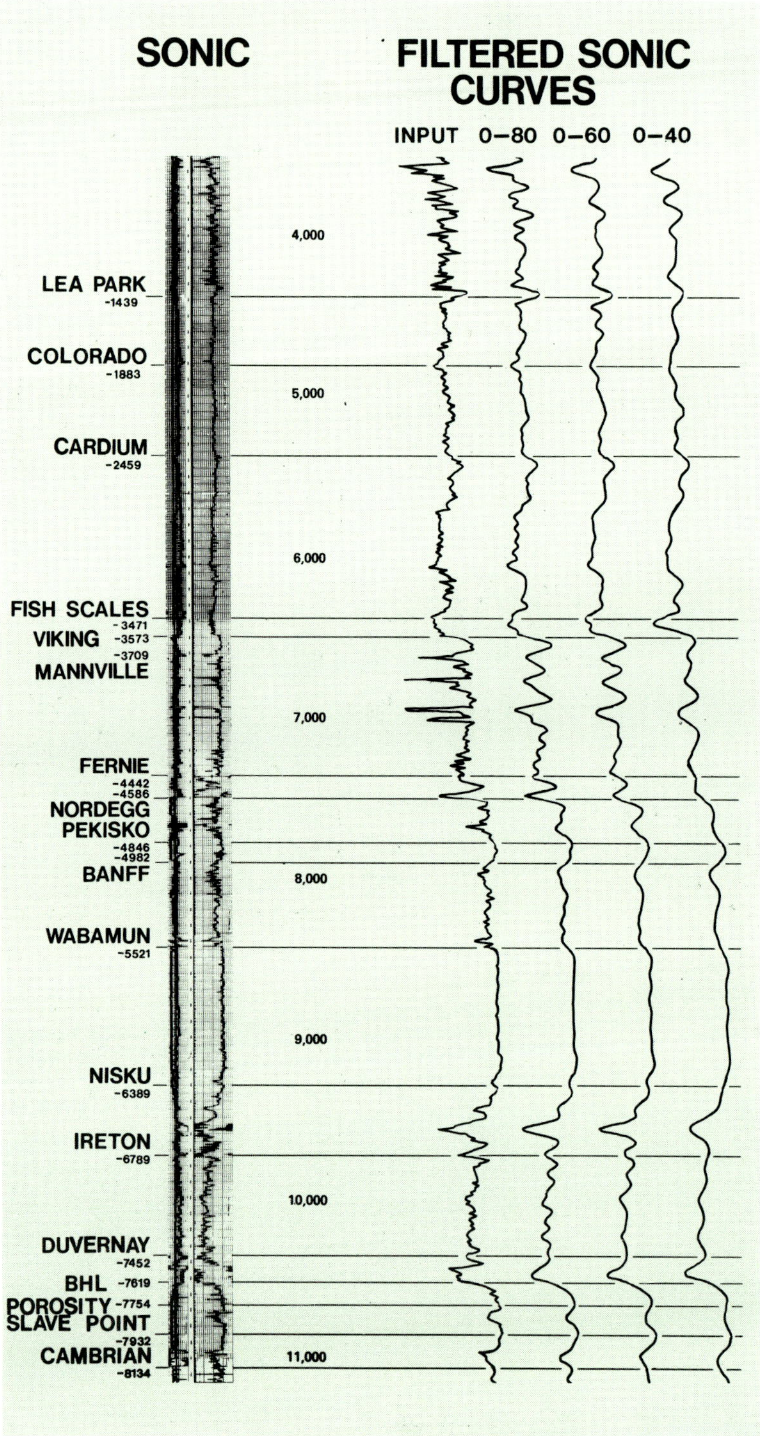

FIGURE 12.11

Loss of high frequency decreases resolution on sonic logs.

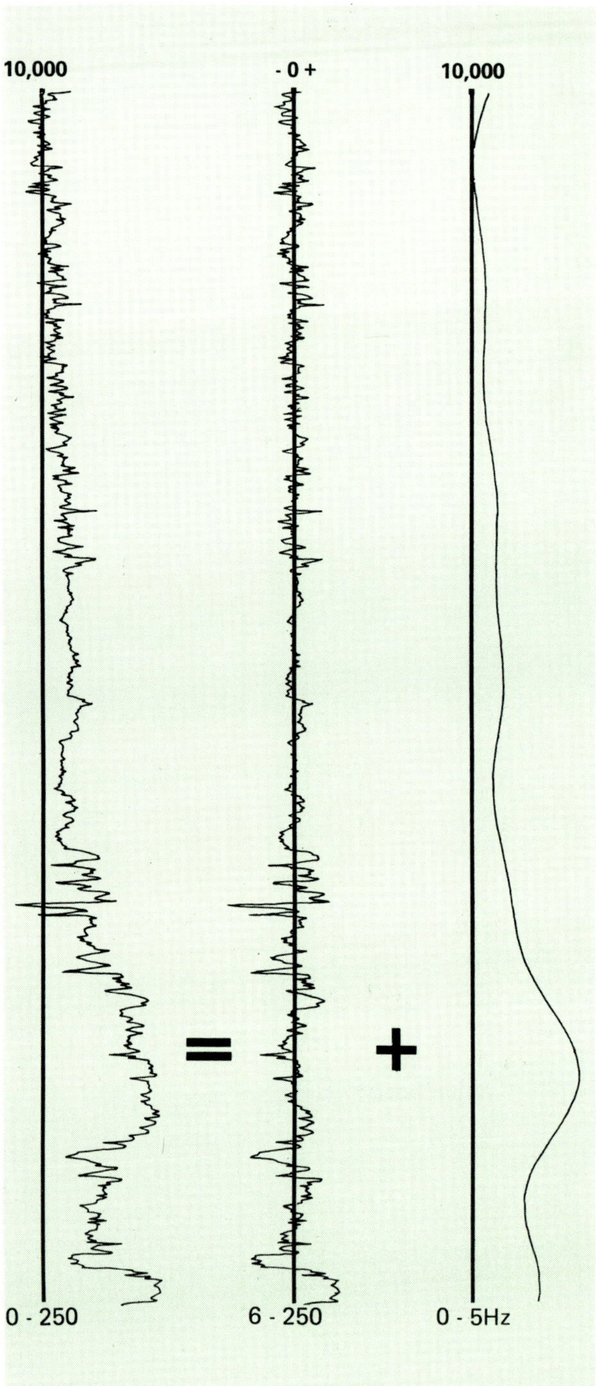

FIGURE 12.12

FIGURE 12.12

The basic velocity structure of a sonic log is provided by the very low frequencies, while the higher frequencies provide the detailed resolution.

true vertical average velocities, and must be corrected accordingly. In near-flat sections of fairly homogeneous material and relatively short recording spreads, the Dix (1955) method is adequate. For dipping sections, or long recording spreads, a ray tracing method, as described by Gangi, and others, is preferable.

The precision of any individual stacking velocity determination is very low. But if statistical surface fitting techniques are applied to the results of several adjacent filtered stacking velocity analyses, the resulting values should be more reliable than the raw data points. Smoothed iso-velocity values may be contoured to develop a continuous velocity section.

The velocity surface derived in this manner will yield a velocity value for any t,x combination. Therefore, the interval velocity may be calculated between each increment of time to yield a continuous interval, or instantaneous, velocity surface.

A vertical profile from the instantaneous velocity surface should represent the low frequency velocity structure of the sonic log at that point. **Figure 12.15** makes this comparison. The sonic log is on the left of the figure. Next is the same log filtered to remove all frequencies above 6 Hz. On the right is the derived velocity profile. With some slight differences, the two match quite well. Hence, the velocity surface provides a reasonably reliable approximation to the very low frequencies that were missing from the seismic data.

Thus, if the field seismic data, above some low frequency cutoff, are broadband wavelet processed to a reasonable approximation of reflection coefficients, corrected for density, reverse modelled, and suitably scaled, the resulting higher frequency velocity coefficients can be used to modulate the lower frequency information derived from velocity data to produce a synthetic sonic log.

At this stage the output is still in the seismic domain, with dimensions of time and velocity. An important final step transforms the dimensions into the geologic domain of depth and transit-time. **Figure 12.5** represents the final output of all these processes and procedures, the synthetic sonic log commonly termed a Seislog.

Quality of the final Seislog will be a direct function of the quality of the seismic data input and the processing applied to it. The acid test of the process is to correlate a sonic log, plotted to the same scale as the Seislog, at a tie point. The quality of correlation provides a direct estimate of the quality of the data and the success of the inversion process.

Inevitably, comparison reveals the seismically derived log to have less resolution than a borehole log. This is to be expected, since the upper limit of the seismic frequencies recorded in the field cannot hope, at the current state of the art, to match those of the sonic log. Nevertheless, resolution is much improved over the conventional seismic section.

Correlation of the sonic to the synthetic log may be improved by high-cut filtering the sonic to approximately the upper frequency limit of the seismic input data. A poor correlation does not necessarily indicate a fault in the basic procedure, since the output is merely a direct transformation of the existing seismic data. It may indicate otherwise unrecognized problems in the basic seismic data, or in the data processing, or in the field instruments.

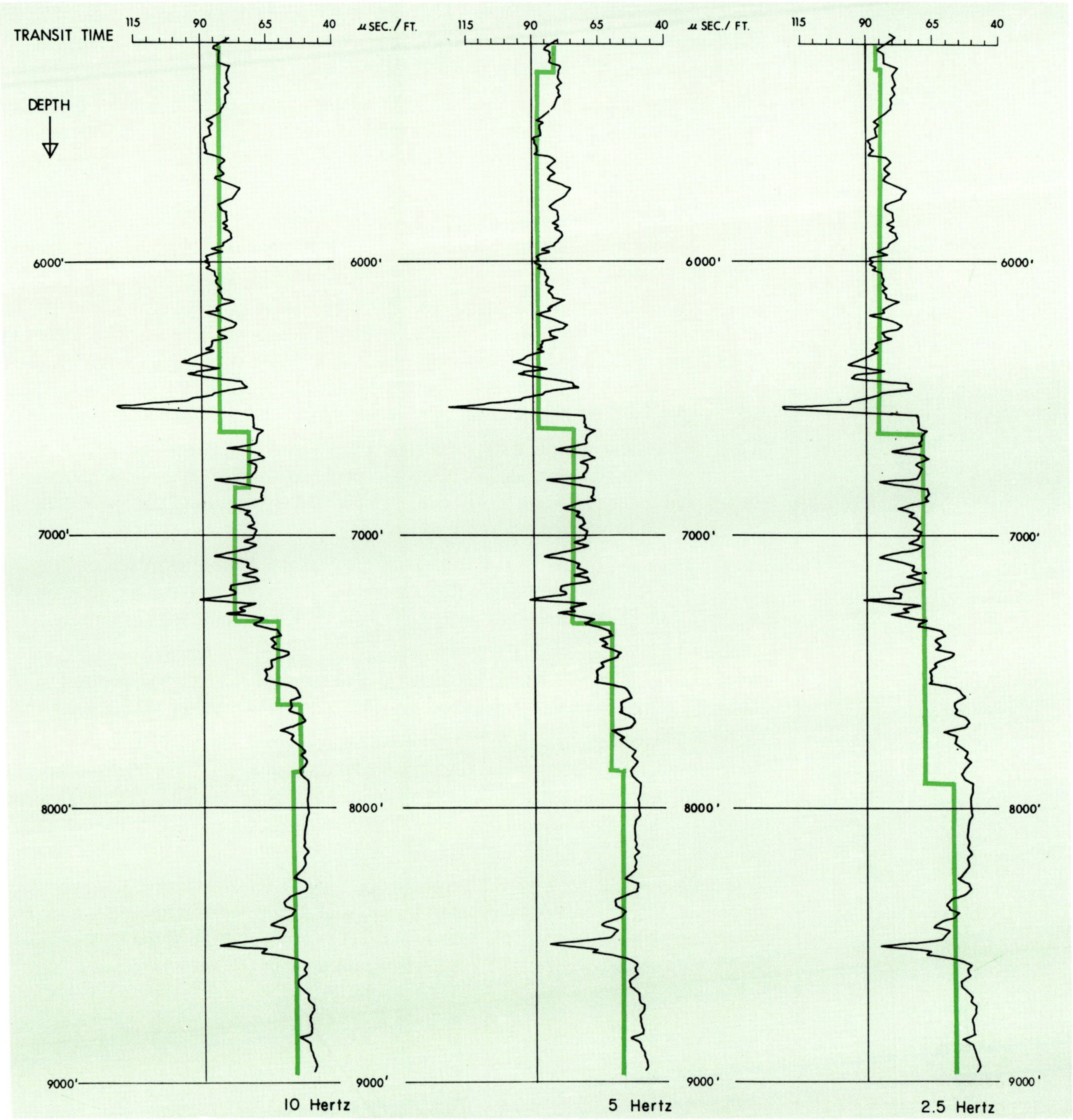

FIGURE 12.13

Comparison of interval velocity to log

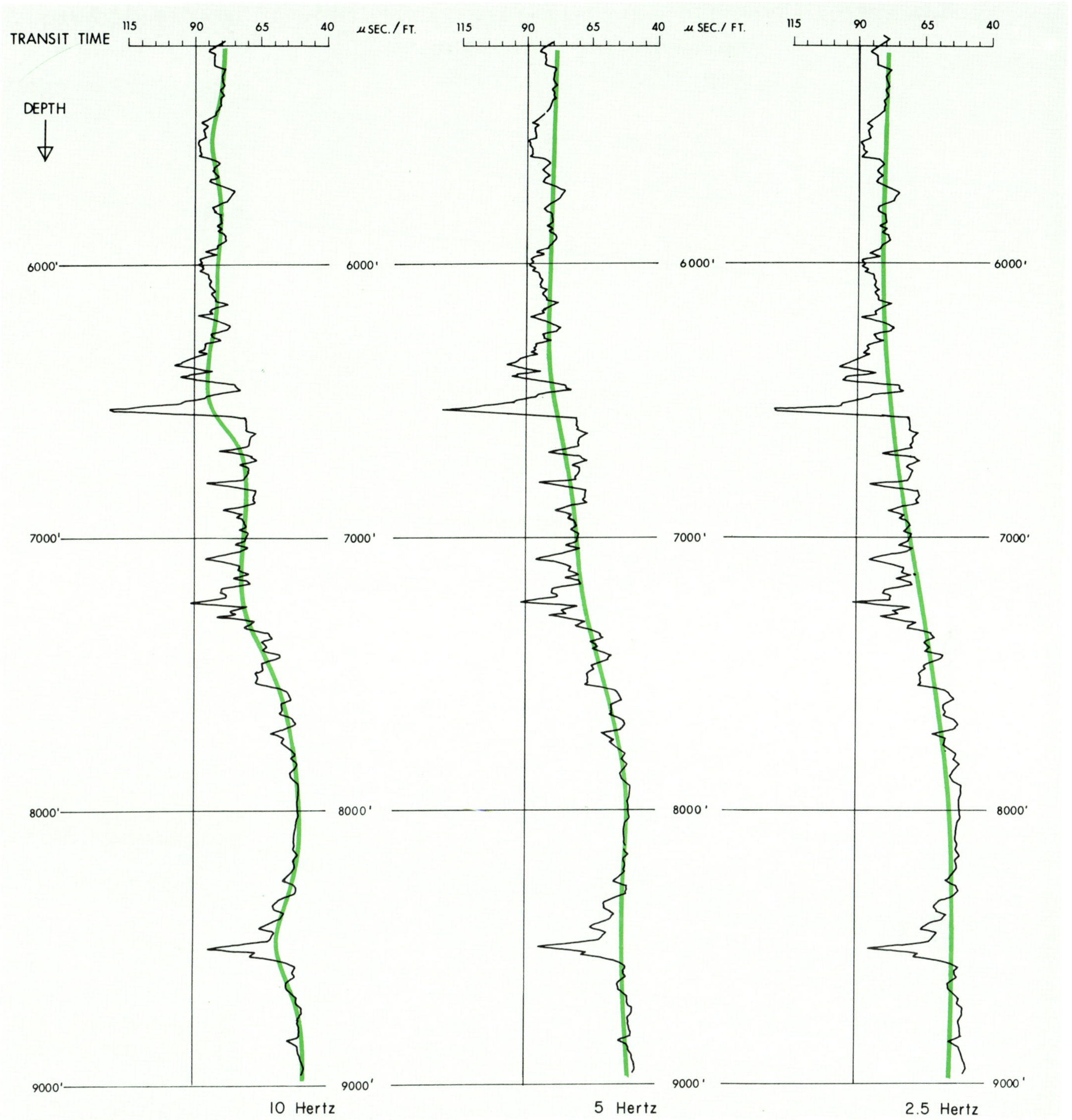

FIGURE 12.14

Smoothed log

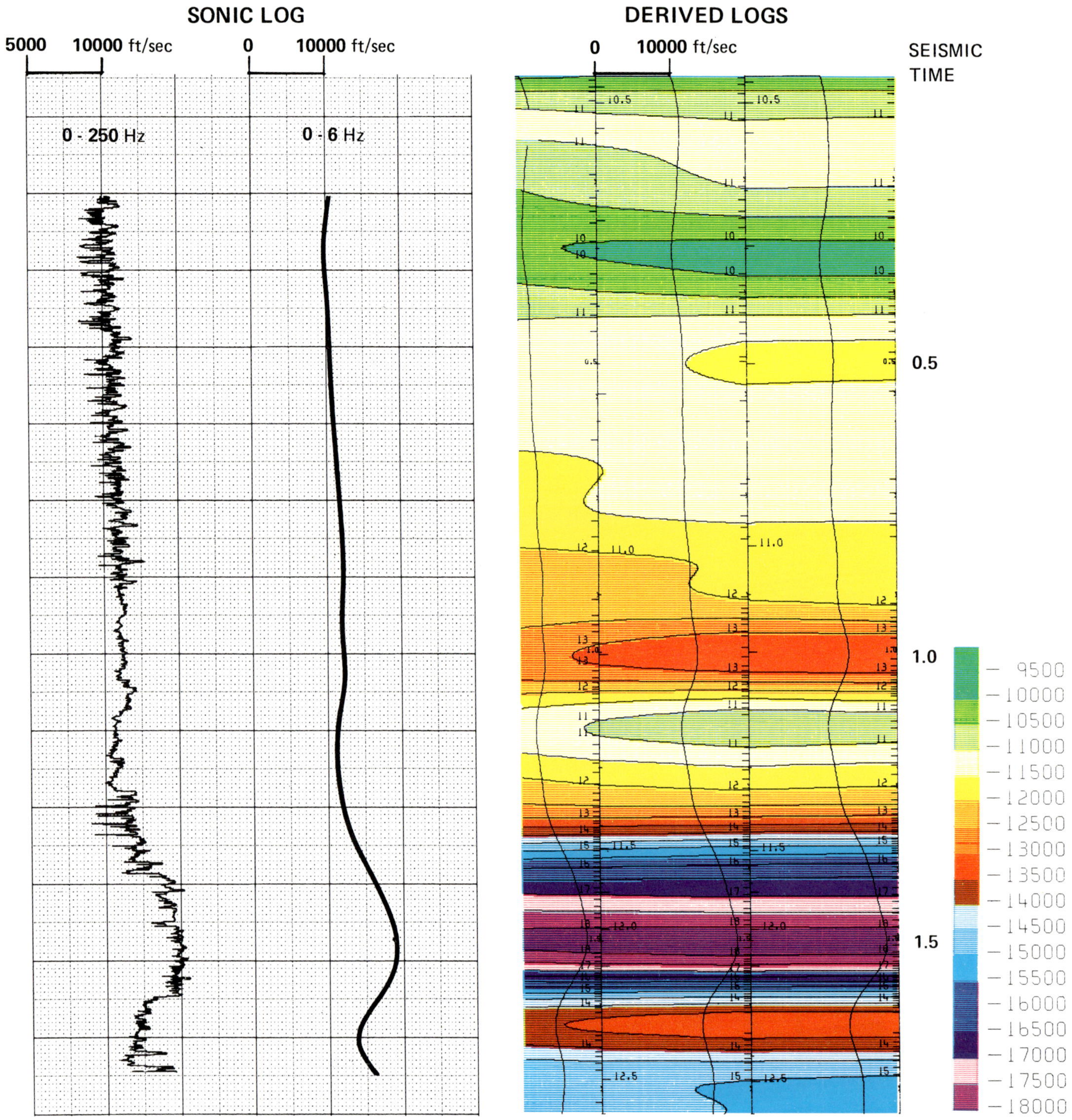

FIGURE 12.15

Comparison of low frequency components from well data to the low frequency of the seismic data.

Display

For an interpreter, transition from the relatively simple conventional seismic section to the information-laden Seislog section is not easy. Most of the information present on the Seislog section differs from on the conventional section and there is a great deal more of it. Conventional seismic display techniques will not be adequate to demonstrate the significant geological data contained in the output. New skills and new techniques must be developed to recognize and utilize the new information, interpret it, and combine it with conventional structural information into a complete and cohesive geological model. Experience with these new techniques and data must be acquired by the interpreter if the full potential of the method is to be realized.

The most important initial concept is to recognize the transformation of the data out of the geophysical domain into the geological and engineering domain. This makes it much easier for the geologist and/or engineer to work with but creates a problem for any geophysicist who attempts to interpret it with the usual seismic techniques. Conventional seismic data measures one basic quantity; reflection travel time. The significance of this information can be expanded by interpreting changes in amplitude and reflection character in terms of geological conditions, but the only quantitative measurements placed on a seismic map are reflection times. Seismic time measurements never correspond directly to geologic depth since they are always distorted by velocity. Most of this distortion should be removed in the inversion process, so the vertical measurements made by Seislog interpreter should more closely approximate the true geologic depth measurements of geologists and engineers. An additional geological parameter, formation transit-time, is also measured. Changes in transit-time within a formation which occur independently of structure likely represent changes in lithology and/or porosity within that interval.

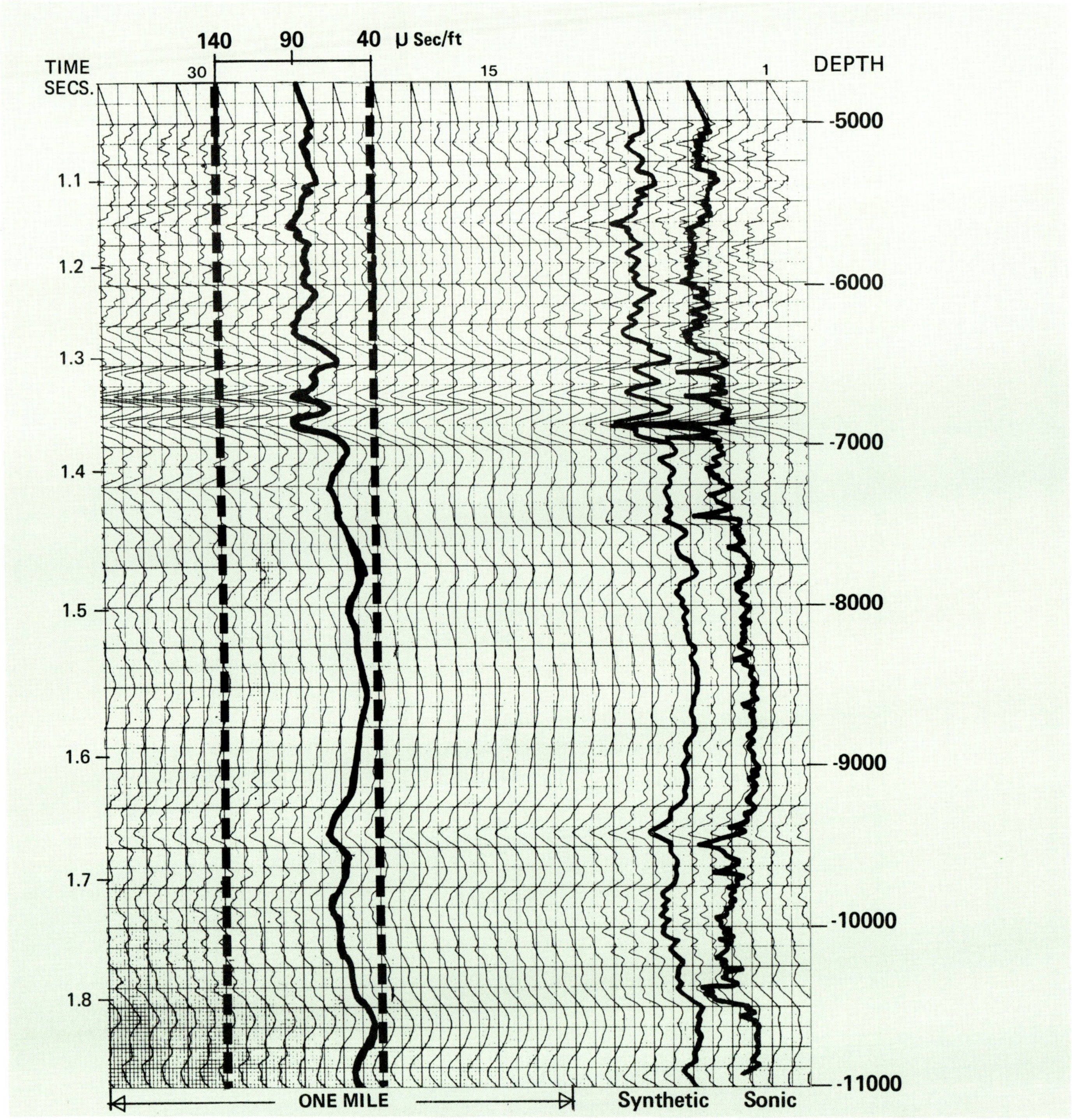

FIGURE 12.16

A continuous Seislog section produced from reflection data. Each synthetic sonic log has a corresponding transit time scale. Comparison of the synthetic to a borehole sonic log is demonstrated on the right. Comparison of the synthetic log to a borehole log is the acid test of the inversion process. Differences between the two are explained in the text.

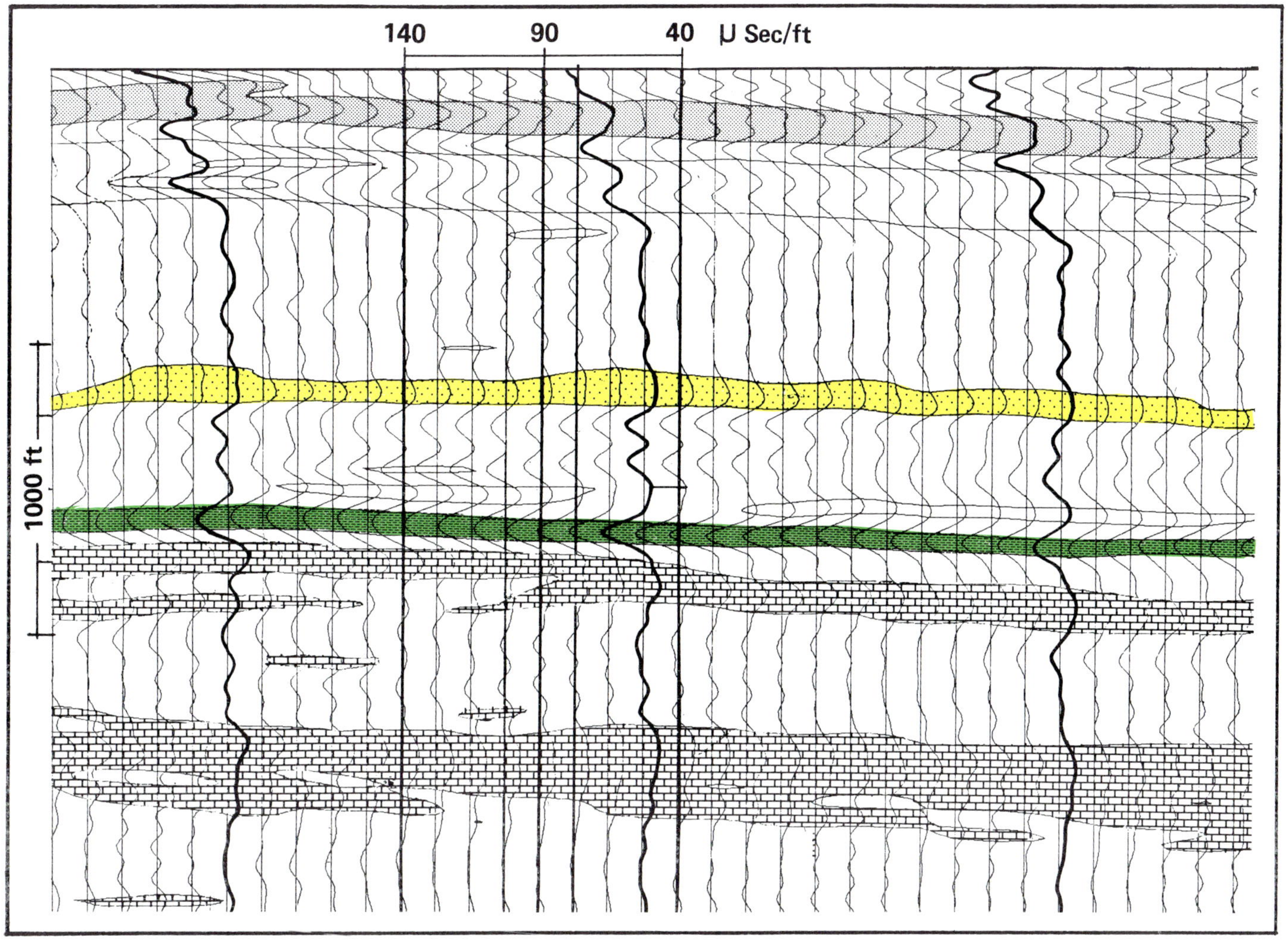

Figure 12.17

Contours of isotransit - time values map contacts between units of contrasting velocity and outline individual rock members in the section.

Normally, the inversion of all of the traces on the seismic section will result in a synthetic log section similar to that shown in **Fig 12.16**. Spacing is close enough to permit detailed correlation from log to log with a high degree of confidence. Lithologic units may be mapped by character correlation, and stratigraphic interpretation could be done in the conventional manner of correlating from unit to unit, in much the same manner as a geologist constructs a subsurface stratigraphic section. Visual correlation of equivalent units is an accepted practice for the interpretation of the Seislog sections, but since each trace has definitive velocity values a more deterministic method can be applied by mapping lines of constant velocity, a process which lends itself readily to automation.

Figure 12.17 is a small portion of a synthetic log section, expanded in scale, and with most of the grid lines removed for clarity. A central log trace has been darkened for emphasis and the appropriate transit-time scale has been superimposed on it. A transit-time scale, similar to that superimposed on the heavy log trace, corresponds to each individual log trace.

Velocities in the upper part of the logs tend to fall to the left of the 65 μsec/ft line. Deeper in the section the curves shift right rather abruptly to track approximately 55 μsec/ft. Evidently some change in geology has taken place, and should be mapped. On a borehole sonic log, the boundary at a major velocity contrast would be a very sharp step, but due to limited frequency bandwidth of the synthetic logs, the exact contact is not so clear. The effect is very similar to borehole logging with long source-receiver spacing. Sharp boundaries logged with long source-receiver spacing are spread over several feet because the velocity of the rock on either side of this boundary is averaged over the length of the tool span. Some procedure is needed to locate the boundary.

FIGURE 12.18

A machine contoured Seislog section. Contours of isotransit-time outline geologic members and map facies changes within the section.

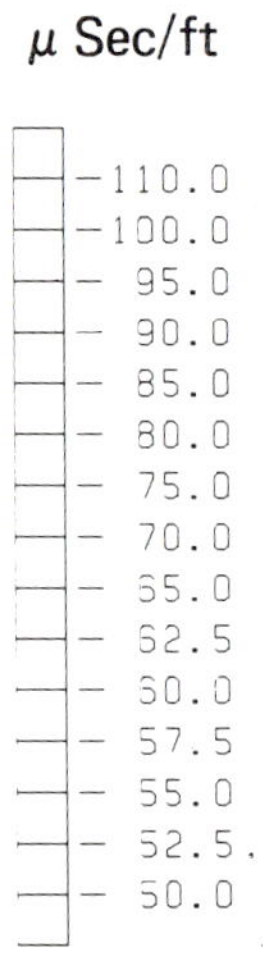

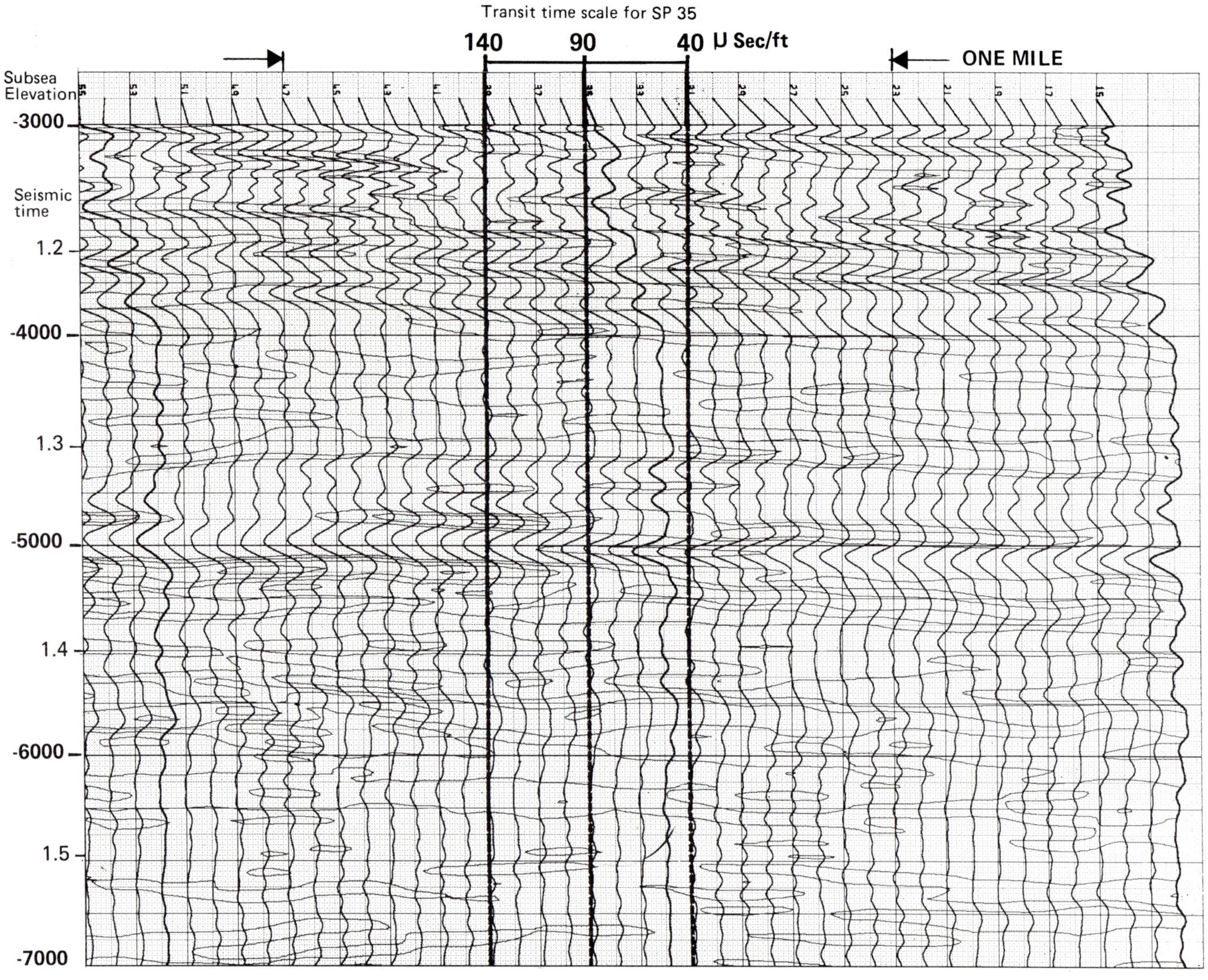

A useful rule of thumb is to place the boundary at the inflection point, or about half way between the limits of deflection. In this case, the corresponding mid-velocity value appears to be about 60 μsec/ft. This value can be marked on all of the synthetic logs and the points joined by a single contour. The resulting isotransit-time (velocity) line maps the geologic boundary very precisely, in this case the contact between Cretaceous clastics and Paleozoic carbonates. In practice, the process can be automated, and a computer driven plotter can draw in the formation top. Contouring in this manner can provide true structural information from which normal seismic time distortion has been eliminated.

Of perhaps greater value for the stratigrapher, the production geologist, and engineer, is the ability to delineate individual rock members in the geologic section. The assumption is made that any given lithologic unit has a characteristic rock velocity which will remain constant until either the rock composition changes, its porosity varies, or there is an appreciable change in depth of burial. A change in rock composition or porosity, which may have geological significance, should be revealed by a change in the transit time. Individual rock members can be outlined by contouring the boundaries at the top and the base of each member having a velocity contrasting with that of adjacent members.

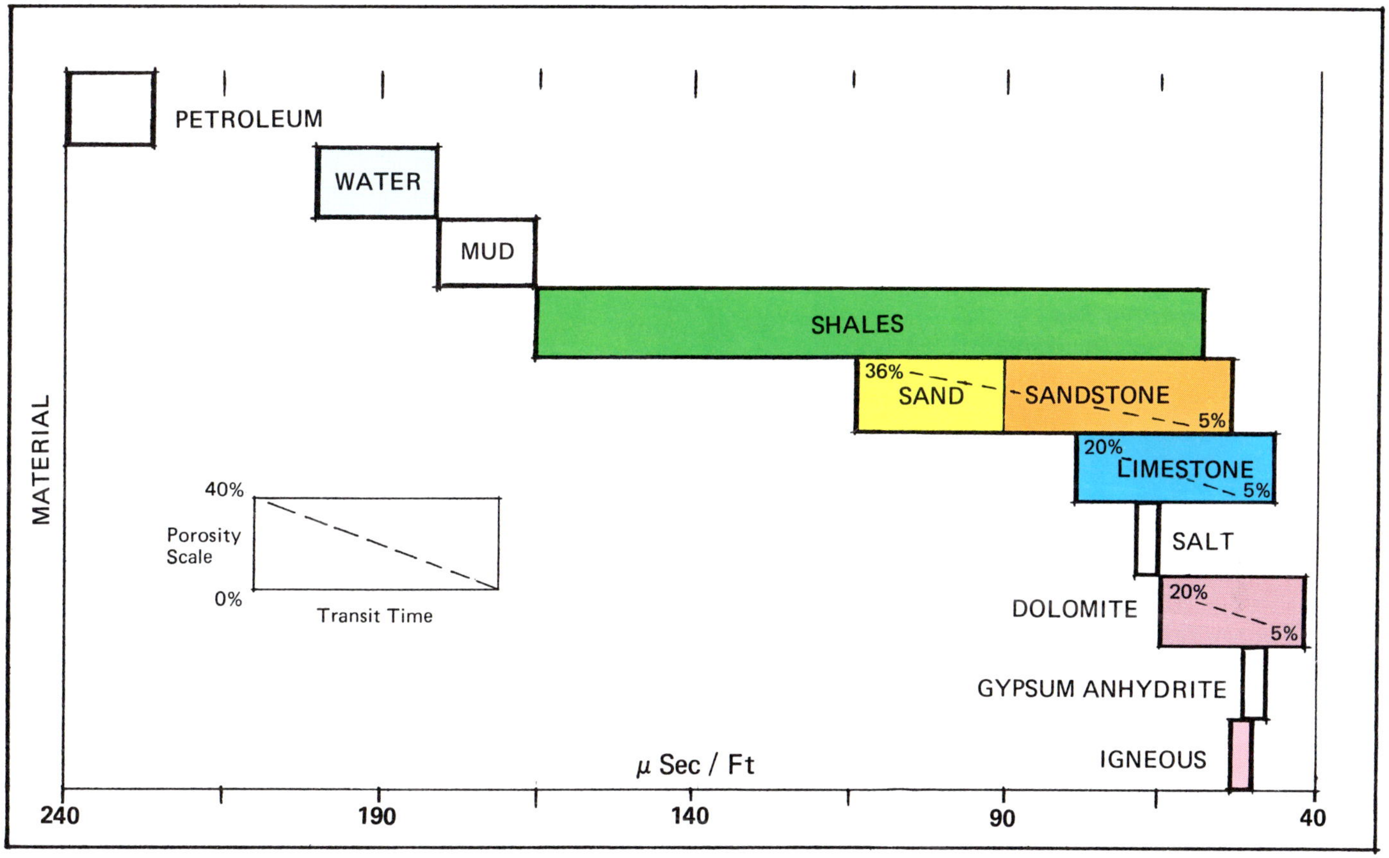

FIGURE 12.19

Normal range of transit time for several materials.

DISPLAY

Bedding is relatively flat in the example, so the influence of increased depth on velocity can be ignored. Below the major boundary just mapped, the logs more or less track the 52.5 μsec/ft reference line. The corresponding velocity for this transit-time is about 19,000 ft/sec, a typical value for carbonate rocks. About half way down the section the log makes a small excursion to the right of the vertical grid line. A continuous contour line has been drawn through the 50 μsec/ft transit-time value on all logs. One contour is at the top, and another at the base, delineating a single, continuous, relatively homogeneous lithologic unit which, under the assumptions made, represents a single rock unit or member in the section.

The top of the unit is rather irregular, a condition which might possibly be attributed to noise, but the base of the unit, no more than 100 ft deeper, is quite smooth. At 50 μsec/ft, an interval of 100 ft corresponds to a two-way travel time of 10 ms. Two or three members this thick would be included within one reflection cycle of conventionally processed data. If noise caused the irregular contour on the surface of the member, it would have affected only the top, but not the base, of this member occupying a fraction of a single reflection cycle on a normal seismic section. Such an occurrence is extremely unlikely. In all probability the contours faithfully outline a single geologic unit within the sedimentary section.

FIGURE 12.20

The velocity within shale varies with depth of burial and geological time.

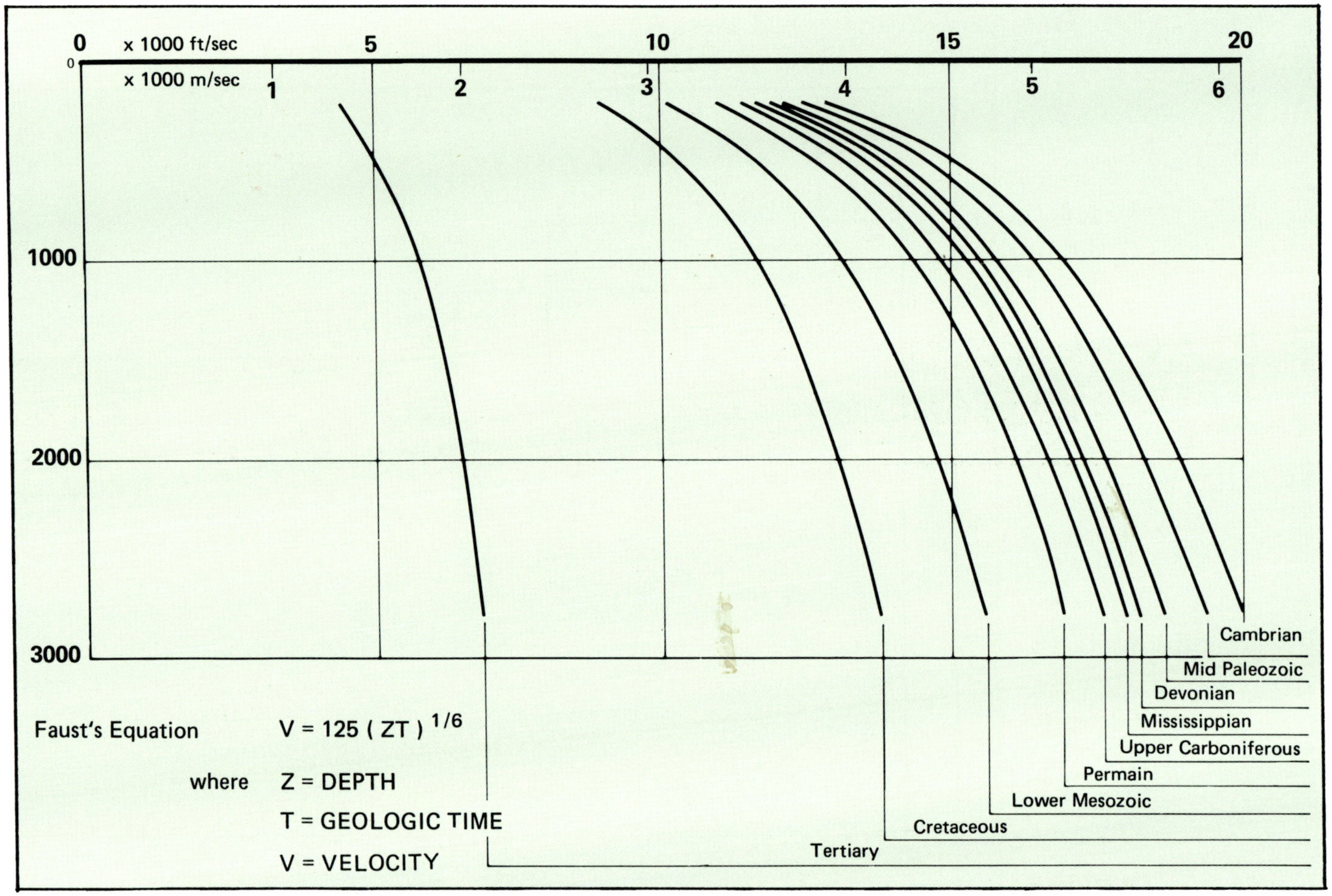

Deeper in the section the log curves shift abruptly left, to a lower velocity. The top and base of a low velocity unit, probably shale, is mapped by two, slightly converging, quite similar contours. Note how closely these two contours track, even across slight vertical deformities, confirming relatively noise-free data mapping small changes in geology very precisely.

Immediately below, a 50 μsec/ft contour carries smoothly across the left half of the section, then rather abruptly drops almost 100 ft probably signifying a lateral facies change, rather than structure. Bedding planes above and below are almost parallel, indicating relatively uniform, flat deposition. On the left hand side of the section, immediately below the shale member, the log response between contours is a consistent, rounded high velocity peak. On the right hand side, at the equivalent depth, it changes. The log trace pulls back to about 52.5 μsec/ft line, suggesting a lateral change to argillaceous material, or possibly increased porosity. The bounding contour follows the continuation of the remaining high velocity material deeper in the section.

For clarity, only a few values were contoured on this illustration. In normal use, the entire section is machine contoured, like **Figure 12.18**, illustrating the potential for automated stratigraphic mapping. Individual lithological units are outlined wherever there is a small velocity contrast between adjacent units. While the Seislog velocities may not match the borehole sonic velocities exactly, the relative change in velocity from trace to trace is normally quite reliable and very sensitive to changing lithology.

Unfortunately, many closely spaced contour lines can be confusing, making it difficult to distinguish individual units. This can be remedied by a further refinement, the addition of a color code. One method might be to color contour lines but a better method is to color the interval between contours. The coding scheme should have some relation to geology.

Figure 12.19 is a compilation of velocity ranges for different rock types, as tabulated in well and logging handbooks. The rather broad ranges overlap substantially from one rock type to another. However, they represent the total known range for each type of rock. Faust has shown that, for a given rock type, velocity varies with depth of burial and age **(Fig 12.20)**. Age appears to have the greatest influence on very young clastics but beyond Tertiary the major factor is depth of burial. The interval velocity of a single sedimentary package within the Miocene was mapped as a function of reflection time over a large marine survey to produce **Figure 8.13** (see: Digital Processing of Geophysical Data – A Review. A Teknica publication). A profile along any horizontal reflection time line would yield a sharp Gaussian curve centered about the heavy, central straight line average. The lateral range of velocity for a formation of fairly uniform lithology is relatively narrow and similar for all depths mapped. In this case, depth of burial has a major influence on velocity. In older rocks the effect is much less, and for old rocks of the same geologic age, the principal source of velocity variations will be rock composition. In more competent rocks, porosity may exert the greatest influence on velocity **(Fig 12.21)**.

Resolution of at least some of these variables is the usual purpose of any study. Obviously the conclusions may be interpretive, with more than one possible solution. In any case, when the depth of a given sedimentary sequence is known, the velocity range becomes restricted to relatively narrow bounds,and hence can be used to predict rock type with some reasonable degree of confidence.

With this in mind, the velocity intervals between contours can be color coded for a given range of depth and age. Under one system blue tones are assigned to transit-times normally corresponding to carbonates; orange and yellow to sandstones; and green to shales. The darker shades of each color represent the cleaner formations, hence a dark blue corresponds to a dense, tight limestone while a lighter tone would correspond to an increase

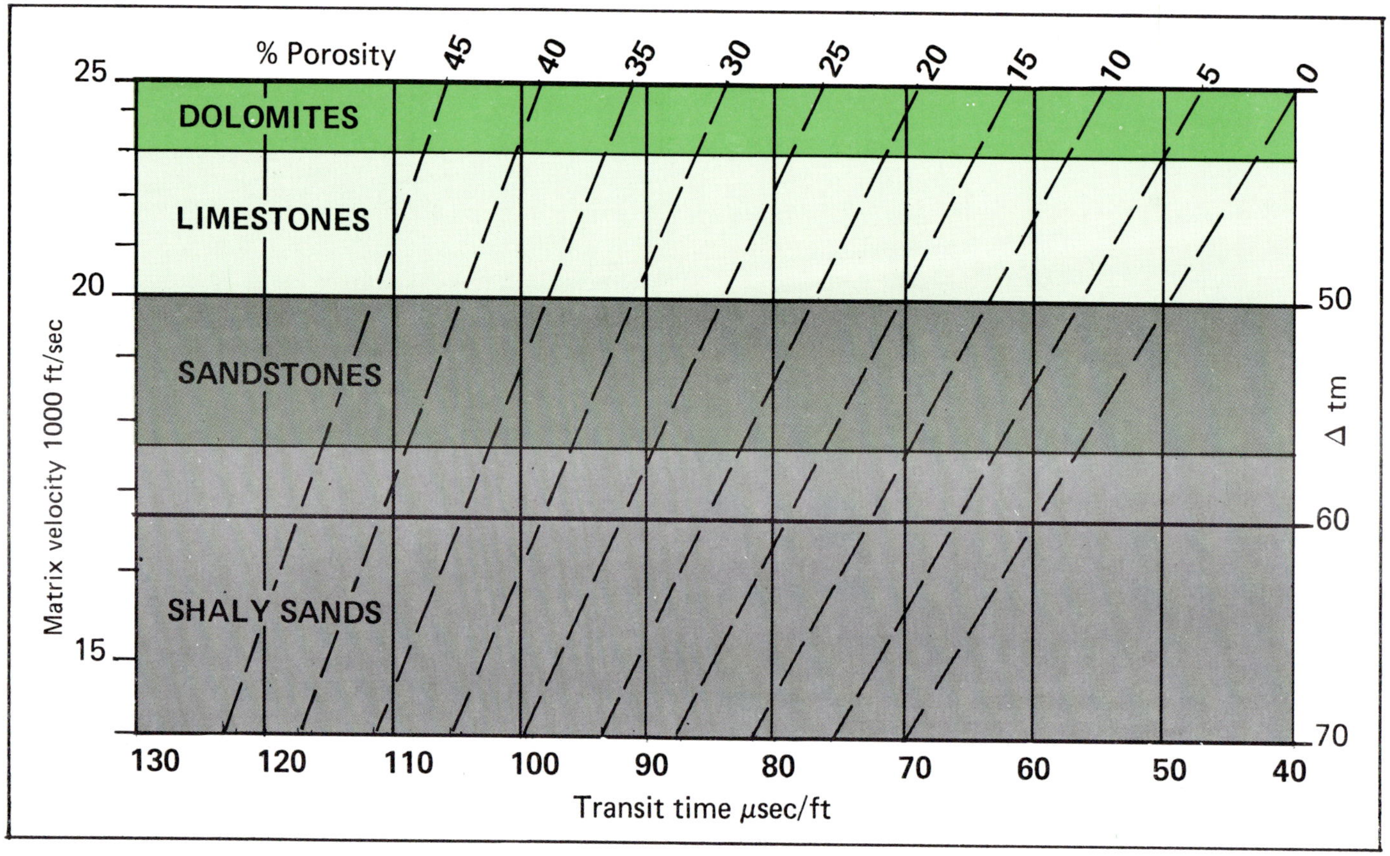

FIGURE 12.21

The transit time for a given rock matrix increases with increasing porosity. For a given porosity, the average transit time will vary with the type of fluid in the pore space.

in porosity or argillaceous content. In practice, the interpreter is given the option of varying the colors to fit any coding scheme.

Let us now attempt to put these considerations and conclusions to practice, by comparing a standard seismic section, processed and displayed in the conventional manner, to the same data, wavelet processed, converted to a synthetic log section, and contoured and color coded using the procedures outlined.

Figure 12.22 is a conventional seismic section of relatively recent vintage. A distinct unconformity separates Lower Cretaceous clastics from Devonian carbonates. One would normally expect to identify this boundary by the strongest reflection, but there are at least three high amplitude reflections near the level of interest. The actual contact, the trough near 1.1 sec, is not particularly clear on the conventional seismic section. Deeper, a Devonian bioherm reef lies buried in the section. An experienced geophysicist would recognize indirect evidence of this, such as drape over the edge, discontinuous reflections, and changes in character, but the outline of the reef mass cannot be clearly defined.

Figure 12.23 is the color coded synthetic log section of the same input data as **Figure 12.22**. The two presentations are derived from the same source data. Due to the difficulties of displaying an entire section at this scale, the synthetic log section includes only a window of the seismic section from 0.7 to 1.4 secs. Within this interval both map the same geologic section, and both examples represent the final output displays from the data processing center that would be delivered to the interpreter for his evaluation.

Each synthetic sonic log has its origin on the 90 μsec/ft baseline of the transit-time scale at the shotpoint. For example, the slightly heavier log trace on the extreme right corresponds to shotpoint 12, which is located on the 90 μsec/ft baseline. The log trace shifts laterally with depth, corresponding to increasing velocity, but its geographical ground position remains at the 90 μsec/ft baseline. Note that the log trace must always be followed back to its origin at the top of the section to obtain its true geographic position on the line.

The vertical dimension has been transformed out of the geophysical time domain into the geological depth domain, with the window extending from about sea level to 5000 ft subsea. The ground surface here has an elevation about 3000 ft. The Seislog section is near true scale, with a vertical exaggeration of about 5:3. The time markers on the section are a convenience only, to permit reference back to the conventional seismic section.

The major unconformity lying at 1.1 sec, is more than half way down the time window from approximately 0.7 sec to 1.4 sec. In depth, the unconformity lies near 2000 ft subsea, less than half of the same interval. The difference is due to the variable scale distortion of the seismic section resulting from the large velocity difference between the Cretaceous clastics above, and the Paleozoic carbonates below, the unconformity.

The abrupt right shift of the logs at the unconformity, and the change in color, corresponds to the contact of two types of rock of different velocity, as defined by the color coding on the right hand side of the section.

The large reef is near -3000 ft, shown by a high velocity carbonate mass in dark blue, surrounded by lighter blue representing lower velocity argillaceous limestones and shaly material surrounding the reef.

The top of the reef has a fair amount of relief, rising from about -3100 ft at the right end of the section, to about -2800 ft near shotpoint 40, and then dropping to about -2900 ft. Facies within the reef appear to represent three distinct types: forereef material from shotpoint 18 to shotpoint 25; the organic reef proper, between shotpoints 25 and 49; and back reef facies beyond shotpoint 49. From shotpoint 53 to 75 the back reef facies grade from dense high velocity material at the top to lower velocity material at the base. Forereef material, between shotpoints 30 and 13, is also of moderately high velocity with alternating zones of high and low velocity. **Figure 12.24** is a full size reproduction of a portion of the section, illustrating the amount of detail presented by the display.

At first glance, the overlying beds appear to drape over the reef, with perhaps as much relief as the reef itself. Closer inspection (which has been confirmed by well control) reveals this is not the case. The overlying material is quite flat. Character correlation of a weak inflection point on most of the log traces, as illustrated by the heavy line, closely parallels the -2750 ft grid line tying the base of the thin, blue, 57.5 μsec/ft lenses together. The light blue material above this line is a halo of dolomitized porosity from which production is obtained.

It is interesting to observe, on **Figure 12.23**, that events much deeper in the section may have had some influence on deposition, eventually creating a condition propitious for reef growth. A distinct contrasting sequence of shale (green-red) and carbonates (blue) near -4500 ft depth, between shotpoints 13 and 40, changes beyond this point to moderately high velocity material having little contrast. This appears possibly to have affected deposition within the overlying carbonate platform, for a definite lateral change occurs near the top of it, at roughly -3700 ft, and reef growth probably initiated shortly thereafter.

It should be remembered that the color code is strictly a function of velocity (transit-time) and indicates lithology only indirectly. For instance, the orange-red colors on the example may be somewhat ambiguous. Although they correspond to sandstones in the Cretaceous clastics, they represent shale in the deeper and much older Paleozoic carbonate section. Since the color coding is based strictly on velocity, such ambiquities must be recognized and considered in the interpretation.

Note that the correlation lines do not always follow color boundaries. The horizon at the top of the Devonian for example, contacts variable color below the line, corresponding to changes in Devonian facies at the contact.

Subsequent interpretation of the synthetic log section would proceed by drawing in lines of correlation on the section, which, depending upon the situation and interpretation, might be made to follow color boundaries on the assumption that they mark rock boundaries. Very often, in a section of substantial lateral lithology change such as this, they would not, but rather would join correlations of log character and inflection points as illustrated by the material overlying the reef. Such lines often map time surfaces, bounding chrono-stratigraphic

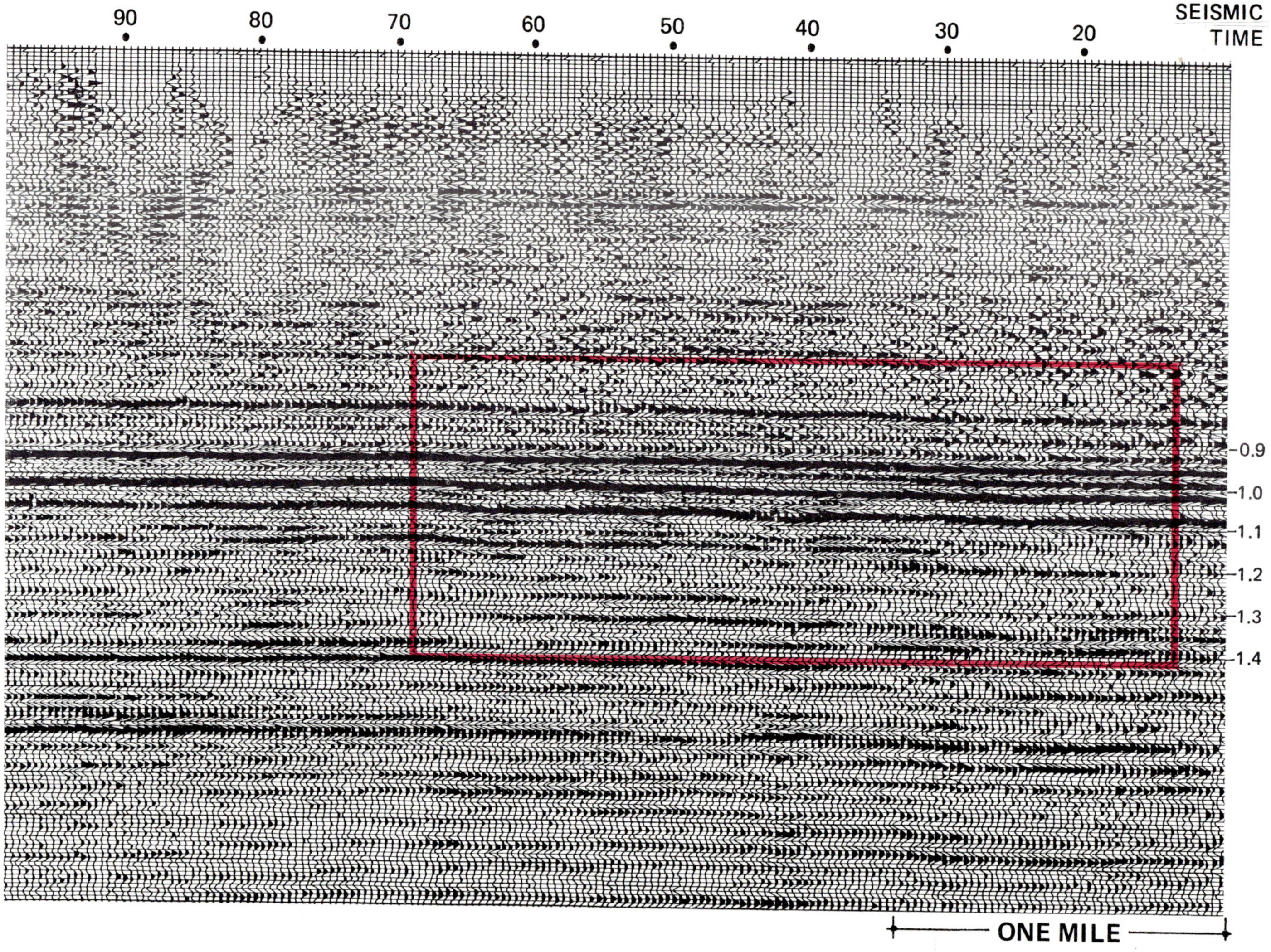

FIGURE 12.22

A conventional seismic section display across a large reef. The section is divided about 1.1 sec between Cretaceous clastics above and mainly Devonian carbonates below. The reef lies about 1.2 sec.

units which may have changing lithology at the boundary and thus feature different color coding at different places.

This completes the review of basic reverse modeling of seismic data to produce the synthetic sonic logs termed Seislogs, and the automated procedures for mapping and color coding of individual velocity units so that they can be interpreted in terms of stratigraphy and estimated rock lithology.

FIGURE 12.23

Inversion of a window of the seismic data. Transit times have been contoured and color coded as indicated by the color bar to the left. Each trace is a synthetic sonic log. The unconformity separating clastics from Carbonates is clearly visible as a distinct color change near 2000 ft. The reef is the irregular dark blue mass near 3000 ft. Both sections are uninterpreted, displaying the seismic data presentation after processing ready for interpretation.

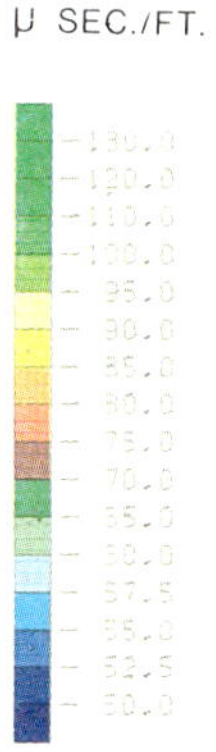

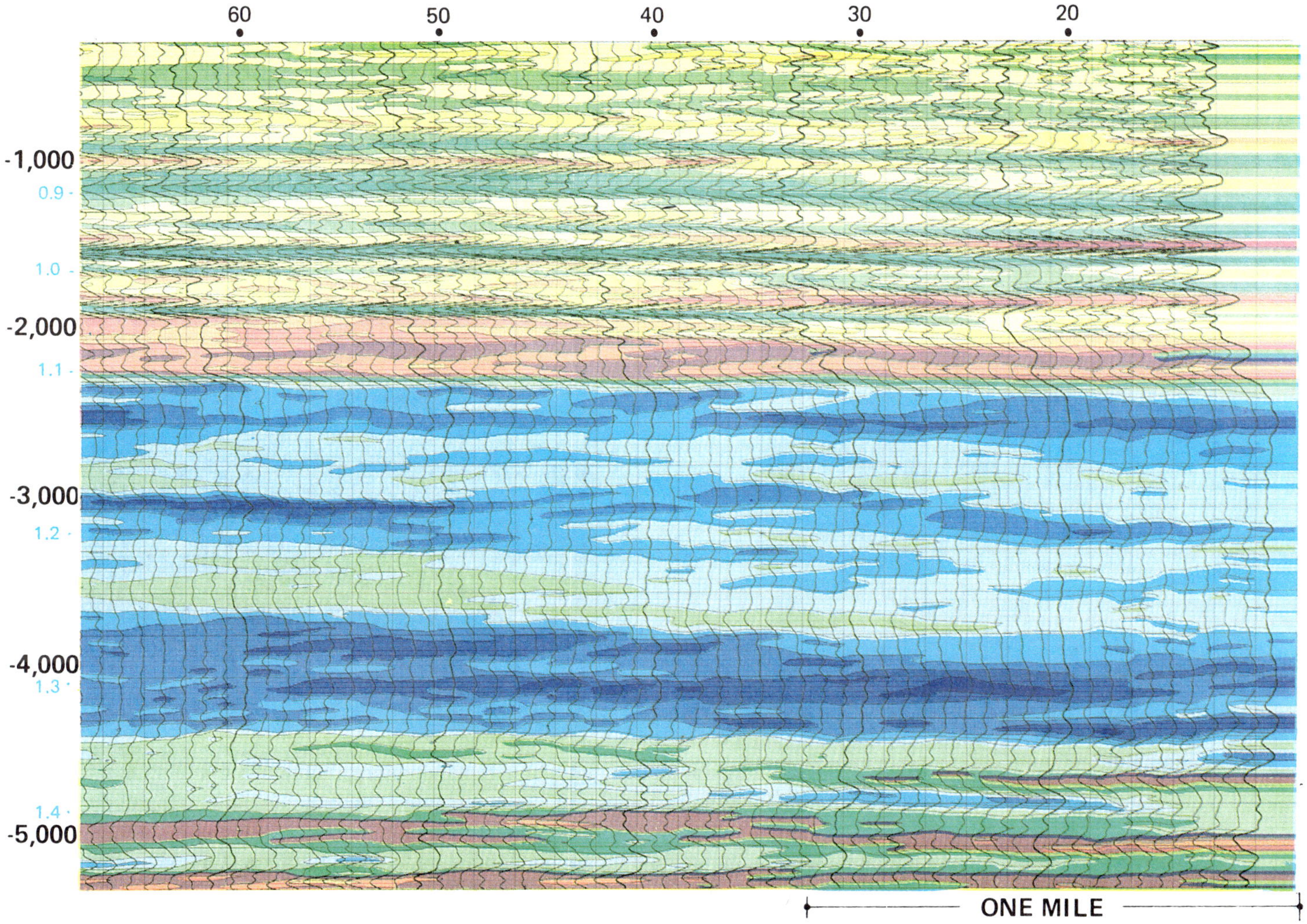

FIGURE 12.24

A full size portion of the Seislog section of Figure 12.23. The heavy black line marks the base of Upper Devonian carbonates. Production is obtained from porosity appearing as the light blue color above the line. The same color below the line corresponds to argillaceous material having the same velocity.

Interpretation

The geophysicist is truly an interpreter of seismic data. This includes not only mapping of the subsurface, but, importantly, recognition and rejection of the many aberrations common to seismic displays, such as diffractions, refractions, multiples, reverberations, and the spatial phenomenon of migration displacement. Increasingly, but certainly not yet completely, these distractions can be removed or compensated for in the data processing sequence, allowing the interpretation to become more analytical in terms of structure and stratigraphy.

Signals on a seismic section have two dimensions: reflection time and amplitude, but only reflection times have sufficient dimensional stability and geological correspondence to yield significant quantitative measurements. Transmission losses, coupling problems, instrumentation, and digital data processes invariably scale reflection amplitudes, often in non-linear manner. Small wonder then, that the majority of seismic interpretation to date has been based on the single reliable parameter of reflection time to produce structure maps.

As a result, most geophysicists are fairly well versed in structural geology. The basic elements of geologic structures are limited to relatively few variations of folding and/or faulting, although combinations of these can be very complex. The processes and features of folding and faulting can be and have been modeled, defined and catalogued. Such geological terms as syncline, anticline, horst and graben instantly call up mental images of structural types in every interpreter's mind.

But geophysicists have not been so well versed in stratigraphic mapping, particularly that related to lateral facies changes, since most information of this type is buried in the amplitude and phase of seismic signals. Although the seismic method is well established and respected for structural mapping, there is much less confidence in its ability to map stratigraphy. To complicate matters, mapping stratigraphy may be a more difficult endeavor than mapping structure. There are perhaps many more types of basic stratigraphic features than structural, and these may be sub-classified further in a great many different ways. The identities of lithology and form, which identify a particular stratigraphic feature and which distinguish it from another, are often gradational from one type to the other, making distinct identities difficult to detect and interpret.

Fortunately, examples of most stratigraphic type features abound on the face of the earth today, in every conceivable variation and at every stage of development. Processes of formation and transition can be readily studied either on site or in a laboratory.

Increasingly, it is recognized that although any given accumulation of petroleum probably exists because of some structural condition which formed a trap influenced the migration of hydrocarbons, the distribution of reserves within the reservoir is more likely dependent upon stratigraphy. In some cases the trap may be purely stratigraphic and have substantial variations in pay thickness and porosity.

INTERPRETATION

Although an oil company exploration department would very likely use conventional reflection seismic surveys to locate a suitable structure to drill an exploratory test, the design and execution of a development drilling program is more often the concern of the production department. The initial study might be based on the scanty subsurface information available on a seismic structure map, and a limited amount of borehole data. The production geologist or engineer might use only the logs from a few exploratory wells and his considerable knowledge of depositional environments to interpolate subsurface stratigraphy. As the drilling program progresses, more stepout locations yield more detailed stratigraphic and structural analyses of the sample and well log information, which would be added on a continuous basis to expand and refine the definition of the reservoir. Unfortunately, such trial and error procedures often fail to provide critical information when it is most needed prior to drilling, and many undesirable situations, such as poor or non-productive locations, may not be detected until after an expensive hole is drilled.

Synthetic sonic logs allow a great deal of this type of information to be obtained in advance, albeit without the precision of borehole logs. The special ability of Seislogs to measure both the thickness of an existing hydrocarbon reservoir, and transit-time variations within it, leads to perhaps the most successful application of the technique. When combined with knowledge, experience and some limited well control, synthetic sonic logs can provide a fairly confident and reliable mapping of the reservoir before the drilling program commences, and can be a valuable aid to maximum exploitation of the reservoir. Under good conditions, synthetic logs allow the physical characteristics of a reservoir to be defined with satisfactory precision prior to drilling, enabling engineers and geologists to anticipate results, not only outlining the reservior limits, in order to avoid dry holes, but also enabling optimum well placement to take advantage of the best porosity and reservoir thickness. This is particularly advantageous for offshore discoveries, where a large and very expensive platform must be located before development drilling can commence.

Similar reasoning can be applied to secondary and tertiary recovery projects, where it is extremely important to locate injection wells in the proper relation to producing wells in order to obtain maximum recovery.

In all of these applications, a relatively modest improvement in pay thickness or production rates obtained from producing wells, or avoiding one dry hole, can provide an appreciable improvement of the economics, and in the recovery, from a given field. For many geologists and geophysicists, reliable prediction of such key factors is a completely new and different experience, providing new opportunities and expanding their scope of endeavor.

Having the information to perform such a study is one thing; interpreting it correctly and extracting all of the useful information is another. The old saw about a geologic study being relatively simple when the interpreter is not hampered by data has a large element of truth. But when confronted with massive amounts of data from stratigraphic survey processes, the interpreter often becomes overwhelmed and is unable to assimilate and digest the significance of all the interacting information elements and their relationships. Past expereince may not be adequate to cope with the new form of data presentation. Training and experience, to recognize the technical merits and the economic benefit to be gained from a procedure, and to evaluate it properly, is necessary. Intuition is no substitute for knowledge and experience.

Case Histories

The following case histories may serve to illustrate some of the comments made and help to develop interpretive knowledge.

SANDSTONES

The problem of interpretation and identification can become very complicated in sandstone reservoirs. With very few exceptions, hydrocarbon reservoirs in carbonate rocks are identified as low velocity intervals in a high velocity matrix. The situation for sandstone reservoirs may be quite different. Whereas non-porous rocks tend to be dense, uniformly homogeneous mass of solid matrix material, sandstone rocks are granular, with interstitial spaces of varying size which can be filled with fluid or other material.

Most frequently material such as silt or shale or clay has a lower velocity than the sand matrix, a situation common to carbonate rocks. Sandstone rocks may include a mixture of sand with clay, silt, or shale, in proportions ranging from a mix containing so little sand the grains are not in contact with each other, to a clean sand, in which the rock matrix is composed entirely of sand grains all in contact, with possibly some other material confined to the interstitial space between the packed sand grains.

The adulterants usually have lower velocity than the sandstone matrix. The exception is cemented material. Most commonly, a progression from silty or shaly material to clean sand would show a regular increase in velocity. Hence, clean reservoir sandstones very often have higher velocities than the non-permeable ones, in which the pore spaces are plugged with fine grained material.

But exceptions are numerous and frequent. The transition from pore space filled with, say clay, to clean porosity filled with a fluid, can cause the average velocity to drop, since fluid is most often of lower velocity than any sedimentary material filling the pore space. More often, this seems not to be the case. Usually, a clean productive sandstone will have a higher average velocity than the dirty, impermeable mixture.

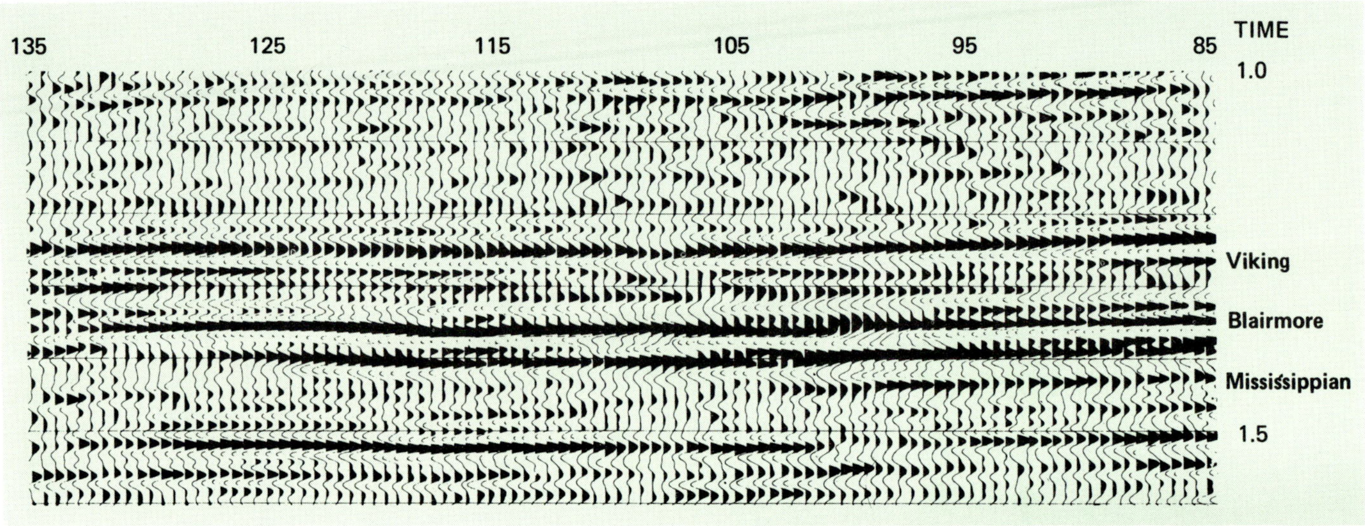

FIGURE 12.25

The conventional seismic section across a local anomaly in a generally flat, parallel bedded Lower Cretaceous sandstone and shale sequence. Some low relief occurs near shot-point 112 below the strong continuous reflection about 1.22 secs.

The maximum response is caused by the presence of gas. Gas filled porosity will almost always lower the average velocity, often by a substantial amount, so that the average velocity through the reservoir rock becomes lower than the shales which may surround it.

Although silts and shales surrounding sandstones are usually of a recognizably lower velocity than the potential sandstone reservoir, cemented and/or highly compacted fine grained clastic material, such as siltstone, may have a higher velocity than sandstone. A similar problem exists in very young rocks, generally no older than Miocene, where density, which always influences seismic reflections, may become the principal influence. In shallow sections, particularly offshore, density may cause a reversal of relationships established for older clastics in the section. Sandstones will appear, on the synthetic sonic logs, to have a lower velocity than shales. It is not uncommon to see a relationship of sandstone units having a higher velocity than shales to be well established below a certain key geological markers, but reverse abruptly above it.

The only general rule for sandstones, but one which for many different reasons, is probably violated as often as it applies, is that clean oil or water filled reservoir sands will have higher velocity than dirty, non-productive sands or shales which may surround them. They may be expected to suddenly switch to the opposite condition when the fluid filling the reservoir space is changed to gas. This is, of course, the principal upon which the well known and often very effective bright spot effect of conventional seismic displays is based.

The conventionally processed seismic data of **Figure 12.25** includes a sequence of Lower Cretaceous sands and shales which extend more or less uniformly over a broad area. They yield, in total, substantial volumes of hydrocarbons, although individual fields tend to be small.

The strong reflection below 1.2 secs is from a distinct continuous shale member which caps a sequence of sands deposited in shallow marine, often near-shore, conditions. Three distinct sand sequences can be recognized in descending order, the Viking, Blairmore and Home; the last resting on a major unconformity; the irregular, eroded surface of Mississippian rocks.

Although a substantial increase in age occurs across the Mississippian unconformity, and emphasis of this contrast might be expected from a transition in lithology from clastics above to carbonates below, the reflection marking the

FIGURE 12.26

The interval from BFS to Elkton is a series of Lower Cretaceous sands and shales, deposited in a shallow marine near shore environment. The Elkton marks an erosional unconformity in which a channel of low relief appears to have influenced subsequent deposition.

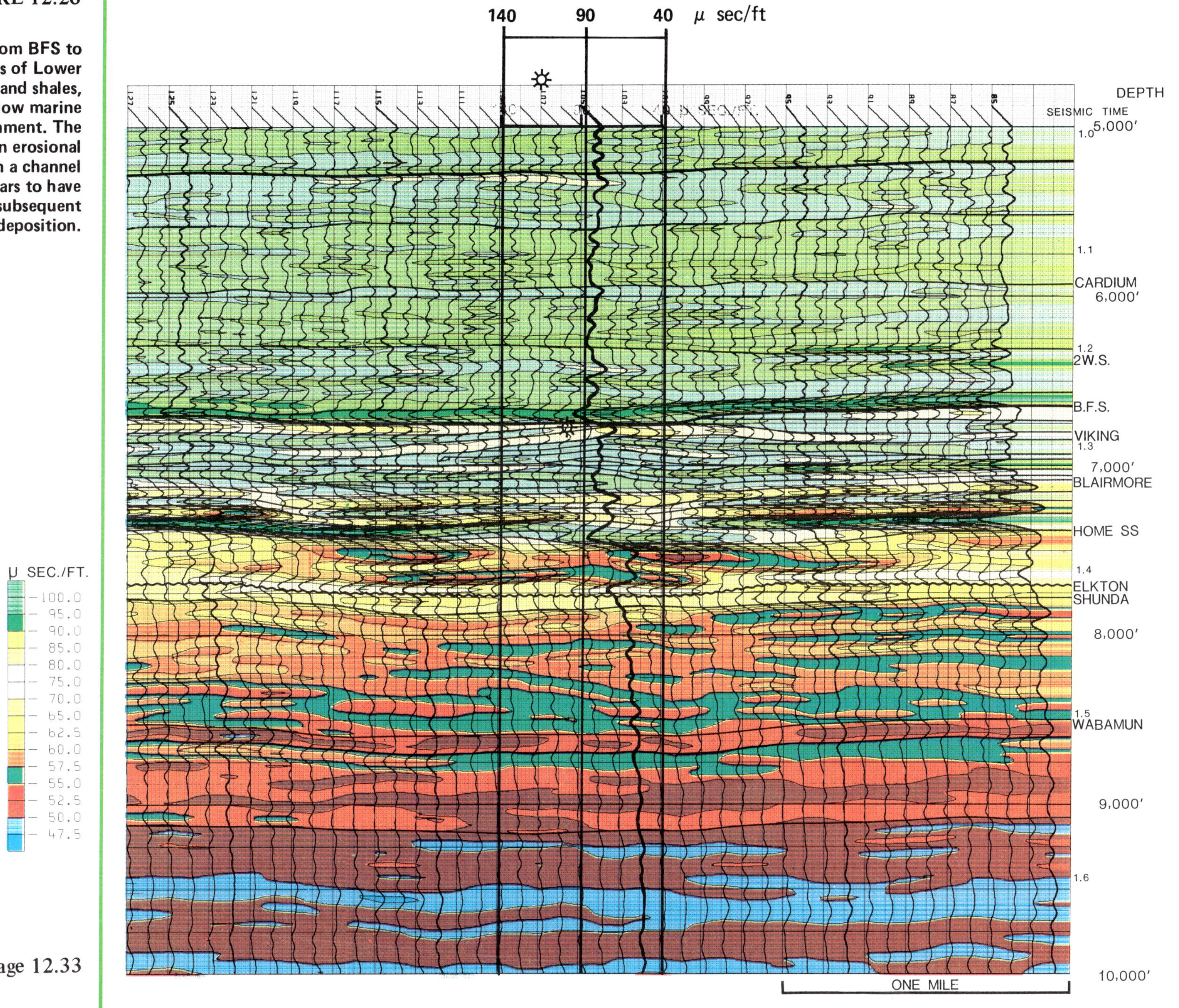

boundary is in fact rather weak. This occurrence, which is not unusual in the area, is the result of unusually high sandstone velocities in the overlying Home group, while the velocity of the subcropping Mississippian rocks is low, due to the presence of some porosity. As a result the velocity contrast across the boundary is much less than would be expected, and a strong reflection is not produced.

The normally flat overlying Viking and Blairmore sandstones appear to be somewhat disturbed in the area, as indicated by the variation in reflection character across the short line segment.

Seislogs, produced by inverting the seismic data of **Figure 12.25**, appear in **Figure 12.26**. (The log for shotpoint 105 has been drawn heavier for easy reference to the transit-time scale appearing at the head of the section). The vertical scale is in depth, with about 3:1 vertical exaggeration. Time markers provide a reference to the seismic section.

Interpretation has been applied by carrying boundary lines, based on character correlation of individual stratigraphic members, across the section. As in all examples, correlations and evaluations are interpretive. The Seislogs merely display a representation of subsurface lithology and stratigraphy based on velocity measured within rock members. The significance of that information is the job of the interpreter, whose opinion and analysis may differ from another, when presented with the same set of information.

The location of the Elkton erosional top, which locally is the Mississippian unconformity, has been confirmed by well control, and correlated on log inflection character.

A shallow downward warp is shown to exist in the entire Mississippian section between shotpoints 97 and 129. This appears to have been accentuated by erosion at the Elkton surface, producing a channel between shotpoints 93 and 123. (The best way to locate the subsurface position of the log trace is to count over to it from the nearest heavy log trace. Every 10th log trace is normally accentuated on the display, and can be readily followed down through the section. Shotpoint 123 may be located at the level of the unconformity by first noting, at the top of the Seislog display, that it is two traces to the right of heavy trace 125. This heavy trace is followed down to the unconformity and counted two traces right). The upper Elkton member is truncated across the channel. (For clarity, refer to the expanded scale view of this portion, in **Figure 12.27.** Due to the limitations of color reproduction, distinct shading of the yellows coded in this range may not always appear). In turn, the presence of the channel appears to have had some influence on the deposition of the overlying Home sand. An unusual amount of high velocity (here glauconitic) material lies over the depression. (Beyond the ends of the display, the Home sandstone is much more uniform in velocity and continuous, with transit-times ranging from 65 to 75 μsec/ft).

A thin, normally continuous shale (light green) at about 7300 ft separates the Home sand from the overlying Blairmore, but the continuity of this shale member has been interrupted across the depression. There is a definite sag of the lower of the two overlying Blairmore sandstone members into a local trough near shotpoint 103. The upper Blairmore sand member has, however, thickened locally by some 25 to 40 ft to fill in the sag of the lower member, leaving the top of the Blairmore fairly level and smooth, obliterating the evidence of the many changes below. It might be expected that subsequent deposition would be smooth and continuous, as it normally is in the overlying sequence, but apparently another local anomalous (possibly related?) depositional cycle has occurred.

The Viking is a sequence that extends more or less uniformly and continuously over a very large area. The sandstones are porous and most commonly water wet, but frequently produce gas (and occasionally oil) if a suitable trap can be located. Over this example a shoal appears to have developed in the lower Viking, piling up the sediments in a symmetrical mound. Apparently the highest member may have been truncated between shotpoints

FIGURE 12.27

The horizontal black lines represent the interpreter's correlation of individual members within the section. As such they are interpretive, and the same data might be interpreted in a different manner by the reader.

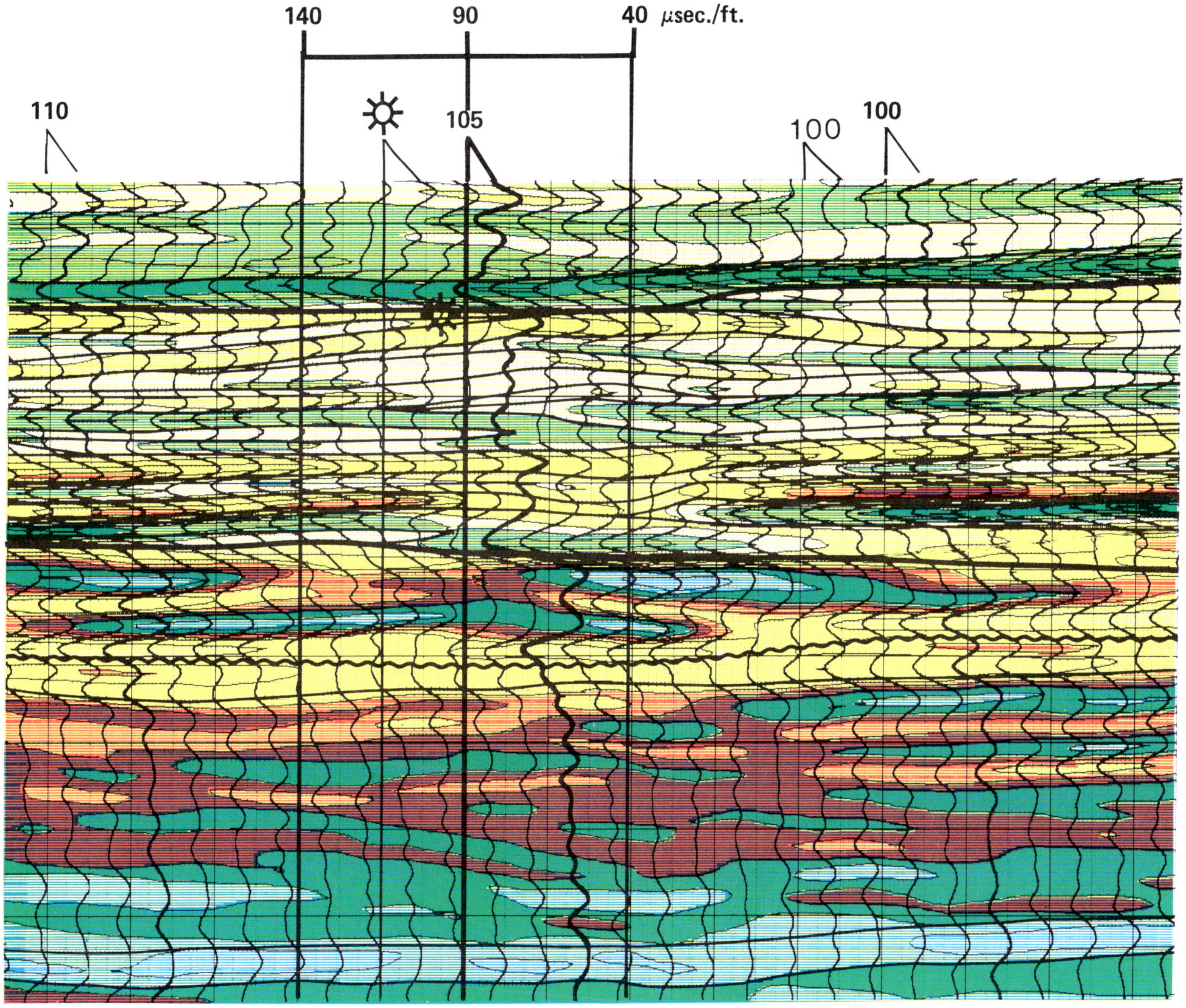

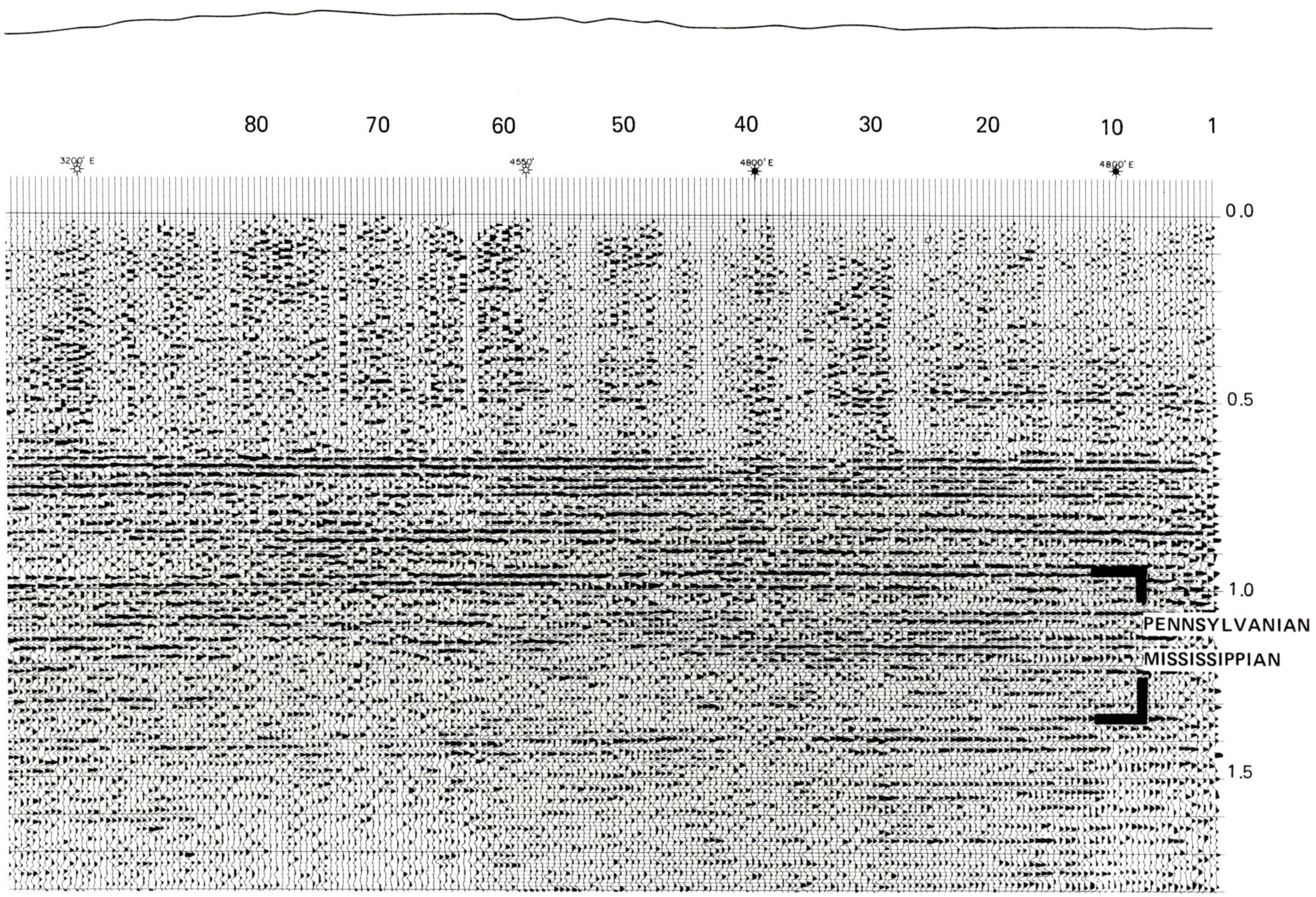

FIGURE 12.28

Conventional seismic section across the producing area. The data appears noisy and reflections are discontinuous but to a large degree this is the nature of the subsurface.

102 and 105, possibly by submarine erosion, since deposition of the overlying sequence is considered to have been part of a continuous succession. In any case the upper Viking member near shotpoint 105 terminates against the overlying shale to resume again at shotpoint 102.

A borehole near shotpoint 106 found gas in the pinchout of the Viking against the overlying shale. Individual members of the Viking at this point thin beyond the limits of resolution carried in the original seismic source.

It may be of some value to compare closely the conventional seismic display with the equivalent data on the Seislog display. The locations of the Viking and Blairmore features on the seismic section do not appear to correspond to the Seislog display, yet the deeper channel feature is in the same location. This emphasizes the point that there is no one-to-one relationship between the conventional seismic data and the inverted seismic trace integration. It is doubtful that this test would have been drilled on the basis of the conventional seismic section.

To this point, the changes in velocity of the members within the section have been related primarily to changes in

FIGURE 12.29

The material below the unconformity is limestone, with some shale members. Variable porosity was developed below the unconformity (wiggle line) which caused the transit-time to vary as a function of porosity. The rate of gas production also varies with the porosity and transit time.

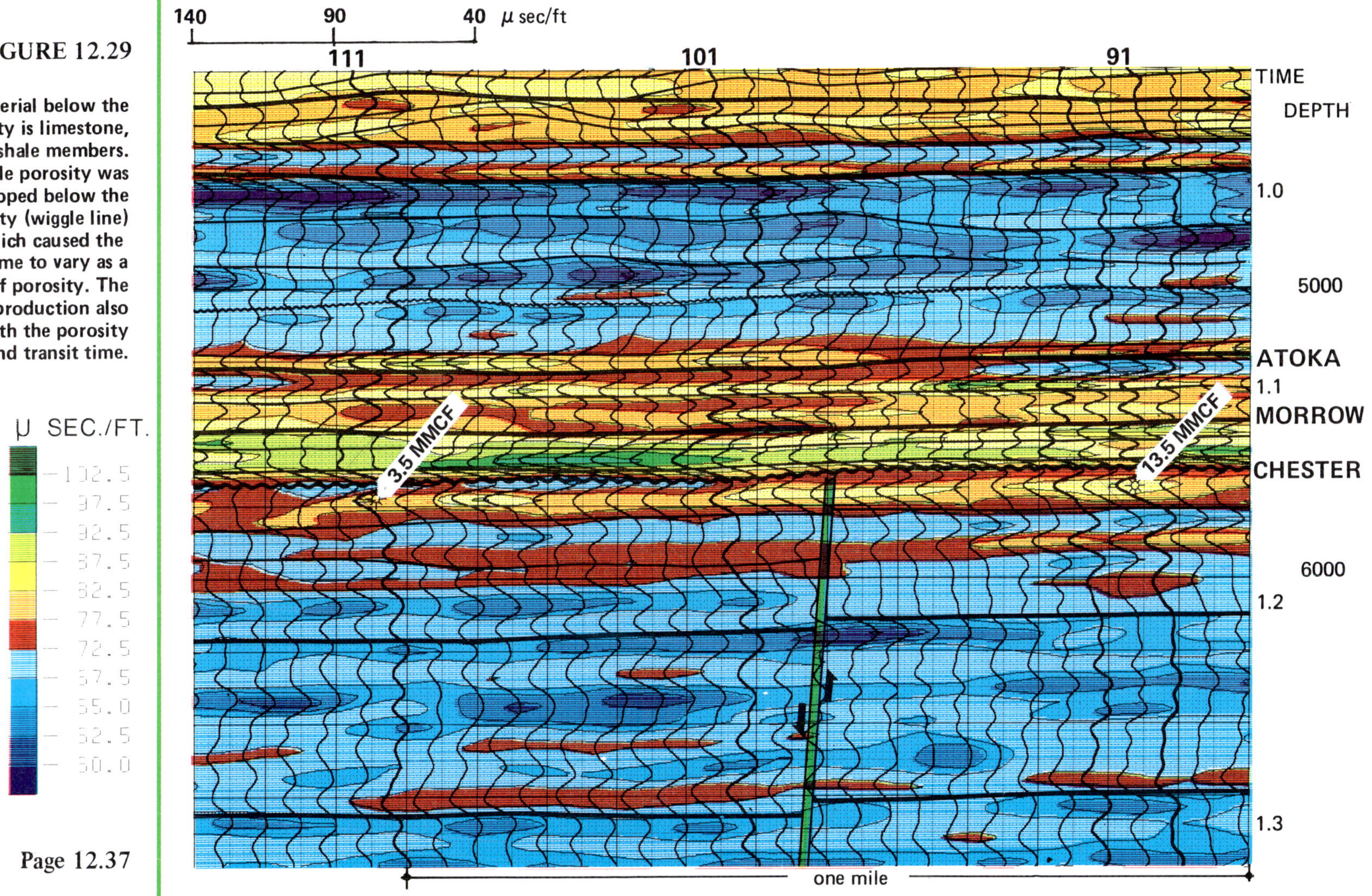

lithology or rock type. Of considerable importance and value is the knowledge that a change in porosity within a given rock member may have the identical response. Hence, a change in color corresponding to a decrease in velocity within a given member may be either a change in velocity or a change in porsity (**Fig 12.28**).

Although this possibility creates the potential for ambiguity and errors in the interpretation, it also creates the opportunity to estimate the extent and distribution of porosity within a newly discovered hydrocarbon reservoir prior to starting the development drilling program.

This is indeed a valuable trait, which opens the door to an entire new field for geophysics, and offers the opportunity to make substantial improvements in the economics of and performance of a field development.

Some control, in the native of sonic logs from boreholes, will be needed to establish the relationships which exist around the borehole and to calibrate the response, but this is usually available from at least the discovery well.

Figures 12.28 to **12.31** illustrate the results of a reservoir study in Oklahoma. The example maps production obtained from a local porosity development in the Chester formation, a Mississippian carbonate. The Chester has been eroded and the exposed reservoir rocks were sealed with a cover of Pennsylvanian Morrow shale. The reservoir is purely stratigraphic, resulting from lateral porosity changes below the Chester subcrop. A few rather poor sands develop in the Morrow, but usually they must be fraced to produce.

The thickness of the Chester pay zone varies irregularly from zero to more than 100 ft. The porosity is also quite variable. Hence the flow rates, obtained from producing wells, can range from non-commercial quantities to very prolific.

Neither porosity nor thickness of the pay zone can be recognized on the conventional seismic data (**Fig 12.28**), which has always been considered to be of poor quality, primarily because of its lack of continuity. In fact, it is difficult to map the Chester unconformity on conventional seismic data. The results of this study suggest that the discontinuous data may be a faithful representation of the subsurface, and contain reliable detailed information.

Over much of the line the Chester is tight and no production is obtained from it. However, northward from about midway along the line, rocks which subcrop at the unconformity, change from dense, tight limestones to an equivalent interval that has porosity.

Porosity in the Chester has a pronounced effect on the transit-time. Up to 25 μsec/ft total transit-time shift can be observed at the zone of interest. The tight, non-productive carbonate has a transit-time of 72.5 μsec/ft or less. Transit-times for porosity vary from 72.5 μsec/ft to 87.5 μsec/ft, and occasionally greater.

Figure 12.29 shows the log response of the Chester and the production obtained from it. On the left, near shotpoint 111, a borehole was completed as a gas well having a flow rate testing 3.5 mmcf of gas per day. The transit-time is shown to be about 82.5 μsec/ft or less, a 10 μsec/ft shift over an interval about 50 ft thick. On the right side of a much better well near shotpoint 91 tested 13.5 mmcf of gas per day from a thicker equivalent interval. Here the transit-time in the porosity is about 87.5 μsec/ft, for a shift of 15 μsec/ft.

Not all the wells are located exactly on the line, but have been projected some varying distance into the line of section. However, the next location lies close to the line so this cannot be used as an excuse to explain the contrary results of the next location, which violates the general relationship found to exist in the area. A well near shotpoint 63, (**Fig 12.30**) which falls in the range above 82.5 μsec/ft, light yellow code, would suggest good porosity, and should have yielded good results by this analysis, but it did not. Extensive testing (which included several failed tests) yielded only very modest results. The well was finally completed as a dual Morrow-Chester gas well for a total of only 1.8 mmcf/day of gas. The transit-time color code would suggest much higher production should have been obtained, say 8 mmcf/day.

Two other wells, at shotpoints 55 and 41 (**Fig 12.31**) produced about what might have been predicted, 5.0 and 3.5 mmcf/day, from porosity corresponding to a transit-time averaging 80 μsec/ft. The lower rate, from a transit-time at shotpoints 41 similar to the well at shotpoint 55, came from a thinner zone.

The final example in the sequence (**Fig 12.32**) contains one very poor well at shotpoint 10. At this location Chester transit-time has dropped below 72.5 μsec/ft, indicating no porosity. Extensive frac treatment was required to obtain 0.8 mmcf/day from a poorly developed Morrow sandstone, with a transit-time of 75 μsec/ft.

FIGURE 12.30

This example of mapping across the reservoir includes one well near shotpoint 61, where results did not agree with the prediction. The color coding indicates good porosity but production from it was poor.

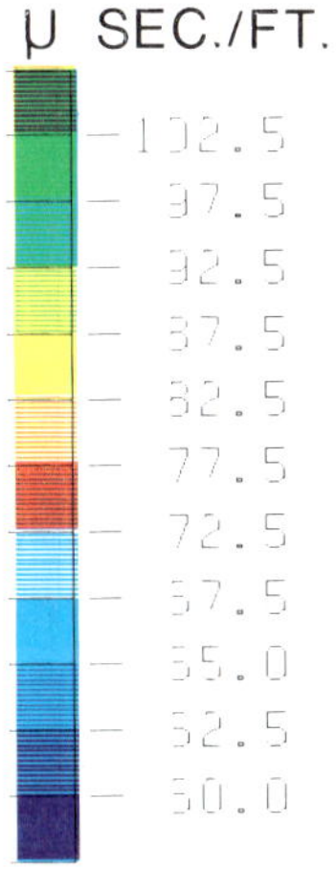

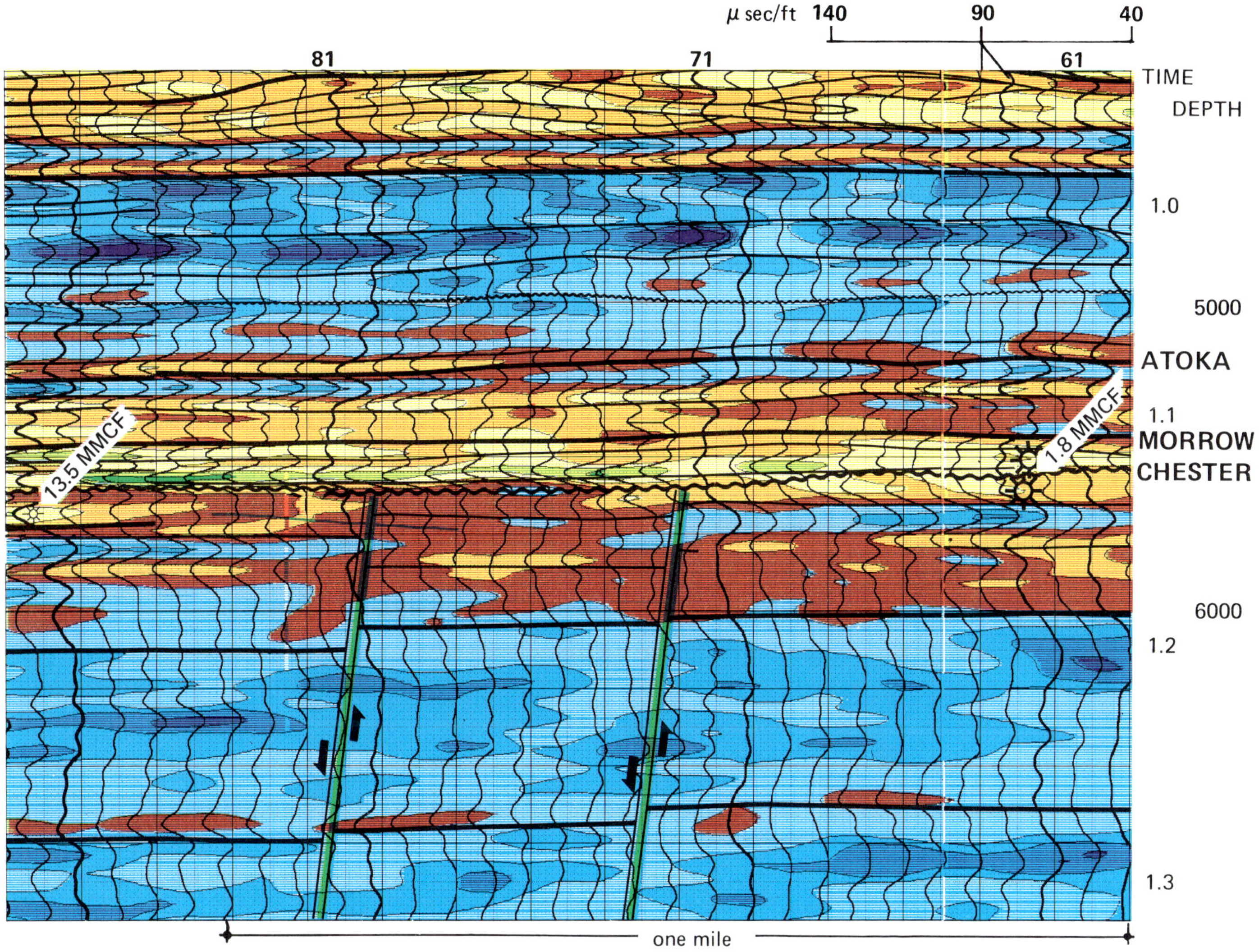

The relationship found between production and Seislog transit-time was as follows:

Transit time μsec/ft	Production, mmcf
72.5 - 77.5	< 2.0
77.5 - 82.5	2.0 - 8.0
82.5 <	8.0 <

The results of the predictions were by no means perfect, with the exceptions amounting to about 10% of the wells in the area. Nevertheless the high correlation from what, at first glance, appears to be poor data, the substantial correlation of results with development drilling, and subsequent experience in similar situations, all confirm the potential for prediction of reservoir production. Under good conditions, a relatively reliable prediction of relative reservoir thickness and porosity can be made prior to drilling.

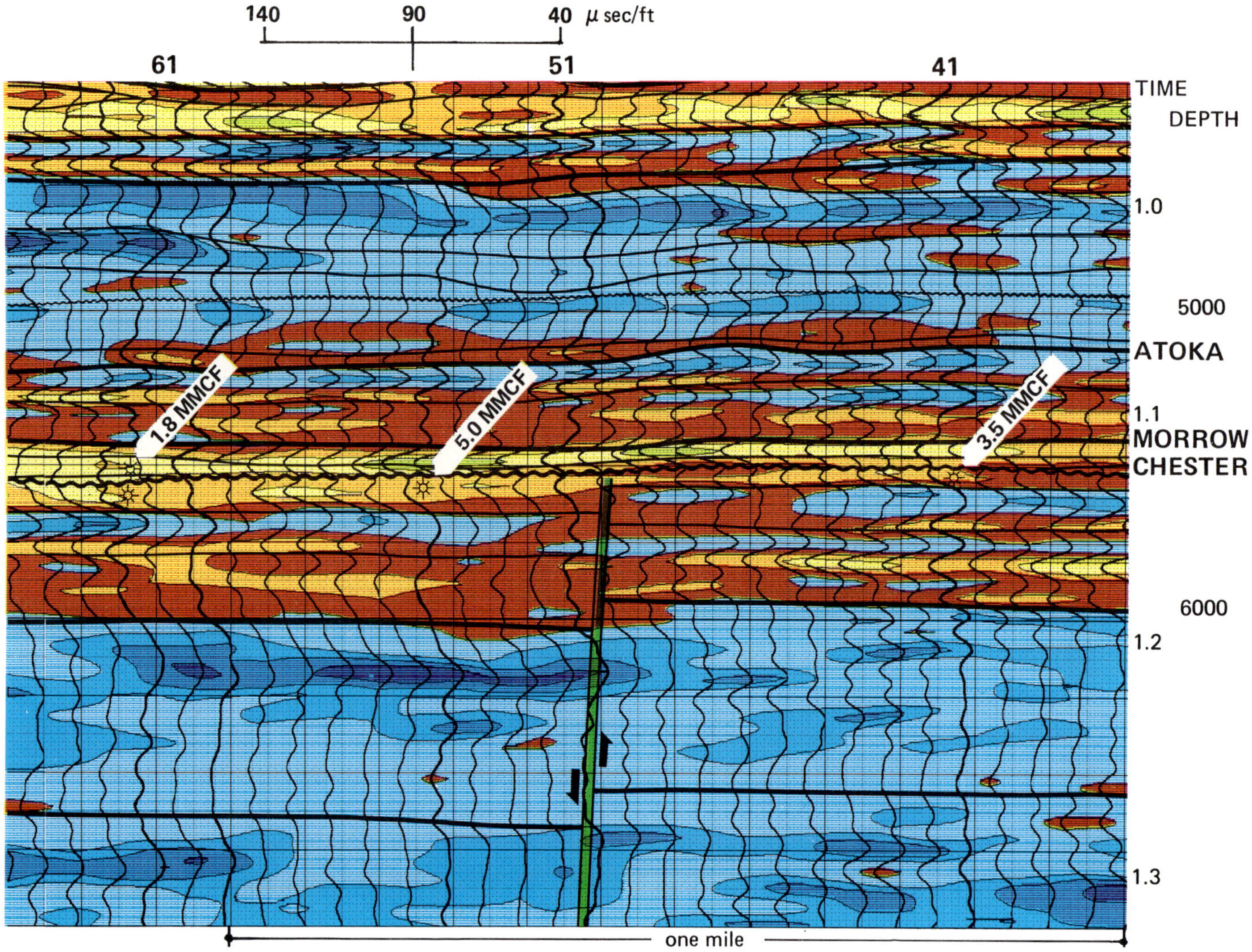

FIGURE 12.31

Production from porosity indicating similar transit time across the fault is reduced on the up-thrown side of the fault because the producing section is thinner. Previous published studies of the area, including drilling, failed to recognize the extensive faulting, which appears as slight misalignments of horizontal correlations of log inflections.

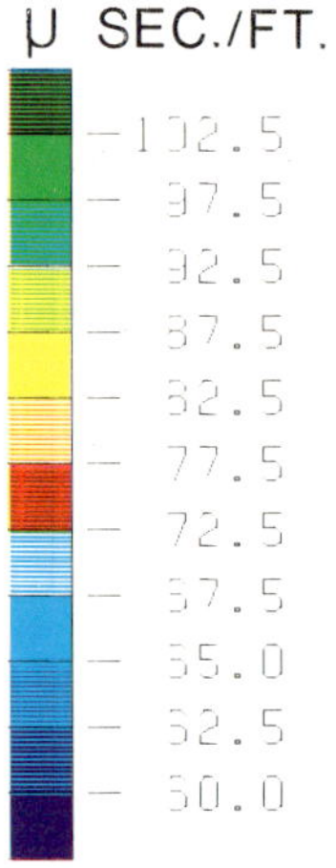

The next example is from a small gas field in central Alberta. Exploratory drilling in the area, completed some years before the inversion technique became available, discovered gas in a thin stratigraphic reservoir. The map (**Fig 12.33**) shows pay thickness, not structure as defined by drilling. Gas is obtained from porosity near the top of a massive Devonian carbonate section. The reservoir is unusual, since production from this part of the sequence is extremely rare. A local facies change exists, the result of a stromatoporoid and algal bank development at the time of deposition. Porosity is found in dolomitized rock, original organic pore space, and from vugs in the algal bank. Porosity development varies with the considerable variations in lithology, and is further complicated by anhydrite and calcite cement filling some of the original pore space.

The area has a gentle southwest regional tilt. At this time of the Seislog study, the field was considered to be a mat-

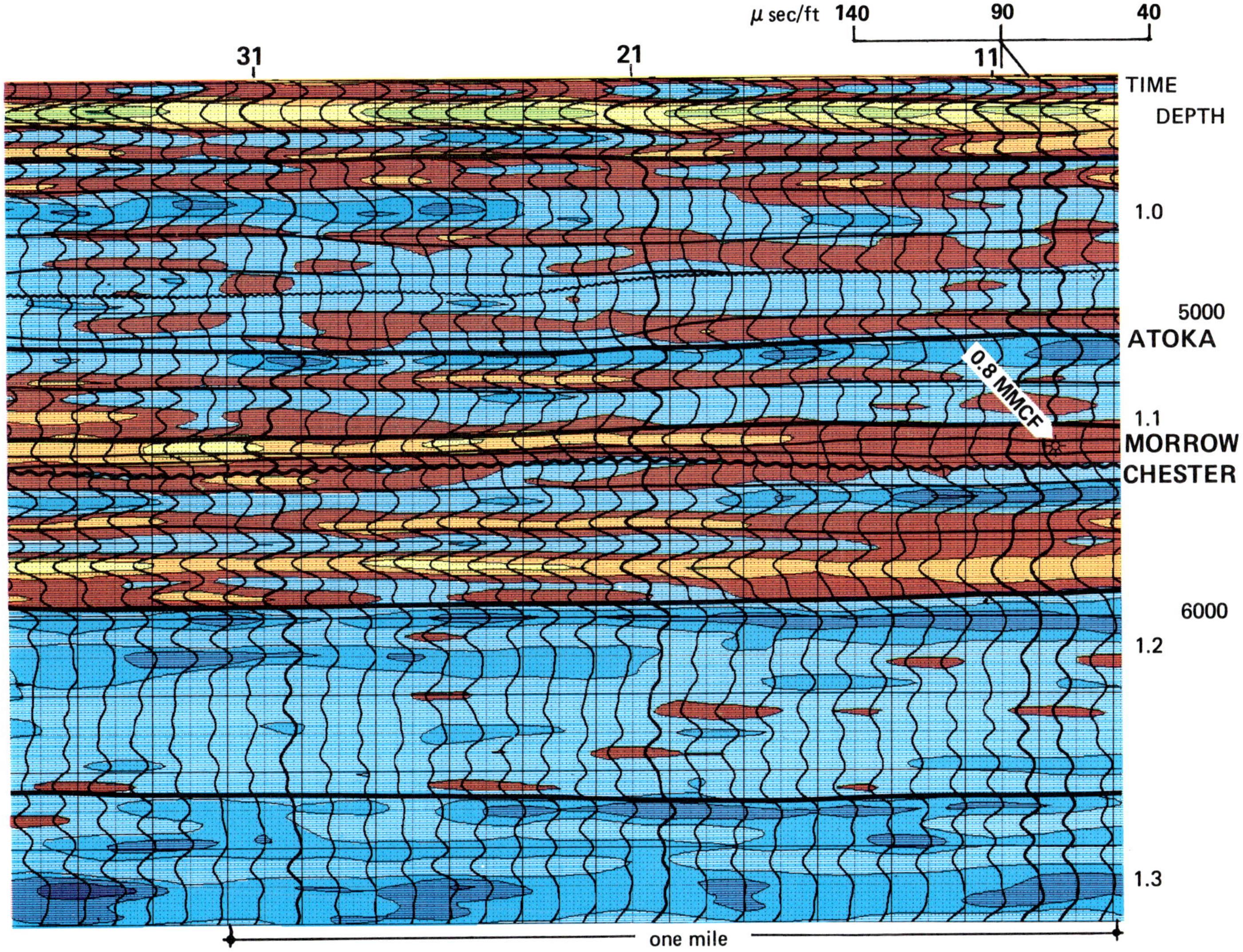

FIGURE 12.32

At the drilled location the colour code below the unconformities indicates a transit-time less than 72.5 µsec/ft, which has been determined to correspond to zero porosity. Marginal production was obtained from an overlying sandstone after a fracturing treatment.

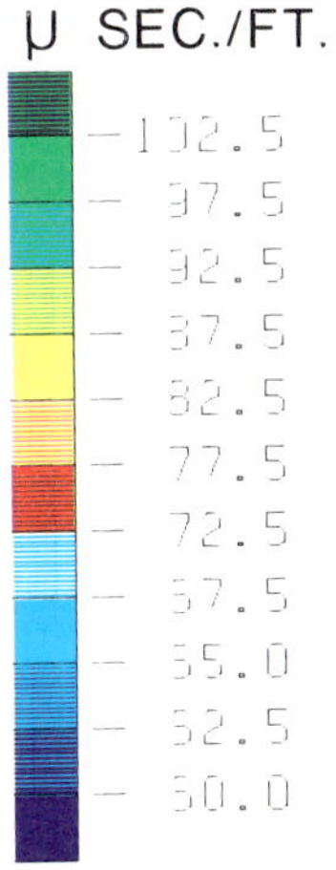

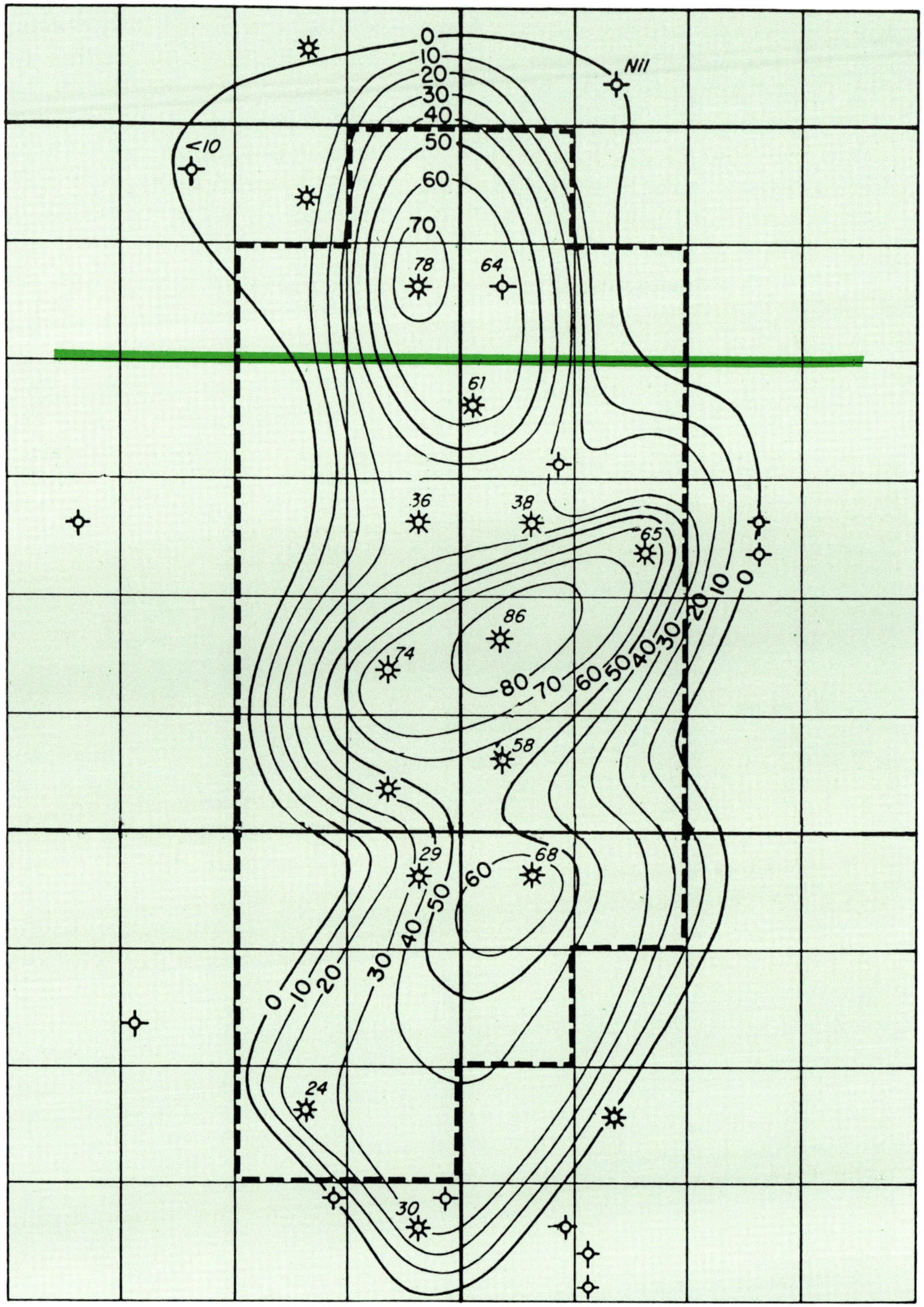

FIGURE 12.33

Porosity map of the gas field. Contours are pay thickness, as interpreted from well data.

~~Orange~~ line indicates Seismic line of Fig 12.34

Green

ure development which had been essentially drilled out. The limits of the field had been defined by a few dry holes. The eastern edge was based on one dry hole to the north, another near the middle, and by one well toward the south which found a very thin pay zone.

The economics of the field are very good but the reservoir could not be readily detected with conventional seismic surveys, so it was decided to shoot a seismic line across the field and invert the data to obtain a model which could be used as a guide for further exploration. The line yielded relatively good data, which is illustrated in **Figure 12.34**. The top of the Devonian carbonate at approximately 1.35 secs is designated the Wabamun. The zone of interest, the Crossfield member, lies immediately below. The field limits originally extended from about shotpoint 85 to shotpoint 55. It will be noticed that considerable variation exists in the character of the seismic reflection at shotpoint 55, clearly corresponds to a change in reflection character, and similar changes may be noted at a number of points.

Inversion of the seismic data yields a synthetic log section (**Fig 12.35**) having a number of distinct features. The Seislog traces have been omitted from the display to more clearly define the stratigraphy. About -3700 ft, an unconformity is clearly demonstrated by a major change in color from predominately orange and yellows, corresponding to Cretaceous clastics, to the light blues, for shaly carbonates of Mississippian age. Deeper in the section at -4800 ft, a continuous low velocity shale, coded green, marks the base of the Mississippian and the contact with the top of the Devonian carbonates. The zone of interest lies within the uppermost 400 ft of the Devonian in the Crossfield member. At the right hand end of the section the Crossfield has a core of high velocity material (indicated by dark blue) with some lower velocity facies (lighter blue) at the top and the base of the interval. Westward, this situation is relatively constant to approximately shotpoint 37. At this point, the response within the Crossfield unit changes. Proceeding westward, the response in the middle of the sequence, which to this point had been high velocity material having a transit-time less than 47.5 μsec/ft changes. Within the mid-Crossfield the log is now deflected in the opposite direction, outlining a lower velocity zone across which transit-time has increased to 50 μsec/ft or more. This anomalous low velocity zone extends from shotpoint 39 to the western edge of the field near shotpoint 85. The transit-time contour, of 50 μsec/ft or greater within this interval, defines the productive porosity within the Crossfield member.

Of special interest is the constriction across a very thin porosity interval near shotpoint 55, the previously established field boundary. Apparently the very thin porosity found in the well at the east side of the original field was considered to be very near the field limit. A subsequent well, located on strike with shotpoint 45, encountered good porosity development and extended the field approximately 1 mile east.

Note how clearly the dimensions of the reservoir are defined when the logs and contours are included in the display (**Fig 12.36**). The space between the 50 μsec/ft contours at the top and the base provides a reliable measurement of the pay thickness at each log, while the variations in transit-time provide a good qualitative, and to a degree quantitative, measure of relative porosity. When pay is thin, the transit-time is not a precise measure of porosity because, due to the limits of resolution, the transit-time response within the very thin bed is averaged with surrounding material. The procedure is relatively simple. The cutoff value of transit-time is determined and a computer program instructed to examine a small window of the section contouring the reservoir. The interval between the two contours defining the top and the base of the reservoir yields pay thickness (**Fig 12.38**). Then, the porosity corresponding to the displacement of the transit-time curve within the reservoir is determined. Integration of the Seislog curve behind the cutoff line can then be carried out and a relative porosity-foot value determined for each trace position. A continuous plot of these findings yields an indication of the reservoir volume along the line (**Fig 12.39**), and, when carried out over a grid of lines can map the entire field. Nevertheless, the zones bounded by the 52.5 μsec/ft contour in the example are zones of superior porosity and will have lower velocity. It is not difficult to imagine how a grid of lines over the entire field area could provide all the information required to construct a detailed map of pay thickness and porosity, **(see Fig 12.28)**. Integration of these two components, calibrated by some scaling factor relating transit-time to porosity, yields a porosity-foot map of the field.

The presence of the reservoir is clearly demonstrated by a change in seismic reflection character on the conventional seismic display, and many geophysicists might be inclined to believe that inversion would not be required,

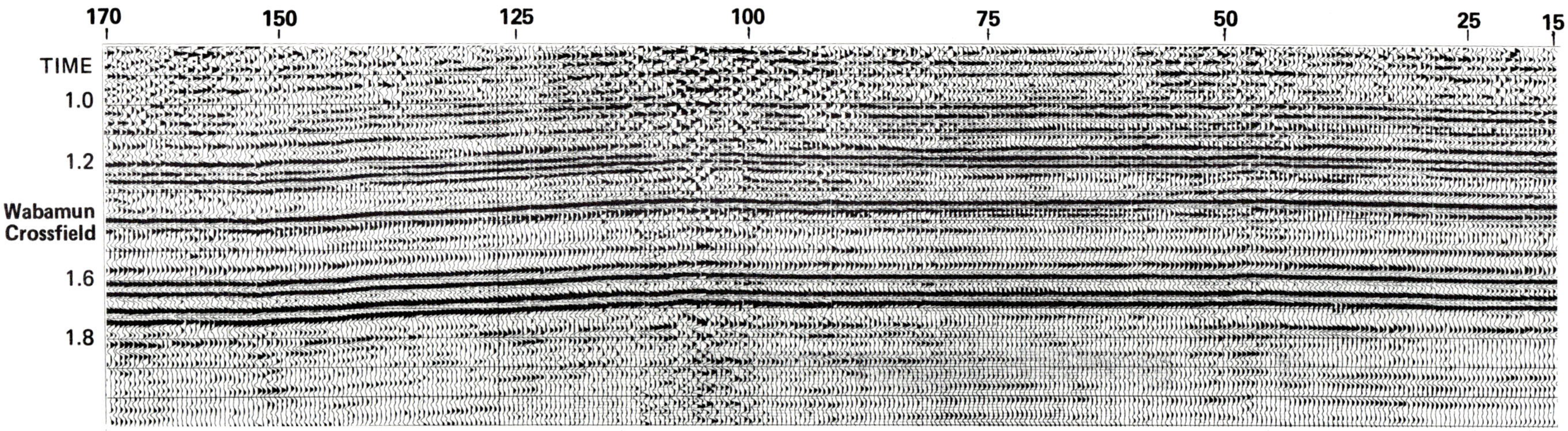

FIGURE 12.34

Seismic line across a gas field. Production is obtained from dolomitized porosity in an algal bank in Devonian carbonate section. The zone of interest is the Crossfield.

since they could detect the gas field on a conventional seismic section. This may be true, but it may be misleading. Also, the definition of the reservoir extent, and detailed measurements of its thickness and porosity, cannot readily be obtained from the conventional seismic section.

Reference to the original seismic section (**Fig 12.34**) shows clearly a change in reflection character east of shotpoint 55 which could be correlated to that of the extended producing area now known to exist, but it would be difficult to conclude from the same data that the eastern limit of the field extends to shotpoint 39.

Furthermore, it is by no means certain that there is a direct correlation of character to porosity in the area. For instance, at the western end of the line reflection character similar to that across the field appears to exist. Yet, examination of the inverted data (**Fig 12.37**) does not necessarily confirm the conclusions that might be drawn from the character changes of the conventional seismic display.

When results are as good as this example, and both the reservoir thickness and transit-time can be measured with reasonable confidence, the method can be extended to predict relative reserves in phase within the reservoir. Within the limits imposed by bandwidths, thickness of the reservoir can be measured with greater confidence than transit-time.

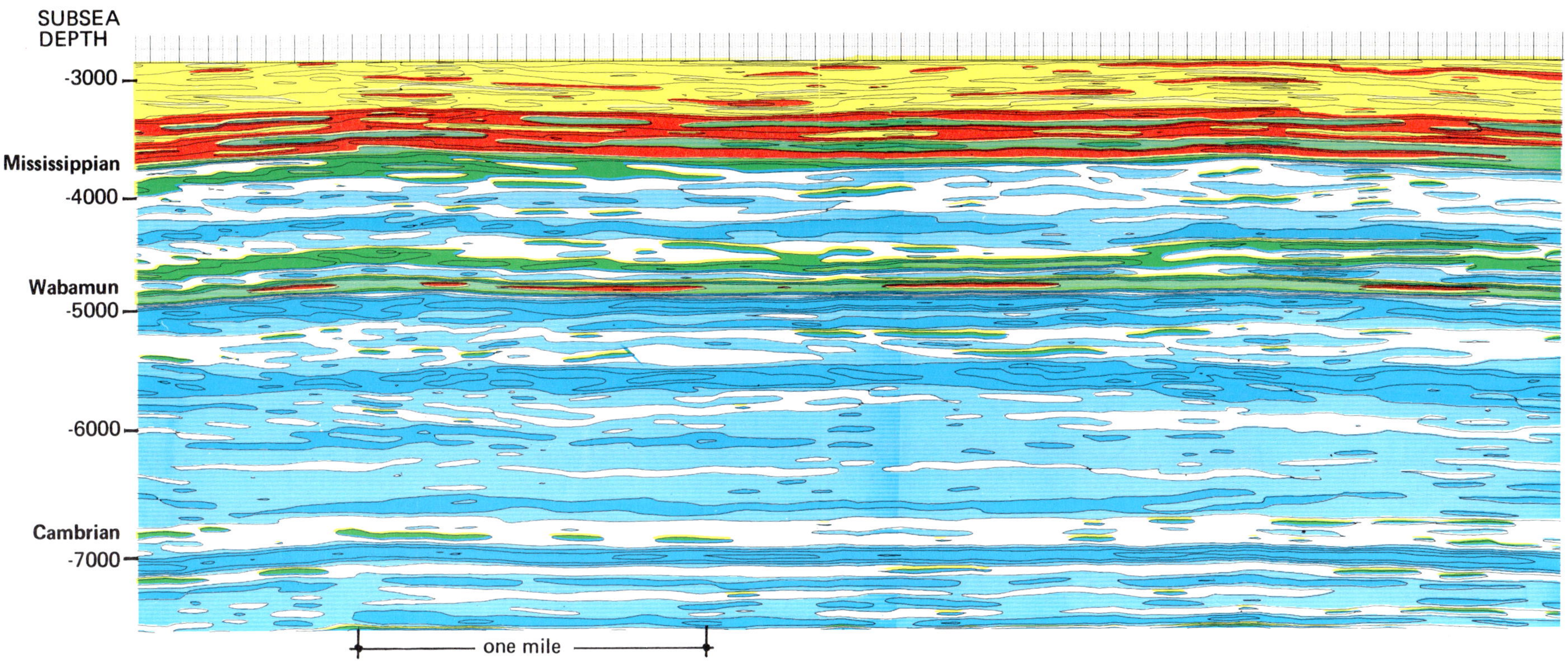

FIGURE 12.35

Inversion of the Seismic data of (Fig 12.34). The Seislog traces have been omitted for clarity. The gas reservoir is the long thin, low velocity interval indicated by the light blue color, and the gas well symbols, at about -5000 ft.

no gas well symbols indicated in Fig 12-35

Although this was a successful study, it illustrates rather clearly some of the difficulties of proper identification and interpretation without the help of additional information, such as at least one sonic log.

The examples illustrate not only the successes, but also point out some of the difficulties, of stratigraphic exploration common to this or any method. In all cases, the traps are purely stratigraphic, with little or no structural element contributing to the accumulation. The reservoir porosity, although correctly defined by a low velocity response, is merely one low velocity interval among many on the section, and it has few, if any, unique characteristics to distinguish it from several other similar appearing intervals which are defined in the subsurface lithology, such as shale, with equally correct response characteristics.

The fundamental source of this difficulty is the fact that although the presence of porosity, corresponding to a reservoir development in a given rock matrix of a defined transit-time, will generate a characteristic and measurable change in transit-time, other changes to the matrix, of no economic importance, can cause a similar response. These include the presence of any lithologic component having a lower velocity than the host rock matrix. In carbonate rocks, these components are most commonly shale or sand facies, and evaporates. The single Seislog curve responds in a similar manner to all. Conventional borehole surveys log several different types of response curves, which provide several combinations of measurements for multiple equations capable of solving the several variable common to the problem.

Refer once again to **Figure 12.35**. Within the section are numerous low velocity elements which could be interpreted as being significant in terms of possible porosity developments. In a pure exploratory situation, with no further detailed information, it would be difficult to detect the one significant anomaly that exists on the section. Under such circumstances, one hint would be provided by the clear updip termination of low velocity zone of the reservoir, which suggests at least that if low velocity did correspond to porosity, rather than shale, it would be effectively sealed by the dense high velocity material above and to the right.

The ability to map out a petroleum reservoir before a substantial development drilling program is undertaken, can yield a substantial improvement in the production economics of the field. Under good conditions, the transit-time response of synthetic sonic logs has been found to be sufficiently reliable to estimate reservoir porosity and thickness and so improve reserves calculations and optimize drilling programs.

FIGURE 12.36

A conventional seismic section across a gas field. Production is obtained from porosity developed in a Devonian Carbonate. The zone of interest is the reflection labelled Crossfield, about 1.38 secs.

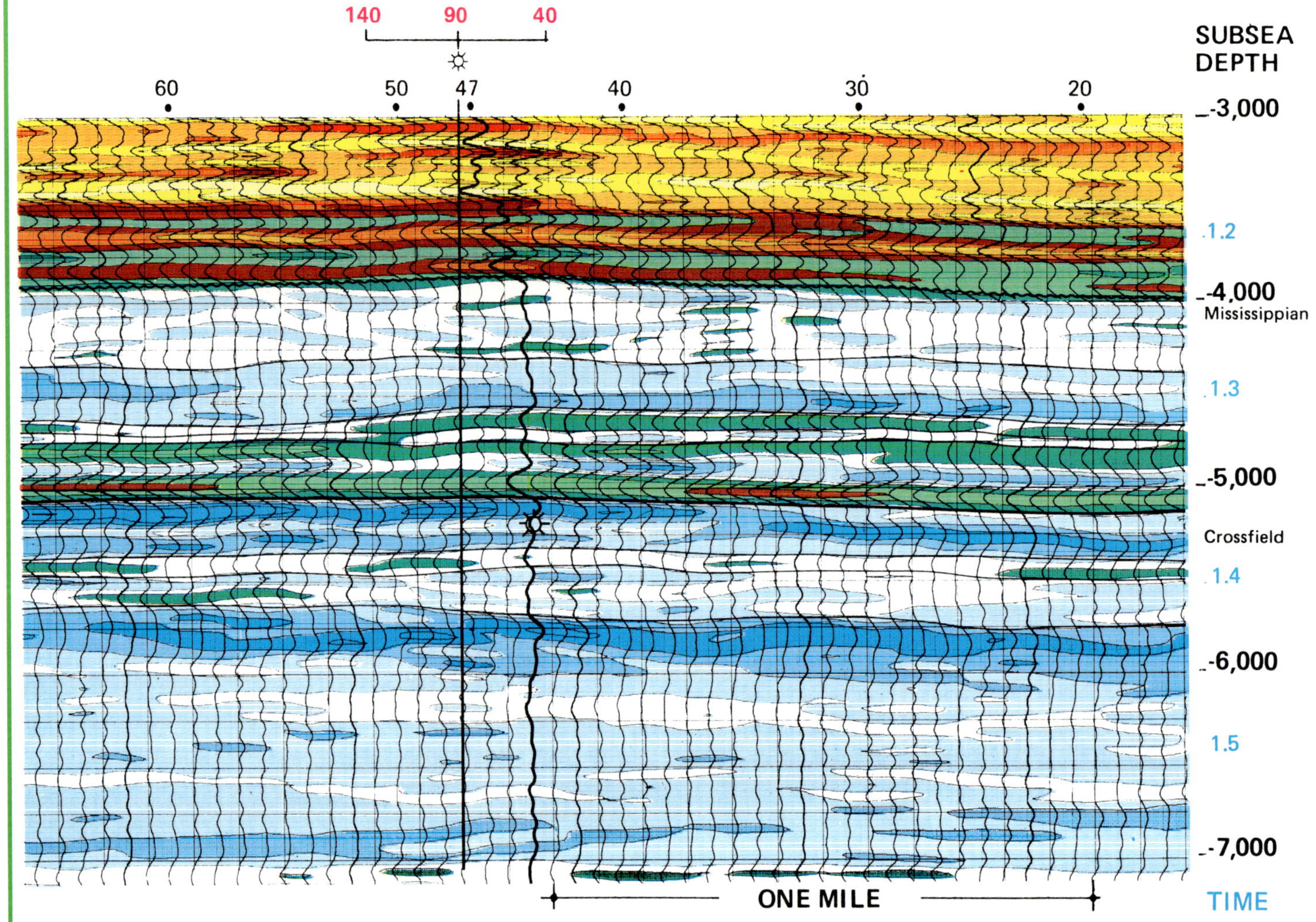

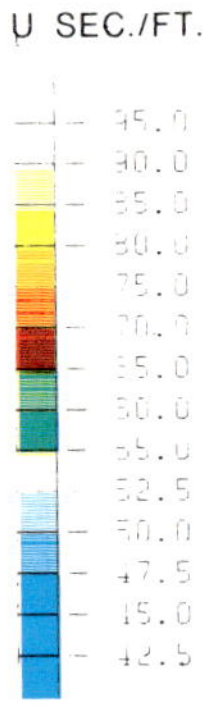

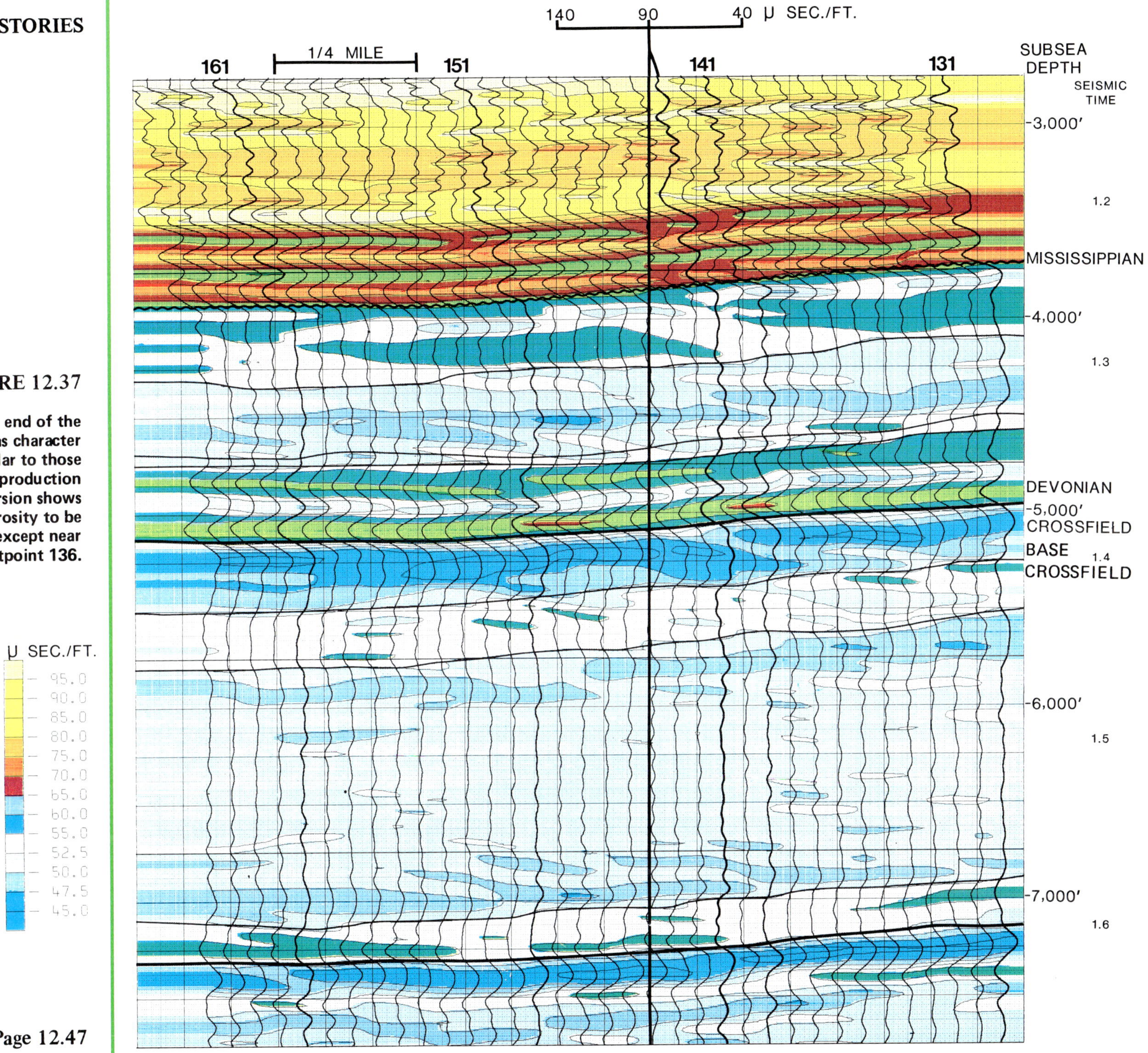

FIGURE 12.37

Although the west end of the seismic line has character changes similar to those associated with production (Fig 12.34) inversion shows insufficient porosity to be present except near shotpoint 136.

Special Considerations and Problems

The inversion of seismic data attempts to produce the best possible approximation to a sonic log, within the constraints imposed by the information content of the seismic field data and the success of the data processing. Sometimes it is very successful, producing logs that match borehole logs with considerable fidelity. But at all times, a limit is imposed by the quality of source data which is, after all, a seismic signal that has travelled through several thousand feet of overburden. In spite of any similarity of the synthetic logs to directly measured borehole data, the user should always remember that the source data is a seismic reflection profile, and the synthetic log output is subject to all of the problems and distortion resulting from wave field propagation that plague all seismic data.

All of the basic considerations for acquisition and processing of high quality conventional seismic data are of at least equal importance to both conventional and synthetic log displays. But the inversion of seismic reflection data is much more demanding (and revealing) of data acquisition and processing quality than conventional output. Conditions which can be overlooked or are ignored, or which may not be recognized on conventional seismic displays, may become extremely critical to the production of good, reliable synthetic logs. These include, but are not limited to, such basic considerations as proper selection of field equipment and procedures, and data processing parameters and routines. Worthy of special emphasis in field operations are: efforts to obtain maximum bandwidth, including both high and low frequencies; the use of small charges (when explosives are the energy source) to minimize distortion of the input waveform; and the avoidance of notch filters where possible. While inferior input data, as in most seismic operations, does not preclude application of a data process to it, the quality of the inversion may be degraded, perhaps to the point where any benefits from the procedure are doubtful.

There is probably nothing to be gained here by reviewing those basic pitfalls and problems common to all seismic operations. Suffice to note that such things as erroneous static and dynamic corrections, improper stacking or incorrect data handling and scaling procedures, will degrade any process output.

SPECIAL CONSIDERATIONS AND PROBLEMS

Some of the problems related specifically to synthetic logs, or having effects which are more critical or become more evident on Seislog displays than on standard seismic sections, have been discussed by Beitzel et al, (1977), Robinson (1978), and Thompson (1979). An inherent advantage of inversion is that the synthetic sonic log can be compared directly to a borehole sonic log at a tie point, so that any differences between them become obvious in a clear and measurable manner. This simple test provides a direct indication of the quality of the seismic source data from which the Seislogs were derived, and the processing applied to it. If a consistent process sequence has been established, and applied with equal precision to all input data, output quality should be a direct measure of the quality of the seismic field data. The results are often surprising, since field data which might be considered poor or noisy on a conventional display may match borehole logs very well, suggesting that the 'noise' or lack of continuity is truly the result of subsurface conditions, while the smooth conventional 'good' data, so dear to many interpreters, may not match the borehole measured conditions very well at all.

It is not uncommon for an inexperienced interpreter to reject the use of synthetic logs because of a poor tie to borehole logs; preferring in such cases to interpret the conventional seismic section. This is his prerogative, but he should realize the conventional seismic data is probably no better than the synthetic logs obtained from the same data and the conventional seismic display only serves to conceal the true nature of the problem. Beitzel et al have provided a good overview of some of these problems and how they affect synthetic logs.

If the discussion of problems and/or constraints were to list the three major items of particular importance to inversion, they would probably include; bandwidth, phase and velocity. All three are related and process considerations for all three are somewhat interdependent.

A primary requisite, for both data acquisition and processing, is to obtain sufficient bandwidth. Loss of any portion of the frequency spectrum distorts the output in a somewhat predictable manner. Frequencies can be lost from either end of the spectrum, or from the middle, and each loss produces a specific type of damage to the output. Of these, loss of frequencies at the high end of the spectrum causes the least amount of harm, resulting primarily in loss of resolution. Fortunately, under good conditions, it is generally possible to obtain a sufficiently high frequency cutoff of the data so that deconvolution will restore a frequency range sufficiently high to permit resolution of any play which is within the range of acceptable economic return. Obviously there will be exceptions to this sweeping generality as there will be to the following, equally general statement: with good data it should be possible to resolve any given member whose thickness is 1% of depth. Much thinner units may be detected, and changes within them mapped correctly, but the absolute resolution may not be reliable.

The difference between detection and resolution may be considered in terms of tool spacing. When a given member is at least equal in thickness to the tool spacing, it will be possible to resolve the bed, but when the bed is thinner than the tool spacing, although it will be detected by the tool, it will not be resolved, since the tool space will straddle the bed and average its properties with some of the surrounding material.

Another problem resulting from a loss of high frequency is the inability to map precisely a very sharp velocity step, such as a shale/limestone contact, which requires a very broad band of frequencies in the log. Obviously, given the relatively narrow frequency range of seismic signal input,the synthesis of logs will yield something less than perfection, but equally obvious, this problem is not fatal. Like the loss of thin bed resolution, loss of sharp contact definition is also similar to the results of logging with a long space between source and receiver. The long spacing averages varying proportions of adjacent material over the interval, spreading its response on the log over the length of the tool spacing.

SPECIAL CONSIDERATIONS AND PROBLEMS

A seismic signal should have at least three octaves of frequency bandwidth in order to construct a reasonable approximation to the sonic log. If less than three octaves in width, or if the upper limit drops much below 60 Hz , then the inversion will yield poor results, unable to map sharp contacts, nor, in response to another phenomenon of narrow bandwidth, able to map intervals of uniform velocity successfully.

At the other end of the spectrum, loss of low frequency information is of much more serious consequence. Lack of the low frequencies is a chronic property inherent in nearly all seismic field data. Low frequencies contain the basic velocity structure of the log. The loss of even a few of the very low frequencies causes dramatic effects. The low frequency cut synthetic log may appear to be merely a phase shifted version of the seismic source trace. As discussed earlier, the low frequencies not present in seismic field data can be replaced, to a degree, from velocity analyses.

low frequencies contain the basic velocity structure of the log.

Finally, and a consideration which will be discussed in more detail later, the spectrum should be continuous, including all frequencies within the band limit. A notch in the spectrum, where a few frequencies have been lost, produces undesirable distortion. Let us now examine these, and some other problems in detail.

Figure 12.16 is a Seislog section resulting from a line of relatively good seismic reflection data. It is an early effort, before the introduction of machine contouring and colouring and prior to some of the current improvements in processing technique. At the right side of the illustration the trace of a synthetic sonic log at a tie location has been accentuated, and beside it, plotted to the same scale, is the real sonic log for comparison. The correlation would be considered good, even by today's standards, yet the synthetic differs from the sonic log in a number of ways.

The most obvious difference, and the simplest to explain, is the difference in resolution. The sonic has considerable fine detail which is not present on the synthetic. Note particularly the response at the level of the thin, very sharp spike above 7000 ft. Such losses in resolution are the natural result of the loss of high frequencies, an effect aggravated by increasing depth. To a degree, weak or attenuated high frequencies can be restored by deconvolution, but only if the remnants of high frequency which remain in the input data are indeed signal and not noise.

Nevertheless, this common problem is not the most common source of local small velocity and depth errors on synthetic logs. **Figure 12.38** reproduces, on the left, the lower portion of a Seislog synthetic and a borehole sonic, similar to those of **Figure 12.16**. Part of the difference between the two logs is merely in resolution. High cut filtering the sonic log to approximately the high frequency limit of the synthetic, yields the closer match on the right. Other variations may appear to be cyclic, which indeed is often the case.

A very sharp velocity step, such as the shale/limestone contact at the top of the Swan Hills member, requires a very broad band of frequencies in the log. Filtering the sonic log has increased the slope of the contact at this level, while the Seislog evidently has a still narrower bandwidth, as demonstrated by its very rounded slope. Obviously, given the relatively narrow frequency range of seismic signal input, the synthesis of logs will yield something less than perfection.

Figure 12.39 illustrates the effect of bandwidth on a spike which might be a reflection coefficient. Note that at four octaves, the polarity of the spike is clearly defined and it is very narrow in time. At two octaves it broadens and there may be some doubt whether polarity is a single positive pulse, or two negative pulses. At the depth

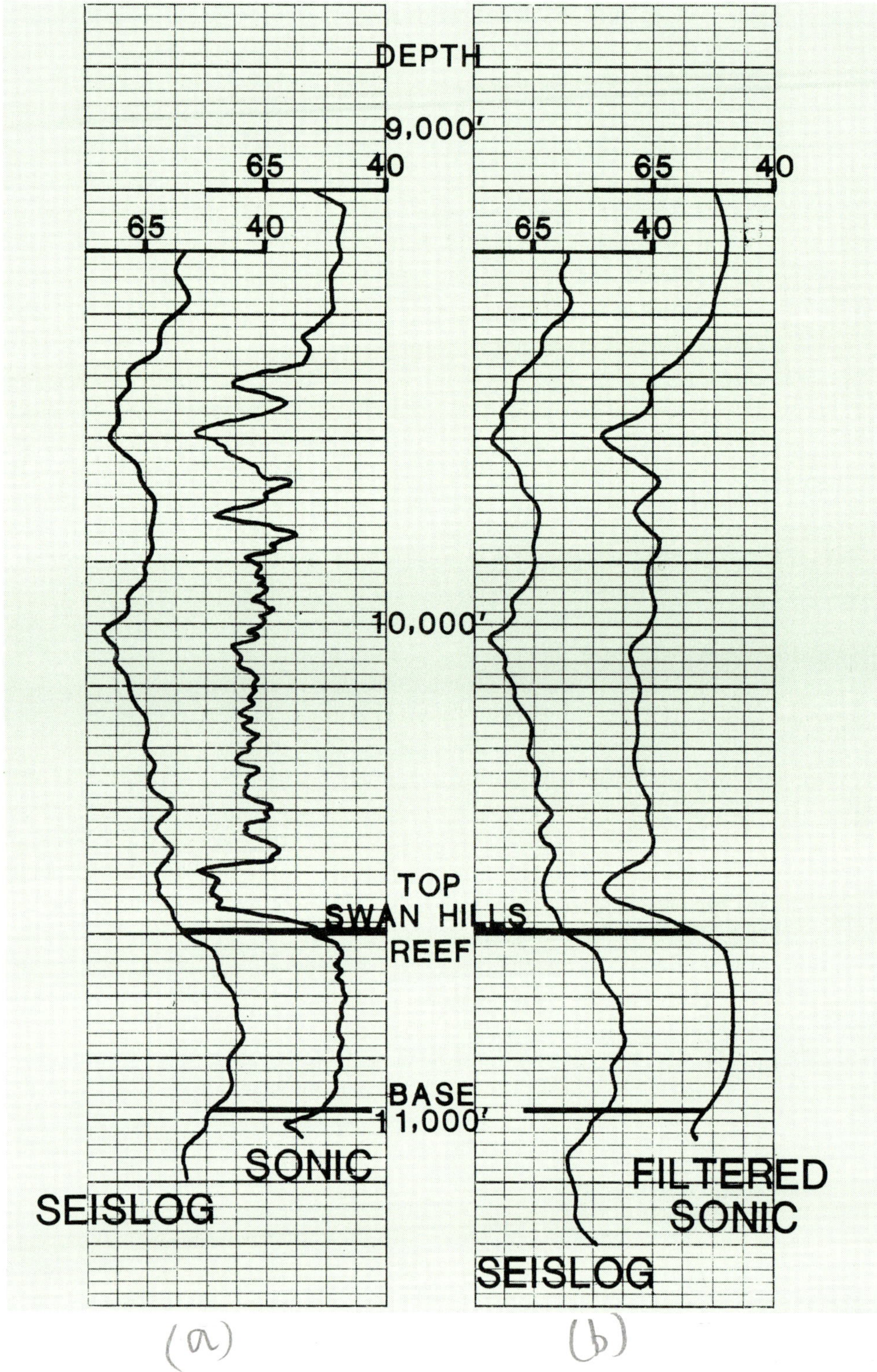

FIGURE 12.38

(a) Comparison between a synthetic log and a borehole log. The major difference is due to higher frequencies in the borehole log.

(b) Filtering the borehole log improves the match. Other differences are due to bandwidth limitations.

of 11,000 ft on **Figure 12.38** the bandwidth is barely sufficient to define the log.

There is also another problem on **Figure 12.38**. There are other differences, not due to missing high frequencies. The upper part of the synthetic log appears to track the borehole log very well. Below 8000 ft not only does resolution degrade, but the deviation between the two logs increases indicating less reliable absolute velocity values of the synthetic log.

The transit time values of the Seislog, which are the reciprocal of absolute velocities do not track the sonic log very well. There are two possible causes for this, both related to low frequency. One is the diminishing reliability of velocity determination with increasing depth. The other is the absence of certain low frequencies. The spectrum should be continuous, having all frequencies within the band limits included.

Construction of a good synthetic log requires missing low seismic frequencies to be replaced by low frequencies extracted from velocity information, but in areas of poor velocity control it may not be possible to develop reliable low frequency information above 5 Hz. Similarly, due to the type of geophones used, or the

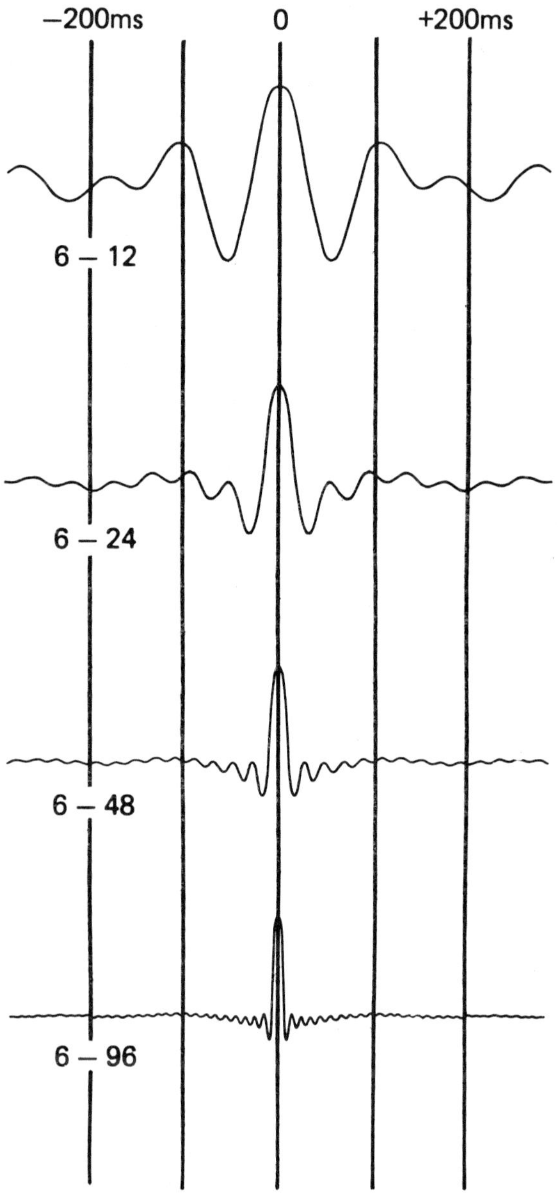

FIGURE 12.39

Seismic definition of a single reflection coefficient diminishes with decreasing bandwidth.

type of energy source, or as the result of a frequency filter applied in the field, the lower end of the useful seismic signal spectrum may be limited to, say, 12 or 15 Hz. When this gap in frquency exists, and it is not infrequent, the spectrum of the synthetic log will have a notch, say from 5 to 12 Hz. **(Fig 12.40)** The effects of this notch on the log can be best illustrated by a model. **Figure 12.41** includes a sonic log **(Fig 12.41a)**, and the same log notch filtered **(Fig 12.41b)** to remove a narrow band of frequencies from 6 to 11 Hz **(Fig 12.41c)**. Although the filtered components of **Fig 12.41c** are plotted to the same scale as the others, the velocity values oscillate about zero, defining positive and 'negative' velocities. The sum of the two signal components will yield the original log.

At a cursory glance the notch filtered version appears similar to the source log, but closer examination reveals several levels where the absolute velocity of the notch filtered log is higher or lower than the original. These variations correspond to the oscillations of the 6-11 Hz components about the central baseline. Summing of these alternating positive and negative velocity values to the notch filtered signal would equal the original log.

The notch is one of the most frequent, and in some cases, most serious problems encountered in the inversion process, for incorrect velocity can lead to erroneous interpretation of lithology. Fortunately, the relative lateral changes in velocity are usually correct, and will still define significant changes in lithology, at the zone of interest.

The type of problem resulting from a notch is very well illustrated by the example of **Figure 12.42**, a comparison between a filtered borehole sonic log and a section of synthetic logs. The sonic log is first shown without filtering, and beside it is plotted the same log, high cut filtered to 75 Hz to make the resolution more compatible with the field recorded data. It has also been filtered from 3 to 10 Hz, a somewhat broader than usual notch filter.

Within the limits of resolution the notch-filtered log matches the unfiltered version fairly well, until it reaches the thick shale zone extending below about 6400 ft. There the notch-filtered log is unable to maintain the integrity of the

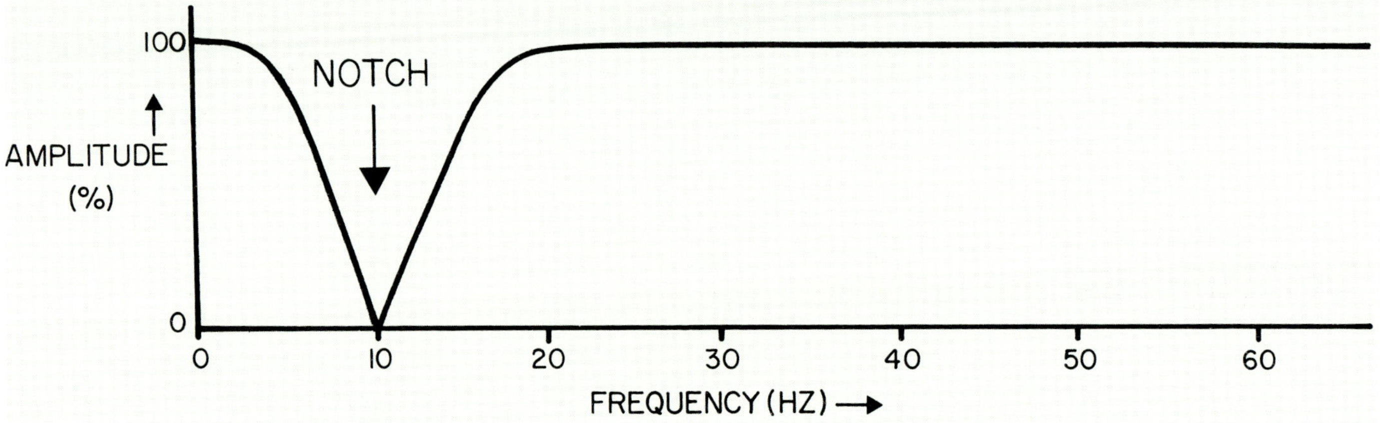

FIGURE 12.40

Model spectrum of synthetic components which fail to cross over in frequency. The notch causes velocity distortion.

shale velocity, and it bulges to the right, producing a false impression of a high velocity sequence within the shale. Note that the relative changes within the shale are preserved, with the higher velocity members showing the appropriate contrast of the proper position.

The Seislog section, which ties to the sonic, shows a very similar response, suggesting that it too, was deficient in the notch frequencies. The shale apparently contains a fairly thick sequence of much sandier facies than is proper. However, note that across the section, there are changes in the individual members. Such relative changes usually correspond faithfully to the changes taking place within the sediments. Hence, even though the absolute values may not be correct, when the low frequency notch is present, relative changes within them may be valid.

Analysis of the source of error in low frequencies determined from velocity data is not simple. There are at least two principal possibilities. One is simply the error inherent in calculation of true velocity from stacking velocities. Derivation of true instantaneous vertical velocities in this manner is fraught with many difficulties. The quality of results is highly dependent upon the data and the software used to extract such velocities. Velocity determination procedures are well and plentifully documentated in the literature. (Al Chalabi, Brown, Cook and Neidell, Robinson, Tanner and others). Velocity measurements may be distorted by a number of causes, including near-surface phenomena and elevation changes. Dip and anistropy distort the apparent velocity. Levin has shown the stacking velocity analyses are weighted by the horizontal component whenever anistropy is present and cannot yield true vertical velocities.

This appears to be the explanation for the fact that velocity determinations through shale sections are frequently too high. Thin streaks of high velocity material may distort the velocity function.

The errors in velocity determination may vary along a seismic line, particularly across an abrupt discontinuity, but there is very little published information about the relationships between velocity and geologic structure. Across a reef front, for example, velocity does not appear to drop abruptly to that of the surrounding shales, but changes gradually as a function of the amount of talus and near reef material contained in the fore-reef facies.

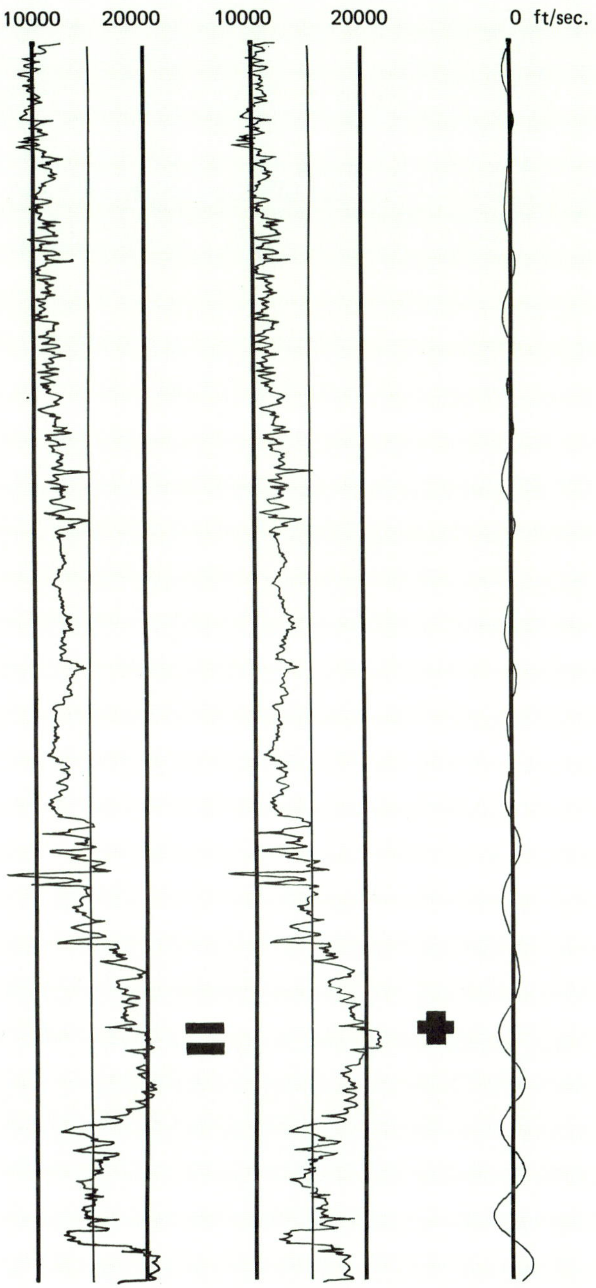

FIGURE 12.41

Loss of a small notch of frequencies (right) caused the log (centre) to oscillate with alternately too high and too low velocities.

Similarly, it appears that velocities may not change abruptly across major faults but pass through a transition zone from one side of the fault to the other. The effect is shown in **Figure 12.45**, where velocity contours flow across a fault into a low velocity graben, rather than breaking sharply across the fault. Yet, undoubtedly, part of this effect in the illustration is simply produced by the mathematical function applied to the velocity surface to smooth out some of the irregularities. Also, it is not possible to obtain a reliable velocity estimate from seismic data near the fault. When crossing the fault with the line, the wide spacing between source and receiver inevitably places the source on one side and the receiver on the other side of the fault for part of the determination, distorting the travel times and hence the velocity.

The fact that velocity, and hence low frequency, may not be completely reliable is an argument used frequently to not use it at all in the calculation of synthetic logs. The problem of imperfect geophysical data is not a new one, and geophysicists accept such conditions as a usual condition of interpretation. They learn to recognize such possibilities but nevertheless use all the data they can. The following example may help support this approach.

Figure 12.44 is a portion of a long seismic section shot across a large fault lying below an unconformity about 1.2 seconds. Flat undisturbed beds lie over numerous faults which evidently cut the deeper section. Displacement across the faulting is not exactly clear due to changes in reflection character, the number and character of reflections and the diffractions. There is substantial interval change across the major fault, near shotpoint 100 or 120. The deepest more or less continuous band of reflections on either side represents the basement, which is covered by a thin layer of Devonian carbonates.

The results of the velocity analysis sequence used to recover the low frequency components are normally plotted as a quality control procedure in the form of a velocity profile or section as in **Figure 12.45**. The

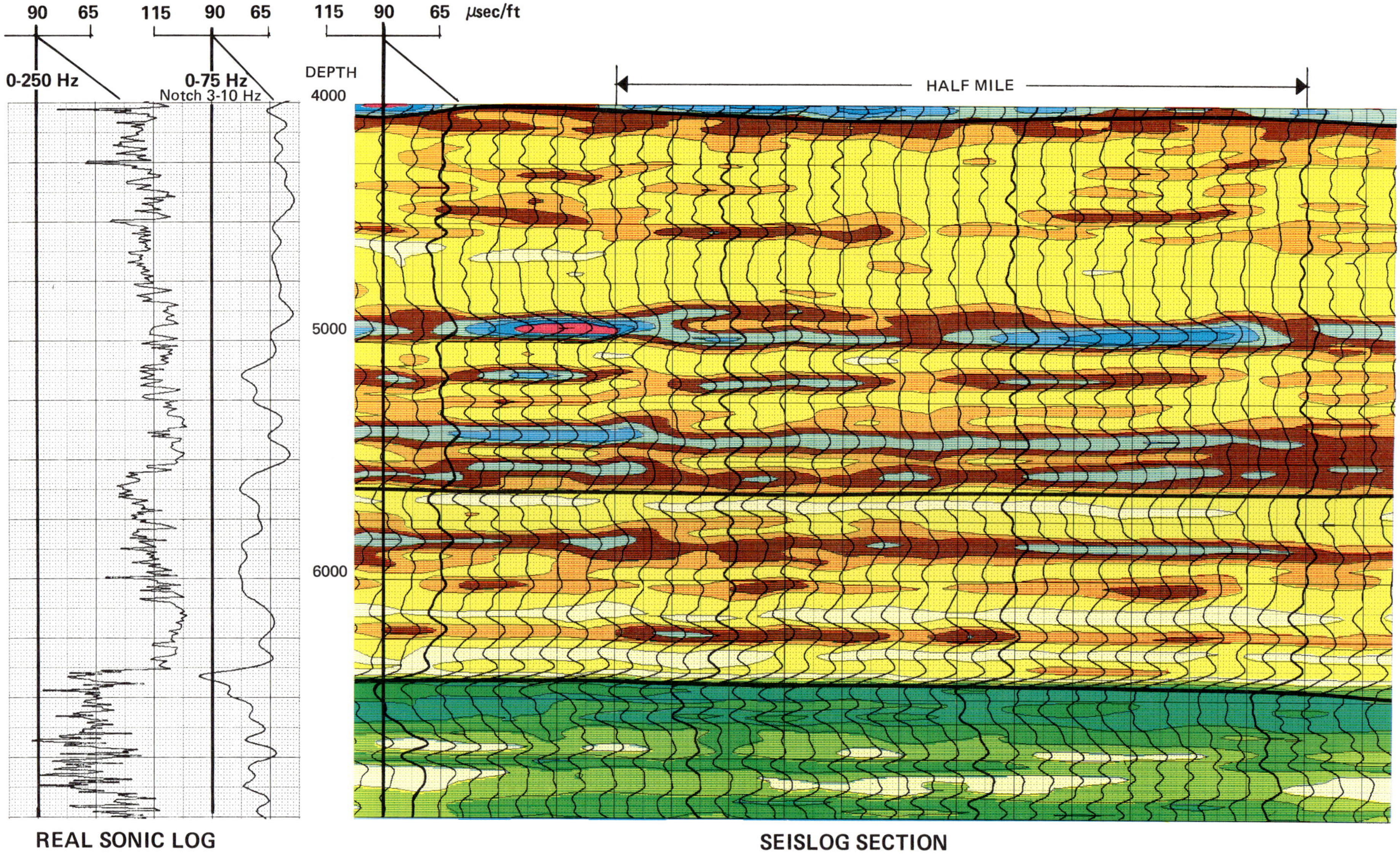

FIGURE 12.42

Note the distorted absolute velocity values on the notch filtered log below 6000 ft. A similar response exists on the Seislog section.

velocities have been contoured, and the interval color coded as indicated by the color bar on the right side of the illustration. Shotpoints and velocity analysis control points are illustrated at the top of the section. Vertical pip bars plot calculated instantaneous velocities on the left side, and average velocities on the right. Also included are time reference markers (see, for example, 0.5 sec) and small pips indicating depth increments of 1000 ft. The vertical profiles are filtered instantaneous velocities which represent the low frequency velocity structure at each location. The vertical dimension is time. Gross indications of the faulting are provided by the results of velocity

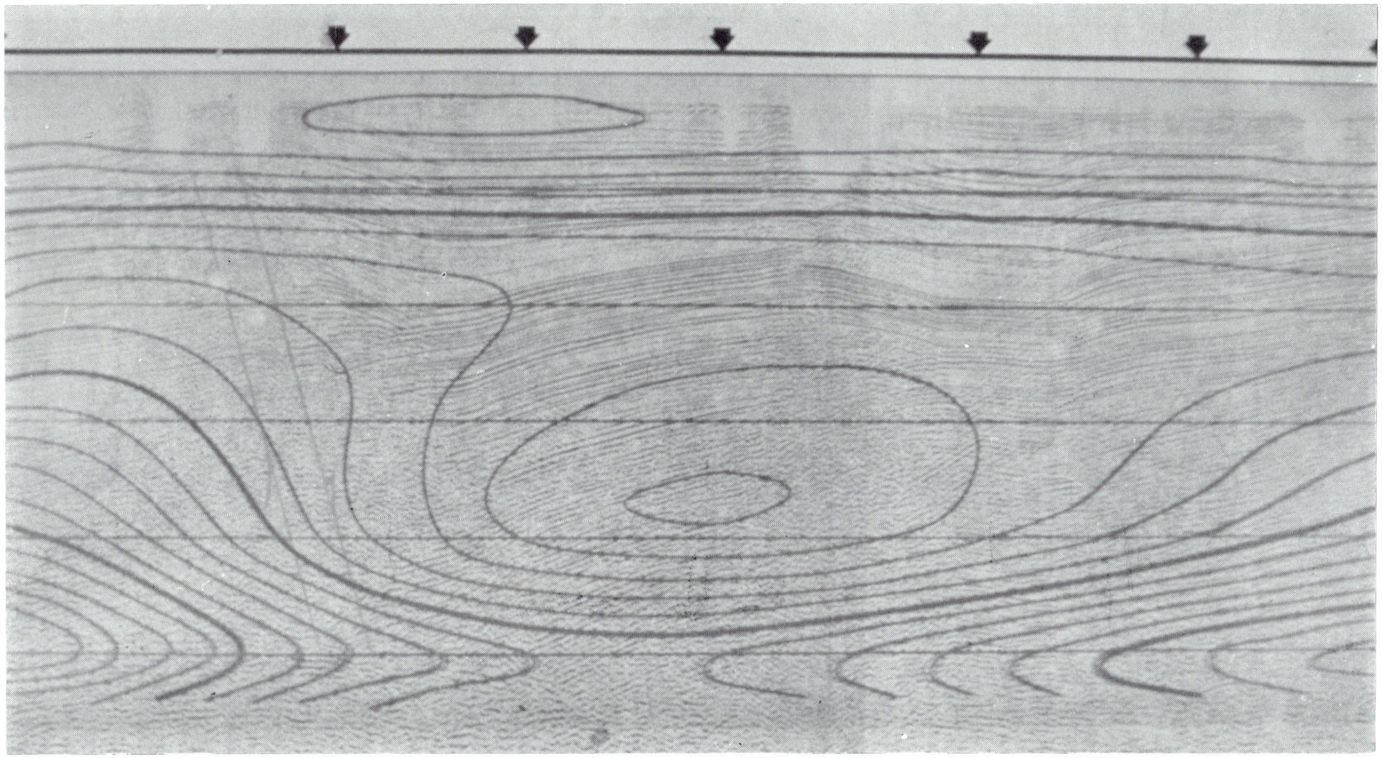

FIGURE 12.43

Velocity contours across a large graben in relatively young sediments. Faulting distorts the determination of velocities, but the gross changes related to the displaced block are evident.

surface mapping.

The most interesting part of the velocity profile is below 1.0 second. The extreme right side and left side of the section can be readily correlated, confirming the displacement of the faulting. The shape of the velocity profile through the thick blue section over the orange, (1.5 sec on the right) can be correlated upward to the left, but the velocity drops slightly, possible due to decreased loading. Between the endings of the well defined velocity components is an indeterminate zone across the faulting, where the displacement and the velocity are not well defined. The slope in the orange coded velocities above suggests much of the faulting takes place about shot-point 111. The high velocity (light blue) material about 1.3 seconds on the right side of the section is not present at the left, suggesting it may have been faulted upward and eroded. Few boreholes were available to verify the velocities and most were shallow.

Confidence in the velocity surface is increased by superimposing it on the seismic section (**Figure 12.46**). The fit to the reflecting horizons is very good, and provides some aid to initial interpretation. A suggestion of an unconformity near 1.2 sec is provided by the wedge of sediments (blue) lying below the continuous strong reflection. The displacement of the major fault is suggested by the offset of the deeper band of color. The wide gaps in color continuity of the deep red unit across the fault are believed to be most likely due to the limits of resolution of the velocity program, but possibly be the result of stratigraphic changes. It is impressive to see how closely the bands of color, corresponding to velocity intervals, follow reflection intervals in the shallow part of the section above 1.0 sec.

SPECIAL CONSIDERATIONS AND PROBLEMS

Seismic data inverted and combined with the low frequency velocity information to yield SEISLOGS.

The fit of the velocity data to the seismic section must provide some degree of confidence in the fit of the data and in the technique of velocity mapping. This may be enhanced when the seismic data is inverted and combined with the low frequency velocity information to yield Seislogs, and the resulting Seislog is correlated to a sonic log, **(Fig 12.47).**

The tie point is a deep stratigraphic test drilled to basement some years ago. Fortunately, it was logged with a sonic.

Although the reader should make his own evaluation and draw his own conclusions regarding the quality of correlation between the Seislog and the sonic the following commentary may be useful.

For some reason, probably due to the limited bandwidth of the field data, the sharp step near the top of the correlation window is not matched by the Seislog. Since this is quite shallow, it would be expected that a better fit would be obtained.

However, although the fit leaves something to be desired at this level, it fortunately is, if anything, somewhat better than might be expected deeper in the section.

There are two major unconformities in the section, one at the top of the Permian, about 3000 ft subsea, (6000 ft drilled depth) and at the top of the Mississippian, near 5000 ft subsea.

The shift to high velocity at the upper unconformity is predicted quite well by the synthetic log as is a similar shift at the top of the Mississippian. Obviously both of these tops can be mapped quite successfully on the seismic section, but the synthetic sonic logs show them both clearly to have high velocity material, probably limestone, at the subcrop, information not evident on the conventional seismic section **(Fig 12.44)** although it is common to the overlay of the velocity profile on the seismic **(Fig 12.46)**.

Not well defined on the seismic section is the sequence of clastic sediments extending a few hundred feet above and below the 6000 ft subsea depth marker. This series of sands and shales generates a series of conventional reflections, but inversion produces the detail of the relative velocity changes which conform very closely to the borehole sonic. Resolution and detail through this part of the section is excellent, and although it starts to deteriorate somewhat in the deeper part of the section, it still matches fairly well at the top of the Mississippian carbonate section near 6200 ft subsea. From this level to the top of the Devonian carbonates below 9000 ft subsea the changes on the sonic log through the massive carbonate section are relatively small, on the order of 2 to 5 μsec/ft. Nevertheless, they are tracked with reasonable fidelity by the Seislogs.

Evident diminished resolving power is noted near 9000 ft subsea, where a fairly well defined shale break, some 300 ft thick lies at the base of the Mississippian, resting on the high velocity Devonian carbonate.

For some reason, although the Seislog recognized a low velocity zone, it shows far less than the proper excursion, agreeing in substance but not in value.

The log indicates a return to high velocity at the top of the Devonian carbonates, near the total depth of the hole. This is matched by the Seislog, which suggests the Devonian carbonates may be as much as 500 ft thick here, before encountering the base of the basement.

Interestingly, the well produces copious quantities of gas from the top of the Devonian section, probably from fracture porosity, since it appears to lie next to a fault below 12,000 ft drilled depth and in nearly the deepest part of the section. It certainly is not a place that one would be inclined to search for production.

The quality of the log fit, in spite of the rough data, and at fairly considerable depth, provides encouragement and

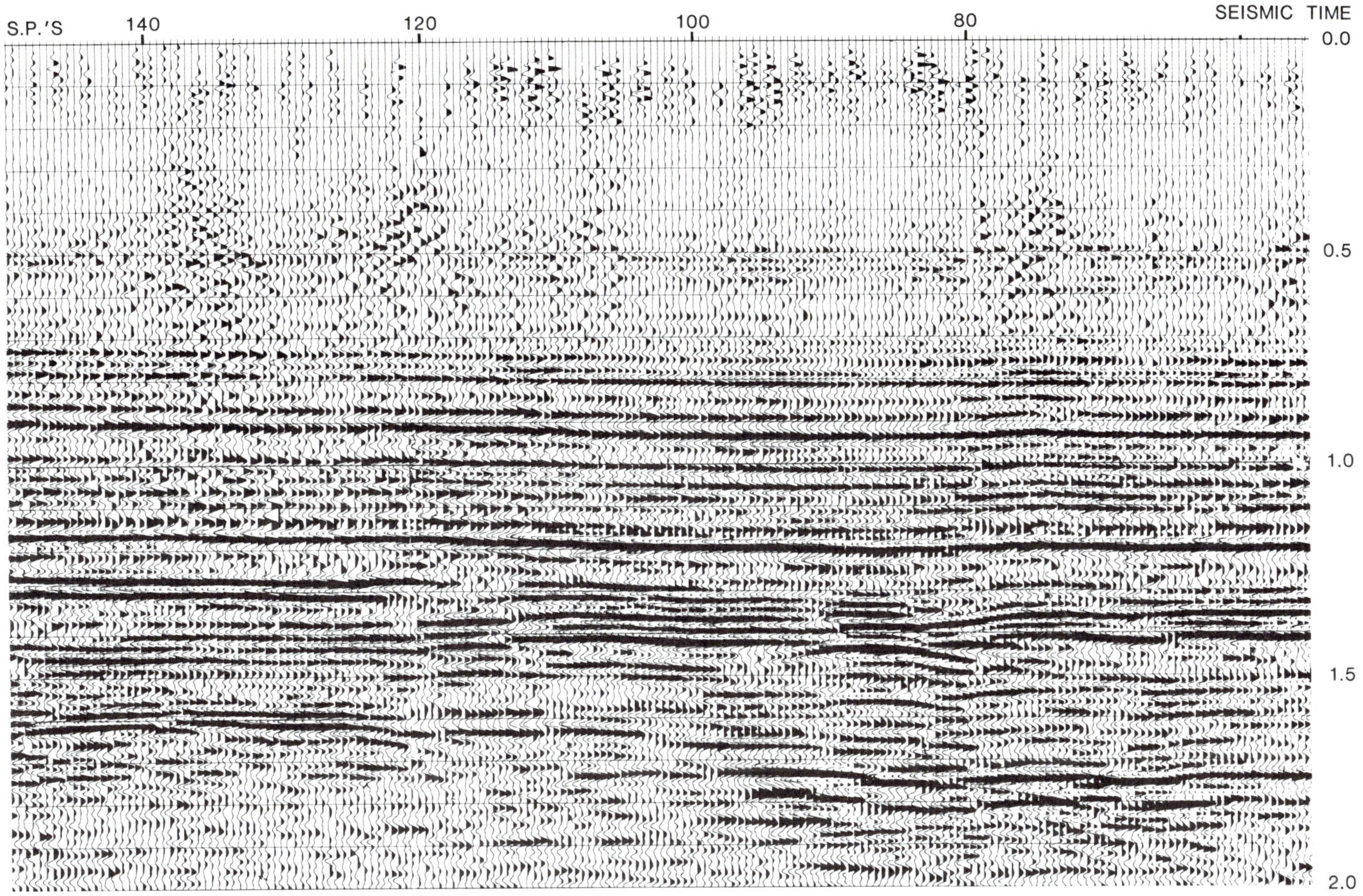

FIGURE 12.44

Conventional seismic section across a heavily faulted area.

confidence in the inversion method, suggesting that it can synthesize borehole sonic logs with a reasonable degree of precision.

A window of detailed seismic data is illustrated in **Figure 12.48** and the corresponding Seislog data in **Figure 12.49**. Firstly note the changed proportions. The Seislog section is near true scale with about 5:8 vertical exaggeration, while the time section, due to the high velociy of the rocks, is compressed.

Much more faulting has been interpreted on the Seislog section than on the conventional seismic section because some very small displacements appear to exist on the Seislog display. Three ages of faulting are shown, old faulting, which cuts the basement and Devonian sediments, but disappears in the Mississippian shale unit at about -9000 ft, extensive faulting cutting the top of the Mississippian carbonates, where it appears to have produced fault block escarpments displaced from 30 to 100 ft; and a few younger faults, which appear to have been initiated at the end of Mississippian carbonate time, and had renewed growth in late Mississippian or early Permian time, but were subsequently eroded before deposition of the overlying sediments.

Faulting must be interpreted to some degree because of the Fresnel effect, which 'smears' the evidence of fault displacement over two or more traces rather than showing a sharp break. Seismic reflections are not returned from a point in the subsurface, but rather, like the beam of a flashlight, spread over a wide circle which increases in diameter

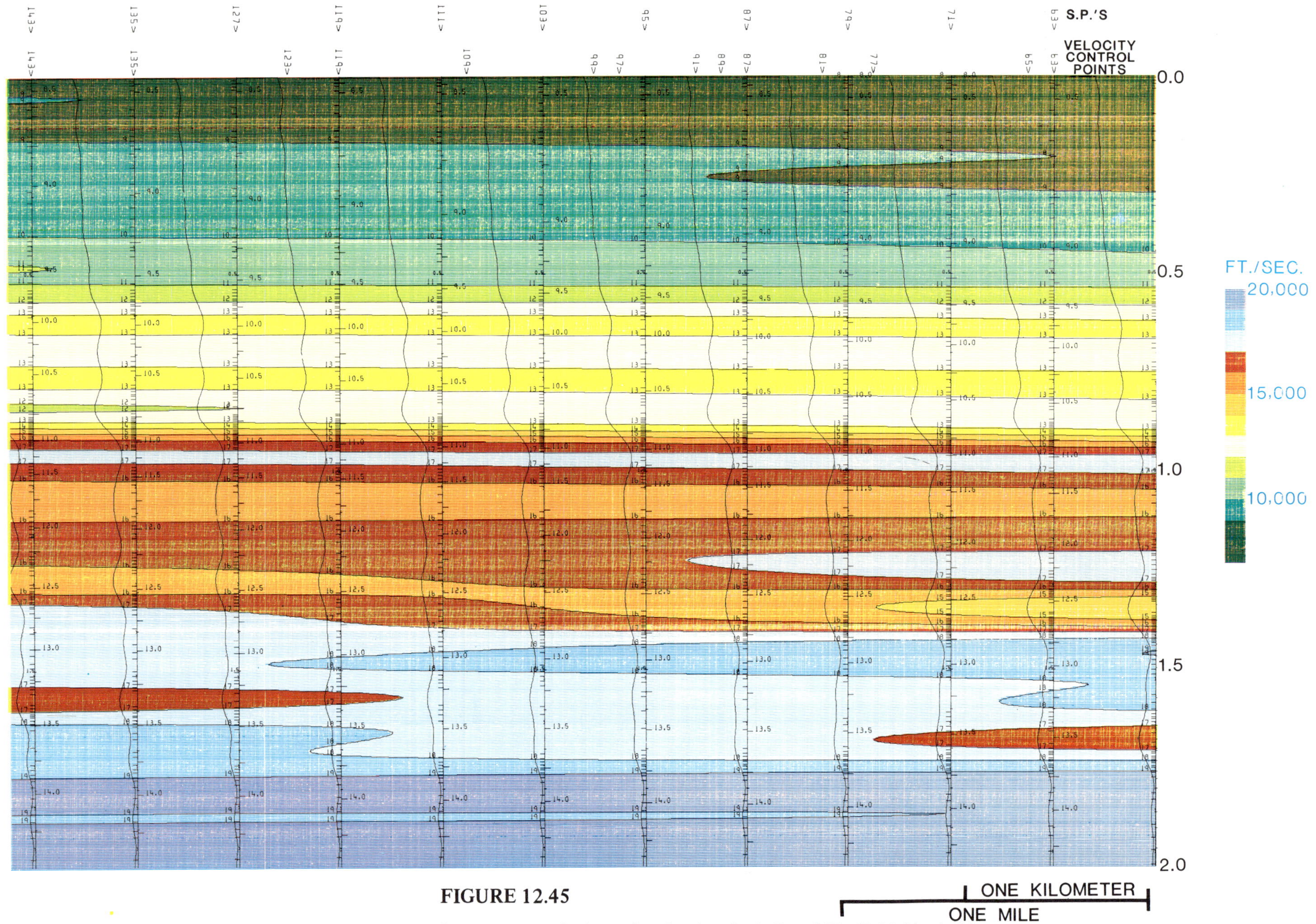

FIGURE 12.45

Instantaneous velocity surface for the seismic line of Fig 12.44. Vertical bars measure instantaneous velocity on the left, average velocity on the right. Line profiles graph instantaneous velocity at 6 Hz.

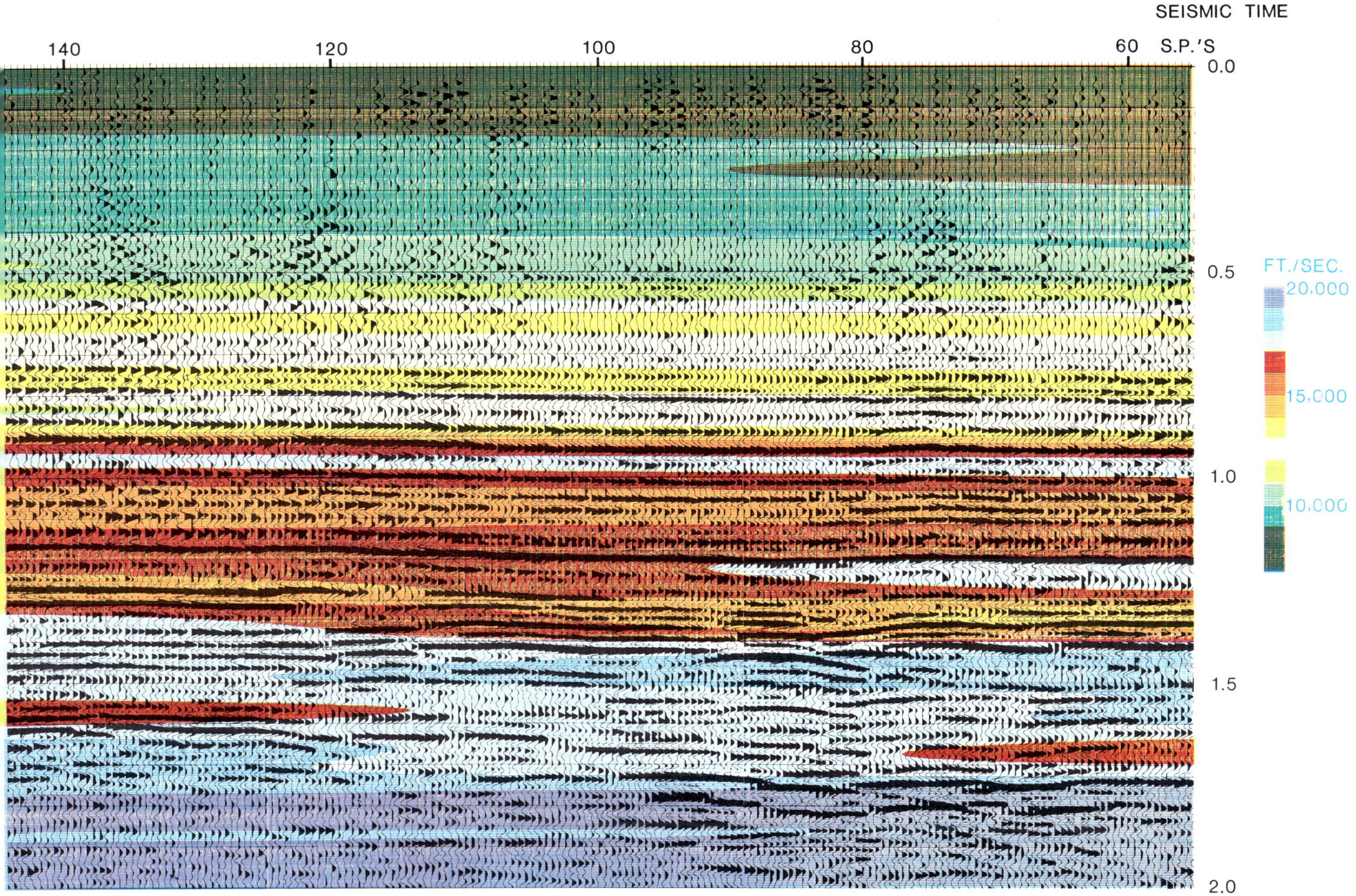

FIGURE 12.46

Superimposing colored velocity code of Fig. 12.45 on the seismic data illustrates the conformity of velocity intervals to seismic reflections.

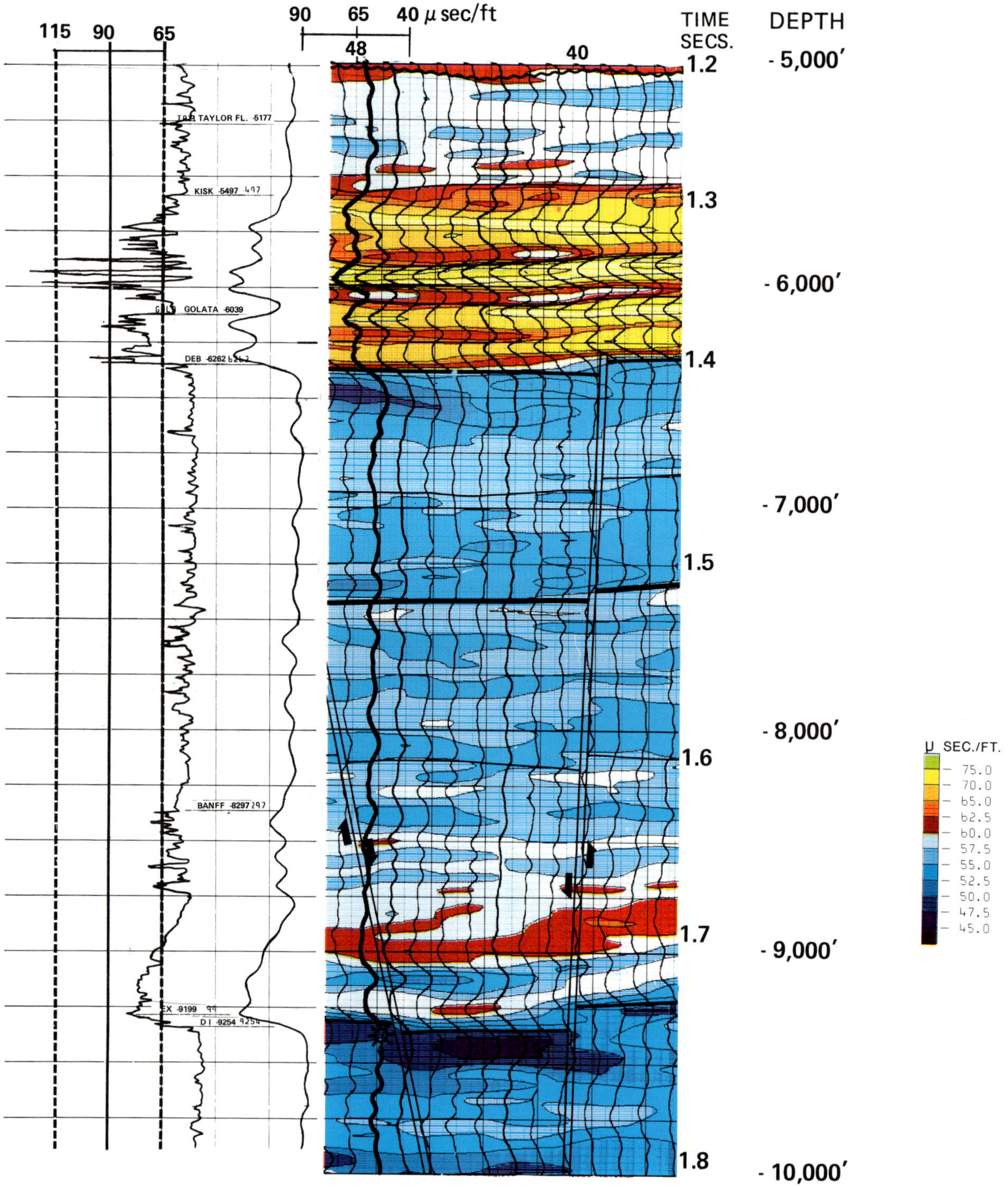

FIGURE 12.47

Comparison of a borehole log high cut filtered to seismic frequencies, with the synthetic log. The match is good, but deteriorates somewhat with increasing depth.

with distance or, in this case, depth **(Fig 12.50)**.

The data have not been migrated. Hence some diffractions may be expected across fault planes, adding to the Fresnel effect.

Faults can be detected by noting the general continuity of Seislog correlations which, in this example, tend to be flat and smooth. Frequently, such correlations are displaced vertically, suggesting a fault. While such evidence may be excessively flimsy if provided by only one horizon, several faults, including those near shotpoints 41, 47, 63, 70, 77, 83 and 91 all have similar displacements at several levels, which can be traced along fault planes to the basement.

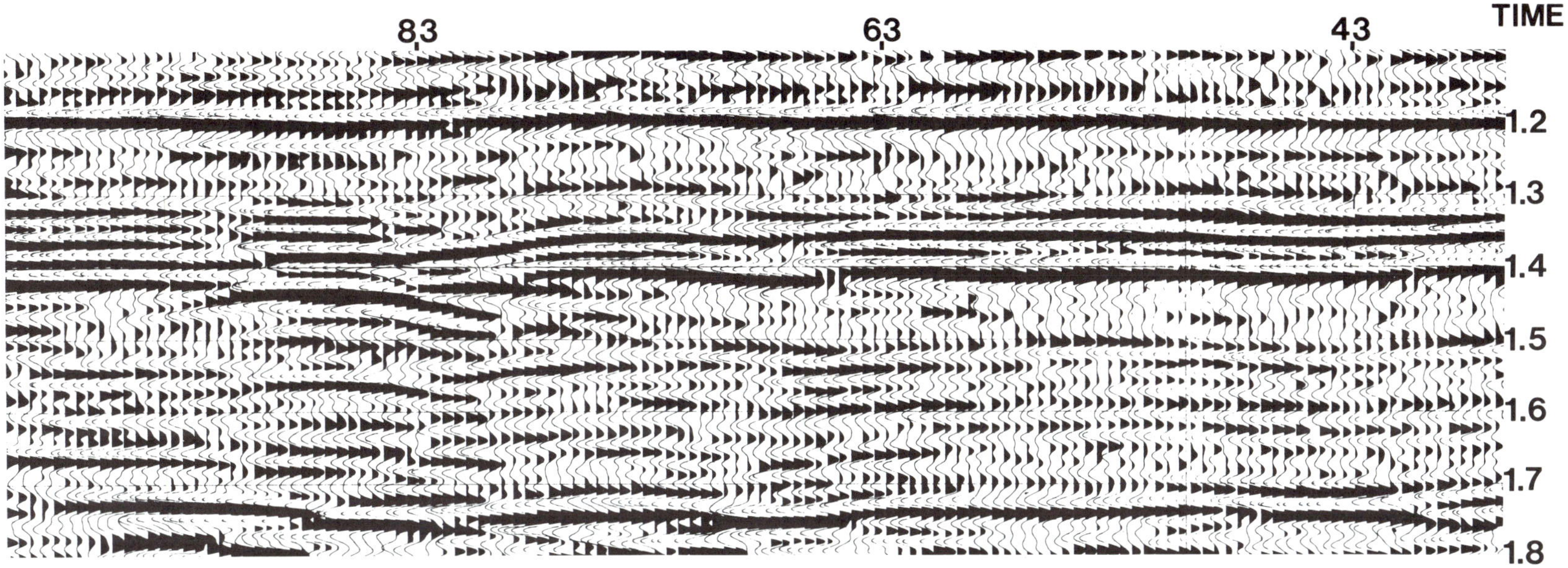

FIGURE 12.48

The seismic section across a faulted zone shows diffractions and the Fresnel effects which obscure the fault trace. Lithology varies substantially within the interval, a condition not evident on the seismic data.

These correlations are worthy of some study, since they are good illustrations of geologic phenomena on Seislog sections, and demonstrate what is probably a near state of the art definition of the subsurface.

Note, the varying types of fault related features near the top of the Mississippian carbonates about 6000 ft subsea. Faults at shotpoints 40 and 63, for example, clearly terminate at the top of the carbonate section (blue), and the resulting scarps are smoothly covered by the earliest members of the overlying clastics. Some detritus may rest on the downthrown side of the scarp, or it may be merely the effects of diffractions and/or the Fresnel zone.

Faulting across the small keystone graben is more complex. The fault on the right side appears to have had some growth through the time of the clastics deposition, since it is displaced appreciably more at the base than at the top. The change in movement appears to have taken place about the time of the well defined central shaly low velocity member, whereas the displacement of the bounding fault on the left side is much more uniform, suggesting it did not have any late movement until after the entire clastic interval was laid down. Meanwhile, the central fault between the two, was displaced in early clastics time, to move no more, since the high velocity member and the middle shale unit above cover smoothly over it, but sag to the right to accommodate growth of the fault.

The detail of this section is very impressive indeed, and the above observations could be expanded upon at great length. It may become obvious to the reader that such detail leads to a substantial increase in the time required

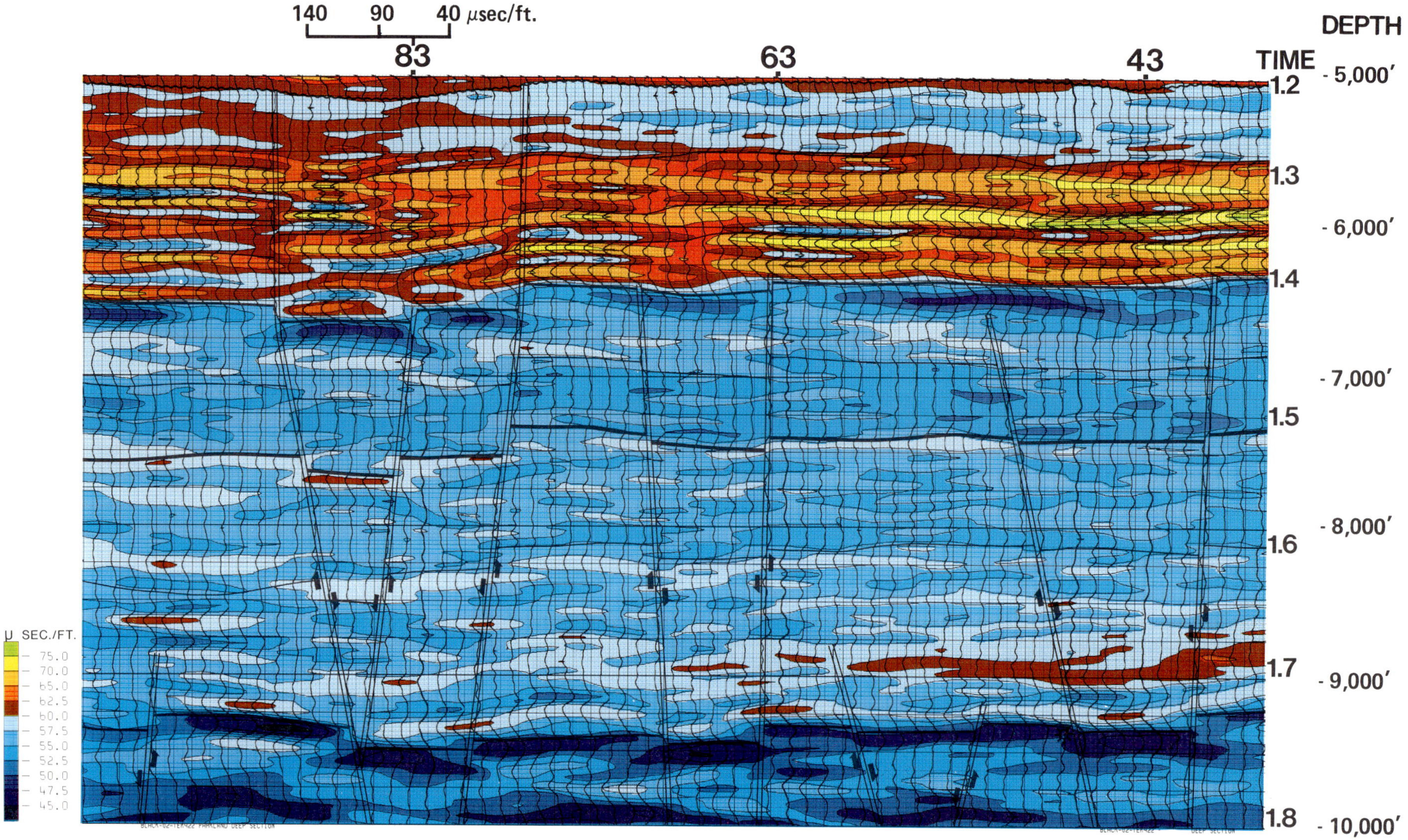

FIGURE 12.49

Extensive faulting can be separated into three stages, or movements, and careful analysis of the intervals provides an indication of when growth occurred.

to interpret a seismic section with any degree of comprehensiveness, compared to the conventional seismic display. The example has served to illustrate some of the difficulty in determining absolute velocity values, particularly at depth on synthetic sonic logs. Stacking velocities do not always yield the desired precision, and there may be some hesitation to use them. However, in spite of their deficiencies the low frequencies determined from velocities provide an important and valuable contribution to the interpretational process, yielding information which is substant-

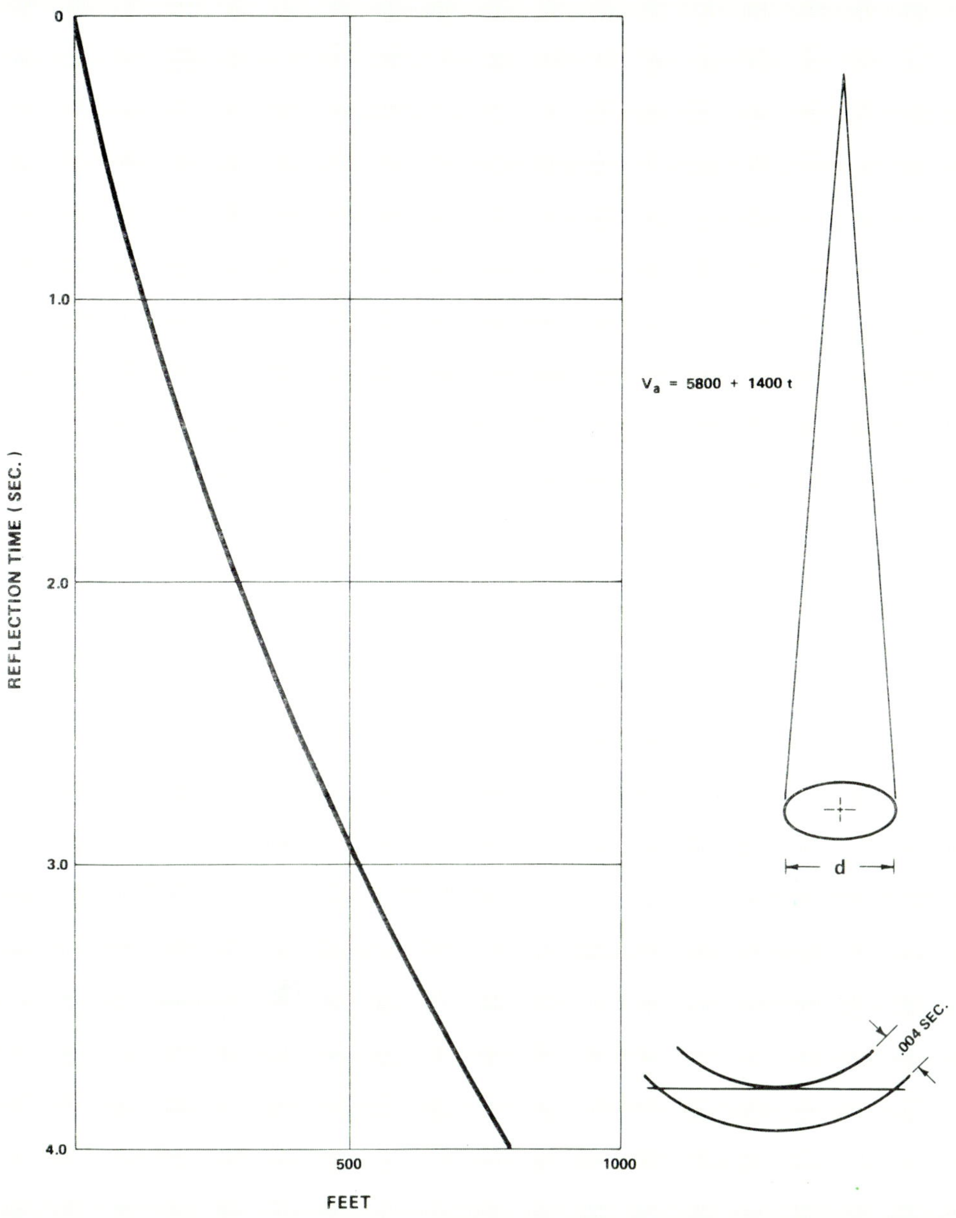

FIGURE 12.50

The heavy line shows the width of the Fresnel zone.

Phase

ially reliable and often offering the only control for evaluation of lithology.

One of the most difficult problems in any seismic data processing operation is the correct determination of the phase, particularly the amount of the phase distortion imposed on the signal by the data acquisition system. In recent years, it has been realized that adequate phase corrections can greatly improve conventional seismic data (Schoenberger). It has also been shown, partly as a result of work with synthetic logs, that old concepts of normal and reverse polarity are totally inadequate to compensate for the wide variation of phase shifts which occur in recording and processing. The most common situation is a phase shift of some number of degrees other than 180°, but different for different configurations of recording instruments. Then seismic sections from different surveys do not tie very well. An attempt to reverse polarity on one will improve the tie at some levels, but make it worse on others. This is the result of inadequate or improper phase matching.

Minimum phase deconvolution has been the industry standard for years, and has served geophysicists and data processors well. It has some very desirable properties for calculation and application, and the resulting sections yield sharp reflections which are easy to map and have a minimum time shift under the circumstances. But the term minimum phase is a mathematical reference which should not be confused with a small phase shift.

Minimum phase is minimum only in a least squares sense, the least amount of phase shifting which will produce a certain type of wavelet. But the actual amount of phase shift is probably close to 90° in many cases, since the typical minimum phase output has a dipole-shaped wavelet, **(Fig 12.51)** which swings from one side to the other of the base line with more or less equal amplitude. A zero (or 180°) phase wavelet is completely symmetrical about a central spike, which represents the band limited version of a reflection coefficient.

It is very difficult to detect modest phase shifts by eye. If one were to correlate the series of wavelets of Fig 12.52, as in carrying a horizon or picking a reflection, the correlation carries smoothly across the illustration without difficultly. Any differences seem to be on the order of minor character changes. But the correlations march across a series of successive phase changes, and the phase has completely reversed across the sequence. A comparison of the top waveform, which is 180° out of phase with the bottom, shows the two to be phase reversed images. Any of the intermediate phase shifts shown (and any other) can and does occur on seismic data.

Geophysicists tend to think in binary terms of positive and negative polarity on their seismic data, and when two surveys do not match, the

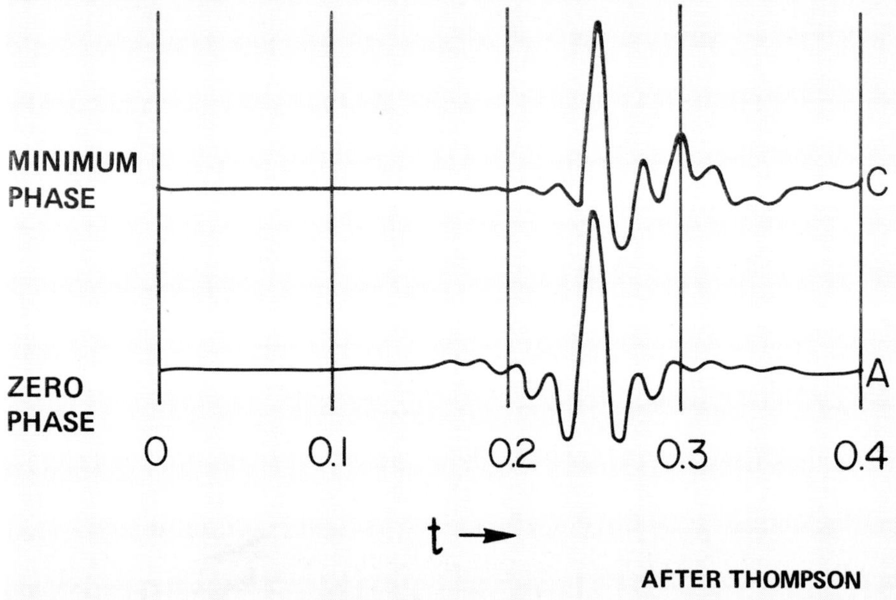

FIGURE 12.51

The minimum phase wavelet has favorable properties for data processing and structural mapping, but the zero phase wavelet is better for stratigraphic resolution and yields better synthetic logs.

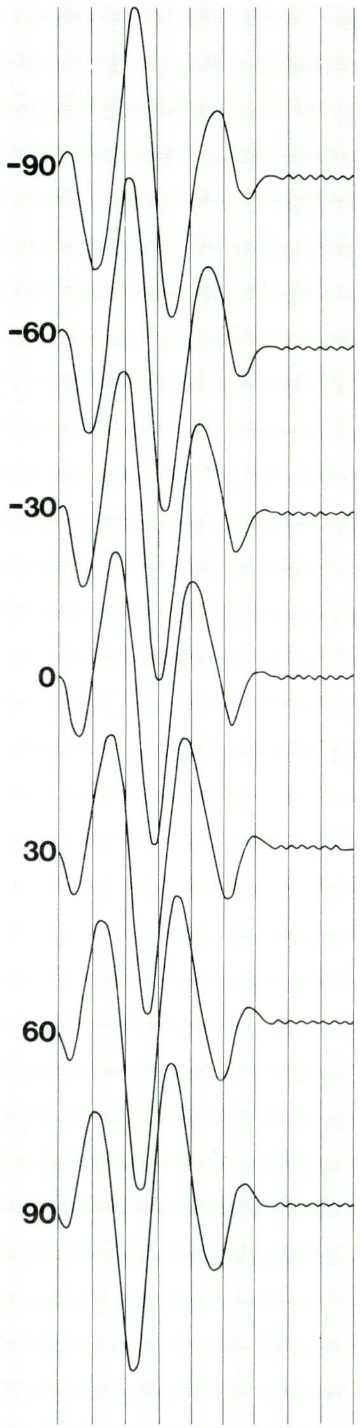

FIGURE 12.52

At first glance these wavelets appear similar, and a reflection appears to follow readily across the record. In fact, each wavelet is phase shifted relative to the others, a characteristic not readily discernible to the eye.

usual recourse is to reverse the polarity on one. The result is often an improvement in match at one level but a deterioration of fit at another. In truth neither is correct, since phase variations between instruments tend to be constant values of a few tens of degrees, like 75° or 50° depending on the particular configuration of geophones, amplifiers, and recorders. There is probably no such thing as 'positive' and 'reverse' polarity, except when data is referred to a polarity reversal of itself. All other comparisons between different field equipment probably have some much smaller intermediate phase difference.

Signature deconvolution seeks to null out all phase distortion, reducing each reflection to the best possible approximation of the zero phase reflection coefficient. Several process methods have been developed for this purpose. None are totally successful in all cases. Their success depends upon how well the assumptions used meet the actual data conditions. Thompson has provided a useful discussion of phase and phase shifts applied to log models.

Constant phase shifts are characterized by a saw-tooth appearance of synthetic logs. **Figure 12.53** compares the inversion of two sets of seismic data at a common tie point. The synthetic log on the left is from the tie line recorded with different instruments. It has been processed with exactly the same parameters as the original line but the results suggest improper phase correction.

A constant phase shift of 50 applied to the synthetic log on the left produced the modified log on the right, which now has a much more satisfactory phase response. A phase shift is a time shift, so a phase error is another possible source of depth error on the log.

In most of these examples, the phase shifts have been constant for all frequencies. Much more complicated non-linear shifts can occur. These are difficult to determine, but can distort the waveform, cause time shifts and modify character of reflections. Minimum phase deconvolution can yield undesireable phase shifts. So-called time-variant applications, in which a seismic trace is segmented into two or more portions and a different operator designed for each, can be worse, compounding the problem by producing different phase shifts (and hence different time shifts) in each segment.

A very poor situation exists when the field data has been recorded through a high-line notch filter, usually 60 Hz. Instrumental filters of this type have a profound effect on the phase of frequencies on either side of the notch. Compounding the distortion resulting from loss of the 60 Hz component, they impose a quick twist of the frequencies around the center value of the notch, in some cases reversing the polarity of those frequencies above the notch. In all cases, the 60 Hz instrumental notch filter creates serious phase distortion that cannot always be rectified in the data process sequence. Whenever possible, use of such filters should be avoided.

To a degree, they can be corrected by convolution with a phase correction operator, provided the amount of phase shift can be determined. Instrument manufactures provide phase shift diagrams with their instruments but these do not always match the observed shift.

Phase correction is an extremely important component of the process sequence, since it determines the form of the signal, and directly influences reflection times. A phase shift is a time shift.

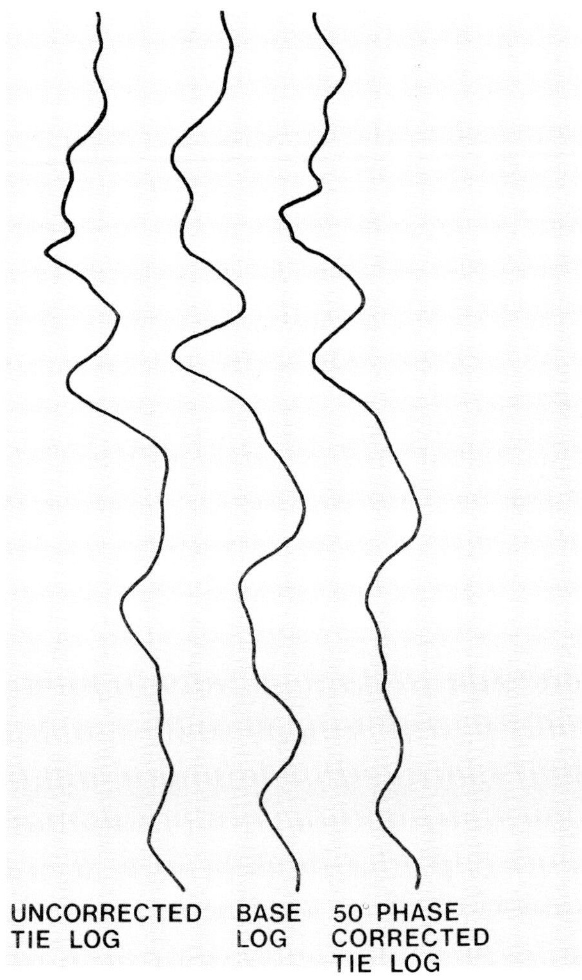

FIGURE 12.53

Tying Seislogs at left and center were shot with different instruments but processed the same. A phase shift of 50° applied to the left trace yields the better tie on the right.

Density

Still another source of local velocity error is inadequate density correction. At times, density may vary independently of rock matrix velocity. The function used for density correction then may be deficient in some lithologies and in excess in others. Precise density correlation first requires the lithology of the rock to be known. But this information is rarely available at the start of a survey, and it varies substantially in the typical geologic column. Fortunately, as has been demonstrated in Figure 12.10, it varies more or less as a linear function of acoustic impedance and can be so corrected.

In very young sections density may influence acoustic impedance more than velocity, contrary to the usual situation in older, more competent rocks. Sandstones usually have higher velocity than surrounding shales, but in the very young rocks, the effect of density will cause the relatively high porosity sandstones to have lower apparent velocities than the surrounding shales. Sometimes the effect of inadequate density correction may be favorable, from an interpretive standpoint, since it tends to emphasize the low velocity response of porosity and heighten the contrast with adjacent rocks.

When both density and sonic borehole logs are available they may be cross multiplied to yield an acoustic impedance log which can then be corrected with the same function as the seismic data hopefully to provide a closer match to the synthetic logs. The typical density log is a curve derived from a radioactive log, by computations, rather than the true density that would be obtained from a down hole gravimeter or from rock samples. As a result it may not yield true density values. Unfortunately, results are often disappointing and it will be found the sonic log alone may provide the best match in many cases.

Sometimes the discrepancies between borehole logs and Seislogs can be traced back to the quality of borehole logs rather than to the seismic data. All logs are subject to errors or distortion of one type or another. Both the sonic and density may suffer severely from borehole washouts.The sonic is susceptible to cycle skipping and mud effects. Undefined or unlabelled scale changes may cause errors in digitizing. Some judicious editing is often necessary before a field log is digitized. Editing logs is, at best, something of an art. Most sonic and density errors are caused by borehole washouts or cavings, which can be recognized on a caliper log. In such cases, the usual practice is to edit with a straight line across the offending portion, but on the other hand, most washouts occur where the formation is a little softer, or more subject to erosion, which usually means a different velocity than the surrounding material.

One test of the sonic log is to compare the elapsed time of the integrated log with seismic reflection times, or still better, with checkshot times when these are available. Usually, if errors exist, the integrated log times will be long.

An important factor can be the open hole condition at the time when the logs were run. The shallow sediments are often younger, less consolidated, and more susceptible to borehole damage. Washouts and sidewall deterioration can happen very quickly from the drilling process. The longer an open borehole is exposed to drilling, the less reliable the logs will be. Comparison of the overlapping portion of successive logging runs in the same borehole, may reveal a surprising amount of difference between the two logs over the same section. Deep holes may have only one logging run. In such cases it is not uncommon to see the correlation between a synthetic log and a borehole log improve with depth, due to improving log quality, when the reverse might be expected due to the broader frequency bandwidth of the seismic signal from the shallow section.

References and Further Reading

Al-Chalabi, M., 1974, An Analysis of Stacking, RMS, Average and Interval Velocities Over a Horizontally Layered Ground: Geophysical Prospecting 22, p. 458-475.

Anderson, Roger Y., 1964, Varve Calibration of Stratification: Kansas Geological Survey Bulletin 169, p.1-20.

Anstey, N.A., 1964, Correction Techniques, A Review: Geophysical Prospecting 12, p. 355-382.

Backus, M.M., 1959, Water Reverberations - Their Nature and Elimination: Geophysics 14, p. 389-426.

Beitzel, J.E., Cone, R.M., Dees, J.L., and Thompson, D.E., 1977, Understanding Seismically Derived Impendance Logs: 47th Meeting of the SEG, Calgary.

Belloti, P., and Giacca, D., 1978, Pressure Evaluation Improves Drilling Programs: Oil and Gas Journal 76, p. 77-85.

Bendat, J.S., and Piersol, A.G., 1966, Measurement and Analysis of Random Data: Wiley.

Blackman, R.B., and Tukey, J.W., 1958, The Measurement of Power Spectra: Dover, New York.

Bracewell, R., 1965, The Fourier Transform and Its Applications: McGraw Hill.

Branisa, F., 1974, Filtering of Well-Log Curves: Geophysics 39, p. 545-549.

Branisa, F., and Nikolic, Z.J., 1981, Evaluation of Pulse Estimation and Seismic Trace Inversion Procedures: 51st Meeting of the SEG, Los Angelas.

Brigham, E.O., 1974, The Fast Fourier Transform: Prentice-Hall.

Brown, R.B., 1963, Smoothing, Forecasting, and Prediction of Discrete Time Series: Prentice-Hall.

Brown, R.J.S., 1969, Normal Moveout and Velocity Relations for Flat and Dipping Beds and for Long Offsets: Geophysics 34, p. 180-195.

Carslaw, H.S., 1930, Fourier's Series and Integrals: Dover, New York.

Chun, J.H., and Jacewitz, C.A., 1981, Fundamentals of Frequency Domain Migration: Geophysics 46, p. 717-733.

Claerbout, J.F., and Doherty, S.M., 1972, Downward Continuation of Moveout Corrected Seismograms: Geophysics 37, p. 741-768.

Claerbout, J.F., 1971, Toward a Unified Theory of Reflection Mapping: Geophysics 36, p. 467-481.

Clement, W.G., 1973, Basic Principles of Two Dimensional Filtering: Geophysical Prospecting 21, p. 125-145.

Cook, E.E., and Neidell, N.S., 1969, The Use and Limitations of Velocity Spectra in Lithological Model Studies: 39th Annual Meeting of the SEG, Calgary.

Cooley, J.W., and Tukey, J.W., 1965, An Algorithm for the Machine Calculation of Complex Fourier Series: Mathematics of Computation 19, p. 297-301.

Cressman, K.S., 1968, How Velocity Layering and Steep Dip Affects CDP: Geophysics 33, p. 399-411.

Deregowski, S.M., 1971, Optimum Digital Filtering and Inverse Filtering in the Frequency Domain: Geophysical Prospecting 19, p. 729-768.

Dix, C.H., 1955, Seismic Velocities from Surface Measurements: Geophysics 20, p. 68-86.

Dobrin, M.B., Ingalls, A.L., and Long, J.A., 1965, Velocity and Frequency Filtering of Seismic Data Using Laser Light: Geophysics 30, p. 1144-1178.

Dresser Atlas, 1971, Log Interpretation: Dresser Industries Inc., Houston.

Dunkin, J.W., and Levin, F.K., 1973, Effect of Normal Moveout on a Seismic Pulse: Geophysics 38, p. 643-656.

Embree, P., Burg, J.,P., and Backus, M.M., 1963, Wide Band Velocity Filtering - The Pie Slice Process: Geophysics 28, p. 948-974.

Faust, L.Y., 1951, Seismic Velocity as a Function of Depth and Geologic Time: Geophysics 16, p. 192-206.

Finnette, J., Nicholich, R., and Sancin, S., 1971, Review of the Basic Theoretical Assumptions in Seismic Digital Filtering: Geophysical Prospecting 19, p. 339-344.

Fitton, J.C., and Dobrin, M.B., 1967, Optical Processing and Interpretation: Geophysics 32, p. 801-818.

Ford, W.T., and Hearne, J.H., 1966, Least Squares Inverse Filtering: Geophysics 31, p. 917-926.

Foster, M.R., Sengbush, R.L., and Watson, R.J., 1964, Suboptimum Filter Systems for Seismic Data Processing: Geophysical Prospecting 12, p. 173-191.

Foster, M.R., and Guinzy, N.J., 1967, The Coefficient of Coherence. Its Estimation and Use in Geophysical Data Processing: Geophysics 32, p. 602-616.

Galbraith, J.N. Jr., and Wiggens, R.A., 1968, Characteristics of Optimum Multi-Channel Stacking Filters: Geophysics 33, p. 36-48.

Gangi, A.F., and Yang, S.J., 1976, Travel Time Curves for Reflection in Dipping Layers: Geophysics 41, p. 425-440.

Gazdag, J., 1978, Wave Equation Migration with the Phase Shift Method: Geophysics 43, p. 1342-1351.

Gentleman, W.M., and Sande, G., 1966, Fast Fourier Transforms for Fun and Profit: AFIPS Proceedings 29, p. 563-578.

George, C.F., Hall, S.D., and Slack, H.A., 1968, Coherence Measurements and Their Application in the Analysis of Seismic Data: 38th Annual Meeting, SEG, Denver.

Gold, B., and Rader, C.M., 1969, Digital Processing of Signals: McGraw Hill.

Grant, F.S., and West, G.F., 1965, Interpretation Theory in Applied Geophysics: McGraw Hill.

Guyer, R.L., 1969, The Robinson-Treitel Reader: Seismography Service Corp., Tulsa.

Guyod, H., and Shane, L.E., 1969, Geophysical Well Logging: Hubert Guyod, Houston.

Hagedorn, J.G., 1954, A Process of Seismic Reflection Interpretation: Geophysical Prospecting 2, p. 85-127.

Herman, A.S., Anania, R.M., Chun, J.H., Jacewitz, C.A., and Pepper, R.E.F., 1982, A Fast 3-D Modelling Technique and Fundamentals of 3-D Frequency Domain Migration: Geophysics 47, p. 1627-1644.

Hilterman, F.J., 1970, Three Dimensional Seismic Modelling: Geophysics 35, p. 1020-1037.

Jones, H.J., and Morrison, J.A., 1954, Cross Correlation Filtering: Geophysics 19, p. 660-683.

Kanasewich, E.R., 1981, Time Sequence Analysis in Geophysics: The University of Alberta Press, Edmonton.

Lacoss, R.T., 1971, Data Adaptive Spectral Analysis Methods: Geophysics 36, p. 661-675.

Lee, Y.W., 1960, Statistical Theory of Communication: Wiley.

Levin, F.K., 1971, Apparent Velocity from Dipping Interface Reflections: Geophysics 36, p. 510-516.

Levin, F.K., 1979, Seismic Velocities in Transversely Isotropic Media: Geophysics 44, p. 918-936.

Letton, III, W., and Bush, A.M., 1969, Time Varying Velocity Filters: 39th Annual Meeting of the SEG, Calgary.

Lindseth, R.O., 1967, The Nature of Digital Seismic Processing: Journal of the CSEG 3, p. 31-111.

Lindseth, R.O., 1972, Approximation of Acoustic Logs from Seismic Traces: Canadian Well Logging Society Journal 5, p. 13-26.

Lindseth, R.O., 1979, Synthetic Sonic Logs – A Process for Stratigraphic Interpretation: Geophysics 44, p. 3-26.

Lines, L.R., and Ulrych, T.J., 1977, The Old and the New in Seismic Deconvolution and Wavelet Estimation: Geophysical Prospecting 25, p. 512-540.

Marschall, R., 1972, Parameter Optimization of Velocity Depth Functions by Use of Root Mean Square Velocities: Geophysical Prospecting 20, p. 700-711.

Mayne, W.H., 1962, Common Reflection Point Horizontal Data Stacking Techniques: Geophysics 27, p. 927-938.

McCollum, E.V., 1952, Quality of Geophysical Measurements: Geophysics 17, p. 56-69.

McLellan, J.H., and Parks, T.W., 1972, Equiripple Approximation of Fan Filters: Geophysics 37, p. 573-583.

Metherell, A.F., El-Sum, H.M.A., and Larmore, L., 1969, Acoustical Holography: Plenum, New York.

Neidell, N.S., and Taner, M.T., 1971, Semblance and Other Coherency Measures for Seismic Data: Geophysics 36, p. 482-497.

O'Brien, J.T., Kamp, W.P., and Hoover, G.M., 1982, Sign Bit Amplitude Recovery with Applications to Seismic Data: Geophysics 47, p. 1527-1539.

Oppenheim, A.V., and Schafer, R.W., 1975, Digital Signal Processing: Prentice-Hall, Englewood Cliffs.

Papoulis, A., 1962, The Fourier Integral and Its Application: McGraw Hill.

Peacock, K.L., and Treitel, S., 1969, Predictive Deconvolution: Theory and Practice: Geophysics 34, p. 155-169.

Peterson, R.A., 1969, Seismography 1970, The Writing of the Earth Waves: Symposium of the Pacific Coast SEG on Seismic Waves from Vibrational Sources.

Pfleuger, J., 1972, Spectra of Water Reverberations for Primary and Multiple Reflections: Geophysics 37, p. 788-796.

Pierce, J.R., 1961, Symbols, Signals and Noise: Harper and Brothers, New York.

Roh-Hsi, P., and Bremaecker, J.C., 1970, Direct Locations of Oil and Gas by the Seismic Reflection Method: Geophysical Prospecting 18, p. 712-727.

Rice, R.B., 1962, Inverse Convolution Filters: Geophysics 27, p. 4-18.

Robinson, E.A., 1957, Predictive Decomposition of Seismic Traces: Geophysics 22, p. 767-778.

Robinson, E.A., 1966, Multichannel Z-Transforms and Minimum Delay: Geophysics 31, p.482-500.

Robinson, E.A., 1967, Predictive Decomposition of Time Series with Application to Seismic Exploration: Geophysics 32, p. 418-484.

Robinson, E.A., 1967, Multichannel Time Series Analysis with Digital Computer Programs: Holden-Day.

Robinson, E.A., and Treitel, S. 1964, Principles of Digital Filtering: Geophysics 29, p. 394-404.

Robinson, E.A., and Treitel, S. , 1967, Principles of Digital Weiner Filtering: Geophysical Prospecting 15, p. 311-333.

Robinson, J.C., 1972, Computer Designed Weiner Filters for Seismic Data: Geophysics 37, p. 235-259.

Robinson, J.C., 1969, HRVA, A Velocity Analysis Technique Applied to Seismic Data: Geophysics 34, p. 330-356.

Robinson, J.C., and Aldrich, C.A., 1972, A Novel High Speed Algorithm for Summational Type Seismic Velocity Analyses: Geophysical Prospecting 20, p. 814-827.

Savit, Carl H., 1960, Preliminary Report: A Stratigraphic Seismogram: Geophysics 25, p. 312-321.

Schlumberger, 1972, Log Interpretation: Volume I Principles: Schlumberger Limited, New York.

Schneider, W.A., and Backus, M.M., 1968, Dynamic Correlation Analysis: Geophysics 33, p. 105-126.

Schoenberger, M., 1974, Resolution Comparison of Minimum-Phase Signals: Geophysics 39, p. 826-833.

Schoenberger, M., and Levin, F.K., 1976, Reflected and Transmitted Filter Functions for Simple Subsurface Geometrics: Geophysics 41, p. 1305-1317.

Schoenberger, M., and Levin, F.K., 1979, The Effect of Subsurface Samplings on One Dimensional Synthetic Seismograms: Geophysics 44, p. 1813-1829.

Sengbush, R.L., and Foster, M.R., 1968, Optimum Multichannel Velocity Filters: Geophysics 33, p. 11-35.

Seriff, A.J., and Kim, W.H., 1970, The Effects of Harmonic Distortion in the Use of Vibratory Surface Sources: Geophysics 35, p. 234-246.

Shanks, J.L., 1967, Recursion Filters for Digital Processing: Geophysics 32, p. 33-51.

Shannon, C.E., 1949, The Mathematical Theory of Communication: University of Illinois.

Sherriff, R.E., 1973, Encyclopedic Dictionary of Exploration Geophysics: SEG, Tulsa.

Sherwood, J.W.C., and Poe, P.H., 1972, Continuous Velocity Estimation and Seismic Wavelet Processing: Geophysics 37, p. 769-787.

Sicking, C.J., 1980, Sampling Requirements for Reflection Seismograms in Geophysical Data Acquisition: University of Texas, Austin.

Siems, L., and Hefer, F.W., 1967, A Discussion on Seismic Binary Gain Switching Amplifiers: Geophysical Prospecting 15, p. 23-34.

Silvia, M.T., and Robinson, E.A., 1979, Deconvolution of Geophysical Time Series in the Exploration for Oil and Natural Gas: Elsevier, Amsterdam.

Smith, M.K., 1958, A Review of Methods of Filtering Seismic Data: Geophysics 32, p. 44-57.

Stoffa, P.L., Buhl, P., and Bryan, G.M., 1974, The Application of Homomorphic Deconvolution to Shallow-Water Marine Seismology: Geophysics 39, p. 401-426.

Stolt, R.N., 1978, Migration by Fourier Transform: Geophysics 43, p. 23-48.

Street, A.V., and Lindseth, R.O., 1974, Continuous Velocity Mapping from Seismic Reflection Data: 36th Meeting of the EAEG, Madrid.

Taner, M.T., Crook, E.E., and Kohler, F., 1968, On Processing Vibroseis Data: 38th Annual Meeting of the SEG, Denver.

Taner, M.T., and Kohler, F., 1969, Velocity Spectra - Digital Computer Derivation and Applications of Velocity Functions: Geophysics 34, p. 858-881.

Taner, M.T., Kohler, F., and Sheriff, R.E., 1979, Complex Trace Analysis: Geophysics 44, p. 1041-1063.

Titchkosky, K.A., 1967, Seismic Deconvolution, Volume II.

Titchkosky, K.A., 1970, The Finite Fourier Transform: Journal of the CSEG 6, p. 7-33.

Treitel, S., 1970, Principles of Multi-Channel Filtering: Geophysics 35, p. 785-811.

Treitel, S., Shanks, J.L., and Frasier, C.W., 1967, Some Aspects of Fan Filtering: Geophysics 32, p. 789-800.

Verma, R.K., and Roy, A., 1970, A Graphical Method for Computing Geophone Group Response: Geophysics 35, p. 704-707.

Ulrych, T.J., 1971, Adaption of Homomorphic Deconvolution to Seismology: Geophysics 36, p. 650-660.

Walter, W.C., and Peterson, R.A., 1976-77, Seismic Imaging Atlas, V. I & II: United Geophysical Corporation, Pasadena.

Wang, R.J., 1969, The Determination of Optimum Gate Lengths for Time Variant Weiner Filtering: Geophysics 34, p. 683-695.

Weiner, N., 1949, Extrapolation, Interpolation, and Smoothing of Stationary Time Series: MIT Press, Cambridge.

Widess, M.B., 1973, How Thin is a Thin Bed?: Geophysics 38, p. 1176-1180.

Wiggens, R.A., 1966, WK Filter Design: Geophysical Prospecting 14, p. 427-440.

Woeber, A.F., and Penhollow, J.O., 1975, Depth Prediction from Velocity Profiles on the Texas Gulf Coast: Geophysics 40, p. 388-398.

Wold, H.O.A., 1964, Econometric Model Building: North Holland Publishing Co., Amsterdam.

Wood, L.C., Heiser, S., Treitel, S., and Rilet, P.L., 1978, The Debubbling of Marine Source Signatures: Geophysics 43, p. 715-729.

Wyllie, M.R.J., Gregory, A.R., and Gardner, G.H.I., 1956, Elastic Wave Velocities in Hetrogeneous and Porous Media: Geophysics 21, p. 41-70.

Wyllie, M.R.J., Gregory, A.R., and Gardner, G.H.I., 1958, An Experimental Investigation of Factors Affecting Elastic Wave Velocities in Porous Media: Geophysics 23, p. 459-493.

ORDER FORM

Additional copies of this Review may be obtained at a price of $36.00 per copy by completing the order form below.

SEG BookMarT

Society of Exploration Geophysicists

SHIPPING INFORMATION:

Name: ______________________________

Address: ______________________________

____

____

____

Member of SEG ☐ Yes ☐ No

ORDER FORM

Mail orders to:

Society of Exploration Geophysicists
P.O. Box 702740
Tulsa, OK 74170-2740 U.S.A.
Attn: BOOK ORDER DEPT.

Number of Copies	Catalog Number	SEG Publication Title	Amount in US $

METHOD OF PAYMENT:

☐ Send statement ☐ VISA
☐ Payment enclosed ☐ MasterCard

Expires ________

Card Account Number

Signature ______________________________

POSTAGE AND HANDLING:

In U.S.A.
$4.00 for initial item
$2.00 for each additional item

Outside U.S.A.
$8.00 for initial item
$4.00 for each additional item

Airmail or special handling available upon request at additional cost.

Subtotal US $	
Add 3¼% on Oklahoma orders; 6¼% on Tulsa orders for Sales Tax	
Postage and handling	
TOTAL AMOUNT DUE US $	

Prices are subject to change without notice.

L87